AF597674

Catalytic Aspects of Metal Phosphine Complexes

Catalytic Aspects of Metal Phosphine Complexes

Elmer C. Alyea, EDITOR
University of Guelph

Devon W. Meek, EDITOR
The Ohio State University

Based on a symposium sponsored by the ACS and CIC Divisions of Inorganic Chemistry at the 1980 Biennial Inorganic Chemistry Symposium, Guelph, Canada, June 5–7, 1980.

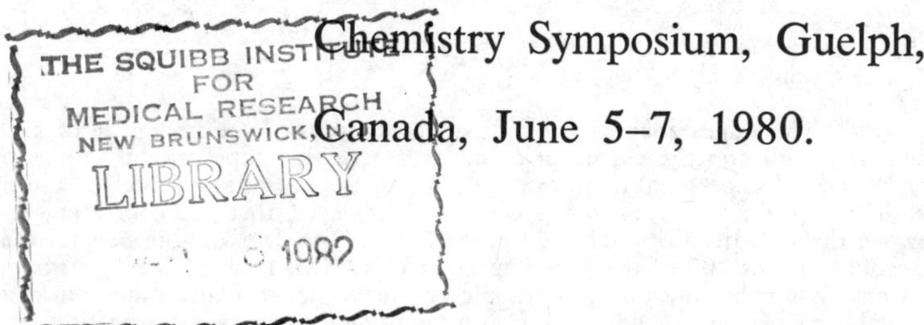

ADVANCES IN CHEMISTRY SERIES **196**

AMERICAN CHEMICAL SOCIETY
WASHINGTON, D. C. 1982

Library of Congress CIP Data

Catalytic aspects of metal phosphine complexes.
(Advances in chemistry series, ISSN 0065-2393; 196)

Includes bibliographies and index.

1. Phosphine—Congresses. 2. Catalysis—Congresses. 3. Legands—Congresses. 4. Transition metal compounds—Congresses. 5. Complex compounds—Congresses.

I. Alyea, Elmer C., 1941– . II. Meek, Devon W. III. American Chemical Society. Division of Inorganic Chemistry. IV. Chemical Institute of Canada. Inorganic Chemistry Division. V. Inorganic Chemistry Symposium (1980: Guelph, Ont.). VI. Series.

QD1.A355 no. 196 [QD181.P1] 540s [660.2'995]
ISBN 0-8412-0601-5 81-12903
AACR2 ADCSAJ 196 1–421 1982

PRINTED IN THE UNITED STATES OF AMERICA

FOREWORD

ADVANCES IN CHEMISTRY SERIES was founded in 1949 by the American Chemical Society as an outlet for symposia and collections of data in special areas of topical interest that could not be accommodated in the Society's journals. It provides a medium for symposia that would otherwise be fragmented, their papers distributed among several journals or not published at all. Papers are reviewed critically according to ACS editorial standards and receive the careful attention and processing characteristic of ACS publications. Volumes in the ADVANCES IN CHEMISTRY SERIES maintain the integrity of the symposia on which they are based; however, verbatim reproductions of previously published papers are not accepted. Papers may include reports of research as well as reviews since symposia may embrace both types of presentation.

CONTENTS

Preface .... **ix**

1. **Applications of P-31 NMR to the Study of Metal–Phosphorus Bonding** **1**
   A. Pidcock

2. **P-31, Sn-119, and Pt-195 NMR Studies on Platinum–Tin Homogeneous Hydrogenation Catalysts** **23**
   K. A. Ostoja Starzewski and P. S. Pregosin

3. **P-31 NMR Studies of Equilibria and Ligand Exchange in Triphenylphosphine Rhodium Complex and Related Chelated Bisphosphine Rhodium Complex Hydroformylation Catalyst Systems** **43**
   R. V. Kastrup, J. S. Merola, and A. A. Oswald

4. **The Effect of Chelating Diphosphine Ligands on Homogeneous Catalytic Decarbonylation Reactions Using Cationic Rhodium Catalysts** **65**
   D. H. Doughty, M. P. Anderson, A. L. Casalnuovo, M. F. McGuiggan, C. C. Tso, H. H. Wang, and L. H. Pignolet

5. **Cobaloximes Containing Phosphorus-Donor Ligands** **85**
   L. G. Marzilli, P. J. Toscano, J. H. Ramsden, L. Randaccio, and N. Bresciani-Pahor

6. **Stereopopulation Control on Cyclometallation Reactions and the Stability of Very Large Chelate Rings** **101**
   B. L. Shaw

7. **Preparation, Structure and Reactions of $Rh_2H_4(P(isopropyl)_3)_4$** **117**
   D. L. Thorn and J. A. Ibers

8. **Splitting of a Water Molecule by Trialkylphosphine Complexes of Platinum and Rhodium and Its Chemical Applications** **135**
   T. Yoshida and S. Otsuka

9. **Tricyclohexylphosphine Complexes of Ruthenium, Rhodium, and Iridium and Their Reactivity Toward Gas Molecules** **145**
   B. R. James, M. Preece, and S. D. Robinson

10. **Phosphido-Bridged Iron Group Clusters** **163**
    A. J. Carty

11. **Chemistry of Tertiary Phosphine Complexes of Rhodium, Iridium, and Platinum** **195**
    D. P. Arnold, M. A. Bennett, G. T. Crisp, and J. C. Jeffery

12. **Palladium–Triarylphosphine Complexes as Catalysts for Vinylic Halide Reactions** **213**
    R. F. Heck

13. **Binuclear Organoplatinum Complexes as Models for Catalytic Intermediates** 231
M. P. Brown, J. R. Fisher, S. J. Franklin, R. J. Puddephatt, and M. A. Thomson

14. **Novel Reactions of Dinuclear Metal Complexes: Reactions and Structures of Palladium Complexes Containing Bis(diphenylphosphino)methane as a Bridging Ligand** 243
A. L. Balch

15. **Hydrogenation Catalysis of Olefins by a Rhodium Hydride Complex of $PhP(CH_2CH_2CH_2PPh_2)_2$: In Situ Generation of RhH(ttp)** 257
J. Niewahner and D. W. Meek

16. **The Preparation, Characterization, and Properties of Highly Soluble Transition-Metal Complexes of Long-Chain Tertiary Phosphines** 273
S. Franks, F. R. Hartley, and J. R. Chipperfield

17. **Homogeneous Catalytic Oxidation with Phosphine-Substituted Complexes of Rhodium Carbonyl Clusters: $Rh_6(CO)_{16}$- and $Re_2(CO)_{10}$- Catalyzed Autoxidation of Ketones and Alcohols** 291
D. M. Roundhill, M. K. Dickson, N. S. Dixit, and B. P. Sudha–Dixit

18. **Metal Complexes of Iminophosphine and Iminoarsine Chelating Agents: Structure, Reactivity, and Stereochemistry** 303
J. E. Hoots, T. B. Rauchfuss, S. P. Schmidt, J. C. Jeffery, and P. A. Tucker

19. **Recent Advances in Polyphosphine Synthesis** 313
R. B. King

20. **Studies of Asymmetric Homogeneous Catalysts** 325
W. S. Knowles, W. C. Christopfel, K. E. Koenig, and C. F. Hobbs

21. **Asymmetric Catalytic Hydrogenation** 337
B. Bosnich and N. K. Roberts

22. **Mechanistic Studies of Rhodium-Catalyzed Asymmetric Homogeneous Hydrogenation** 355
J. M. Brown, P. A. Chaloner, and D. Parker

23. **Advances in Asymmetric Hydrocarbonylation** 371
G. Consiglio and P. Pino

24. **P-31 NMR Comparisons of the Crystalline and Solution States of Rhodium(I) Diphosphine Catalysts** 389
G. E. Maciel, D. J. O'Donnell, and R. Greaves

Appendix 409

Index 413

PREFACE

A major contribution to the efficient use of the world's limited petrochemical and fossil fuel resources could be made by minimizing the energy requirements of industrial processes. This objective is being pursued by the development of molecular catalysts, often involving metal phosphine complexes, to carry out extremely specific chemical transformations. The purpose of the symposium was to stimulate exchanges of ideas and information between academic and industrial chemists concerned with catalytic systems that involve phosphine ligands. Attention was focused on the most relevant aspects of the theme by the following subtopics: P-31 NMR and the nature of the metal–phosphorus bond, the chemistry of bulky phosphine ligands, polydentate phosphines, the reactivity of transition-metal phosphorus compounds, and asymmetric synthesis. These areas were well represented in the five sessions of oral papers at the symposium and their sequence is maintained in this volume.

The symposium program included 30 talks and 23 poster presentations. Due to space limitations, this volume contains 24 of the oral papers, whereas the authors, affiliations, and titles of the poster presentations are listed in the appendix. Sixteen of the oral papers at the symposium were invited by the Organizing Committee, with the remaining 37 papers being selected from those submitted for presentation. The 30 speakers at the symposium included five persons from industry and 12 speakers from outside North America. The symposium speakers attracted 170 participants from the United States, Canada, England, France, Germany, Switzerland, Italy, Hungary, Venezuela, Australia, and Japan.

We would like to acknowledge the assistance of several colleagues from The Guelph–Waterloo Centre for Graduate Work in Chemistry (GWC^2)—R. J. Balahura, A. J. Carty, P. Chieh, H. C. Clark, R. G. Goel, P. M. Henry, R. McCrindle, and G. Rempel—who served on the Organizing Committee for planning and arranging the scientific program. In addition to the financial support of 22 Canadian and U.S. companies, major financial contributions were received from the Natural Sciences and Engineering Research Council of Canada, the Petroleum Research Fund administered by the American Chemical Society, and the United States Army Research Office. Finally, we are pleased to acknowledge the financial contributions from the Universities of Waterloo and Guelph, the expertise of the University of Guelph Conference Office, and the assistance of M. J. Alyea in organizing the nonscientific program of the symposium.

Elmer C. Alyea
GWC[2]
University of Guelph
Guelph, Ontario N1G 2WI

Devon W. Meek
Department of Chemistry
The Ohio State University
Columbus, OH 43210

August, 1980

1

Applications of P-31 NMR to the Study of Metal–Phosphorus Bonding

ALAN PIDCOCK
School of Molecular Sciences, University of Sussex, Brighton BN1 9QJ, U.K.

Evidence is presented that $^1J_{M-P}$ *is Fermi dominated and a molecular orbital (MO) expression derived from that term is used to discuss the relation between* $^1J_{Pt-P}$ *and the Pt–P length for platinum(II) and (IV) complexes, the dependence of coupling constants on the nature of ligands in trans and cis relationship to the detector ligand, and the dependence on* $^1J_{M-P}$ *on groups attached to phosphorus (M = platinum(II), tungsten(0), and hydrogen). New results are included for the following series of complexes:* cis-*[PtPh(SnX₃)(PPh₃)₂],* cis- *and* trans-*[PtCl(SnX₃)(PPh₃)₂],* cis-*[PtCl₂(PX₃)(PEt₃)], and* trans-*[PtCl₂(PX₃)(PCy₃)](Cy = cyclohexyl).*

The characterization and assignment of structures of coordination compounds with phosphorus ligands is aided strongly by P-31 NMR spectra especially when coupling constants between the metal and phosphorus $^1J_{M-P}$ or between phosphorus and other ligand donor atoms $^2J_{P-L}$ can be measured (*1*, *2*). The coupling constants vary considerably in magnitude, and when trends or correlations have been established the coupling constants can provide important structural information beyond that derived from the symmetry of the spectra. The coupling constants (particularly those involving directly bonded atoms) also may be used to investigate bonding in coordination compounds, and the variation in the magnitude of the couplings $^1J_{M-P}$ between a metal in a single oxidation state and a given phosphorus ligand (e.g., from ca. 1500 Hz to 5500 Hz for platinum(II) and PR_3) is a reflection of the wide variation in bond strengths in coordination compounds. In studies of homogeneous catalysis there are many examples of complex identification via coupling constants involving phosphorus, and a greater understanding of the lability of ligands also has come through greater knowledge of ground-state bonding patterns gained by measuring coupling constants (*3*).

0065–2393/82/0196–0001$05.50/0

This chapter emphasizes the establishment of trends in coupling constants between transition metals and phosphorus and the interpretation of such trends in terms of the appropriate molecular orbital (MO) treatments that are currently available.

Framework for Interpretation

Interpretations of coupling constants generally are founded on the assumption that the coupling derives essentially from the Fermi contact term. This assumption is supported by the existence of a number of correlations between $^1J_{M-P}$ and other couplings for which non-Fermi terms are expected to be smaller (*4*). Examples of correlations that extend over a sizeable range of magnitude of $^1J_{M-P}$ are those between $^1J_{Pt-P}$ for the phosphonato ligands in *trans*-[Pt*X*{P(O)(OPh)$_2$}(PBu$_3$)$_2$] and $^2J_{PT-C-H}$ for the methyl ligands in *trans*-[Pt*X* (CH$_3$)(PEt$_3$)$_2$] (*X* = Cl, Br, I, ONO$_2$, N$_3$, NCO, NCS, NO$_2$, and CN) (*5*), and between $^1J_{Pt-P}$ for PBu$_3$ in *trans*- [PtCl$_2$L(PBu$_3$)] and $^1J_{Pt-N}$ in *trans*- [PtCl$_2$L(NH$_2$C$_6$H$_{13}$)] (L = PR$_3$, AsR$_3$, Me$_2$SO, NH$_2$R) (*6*) (*see* Figures 1 and 2). These and other correlations involving direct coupling to hydride ligand $^1J_{Pt-H}$ and indirect couplings to fluorine have been examined by Dixon et al. (*7*), who have shown that the nonzero intercepts of some of the correlations are consistent with Fermi-dominated couplings.

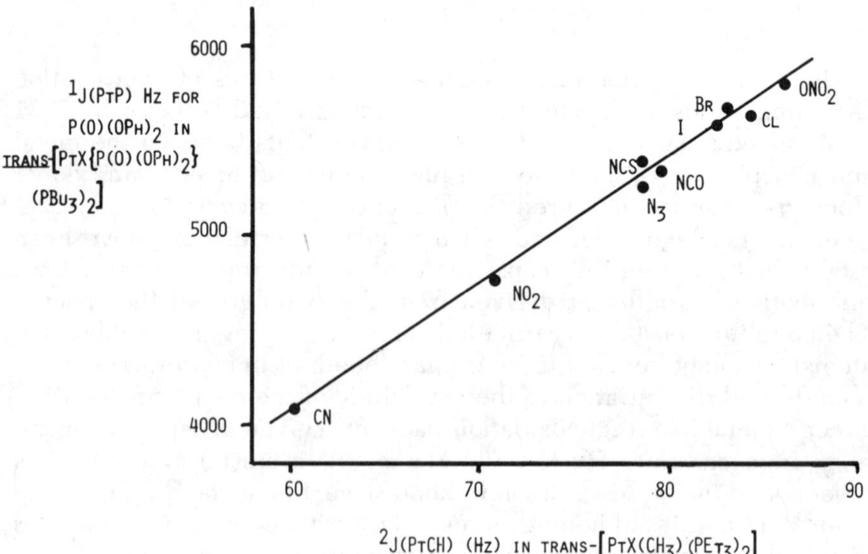

Figure 1. Graph of $^1J_{Pt-P}$ *for the phosphonato ligand in* trans-*[PtX{P(O)(OPh)$_2$}(PBu$_3$)$_2$] against* $^2J_{Pt-C-H}$ *for* trans-*[Pt(CH$_3$)X(PEt$_3$)$_2$]* (5)

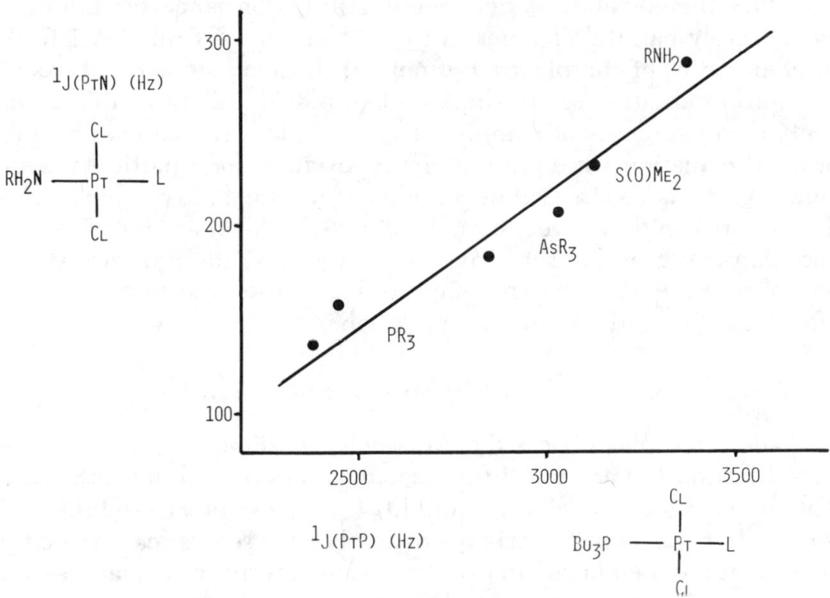

Figure 2. Graph of $^1J_{Pt-N}$ *for* trans-*[PtCl$_2$(NH$_2$R)L] against* $^1J_{Pt-P}$ *for* trans-*[PtCl$_2$(PBu$_3$)L]* (6)

The evaluation of the Fermi contact term in valence bond or MO theory involves several simplifying assumptions (such as the average excitation energy approximation) especially for complexes of low symmetry (*1*), but the form of the Fermi term is such that the couplings must depend strongly on the s orbital content of the bond between the coupled atoms. A convenient form of the MO expression (adapted from Pople and Santry (*4*)) is given in Equation 1; the constant C involves the magnetogyric ratios of the coupled nuclei and other constants, $^3\Delta E$ is an average excitation energy, $|S_M(0)|$ and $|S_P(0)|$ are the magnitudes of the valence state s-orbital basis functions evaluated at the parent nuclei, and $P'_{S_mS_p}$ is the MO bond order between the s orbitals of

$$^1J_{M-P} = C(^3\Delta E)^{-1}|S_M(0)|^2|S_P(0)|^2(P'_{S_MS_P})^2 \qquad (1)$$

the coupled atoms. The term $(P'_{S_MS_P})^2$ increases directly with α_P^2, the s character of the donor orbital of the phosphorus ligand, and with α_M^2, the s character of the acceptor orbital of the metal (M). Of course analogous expressions can be written for other couplings such as $^1J_{M-C}$ and $^1J_{M-N}$, and the magnitudes of indirect couplings such as $^2J_{M-C-H}$ often are discussed in terms of an expression of similar form, where C then includes factors affecting the transmission of coupling through the intervening bonds.

Thus, the correlations between coupling constants (*see* Figures 1 and 2) imply parallel changes in the parameters of Equation 1 in the different series of complexes, but neither the accurate correlations nor the more qualitative establishment of similar trends in coupling constants in related sets of complexes gives a clear indication of which terms in Equation 1 are the important variables for a particular set of coupling constants. In a set of complexes the site and type of variation of structure can determine to some extent which variables of Equation 1 are dominant for that set. However, if possible, attempts also should be made to identify by experiment or calculation the important variables for each kind of structure variation.

Metal Oxidation State and Phosphorus Ligand Constant

Where the phosphorus ligand is held constant, it is normal to assume that the s character of the phosphorus donor orbital (α_P^2), which contributes directly to $(P'_{S_M S_P})^2$ and $|S_P(0)|^2$, does not vary significantly as other ligands on the metal are changed (*8*). It seems reasonable that changes at the metal principally should affect terms in Equation 1 that are associated with the metal and that variations induced in the state of the phosphorus donor should be small in comparison. However, it is clear from studies of coupling constants within the phosphorus donor that some variations in the phosphorus donor do occur. Thus, $^1J_{P-C}$ for PBu_3 is ca. 38 Hz when trans to nitrogen or chlorine in dichloroplatinum(II) complexes and is ca. 32 Hz when trans to PR_3 (*9*). Withdrawal of σ-electrons from PBu_3 by platinum is stronger for PBu_3 trans to nitrogen or chlorine, and the simplest interpretation of the results is that $\alpha_P^2|S_P(0)|^2$ for P–C bonds is larger in these complexes. An increase in α_P^2 (for the P–C bonds) is expected from Bent's rules (*10*) since a stronger acceptor should involve larger p- and smaller s character in the phosphorus orbital of the Pt–P bond, and there should be a complementary change in the orbitals of the P–C bonds. Electron withdrawal from PBu_3 also is expected to contract the radial parts of the phosphorus (basis) orbitals and increase $|S_P(0)|^2$ (*11*). Although these two factors affect ^1J(PC) in the same way, their effects on the phosphorus donor orbital used in the bond to platinum tend to cancel in $^1J_{M-P}$: $|S_P(0)|^2$ increases and α_P^2 decreases as the acceptor strength of platinum increases. This implies that the product $\alpha_P^2|S_P(0)|^2$ for the PtP bonds may not vary significantly in comparison with changes in the platinum parameters for a range of complexes of a given phosphorus donor.

The values of $^1J_{Ag-P}$ for complexes $[Ag(PBu_3)_n]^+$ ($n = 2, 3, 4$), which are in common with several other coupling constants between directly bonded atoms (*12*), depend on the coordination number n in a manner that suggests dominance of the s character of the silver hybrid orbital α_{Ag}^2 (which contributes directly to the $(P'_{S_{Ag} S_P})^2$ of Equation 1).

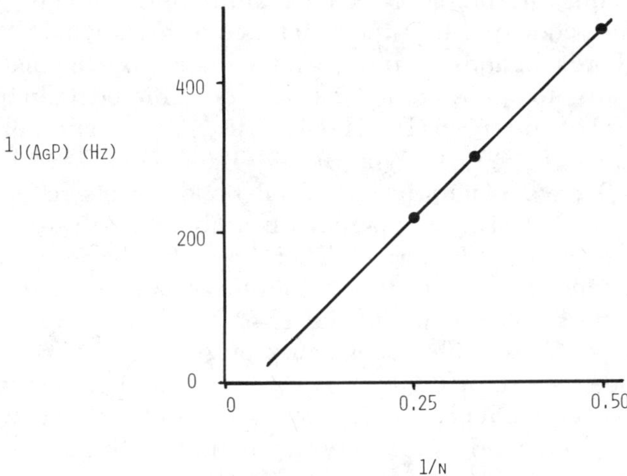

Figure 3. Graph of $^1J_{Ag-P}$ *for* $[Ag(PBu_3)_n]^+(BF_4)^-$ (12) *against* $1/n$, *the s character of the idealized* sp^{n-1} *hybrid orbital of silver. Results for other silver(I) complexes are similar* (13, 14).

Thus the graph of $^1J_{Ag-P}$ against $1/n$, the idealized s character of the silver sp^{n-1} hybrid orbitals, is accurately linear (*see* Figure 3). Results of this kind (especially for $^1J_{C-H}$) probably have been preeminent in leading to the general acceptance of the idea that coupling constants between directly bonded atoms accurately reflect s-orbital contents of bonds. However, in terms of the MO expression (*see* Equation 1) (and especially in terms of more accurate treatments involving several terms with different $^3\Delta E$), such results appear almost coincidental, since an accurate correlation appears to require a rather improbable constancy in the product $|S_M(0)|^2/^3\Delta E$ and in the relative contributions of the metal and phosphorus atomic orbitals to the MOs. Although the fact that the correlation between $^1J_{Ag-P}$ and $1/n$ does not pass through the origin (*see* Figure 3) indicates some failure of this simple framework of interpretation, it seems probable that some fortuitous cancellation of changes in $|S_M(0)|^2$ and $^3\Delta E$ is involved also. It should not be overlooked that Equation 1 itself and the simple interpretation in terms of hybridization involve substantial approximations.

However, where the energies of the valence-state orbitals of metals are reasonably similar, we can expect that the coefficients of the atomic orbitals (which affect $P'_{S_M S_P}$) can undergo large numerical changes while the energy differences between bonding and antibonding orbitals (e.g., $^3\Delta E$) change only fractionally. Therefore changes in $^1J_{M-P}$ in a series of complexes of a given phosphorus donor, metal, oxidation state, and coordination number may be dominated by changes in the s component of the M–P bonds rather than by $|S_M(0)|^2$

or $^3\Delta E$. This interpretation can be supported strongly for square-planar and octahedral transition metal complexes where the binding of phosphorus ligands is influenced strongly by the nature of the ligand trans to phosphorus, as in the chloro(trialkylphosphine) complexes of platinum(II) $[PtCl_{4-n}(PR_3)_n]^{(n-2)+}$ and platinum(IV) $[PtCl_{6-n}(PR_3)_n]^{(n-2)+}$ (n = 1, 2, 3) (*see* Table I) (*15*). The complexes with $n = 3$ have nonequivalent Pt–P bonds (trans to chlorine and phosphorus) that are associated with a common $|S_{Pt}(0)|^2$ term and probably similar $^3\Delta E$ terms. Therefore the large differences in coupling constants within these complexes must derive from differences in the s components of the Pt–P bonds trans to chlorine and phosphorus. These differences must originate primarily in the α^2_{Pt} terms, since the product $\alpha^2_P|S_P(0)|^2$ is expected to be nearly invariant. The strong dependence of $^1J_{Pt-P}$ on the nature of the trans ligand derives from changes in the hybridization of platinum.

Where coupling constants from different complexes are compared, it is not possible to eliminate $|S_M(0)|/^3\Delta E$ as an important variable. However the fact that $^1J_{Pt-P}$ trans to a given ligand (phosphorus or chlorine, Table I) is similar in all of the complexes of a given oxidation state suggests that the $(P'_{s_{Pt}s_P})^2$ terms depend essentially on the trans ligand and that other terms of Equation 1 are similar even in different complexes of a given oxidation state. This argument, which receives some support from bond-length measurements for the platinum(II) complexes (*see* Table II), presumes that the bonds to PR_3 trans to a given ligand are similar in the different complexes. The lengths of the Pt–P bonds trans to chlorine are not significantly different in *cis*-$[PtCl_2(PEt_3)_2]$ (*16*) and $[PtCl(PEt_3)_3]^{2+}$ (*17*), where the coupling constants are very similar. The Pt–P bond is shorter in $[PtCl_3(PEt_3)]^-$ (*18*), where the coupling is ca. 200 Hz larger, suggesting that this bond has a slightly larger *s* component, which is reflected in the shorter bond and in the larger coupling constant. The smaller coupling constants for Pt–P bonds trans to phosphorus in $[PtCl(PEt_3)_3]^+$ and *trans*-$[PtCl_2(PEt_3)_2]$ (*19*) likewise correlate with bonds that are longer than for Pt–P trans to chlorine (*see* Table I).

Correlation of $^1J_{M-P}$ with Bond Length

The substantially larger values of $^1J_{M-P}$ for phosphorus trans to chlorine compared with phosphorus trans to phosphorus correlate with shorter M–P bonds trans to chlorine for a number of metals and oxidation states: tungsten(IV), rhodium(I) and (III), platinum(II) and (IV), and linear mercury(II) (*15*). By analogy with the discussion of the results for the platinum(II) complexes, this indicates the dominance of the $(P'_{s_Ms_P})^2$ term in Equation 1 for couplings with a variety of M, but as discussed earlier it is difficult to determine the extent of variation of

Table I. $^{1}J_{Pt-P}$ (in Hertz) and Pt–P Bond Lengths (in Angstroms) in Trialkylphosphine Complexes of Platinum(II) and (IV)

	$[PtCl_3P]^-$	cis-$[PtCl_2P_2]$	$[PtClP_3]^+$	trans-$[PtCl_2P_2]$
$^{1}J_{Pt-P}$	3704	3508	3474, 2261	2400
l_{Pt-P}	2.215(4) (*18*)	2.258(2) (*16*)	2.251(3), 2.354(3) (*17*)	*ca.* 2.315 (*19*)
	$[PtCl_5P]^-$	*cis*-$[PtCl_4P_2]$	*mer*-$[PtCl_3P_3]^+$	*trans*-$[PtCl_4P_2]$
$^{1}J_{Pt-P}$	2085	2070	2049, 1374	1475
l_{Pt-P}		2.339(4) (*20*)		2.393(5) (*21*)

$|S_M(0)|^2$ and $^3\Delta E$ either for these systems or in comparisons of coupling constants in different complexes of a single metal. If variations of coupling constants $^1J_{M-P}$ derive solely from changes in $(P'_{s_M s_P})^2$, and the M–P bond lengths correlate with this term also, a correlation between $^1J_{M-P}$ and the M–P length is to be expected for complexes of a given M, oxidation state, and phosphorus ligand. It is necessary to restrict comparisons to complexes of a single phosphorus ligand (or similar phosphorus ligands) because, for example, the bond lengths are unlikely to be equally sensitive to changes in the s character of the metal and phosphorus orbitals, and the term $|S_P(0)|^2$ is likely to depend on the nature of the groups on phosphorus. When this correlation first was proposed (*15*), the results for platinum(II) trialkylphosphine complexes were such that a satisfactory correlation could be constructed, but the form of the correlation was ill-defined because the uncertainties in the bond lengths were not small compared with the total range of lengths. Subsequently, accurate bond lengths and coupling constants have become available for a large number of platinum(II) complexes of trialkylphosphines, and the results are plotted in Figure 4 (*22*).

No simple correlation is capable of fitting the results to within the experimental errors, and within the framework of interpretation that is suggested, it must be concluded that either there are significant variations in $|S_M(0)|^2/^3\Delta E$ or that the correlation between $(P'_{s_{Pt} s_P})^2$ and bond length is poor. However, the overall trend in the results in which increasing values of $^1J_{Pt-P}$ are associated with shorter bonds is likely to derive from the sensitivity of both $^1J_{Pt-P}$ and the bond lengths to the s-orbital bond order. The wide range of $^1J_{Pt-P}$ found for the short Pt–P bonds suggests that the bond lengths between platinum(II) and trialkylphosphines may reach a lower limit of ca. 2.21 Å and that the bond lengths therefore become virtually independent of $(P'_{s_{Pt} s_P})^2$ (and $^1J_{Pt-P}$) in this region. It is possible that more accurate correlations between $^1J_{M-P}$ and the M–P length may be found for more restricted sets of complexes. In Figure 4 most of the points correspond to complexes with phosphorus trans to phosphorus or phosphorus trans to chlorine so that within these sets the ligands cis to phosphorus vary. Calculations by Shustorovich, who treats the energy difference between L and X in MX_nL as a perturbation, suggest that a correlation between $^1J_{M-P}$ and the M–P length is to be expected when the ligands trans to phosphorus are varied, but that a correlation is not expected when ligands cis to phosphorus are varied (*25, 26*). Although the perturbation must be much more complicated than the single energy difference parameter used in these calculations, it is clearly desirable that a series of complexes is examined in which only the ligand trans to PR_3 varies. The calculations and the broad agreement found between trans influence

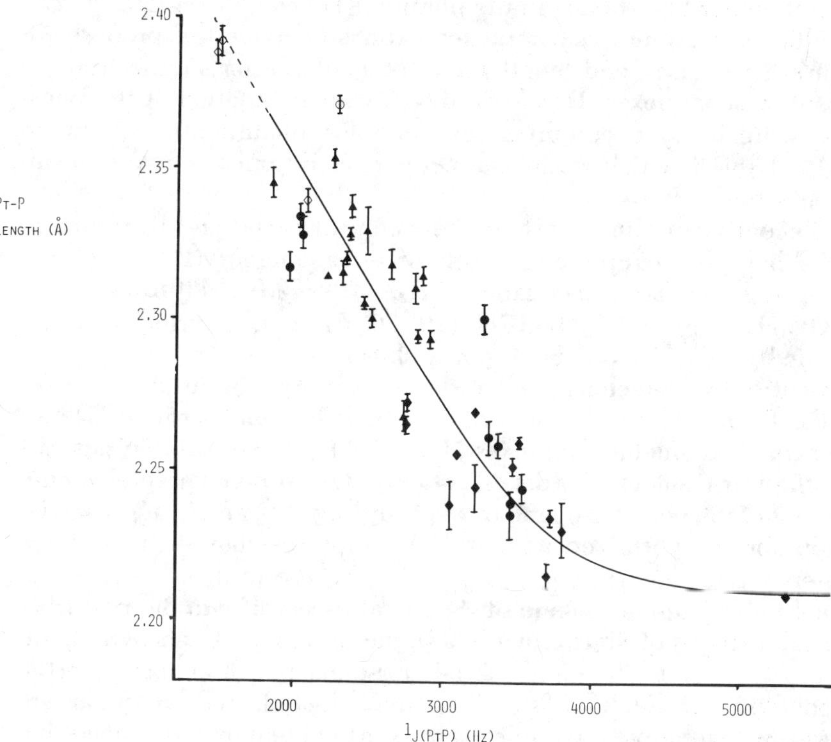

Figure 4. Graph of Pt–P length (shown with standard deviation limits) against $^1J_{Pt-P}$ *for platinum(II) complexes of trialkylphosphines:* ▲, *phosphorus trans to phosphorus;* ◆, *phosphorus trans to chlorine;* ●, *phosphorus trans to other donor atoms;* ○, trans-*[PtI$_2$(PCy$_3$)$_2$]. For discussion see Ref. 23; the point with* $J_{Pt-P} > 5000$ *Hz is for [Pt$_2$(μ-Cl)$_2$(COEt)$_2$(PMe$_2$Ph)$_2$]* (24). *Some platinum(IV) complexes have been included,* ◇.

series based on $^1J_{Pt-P}$ (or other coupling constants) and other physical parameters (3) provide some grounds for expecting that a more accurate correlation between $^1J_{M-P}$ and the M–P length then will emerge.

Results for a few platinum(IV) complexes have been included in Figure 4. We have known for some time that coupling constants $^1J_{Pt-P}$ are smaller in platinum(IV) than in related platinum(II) complexes and that the ratio of the couplings was ca. 0.6, close to the ratio 0.67 expected for idealized d^2sp^3 and dsp^2 hybridization schemes (8). However, the cancellation of $|S_{Pt}(0)|^{2/3}\Delta E$ terms for the two oxidation states appears improbable, as does the correlation with idealized s-orbital contents for such asymmetric complexes. Perhaps of more significance is the apparent conformity of the platinum(IV) results with the platinum(II) correlation (*see* Figure 4) which highlights the presence of longer Pt–P bonds in the platinum(IV) complexes (*20, 21, 27*). Results for platinum(0) complexes cannot be included properly in Figure 4 because bond lengths are not available for simple trialkylphosphine complexes. However, despite some variation of the phosphorus ligand, the coupling constants for platinum(0) complexes $[Pt(PX_3)_n]$ vary with *n* in the manner expected from a simple hybridization scheme (*28*).

Recently coupling constants between the metal and ligand atom nuclei have been reported for sets of linear mercury(II) complexes containing the detector ligands CH_3, CF_3, PMe_3, $P(O)Bu^t_2$, P(O)-$Ph(OC_4H_9$ – n) and $P(O)(OEt)_2$ (*29, 30, 61*) and a variety of other anionic ligands. The orders of trans influence found were almost independent of the detector ligand and the coupling measured, and were similar to the trans influence orders established for platinum(II) and other transition-metal complexes (*3, 7, 31–36*). As we pointed out earlier, the trans influence order is not expected to be completely independent of the detector ligand or the coupling (*29*), and in addition the results for sp-hybridized mercury(II) complexes show that the trans influence phenomenon does not require the use of d_σ orbitals in the formal hybridization scheme of the metal. According to the perturbation calculations of Shustorovich (*25*), the d_σ and s orbitals, which are the most important bonding orbitals for metals such as platinum(II), respond in parallel to a changing trans ligand, and in linear sp-hybridized mercury(II) complexes the s-orbital behavior is expected to closely parallel that in earlier transition metal complexes. Thus, in all of these complexes the metal s orbital is either dominant or very important in the bonding and is the principal contributor to the NMR coupling constants.

Dependence of $^1J_{M-P}$ on cis Ligands

While it appears well established that changes in the σ-bonding are at least mainly responsible for the effects of trans ligands on $^1J_{M-P}$ and other parameters of metal–ligand bonds, the relative importance of σ, π, and steric contributions to the effects of cis ligands remains unclear. Although in some systems, such as in the complexes $[PtCl_{4-n}(PR_3)_n]^{(n-2)+}$ and $[PtCl_{6-n}(PR_3)_n]^{(n-2)+}$ which were discussed earlier, change of the cis ligands has relatively little effect on $^1J_{M-P}$ or on the M–P lengths, considerable variations in $^1J_{M-P}$ and other couplings can be induced by cis ligands. More recently changes in bond lengths induced by cis ligands have been studied systematically in *cis*-$[PtCl_2L(PR_3)]$ complexes (*11, 37, 38*).

There are good correlations between the values of $^1J_{Pt-P}$ for the phosphine ligands in *trans*-$[Pt(CH_3)X(PEt_3)_2]$, *trans*-$[Pt\{P(O)(OPh)_2\}$-$X(PBu_3)_2]$, *trans*-$[Pt(SCF_3)X(PEt_3)_2]$, and *trans*-$[Pt(C_2F_3)X(PEt_3)_2]$ for halide and pseudohalide ligands *X*, and between the latter two series for a wider range of anionic ligands (*5, 32*). The order of decreasing cis influence (as measured by $^1J_{Pt-P}$ for the C_2F_3 complexes was CO > $P(OPh)_3$ > $P(OMe)_3$ > $P(OEt)_3$ > PPh_3 > CN > PEt_3 > pyridine > I > SCF_3 > Br > Cl > N_3 > NO_2 > ONO_2, with $^1J_{Pt-P}$ changing from 2021 Hz (*X* = CO) to 2603 Hz (*X* = ONO_2). The results for the SCF_3 complexes indicate that the series continues to groups of even lower cis influence: CF_3 > H > Ph > CH_3 (*32*). With some minor variations, similar series have been found for $^1J_{Pt-P}$ cis to *X* in [Pt(CH_3)*X* (dppe)] (dppe = $Ph_2PCH_2CH_2PPh_2$) (*33*) and in $[PtX(ttp)]^+$ [ttp = PhP(CH_2-$CH_2CH_2PPh_2)_2$] (*34*). Although there is a good measure of consistency in the behavior of $^1J_{Pt-P}$ cis to *X* in these series of complexes, the correlation between $^1J_{Pt-P}$ and the Pt–P length in the series *cis*-[$PtCl_2L(PEt_3)$] is relatively poor (*see* Figure 5). This may be because the total overlap population (and hence the bond length) cis to the ligand does not correlate with the s-orbital bond order (and hence $^1J_{Pt-P}$) as suggested by Shustorovich (*26*), but it would be unwise to attach significance to the weakness of the correlation for varying cis ligands until a better correlation is obtained for varying trans ligands.

Although the calculations of Shustorovich do not suggest a general relationship between couplings cis and trans to a varying ligand *X* (*26*),

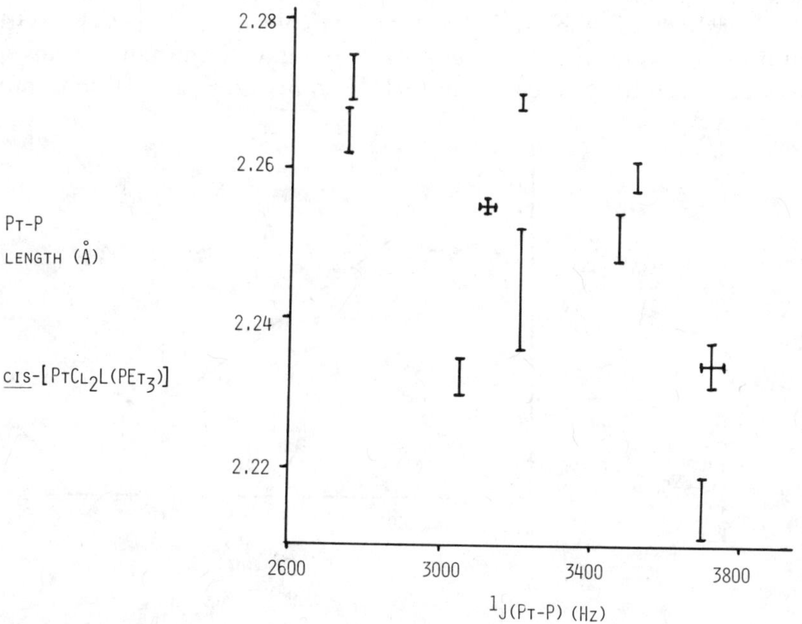

Figure 5. Graph of Pt–P (shown with standard deviation limits) against $^1J_{Pt-P}$ *for the* cis-[$PtCl_2L(PEt_3)$] *complexes*

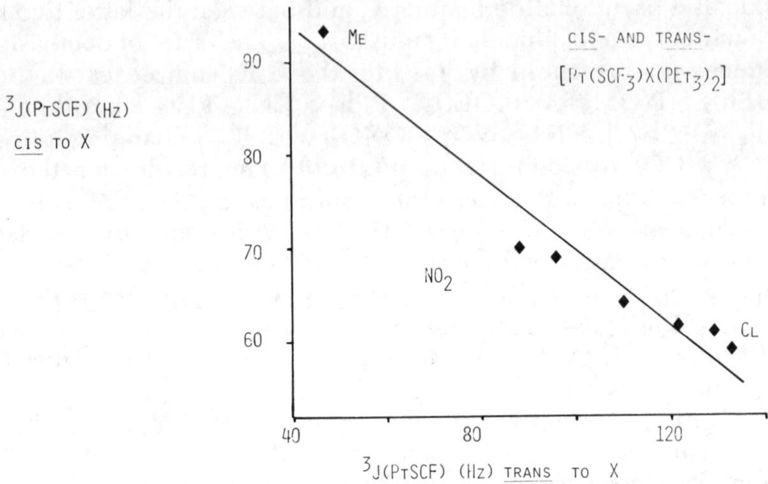

Figure 6. Graph of $^3J_{PtSCF}$ *cis to* X *in* trans-*[Pt(SCF₃)X(PEt₃)₂] against* $^3J_{PtSCF}$ *trans to* X *in cis-[Pt(SCF₃)X(PEt₃)₂]* (7, 39)

the coupling $^3J_{Pt-SCF}$ *cis* to X in *cis*-[Pt(SCF$_3$)X(PEt$_3$)$_2$] is a linear decreasing function of $^3J_{Pt-SCF}$ trans to X in *trans*-[Pt(SCF$_3$)X(PEt$_3$)$_2$] (39). Subsequent results have revealed a similar but slightly less precise correlation for the direct couplings $^1J_{Pt-P}$ (*see* Figures 6 and 7) (32).

In studies of the reactions of organotin compounds with platinum(0) and platinum(II) complexes, we have obtained a series of *cis*-bis(triphenylphosphine)platinum(II) complexes in which one

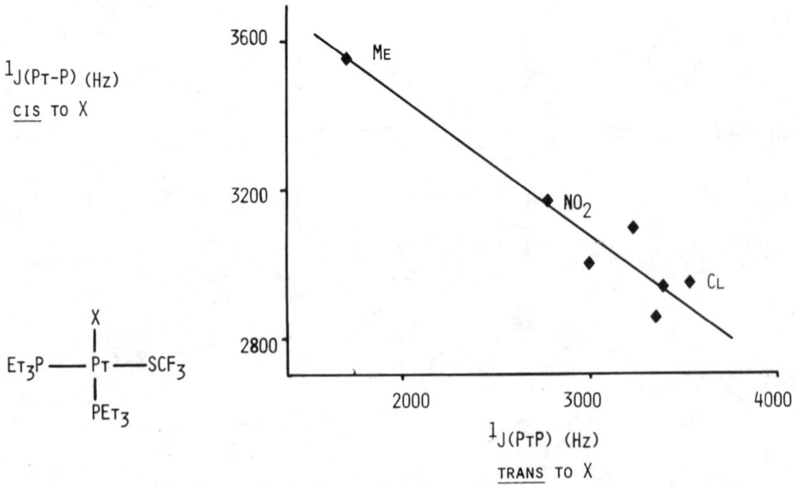

Figure 7. Graph of $^1J_{Pt-P}$ *cis to* X *against* $^1J_{Pt-P}$ *trans to* X *in* cis-*[Pt(SCF$_3$)X(PEt$_3$)$_2$]* (32)

anionic ligand varies through a range of carbon or tin donors (*40, 41, 42*). Linear relationships were found between $^{1}J_{Pt-P}$ cis to the carbon ligand in the two series *cis*-[PtCl(CX_n)(PPh_3)$_2$] and *trans*-[PtCl(CX_n)(PPh_3)$_2$] and *cis* to the tin ligand in the two series *cis*-[PtCl(SnX_3)(PPh_3)$_2$] and *trans*-[PtCl(SnX_3)(PPh_3)$_2$] (*see* Figures 8 and 9). Furthermore, in complexes where the PPh_3 ligand, which is cis to carbon or tin, is also trans to chlorine, $^{1}J_{Pt-P}$ cis to carbon or tin is a linear decreasing function of $^{1}J_{Pt-P}$ trans to carbon or tin (*see* Figures 10 and 11). For the complexes *cis*-[PtPh(SnX_3)(PPh_3)$_2$], where the PPh_3 ligand cis to tin is trans to the phenyl ligand, there is no discernible correlation between $^{1}J_{Pt-P}$ cis and trans to tin (*see* Figure 12). This was also the case for $^{1}J_{Pt-P}$ cis and trans to varying carbon ligands in [$PtCH_3$(CX_n)(dppe)] (*33*), where again the phosphorus ligand is trans to a carbon σ donor. For the *cis*-[PtPh(SnX_3)(PPh_3)$_2$] complexes (*see* Figure 12) $^{1}J_{Pt-P}$ trans to tin varies by a similar amount to $^{1}J_{Pt-P}$ trans to tin in *cis*-[PtCl(SnX_3)(PPh_3)$_2$] (*see* Figure 9), but $^{1}J_{Pt-P}$ cis to tin in the former complexes varies by a much smaller amount than for the corresponding ligand in the latter complexes. This and the results of Appleton and Bennett (*33*) suggest that when PR_3 is trans to a carbon σ donor, the bonds along this axis of the molecule are much more resistant to perturbation than when PR_3 is trans to PR_3 or chlorine. The amount of scatter on the correlations is numerically similar in each

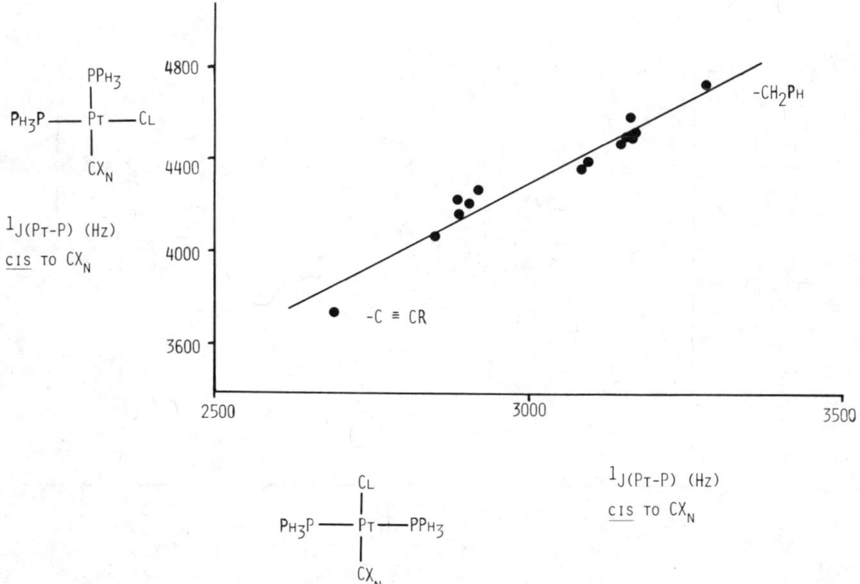

Figure 8. Graph of $^{1}J_{Pt-P}$ *cis to* CX_n *in* cis-*[PtCl(*CX_n*)(*PPh_3*)*$_2$*] against* $^{1}J_{Pt-P}$ *cis to* CX_n *in* trans-*[PtCl(*CX_n*)(*PPh_3*)*$_2$*] (*41*)*

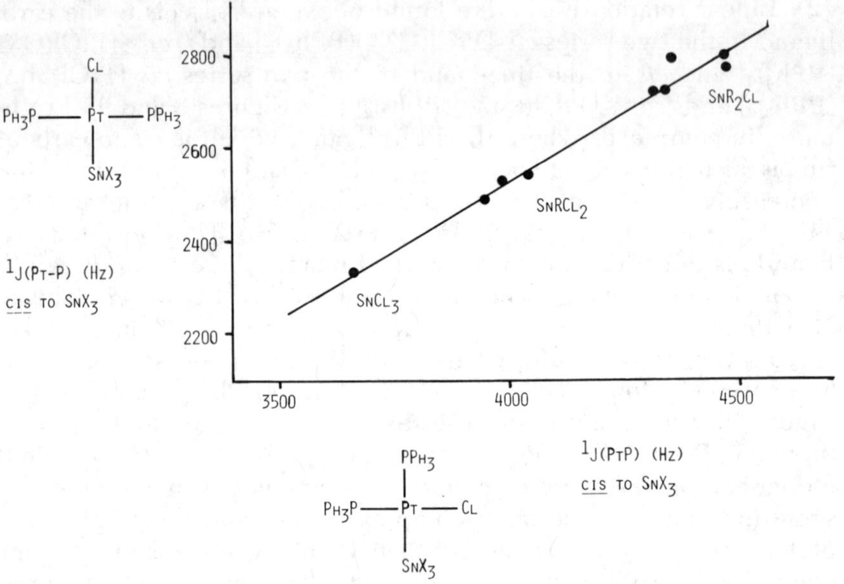

Figure 9. Graph of $^1J_{Pt-P}$ cis to SnX_3 in trans-*[PtCl(SnX$_3$)(PPh$_3$)$_2$] against $^1J_{Pt-P}$ cis to SnX_3 in* cis-*[PtCl(SnX$_3$)(PPh$_3$)$_2$]* (41)

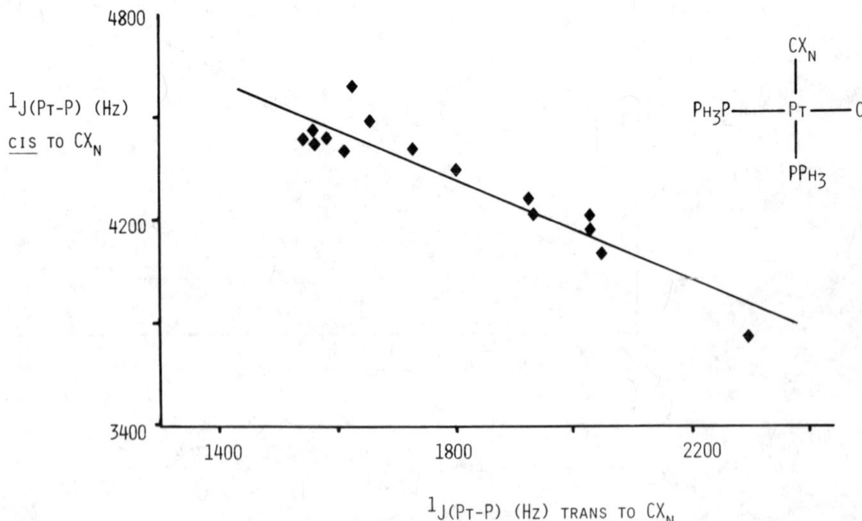

Figure 10. Graph of $^1J_{Pt-P}$ cis to CX_n against $^1J_{Pt-P}$ trans to CX_n in cis-*[PtCl(CX$_n$)(PPh$_3$)(PPh$_3$)$_2$]* (41)

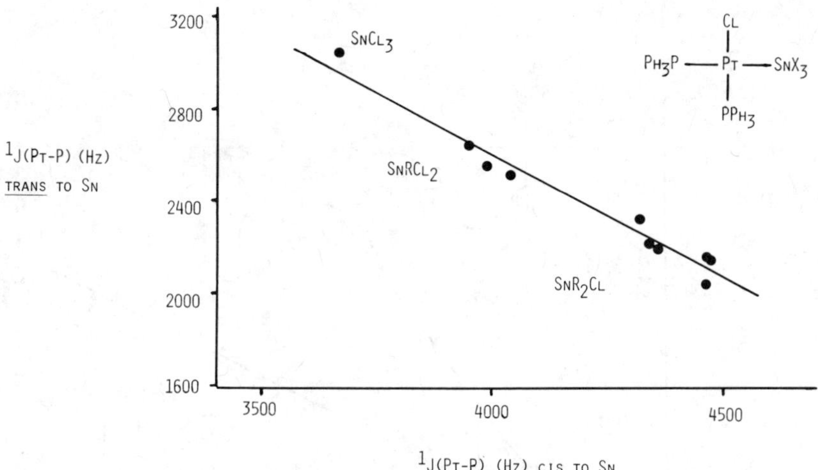

Figure 11. Graph of $^1J_{Pt-P}$ trans to SnX_3 against $^1J_{Pt-P}$ cis to SnX_3 in cis-*[PtCl(SnX_3)(PPh_3)$_2$] (41)*

case, but it is sufficient to obscure any correlation where the overall changes induced by cis ligands are small.

The results for the tin-containing complexes (*41, 42*) also permit the examination of the relationship between $^1J_{Pt-P}$ trans to the tin ligand and the indirect coupling constant $^2J_{P-Pt-Sn}$. There are accurate linear relationships for two series of complexes (*see* Figure 13) that

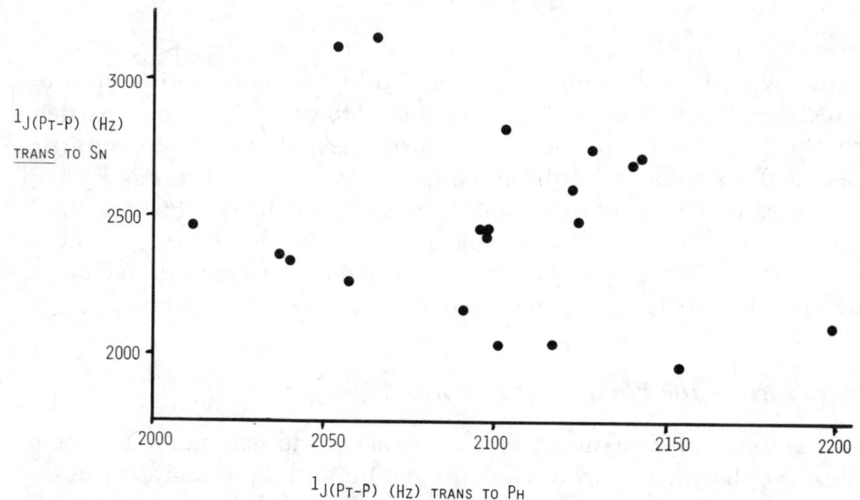

Figure 12. Plot of $^1J_{Pt-P}$ trans to SnX_3 against $^1J_{Pt-P}$ cis to SnX_3 in cis-*[PtPh(SnX_3)(PPh_3)$_2$] (41)*

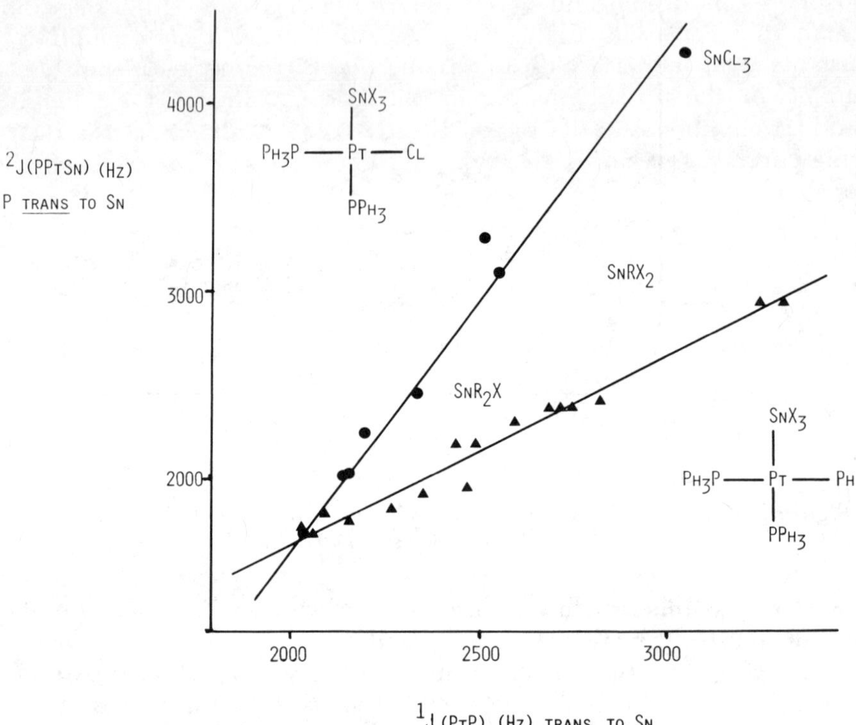

Figure 13. Graphs of $^2J_{P-Pt-Sn}$ *for* PPh_3 *trans to tin against* $^1J_{Pt-P}$ *for* PPh_3 *trans to tin for the* cis-*[PtCl(SnX₃)(PPh₃)₂]* and cis-*[PtPh(SnX₃)(PPh₃)₂] complexes (41)*

extend over sizeable ranges of magnitude. In these series of complexes, the principal variables are expected to be $(P'_{S_{Pt}S_P})^2$ (and particularly the contribution of the platinum orbitals) for the Pt–P coupling, and $|S_{Sn}(0)|^2\alpha^2_{Sn}$ for the indirect coupling. As the electronegativity of the groups on tin increases $|S_{Sn}(0)|^2$ and α^2_{Sn} for the Sn–Pt bond both increase and the trans influence of the tin ligand decreases leading to the observed correlation. Perhaps further correlations will be found when results for $^1J_{Pt-Sn}$ for these complexes become available.

Dependence on the Phosphorus Ligand

The terms of Equation 1 that are expected to be changed most as groups on phosphorus are varied are $|S_P(0)|^2$ and the s character of the phosphorus donor orbital α^2_P, which contributes directly to $(P'_{S_MS_P})^2$. Both of these terms are expected to increase with increasingly electronegative substituents on phosphorus—$|S_P(0)|^2$ by contraction of the phosphorus 3s orbital and α_P^2 by rehybridization of the phosphorus orbitals in accord with Bent's rules (*10*). In most instances coupling constants $^1J_{M-P}$ increase with increasing electronegativity of the substituents on the phosphorus donor, but the opposite trend has been noted for complexes of the ligands PPh_nR_{3-n} (R = alkyl and n = 1-3) in mercury(II) complexes $[HgX_2L_2]$ (*43*). In coordinatively saturated cadmium(II), indium(III), and tin(IV) complexes the M–P bonds are rather long and, therefore, presumably much weaker than for the transition metals (*44, 45, 46*). It seems that the anomalous trend in coupling constants for $[HgX_2L_2]$ may be common to such complexes and may arise from increasing steric pressure on the relatively weak M–P bonds as the number of phenyl substituents on the phosphorus increases.

Coupling constants are much larger for phosphite than for phosphine ligands and the ratio of the coupling constants in otherwise similar complexes lies in the 1.5–1.8 range for a variety of acceptors (e.g., rhodium(I) and (III), platinum(II), silver(I), and BH_3) (*1, 2*). Approximate constancy of the ratio is expected if changes in $|S_P(0)|^2\alpha_P^2$ are dominant. For complexes of a given metal and oxidation state, the ratio of coupling constants between phosphite and phosphine complexes is not rigorously constant. However for platinum(II) complexes we have found that for $P(OPh)_3$ and PPh_3 complexes, and a variety of other pairs of phosphorus donors, that there are accurate linear relationships between the coupling constants for complexes of different phosphorus donors (e.g., *see* Figure 14) (*47*). Where sufficient results are available such linear correlations are best limited to sets of complexes of a given stereochemistry. Thus for complexes of PPh_3 and PEt_3 the three different correlations for *trans*-$[PtAB(PR_3)_2]$, *cis*-$[PtAB(PR_3)_2]$, and $[PtABC(PR_3)]$ complexes are given by Equations 2 (correlation coefficient r = 1.00, 17 points), 3 (r = 1.00, 14 points), and 4 (r = 0.99, 10 points).

$$^1J_{Pt-PPh_3\,trans} = 1.155\ ^1J_{Pt-PEt_3\,trans} - 87 \quad (2)$$

$$^1J_{Pt-PPh_3\,cis} = 1.125\ ^1J_{Pt-PEt_3\,cis} - 215 \quad (3)$$

$$^1J_{Pt-PPh_3} = 1.120\ ^1J_{Pt-PEt_3} - 378 \quad (4)$$

The gradient is larger for Equation 2 than for Equations 3 and 4 because the trans influence of PPh_3 is lower than that of PEt_3. These relationships are sufficiently accurate to make a useful calculation of an unknown coupling constant when the result for an analogous complex with a different phosphorus donor is available.

Results also have been obtained for *cis*-$[PtCl_2L(PEt_3)]$ and *trans*-$[PtCl_2L(PCy_3)]$ (Cy = cyclohexyl) complexes, where L in the cis complexes varies over the full range of phosphorus donors from $PHPh_2$ to PF_3 and for the trans complexes from PR_3 to $P(OR)_3$ (*22*). The coupling

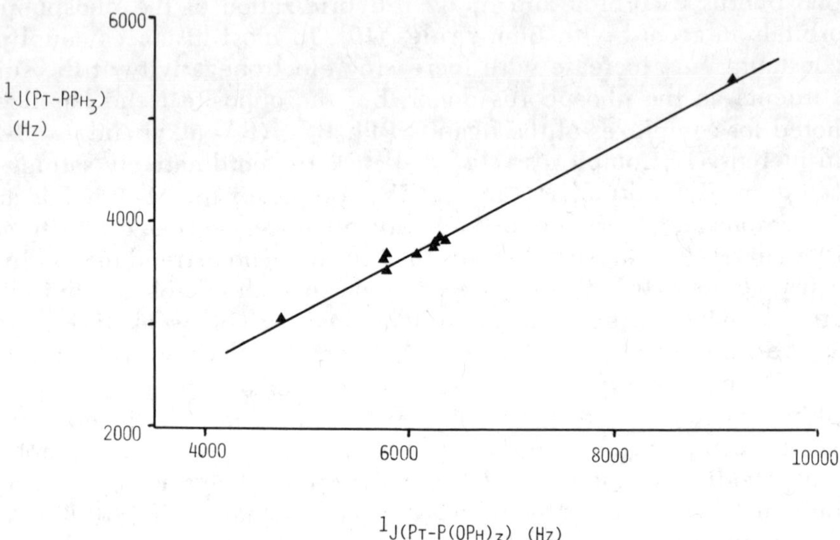

Figure 14. Graph of $^1J_{Pt-P}$ for bonds to PPh_3 in platinum(II) complexes against $^1J_{Pt-P}$ for bonds to $P(OPh)_3$ in otherwise similar complexes: $^1J_{Pt-PPh_3} = 0.530\ ^1J_{Pt-P(OPh)_3} + 517$ (correlation coefficient 0.99) (41)

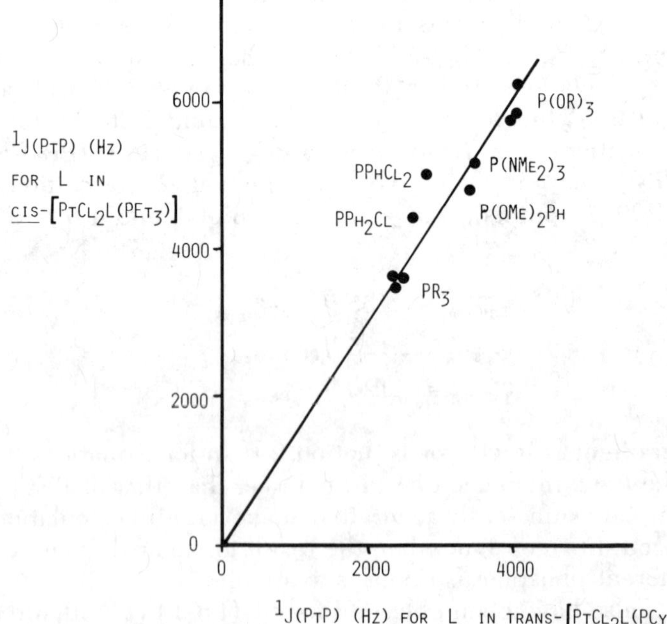

Figure 15. Graph of $^1J_{Pt-P}$ for bonds to the ligand L in cis-[$PtCl_2L(PEt_3)$] against $^1J_{Pt-P}$ for bonds to the ligand L in trans-[$PtCl_2L(PCy_3)$] (Cy = cyclohexyl) (22)

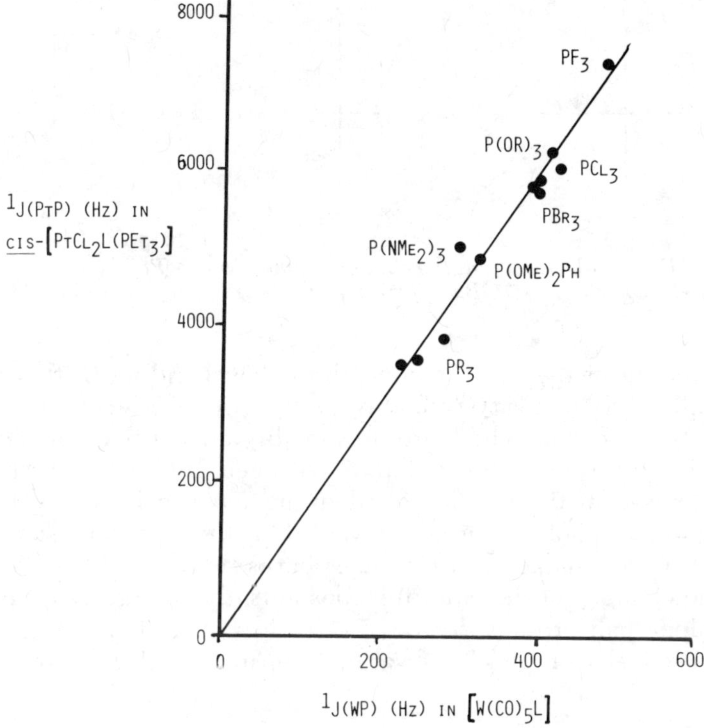

Figure 16. Graph of $^1J_{Pt-P}$ for bonds to the ligand L in cis-$[PtCl_2L(PEt_3)]$ against $^1J_{W-P}$ for bonds to the ligand L in $[W(CO)_5L]$ (22)

constants to the changing phosphorus donor in the two series are related linearly (*see* Figure 15), and there is an accurate linear relationship between the results for the cis platinum(II) complexes and the tungsten(0) $[W(CO)_5L]$ complexes (21, 48, 49, 50, 51) over the full range of phosphorus donors (*see* Figure 16). These correlations principally reflect the dominance of the phosphorus-centered terms $|S_P(0)|^2\alpha_P^2$ in ^1J (ML), but it appears that any induced changes in the terms associated with the metal must be closely parallel in the cis-platinum(II) and the tungsten(0) complexes. That changes in the metal orbitals occur is evident from changes in the bond lengths to the remaining ligands for both platinum(II) complexes (*see* Figure 17) and for chromium(0) complexes, which are presumably very similar to the analogous tungsten(0) complexes (53).

In recent years the availability of strongly acidic solvents has facilitated the determination of $^1J_{H-P}$ for a full range of protonated phosphorus donors $[HL]^+$ (*54, 55, 56*) and the graph of $^1J_{H-P}$ against $^1J_{Pt-P}$ for the cis-platinum(II) complexes is given in Figure 18. The curvature of the graph is opposite to that which is expected from synergic

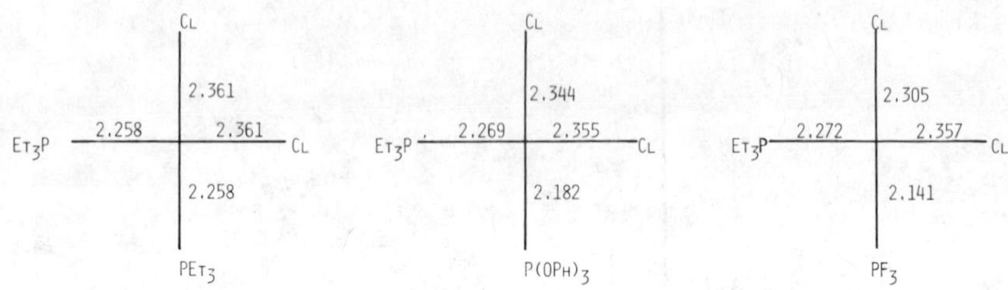

Figure 17. Pt–L bond lengths (in angstroms) in cis-*[$PtCl_2(PEt_3)_2$] (16),* cis-*[$PtCl_2(PEt_3)$ {$P(OPh)_3$}] (52), and* cis-*[$PtCl_2(PEt_3)(PF_3)$] (11)*

strengthening of the Pt–P σ bond through π bonding in the Pt–PF_3 system. The nonlinearity probably derives from induced changes in the platinum orbitals which are presumably greater than any changes induced in the 1s acceptor orbital of hydrogen. It is not surprising that some changes in the orbitals of platinum have to be invoked since three groups on phosphorus are varied in these systems and large changes in the character of the phosphorus donor orbital are evident from the changes in the coupling constants, the energies of the phosphorus lone-pair orbitals (*57, 58, 59*), and in hyperfine coupling constants in cobalt complexes of various phosphorus donors (*60*). Since

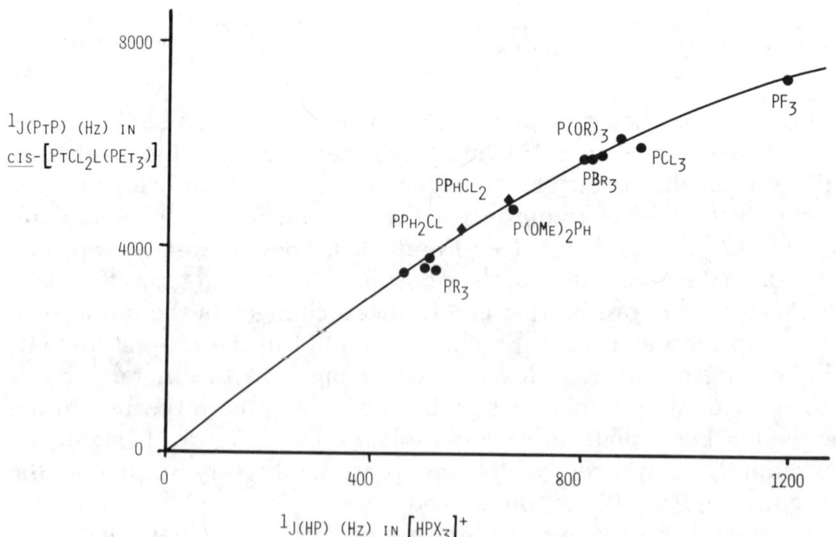

Figure 18. Graph of $^1J_{Pt-P}$ for bonds to the ligand L in cis-[$PtCl_2L(PEt_3)$] against $^1J_{H-P}$ for [HL]$^+$. Values of $^1J_{H-P}$ for L = PPh_2Cl and $PPhCl_2$, ◆, may be unreliable (55)

$^1J_{M-P}$ and the hyperfine coupling constants (*60*) depend on the product $\alpha_P^2|S_P(0)|^2$, it is not possible to determine which of the terms α_P^2 or $|S_P(0)|^2$ is the more sensitive to the groups attached to phosphorus. We expect that the coefficients of the orbitals α_P^2 vary more than the electron density $|S_P(0)|^2$, and it is probable that α_P^2 varies more than is implied by the hybrid orbitals constructed to follow known XPX bond angles. Thus it is more likely that the changes in $^1J_{M-P}$ derive more than α_P^2 than from $|S_P(0)|^2$, but since the terms change in the same way for M–P bonds, discussion in terms of the product $\alpha_P^2|S_P(0)|^2$ is more appropriate at present.

Literature Cited

1. Nixon, J. F.; Pidcock, A. *Ann. Rev. NMR Spectrosc.* **1969,** *2,* 346.
2. Pregosin, P. S.; Kunz, R. W.; "^{31}P and ^{13}C NMR of Transition Metal Phosphine Complexes"; Springer–Verlag: Berlin, 1979.
3. Appleton, T. G.; Clark, H. C.; Manzer, L. E. *Coord. Chem. Rev.* **1973,** *10,* 335.
4. Pople, J. A.; Santry, D. P. *Mol. Phys.* **1964,** *8,* 1.
5. Allen, F. H.; Pidcock, A.; Waterhouse, C. R. *J. Chem. Soc.* (*A*) **1970,** 2087.
6. Motschi, H.; Pregosin, P. S.; Venanzi, L. M. *Helv. Chim. Acta* **1979,** *62,* 667.
7. Dixon, K. R.; Moss, K. C.; Smith, M. A. R.; *J. Chem. Soc. Dalton* **1975,** 990.
8. Pidcock, A.; Richards, R. E.; Venanzi, L. M. *J. Chem. Soc.* (*A*) **1966,** 1707.
9. Balimann, G.; Pregosin, P. S. *J. Mag. Res.* **1976,** *22,* 235.
10. Bent, H. A. *Chem. Rev.* **1961,** *61,* 275.
11. Hitchcock, P. B.; Jacobson, B.; Pidcock, A. *J. Chem. Soc. Dalton* **1977,** 2043.
12. de Bie, M. J. A., unpublished data.
13. Colquhoun, I. J.; McFarlane, W. *Chem. Comm.* **1980,** 145.
14. Alyea, E. C.; Dias, S. A.; Stevens, S. *Inorg. Chim. Acta* **1980,** *44,* L203.
15. Mather, G. G.; Pidcock, A.; Rapsey, G. J. N. *J. Chem. Soc. Dalton* **1973,** 2095.
16. Manojlović–Muir, Lj.; Muir, K. W.; Solomun, T. J. *Organometal. Chem.* **1944,** *142,* 265.
17. Mazid, M. A.; Russell, D. R. *J. Chem. Soc. Dalton* **1980,** 1737.
18. Bushnell, G. W.; Pidcock, A.; Smith, M. A. R.; *J. Chem. Soc. Dalton* **1975,** 572.
19. Messmer, G. G.; Amma, E. L. *Inorg. Chem.* **1966,** *5,* 1775.
20. Hitchcock, P. B.; Jacobson, B.; Pidcock, A. *J. Organometal. Chem.* **1977,** *136,* 397.
21. Aslanov, L.; Mason, R.; Wheeler, A. G.; Whimp, P. O.; *Chem. Comm.* **1970,** 30.
22. Jacobson, B. Ph. D. Thesis, Sussex, Brighton, U.K., 1977.
23. Hitchcock, P. B.; Jacobson, B.; Pidcock, A. *J. Chem. Soc. Dalton* **1977,** 2038.
24. Anderson, G. K.; Cross, R. J.; Manojlović-Muir, Lj.; Muir, K. W.; Solomun, T. *J. Organometal. Chem.* **1979,** *170,* 385.
25. Shustorovich, E. *J. Amer. Chem. Soc.* **1979,** *101,* 792.
26. Shustorovich, E. *Inorg. Chem.* **1979,** *18,* 1039.
27. Muir, K. W.; Walker, R. *J. Chem. Soc. Dalton* **1975,** 272.
28. Ref. 2, p 92.
29. Goggin, P. L.; Goodfellow, R. J.; McEwan, D. M.; Griffith, A. J.; Kessler, K. *J. Chem. Res.* (*M*) **1979,** 2315.

30. Eichbichler, J.; Peringer, P. *Inorg. Chim. Acta* **1980,** *43,* 121.
31. Appleton, T. G.; Chisholm, M. H.; Clark, H. C.; Manzer, L. E. *Can. J. Chem.* **1973,** *51,* 2243.
32. Cairns, M. A.; Dixon, K. R.; Rivett, G. A. *J. Organometal. Chem.* **1979,** *171,* 373.
33. Appleton, T. G.; Bennett, M. A. *Inorg. Chem.* **1978,** *17,* 738.
34. Tau, K. D.; Meek, D. W. *Inorg. Chem.* **1979,** *18,* 3574.
35. Miyamoto, T. *J. Organometal. Chem.* **1977,** *134,* 335.
36. Mather, G. G.; Rapsey, G. J. N.; Pidcock, A. *Inorg. Nucl. Chem. Lett.* **1973,** *9,* 567.
37. Manojlović-Muir, Lj.; Muir, K. W.; Solomun, T. *J. Organometal. Chem.* **1977,** *142,* 265.
38. Russell, D. R.; Tucker, P. A.; Wilson, S. *J. Organometal. Chem.* **1976,** *104,* 387.
39. Dixon, K. R.; Moss, K. C.; Smith, M. A. R.; *Inorg. Nucl. Chem. Lett.* **1974,** *10,* 373.
40. Eaborn, C.; Odell, K. J.; Pidcock, A. *J. Chem. Soc. Dalton* **1978,** 357.
41. Butler, G.; Eaborn, C.; Pidcock, A. *J. Organometal. Chem.* **1979,** *181,* 47.
42. Butler, G.; Eaborn, C.; Pidcock, A. *J. Organometal. Chem.* **1980,** *185,* 367.
43. Grim, S. O.; Lui, P. J.; Keiter, R. L. *Inorg. Chem.* **1974,** *13,* 342.
44. Cameron, A. F.; Forrest, K. P.; Ferguson, G. *J. Chem. Soc.* (*A*) **1971,** 1286.
45. Veidis, M. V.; Palenik, G. J. *Chem. Comm.* **1969,** 586.
46. Mather, G. G.; McLaughlin, G.; Pidcock, A. *J. Chem. Soc. Dalton* **1973,** 1823.
47. Butler, G. Ph.D. Thesis, Sussex, Brighton, U.K., 1978.
48. Verkade, J. G. *Coord. Chem. Rev.* **1972,** *9,* 1.
49. Grim, S. O.; Ferrenco, R. A. *Inorg. Nucl. Chem. Lett.* **1966,** 2, 1846.
50. Grim, S. O.; McFarlane, W.; Wheatland, D. A. *Inorg. Nucl. Chem. Lett.* **1966,** *2,* 49.
51. Grim, S. O.; McAllister, P. R.; Singer, R. M. *Chem. Comm.* **1969,** 38.
52. Caldwell, A. N.; Manojlović-Muir, Lj.; Muir, K. W. *J. Chem. Soc. Dalton* **1977,** 2265.
53. Plastas, H. J.; Stewart, J. M.; Grim, S. O. *J. Amer. Chem. Soc.* **1969,** *91,* 4326.
54. Olah, G. A.; McFarlane, W. *J. Org. Chem.* **1969,** *34,* 1832.
55. McFarlane, W.; White, R. F. M. *Chem. Comm.* **1969,** 744.
56. Verkade, J. G.; Vande Griend, L. J. J. *J. Amer. Chem. Soc.* **1975,** *97,* 5958.
57. Hillier, I. H.; Saunders, V. R. *Chem. Comm.* **1970,** 316.
58. Hillier, I. H.; Saunders, V. R. *J. Chem. Soc.* (*A*) **1971,** 664.
59. Guest, M. F.; Hillier, I. H.; Saunders, V. R. *J. Chem. Soc., Faraday Trans 2* **1972,** *68,* 114.
60. Wayland, B. B.; Abd–Elmageed, M. E. *J. Amer. Chem. Soc.* **1974,** *96,* 4809.
61. Eichbichler, J.; Peringer, P. *J. Inorg. Nucl. Chem.,* in press.

RECEIVED July 15, 1980.

P-31, Sn-119, and Pt-195 NMR Studies on Platinum–Tin Homogeneous Hydrogenation Catalysts

K. A. OSTOJA STARZEWSKI and P. S. PREGOSIN
Laboratorium für Anorganische Chemie, ETH-Zentrum, Universitätstrasse 6, CH-8092 Zürich, Switzerland

P-31, Sn-119, and Pt-195 NMR spectroscopic methods have been used to identify the products of the reaction of $SnCl_2$ with phosphine complexes of platinum(II). These complexes are of the type $[Pt(SnCl_3)L(PR_3)_2]$, L = Cl^-, $SnCl_3^-$, H^-, or alkyl, and are of interest, spectroscopically, in that the one-bond platinum–tin coupling $^1J(^{195}Pt, ^{119}Sn)$ can exceed 30,000 Hz and therefore represents the largest known nuclear spin–spin coupling constant. The two-bond interaction in the poly-$SnCl_3$ complex $^2J(^{119}Sn, ^{117}Sn)$ at > 18,000 Hz is also exceedingly large.

Platinum phosphine complexes are recognized to function as homogeneous hydrogenation and hydroformylation catalysts when combined with an excess of a co-catalyst such as tin(II) chloride (*1*, *2*, *3*). Specifically, α-olefin hydroformylation proceeds both with high yield and selectivity using the combination $[PtCl_2(PPh_3)_2]$ + $SnCl_2$ as shown in Equation 1 (*4*):

$$CH_3(CH_2)_4CH{=}CH_2 \xrightarrow[\substack{PtCl_2(PPh_3)_2 + n\text{-}SnCl_2 \\ 66^\circ}]{H_2/CO} 90\%\ \text{1-octanal} \qquad (1)$$

The homogeneous hydrogenation and hydroformylation reactions are known to involve metal hydride, metal olefin, and metal alkyl coordination complexes (*5*); however, there is relatively little in the literature concerning the function of $SnCl_2$ in this connection. The work that has been done suggests (*6*, *7*, *8*, *9*) that $SnCl_2$ reacts with platinum complexes containing a Pt–Cl bond according to Equation 2.

0065–2393/82/0196–0023$05.00/0

$$Pt\text{–}Cl + SnCl_2 \rightarrow Pt\text{–}SnCl_3 \quad (2)$$

In support of the direct platinum–tin bond are crystallographic structural determinations for $[Mn(SnCl_3)(CO)_5]$ (*10*), $[Pd(\pi\text{-allyl})(SnCl_3)(PPh_3)]$ (*11*), $[Au(SnCl_3)(PMe_2Ph)_2]$ (*12*), and the cyclooctadiene cluster $[Pt_3(SnCl_3)_2(C_8H_{12})_3]$ which has a triangle of platinum atoms capped above and below by two $SnCl_3^-$ groups (*13*). In addition to the metal–metal bond there are two further possible interactions of the $SnCl_3^-$ ligand with the platinum. The most trivial is nonassociative, with the $SnCl_2$ serving as a chloride extractor forming the trichlorostannate counterion, and indeed this has been found in the solid state for $[CoCl(dppe)_2]SnCl_3$ (dppe = 1,2-bis(diphenylphosphino)ethane (*14*). The last and most intriguing option, stems from a recent structural determination of $[Ag(SnCl_3)(PP)]$ where PP = 2,11-bis(diphenylphosphinomethyl)benzo[*c*]phenanthrene, a chelating diphosphine that may span the trans positions of a square planar complex (*15*). In this molecule, abbreviated as **I**, the $SnCl_3^-$ coordinates to the transition metal via the halogen.

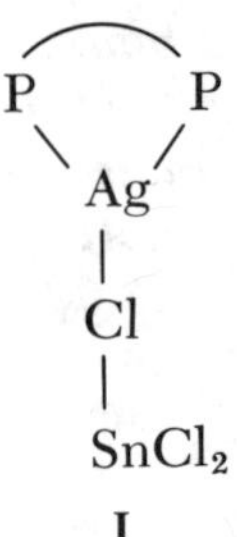

I

The problem of establishing the solution structure of platinum–tin complexes is complicated by the lability of this system. We have found (*16, 17*) (and will refer to this later) that these complexes are often dynamic on the NMR time scale. Despite these difficulties it is possible to characterize such molecules using NMR methods.

This means that we use (a) P-31 NMR both qualitatively, for information on the orientation of the tertiary phosphine ligands, and analytically to determine the number of complexes in solution (this is a very receptive nucleus and 1–2% of impurities often are detected readily); (b) Sn-119 (**I** = $\frac{1}{2}$, natural abundance = 8.6%) NMR as a probe for the identity of the trichlorostannate moiety; and (c) Pt-195 (**I** = $\frac{1}{2}$, natural abundance = 33.7%) NMR for multiplicity data concerned with the number of coordinated phosphines (and NMR spins in general). The value of H-1 NMR in metal hydride chemistry is so well established (*18*) that no further justification is required here.

We begin by considering the products of Equation 3 and follow by discussing the complexes that result from adding molecular hydrogen and eventually an acetylene.

$$PtCl_2P_2 \xrightarrow{SnCl_2} ? \xrightarrow{SnCl_2} ? \qquad (3)$$

Experimental

P-31, Sn-119, and Pt-195 NMR spectra were measured using Bruker HX-90 (^{31}P only) and WM-250 (101.27, 93.28, and 53.77 MHz) spectrometers. The samples were measured as solutions in rotating 10-mm tubes using ~40°–45° pulse angles for ^{31}P, ^{119}Sn, and ^{195}Pt with acquisition times of ~0.7, 0.2, and 0.2 s, respectively. Spectral widths were routinely (WM-250) 10,000, 50,000 and 50,000 Hz for these same nuclei. Chemical shifts are in parts per million relative to external H_3PO_4, Me_4Sn, and Na_2PtCl_6 and ±1 ppm for the two metals. A positive sign indicates a shift to a lower field (higher frequency) relative to the reference. Coupling constants are in hertz and are ±3 Hz for ^{31}P and ±12 Hz for ^{119}Sn and ^{195}Pt. The variable-temperature spectra were measured using the commercially provided controller whose accuracy is ~ ±1°C. The temperatures and solvents for the individual measurements are given in the tables. Sample concentrations were of the order of 5×10^{-2} *M*.

Results and Discussion

[PtCl(SnCl$_3$)P$_2$] Complexes. The reaction of *cis*-[$PtCl_2(PEt_3)_2$] (or the tripropyl or tri-*n*-butyl analogs) in either $CDCl_3$ or CD_2Cl_2 does not afford simple insertion of $SnCl_2$ into a Pt–Cl bond to yield the cis isomer, but gives directly *trans*-[$PtCl(SnCl_3)(PEt_3)_2$]. The trans isomer is suggested by the appearance of a single ^{31}P resonance, $\delta = 13.6$, flanked symmetrically by ^{195}Pt ($^1J(^{195}Pt, ^{31}P) = 2042$ Hz) and $^{117,119}Sn$ ($^2J(^{117}Sn, ^{31}P) = 227$ Hz, $^2J(^{119}Sn, ^{31}P) = 237$ Hz) satellite lines. (In the proceeding pages only values for the ^{119}Sn isotope will be given). The ^{119}Sn spectrum reveals triplet structure from the two PEt_3 groups and ^{195}Pt satellites whose spacing is strongly suggestive of a one-bond coupling constant; the ^{195}Pt spectrum shows a 1 : 2 : 1 structure confirming the presence of two equivalent PEt_3 groups ($\delta\ ^{195}Pt = -4790$ relative to Na_2PtCl_6). We reserve further comment on these coupling constants until the descriptive chemistry is finished. The same complex is obtained directly from *trans*-[$PtCl_2(PEt_3)_2$]. The analogous dichloride complex containing the trans spanning chelate, Complex **II**, which is electronically similar to PEt_3, cleanly affords *trans*-[$PtCl(SnCl_3)$-(Et_2PPEt_2)] (*19*).

The complexes *cis*-[$PtCl(SnCl_3)P_2$] may be obtained starting from either a cis–bis-phosphine complex of a tertiary aryl phosphine or a chelating diphosphine, e.g. PPh_3 or DIOP complexes (*20*), (although careful examination of the P-31 spectra of the monodentate systems suggests that the trans isomer is also present).

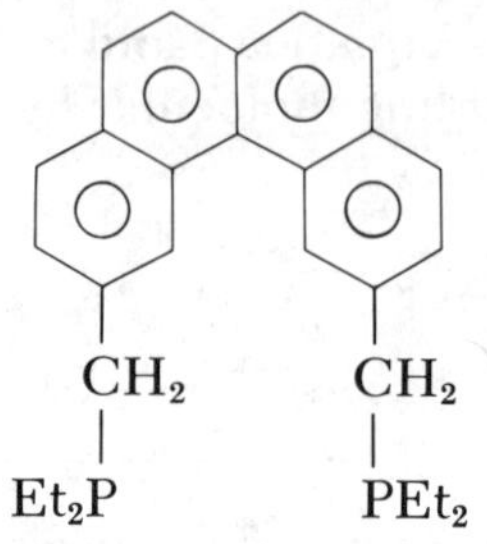

$\equiv Et_2PPEt_2$ = bis(diethylphosphinomethyl)benzo[*c*]-phenanthrene

CH_2 CH_2

Et_2P PEt_2

II

[$Pt(SnCl_3)_2P_2$] Complexes. Further addition of $SnCl_2$ to a solution containing *trans*-[$PtCl(SnCl_3)(PEt_3)_2$] leads to the equilibrium shown in Equation 4. Unfortunately, $SnCl_2$ is only sparingly soluble in chlorinated hydrocarbons; however, sufficient tin(II) dichloride can be dissolved in acetone such that the equilibrium lies far to the right.

$$trans\text{-}[PtCl(SnCl_3)(PEt_3)_2] \xrightleftharpoons{SnCl_2} trans\text{-}[Pt(SnCl_3)_2(PEt_3)_2] \quad (4)$$

At this point it is worth mentioning that there are two ways to detect the presence of a poly-trichlorostannane derivative using NMR spectroscopy: (1) via the relative intensities of the ^{117}Sn and ^{119}Sn satellites in the P-31 (or Pt-195) NMR spectrum and (2) from the observation of a $^2J(^{117}Sn, ^{119}Sn)$ coupling constant in the Sn-119 spectrum.

Point 2 is the more obvious and can be demonstrated for a complex such as **III.** Clearly the two magnetically inequivalent tin atoms

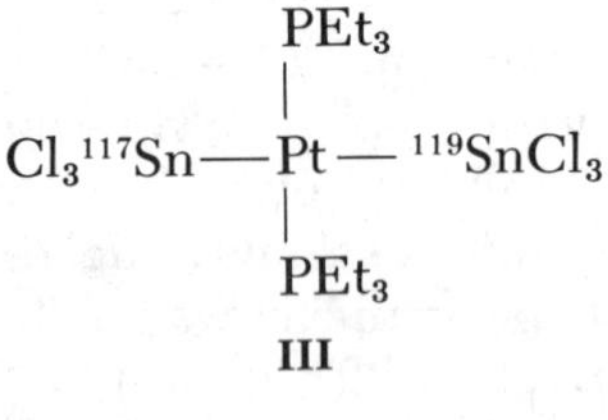

PEt_3

$Cl_3{}^{117}Sn—Pt—{}^{119}SnCl_3$

PEt_3

III

will superimpose an additional simple coupling pattern on the Sn-119 spectrum, thereby unequivocally establishing the presence of more than one $SnCl_3^-$ ligand (*see* Figure 1).

Point 1 is less obvious but equally important. In the P-31 spectrum of a mono-tin complex, the ratio of the ^{119}Sn (or ^{117}Sn) satellites to the main-band (^{31}P not coupled to NMR-active tin) will be approximately 1:20 due to the relatively dilute spin $I = \frac{1}{2}$ tin isotopes ([~84% ^{31}P not coupled to $I = \frac{1}{2}$ Sn]/[~8% ^{31}P coupled to e.g. ^{119}Sn, divided by 2 due to coupling]). In a complex containing two $SnCl_3$ groups there are molecules having only one ^{117}Sn or ^{119}Sn (~13%) and a few (<3%)

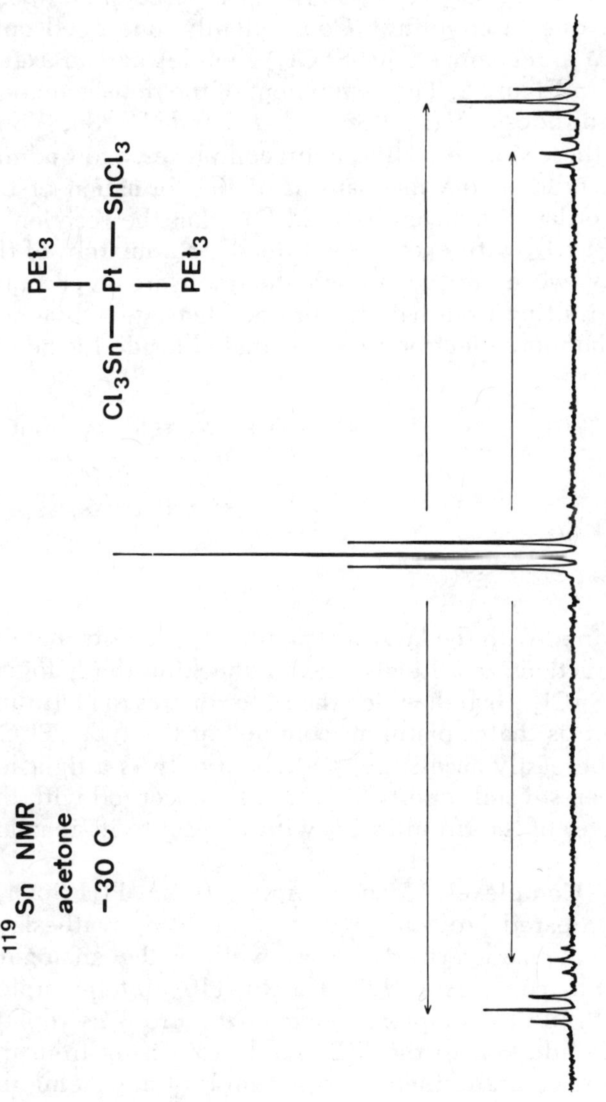

Figure 1. Sn-119 NMR spectrum of trans-[Pt(SnCl$_3$)$_2$(PEt$_3$)$_2$]. The main triplet due to ^{31}P coupling is flanked symmetrically by ^{195}Pt satellites (main-band/satellite ≅ 4:1) and ^{117}Sn satellites (main-band/satellite ≅ 20:1). Note that $^1J(^{195}Pt, ^{119}Sn)$ is only slightly larger than $^2J(^{119}Sn, ^{117}Sn)$. This is clearly a poly-tri-chlorostannate complex; however, a closer examination of the relative integrals (main-band/satellite) suggests a PtP_2Sn_3 type complex in this solvent.

having two active tins (this latter isotopomer affords only weak signals). The tin-satellite, main-band ratio for the former isotopomer is ~1:10, twice as large as in the compound having only a single tin ligand ([~70% ^{31}P not coupled to I = $\frac{1}{2}$ Sn]/[~13% ^{31}P coupled to e.g. ^{119}Sn, divided by 2 due to coupling). Consequently, one need only inspect the spectrum to recognize a bis-$SnCl_3^-$ complex and an example of this is shown in Figure 2. The assignment of the trans geometry is based on the magnitudes of ^2J ($^{117,119}Sn$, ^{31}P) (*16*) and ^2J (^{119}Sn, ^{117}Sn).

Before leaving these simple platinum–tin complexes, a few admittedly speculative words on the mechanism of the formation of the Pt–Sn bond are in order. We have observed (*21*) that the reaction of *sym-trans*-$[Pt_2Cl_4(PEt_3)_2]$ with excess $SnCl_2$ leads to a mixture of the Complexes **IV** and **V** (we are not certain whether these are cis or trans), with no products resulting from halogen bridge cleavage. Obviously the $SnCl_2$ prefers the more electron-rich terminal chloride ligand.

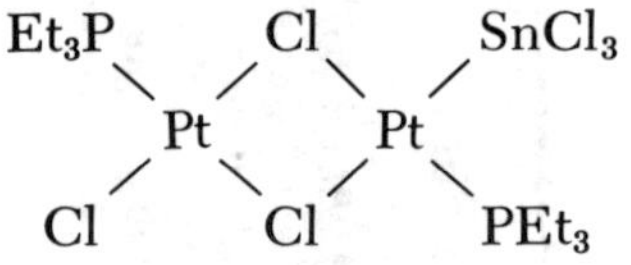

```
Et3P      Cl      SnCl3          Et3P      Cl      SnCl3
    \    /  \    /                   \    /  \    /
     Pt      Pt                       Pt      Pt
    /   \  /   \                     /   \  /   \
  Cl     Cl     PEt3            Cl3Sn     Cl     PEt3
         IV                                V
```

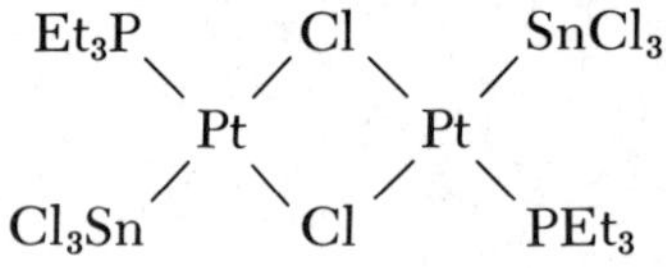

This fact, combined with the known structure, **I**, suggests that the tin(II) dichloride functions as a Lewis acid in attacking the halogen, thus producing the $SnCl_3^-$ ligand, which then coordinates to platinum. The implication here is that a platinum complex of the type $[Pt_2Cl_3(PEt_3)_2]SnCl_3$ may be easily accessible, perhaps if only as a tight ion pair. There have been several reports (*22, 23, 24*) concerned with the electrophilic character of the tin in $SnCl_2$ with respect to its reaction with ligands.

$[PtH(SnCl_3)P_2]$ Complexes. The complex *trans*-$[PtH(SnCl_3)(PEt_3)_2]$ has been prepared previously (*9*) and we have synthesized (*see* Equation 5) and characterized this as well as the analogous PPh_2CH_2Ph and PPh_3 complexes (*17*). The Sn-119 proton-coupled NMR spectrum of the PPh_3 complex, shown in Figure 3, is replete with information. In addition to the ^{195}Pt satellites arising from the metal–metal coupling constant, there is triplet multiplicity stemming from the equivalent tertiary phosphines and a surprisingly large coupling due to the trans hydride. These observations are once again consistent with a structure in which the tin atom occupies a position within the square planar coordination sphere of the platinum.

$$trans\text{-}[PtHClP_2] + SnCl_2 \xrightarrow[CD_2Cl_2]{} trans\text{-}[PtH(SnCl_3)P_2] \qquad (5)$$

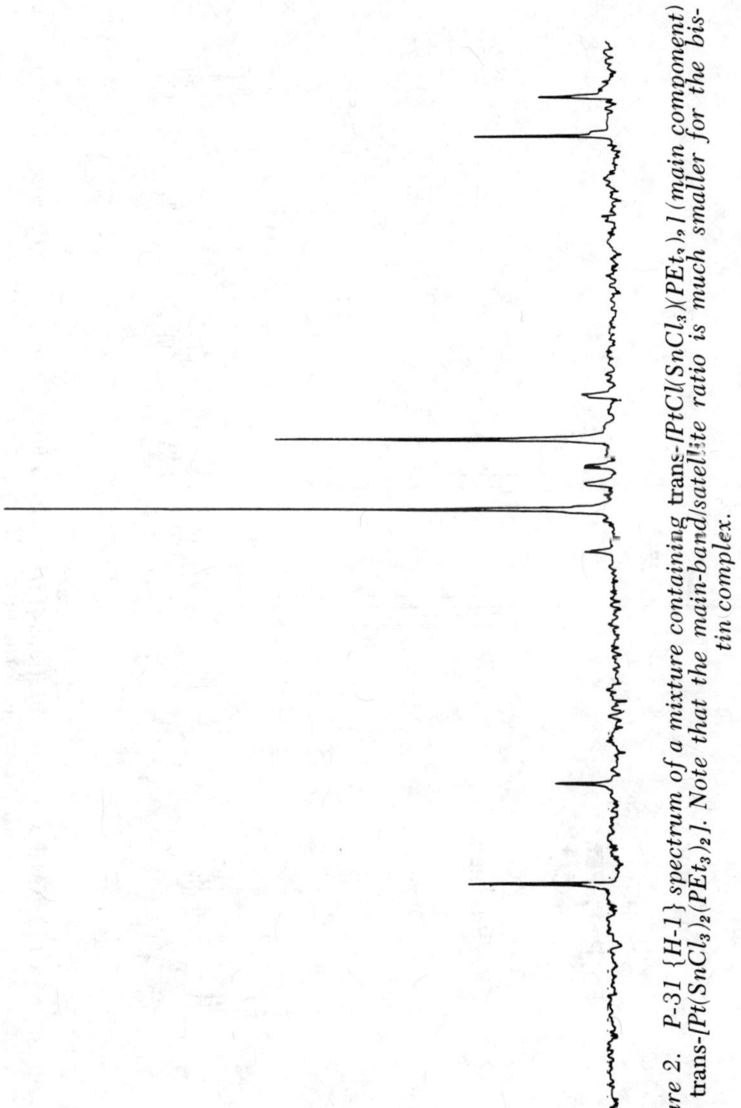

Figure 2. P-31 {H-1} spectrum of a mixture containing trans-*[PtCl($SnCl_3$)(PEt_3)$_2$] (main component) and* trans-*[Pt($SnCl_3$)$_2$(PEt_3)$_2$]. Note that the main-band/satellite ratio is much smaller for the bis-tin complex.*

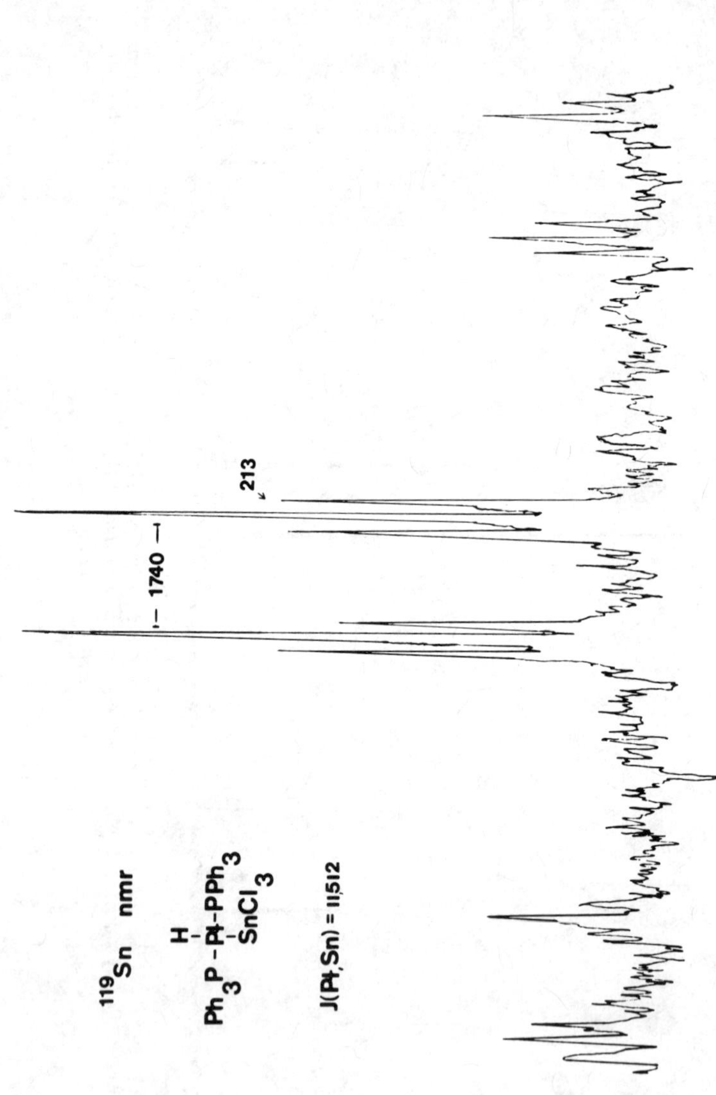

Figure 3. *Sn-119 NMR spectrum of* trans-*[PtH(SnCl$_3$)(PPh$_3$)$_2$] showing ^{195}Pt satellites, ^{31}P triplet multiplicity, and the large ^{1}H hydride doublet splitting*

Unlike Equation 5, the hydrogenation reaction begins from $[PtCl_2P_2]$, which in the presence of an excess of $SnCl_2$, gives poly-tin complex. We find that reaction of *cis*-$[PtCl_2(PPh_3)_2]$ with 2 mol of $SnCl_2$ under 1 atm H_2 in $CHCl_3$, followed by addition of tetramethylammonium chloride, gives good yields of *trans*-$[PtHCl(PPh_3)_2]$ (*25*) (*see* Equation 6):

$$cis\text{-}[PtCl_2(PPh_3)_2] + 2SnCl_2 \xrightarrow[\text{CDCl}_3,\text{ room temperature}]{\text{1) H}_2 \quad \text{2) Me}_4\text{NCl}} trans\text{-}[PtHCl(PPh_3)_2] + (Me_4N)SnCl_3 \quad (6)$$

Consequently, the presence of the $SnCl_2$ promotes the reaction between molecular hydrogen and the platinum complex. Further, with PEt_3 as tertiary phosphine, we have observed the P-31 spectrum of *trans*-$[PtH(SnCl_3)(PEt_3)_2]$ in a similar reaction mixture ($\sim -100°C$ in methanol (*21*)). It seems likely that the hydrido–tin complex is formed from the reaction with hydrogen and is converted to the hydrido–chloride, as in Equation 6, when the sparingly soluble $(Me_4N)SnCl_3$ is precipitated.

In independent experiments we have shown that *trans*-$[PtH(SnCl_3)(PPh_3)_2]$ and *trans*-$[PtH(SnCl_3)(PPh_2CH_2Ph)_2]$ react with $Et_2OOC{-}C{\equiv}C{-}COOEt$, at room temperature, to afford *trans*-$[Pt(EtOOC{=}CHCOOEt)(SnCl_3)P_2]$ complexes, both of which have the tin coordinated directly to platinum. The choice of acetylene is somewhat specific; however, *trans*-$[Pt(C_6H_5)(SnCl_3)(PEt_3)_2]$ and *trans*-$[Pt(p\text{-}CH_3C_6H_4)(SnBr_3)(PEt_3)_2]$ (*21*) as well as *trans*-$[Pt(SnCl_3)(m\text{-}FC_6H_4)(PEt_3)_2]$ (*9*) are all stable molecules suggesting that a carbon ligand trans to an SnX_3 moiety is acceptable.

NMR Coupling Constants. OVER ONE BOND. Throughout this study we have relied heavily on the observation and empirical interpretation of spin–spin coupling constants. Several of these, such as $^1J(^{195}Pt, ^{31}P)$ and $^2J(^{31}P, ^{31}P)$, have been investigated in some detail (*26*); others, such as $^2J(^{117,119}Sn, ^{31}P)$, have been investigated only sparingly (*27, 28, 29*) or, as in the case of $^1J(^{195}Pt, ^{119}Sn)$, not at all. We especially were interested in this last value as, once characterized, it would prove an empirical tool for establishing the presence of a platinum–tin bond.

The first complex that we satisfactorily characterized was *trans*-$[PtCl(SnCl_3)(PEt_3)_2]$ (*16*). The ^{31}P and ^{195}Pt data (we had no ^{119}Sn capability at that time) strongly suggested a Pt–Sn bond, although the Pt-195 spectrum showed no resonances other than the 1:2:1 triplet due to ^{31}P coupling, and some weak, out-of-phase signals. These were weak and out of phase because the value $^1J(^{195}Pt, ^{119}Sn)$, 28,954 Hz, was so large that the ^{119}Sn satellites lay outside of our spectral width! Fortunately we had access to a more sophisticated computer and, eventually, could characterize properly a full spectrum. To our knowledge this represented (*30*) (*see* Structures **IV** and **V**) the largest nuclear spin–spin coupling ever observed, and in Tables I and II we list both the coupling constant and chemical shift data for our platinum–tin complexes. (We have learned recently that the anionic complexes $PtCl_2(SnCl_3)_2^{2-}$ and $Pt(SnCl_3)_5^{3-}$ have been measured and show $^1J(^{195}Pt, ^{119}Sn)$ values of 27,635 and 16,024 Hz, respectively (*31*)). From our values $^1J(^{195}Pt, ^{119}Sn)$ we observe that (a) there is a change of approximately a factor of six as a function of the trans ligand and (b) that these changes are reminiscent of a trans influence (*32*) series, in that the poorer ligands, e.g. Cl^-, are associated with larger, one-bond coupling constants. This latter point suggests that we will, in part, be able to interpret changes in this NMR parameter in terms of our knowledge of the factors affecting the trans influence (*33*). To appreciate the absolute magnitude of this coupling constant it is useful to consider the Fermi contact expression (*see* Equation 7) as given by Pople and Santry (*34, 35*). The symbols γ represent the gyromagnetic ratios of the nuclei, the terms $|\Psi_{ns}(0)|^2$ are the valence s-electron densities at the nuclei, A and B and $\pi_{A,B}$ is the mutual polarizability. The $\pi_{A,B}$ expression contains the s coefficients of the atomic orbitals used in the linear combinations that make up the occupied and unoccupied molecular orbitals, as well as a difference term, $(\epsilon_j - \epsilon_i)$ where ϵ_i is the energy of an occupied ϵ_j and an unoccupied molecular orbital.

$$^1J(A,B) \sim \gamma_A \gamma_B |\Psi_{ns_A}(0)|^2 |\Psi_{ns_B}(0)|^2 \pi_{A,B} \quad (7)$$

$$\pi_{A,B} = \sum_i^{occ} \sum_j^{unocc} (\epsilon_j - \epsilon_i)^{-1} C_{iA} C_{iB} C_{jA} C_{jB}$$

One can ask which of the three terms in Equation 7 would differ markedly if we compare $^1J(^{195}Pt, ^{119}Sn)$ with, e.g., $^1J(^{195}Pt, ^{31}P)$. The ratio $\gamma_{^{119}Sn}/\gamma_{^{31}P}$ is equal to 0.92 and therefore is not an important parameter. The ratio $|\Psi_{Sn}(0)|^2/|\Psi_P(0)|^2$ is of the order of 2 (*26*), but it may have a pronounced dependence on the charge of the ligand atom. It is difficult to know if this factor of 2 suffices to explain differences between these two one-bond coupling constants. Multiplication by 2 for Complexes **VI** and **VII** does not achieve the value for $^1J(^{195}Pt, ^{119}Sn)$ in

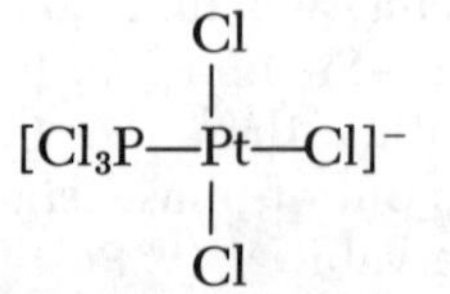

$[Cl_3P—Pt(Cl)_2—Cl]^-$ (Cl above and below Pt)

$^1J(^{195}Pt,^{31}P) = 6182$ Hz (*35*)

VI

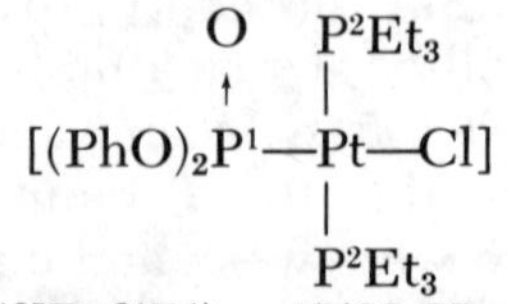

$[(PhO)_2P^1(\rightarrow O)—Pt(P^2Et_3)_2—Cl]$ (P^2Et_3 above and below Pt)

$^1J(^{195}Pt,^{31}P^1) = 5569$ Hz (*36*)

VII

Table I. P-31 NMR Data[a]

		$\delta^{31}P$	$(\Delta\delta)^b$	$J_{^{195}PtP}$	$(\Delta J)^b$	$J_{^{119}SnP}$
trans-$[PtCl(SnCl_3)(Et_4PP)]^c$	CD_2Cl_2/RT	+17.1	(−1.1)	2102	(−633)	224,214
trans-$[PtCl(SnCl_3)(PEt_3)_2]$	CD_2Cl_2/−30°C	+13.6	(+1.4)	2041	(−356)	241,229
trans-$[PtH(SnCl_3)(Et_4PP)]^c$	CD_2Cl_2/−80°C	+18.8	(−7.2)	2372	(−363)	214,205
trans-$[PtH(SnCl_3)(PEt_3)_2]$	$CDCl_3$/−50°C	+20.5	(−2.3)	2363	(−347)	216 av.[d]
trans-$[PtH(SnCl_3)(PBzPh_2)_2]$	$CDCl_3$ CH_2Cl_2 /−60°C	+24.6	(−2.7)	2713	(−339)	197 av.[d]
trans-$[PtH(SnCl_3)(PPh_3)_2]$	$CDCl_3$ CH_2Cl_2 /−60°C	+28.5	(+0.1)	2650	(−369)	213,204
trans-$[PtC^1(SnCl_3)(PEt_3)_2]^e$	$CDCl_3$/RT	+ 9.8	(+4.0)	2493	(−301)	241,229
trans-$[PtC^2(SnCl_3)(PBzPh_2)_2]^e$	$CDCl_3$/RT	+10.1	(−7.9)	2503	(−311)	265,253
trans-$[PtC^2(SnCl_3)(PPh_3)_2]^e$	$CDCl_3$/RT	+18.3	(−5.4)	2640	(−297)	262,250
trans-$[PtC^3(SnCl_3)(PPh_3)_2]^e$	$CDCl_3$/RT	+18.0	(−5.3)	2625	(−294)	262,250

[a] See experimental for details.

[b] The Δ sign refers to the following difference: [($SnCl_3$ complex) − (Cl complex)], e.g. (values for $[PtCl(SnCl_3)(PEt_3)_2]$) − (values for $[PtCl_2(PEt_3)_2]$).

[c] Et_4PP = Complex II.

[d] Average of ^{117}Sn and ^{119}Sn.

[e] C^1 = phenyl; C^2 = $EtO_2C = CHCO_2Et$; C^3 = $MeO_2C = CHCO_2Me$.

Table II. Pt-195, Sn-119, and H-1 Data[a]

		$\delta^{195}Pt$	$(\Delta\delta)$	$\delta^{119}Sn$	$J_{^{195}Pt^{119}Sn}$
trans-[$PtCl(SnCl_3)(Et_4PP)$]	CD_2Cl_2/RT	−4749	(−817)	−197	29242
trans-[$PtCl(SnCl_3)(PEt_3)_2$]	CD_2Cl_2/−40°C	−4780	(−864)	−201	28954
trans-[$PtH(SnCl_3)(Et_4PP)$]	CD_2Cl_2/−80°C	−5201	(−333)	+177	9424
trans-[$PtH(SnCl_3)(PEt_3)_2$]	CD_2Cl_2/−70°C	−5302	(−426)		9067
trans-[$PtH(SnCl_3)(PBzPh_2)_2$]	CD_2Cl_2/−70°C	−5322	(−428)	+160	10955
trans-[$PtH(SnCl_3)(PPh_3)_2$]	CD_2Cl_2/−70°C	−5195	(−354)	+130	11512
trans-[$PtC^1(SnCl_3)(PEt_3)_2$]	$CDCl_3$/RT	−4799	(−528)	+ 92	6745
trans-[$PtC^2(SnCl_3)(PPh_3)_2$]	CD_2Cl_2/RT	−4771	(−469)		11320
trans-[$PtC^4(SnBr_3)(PEt_3)_2$]	$CDCl_3$/RT			+ 58	5780

		δ^1H[b]	$(\Delta\delta)$	$^1J_{^{195}PtH}$	(ΔJ)	$^2J_{^{31}PH}$ ΔJ	$^2J_{^{119}SnH}$
trans-[$PtH(SnCl_3)(Et_4PP)$]	CD_2Cl_2/−80°C	−9.4	(+8.0)	1465	(+111)	10(−4)	1831
trans-[$PtH(SnCl_3)(PBzPh_2)_2$]	CD_2Cl_2/−70°C	−9.7	(+7.3)	1308	(+ 48)	5(−7)	1731
trans-[$PtH(SnCl_3)(PPh_3)_2$]	CD_2Cl_2/−70°C	−8.8	(+7.3)	1294	(+ 90)	9(−4)	1740
trans-[$IrH(SnCl_3)Cl(CO)(Ph_4PP)$]	$CDCl_3$/RT	−9.4	(+4.6)	—	—	10(−2)	1570

[a] For abbreviations and definitions, *see* Table I; $C^4 = p - CH_3C_6H_4$.
[b] Relative to tetramethylsilane.

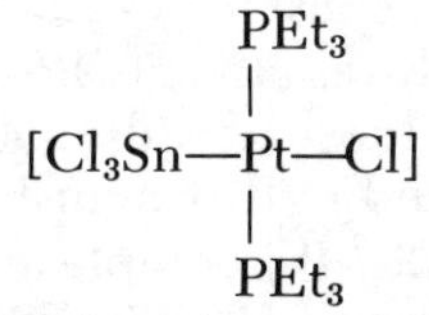

$$\begin{array}{c} PEt_3 \\ | \\ [Cl_3Sn—Pt—Cl] \\ | \\ PEt_3 \end{array}$$

$^1J(^{195}Pt,^{119}Sn) = 28954$ Hz

VIII

Complex **VIII**; although the same operation (plus a second factor of ~2 due to the presence of ethyl groups instead of chlorine (*39*)) seems satisfactory when comparing Complexes **IX** and **X**. Perhaps the π_{AB} term is more important in one case than in another and/or a comparison with a ^{31}P coupling is unsuitable. The trichlorostannate group is thought to be a good π-acceptor due to the electron-withdrawing properties of the halogens, and this conceivably could favor larger s coefficients in a molecular orbital involved with the platinum–tin bond.

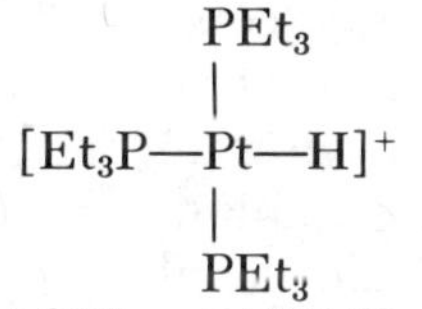

$$\begin{array}{c} PEt_3 \\ | \\ [Et_3P—Pt—H]^+ \\ | \\ PEt_3 \end{array}$$

$^1J(^{195}Pt,^{31}P) = 2,515$ Hz (*37*)

IX

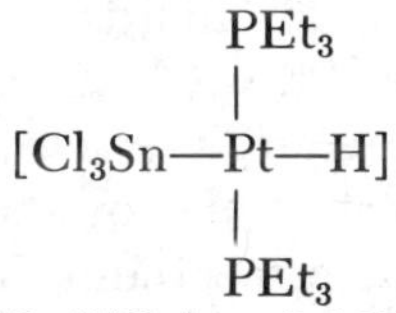

$$\begin{array}{c} PEt_3 \\ | \\ [Cl_3Sn—Pt—H] \\ | \\ PEt_3 \end{array}$$

$^1J(^{195}Pt,^{119}Sn) = 9,067$ Hz

X

In summation, the $^1J(^{195}Pt, ^{119}Sn)$ can be used to assign structure where the mode of binding of the tin is in question and large one-bond platinum–tin J values are to be expected and seem manageable within the context of the Fermi contact term.

OVER TWO BONDS. As mentioned, the splitting $^2J(^{119}Sn, ^1H) = 1740$ Hz (*see* Figure 3) is somewhat large for a two-bond trans coupling constant (*40*) and in fact is the largest known two-bond coupling involving a proton (*17*).

This kind of coupling-constant information can have diagnostic value for other transition-metal hydride complexes. The Vaska analog *trans*-[IrCl(CO)PP] reacts with $SnCl_2$ to give an iridium hydride complex whose structure may be either **XI** or **XII** (*41*). The observation of a $^2J(^{119}Sn, ^1H)$ coupling constant of 1570 Hz suggests that Structure **XI** is correct and vibrational spectroscopic data support this assignment.

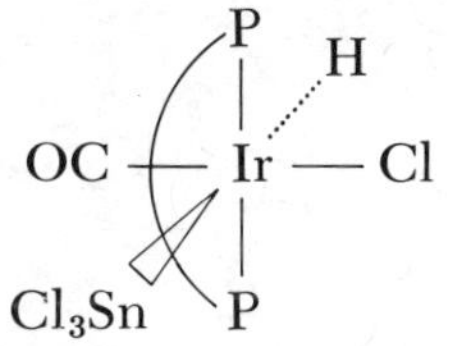

$$\begin{array}{c} P \quad H \\ | \\ OC — Ir — Cl \\ | \\ Cl_3Sn \quad P \end{array}$$

XI

$$\begin{array}{c} P \quad H \\ | \\ OC — Ir — Cl \\ | \\ Cl_3Sn \quad P \end{array}$$

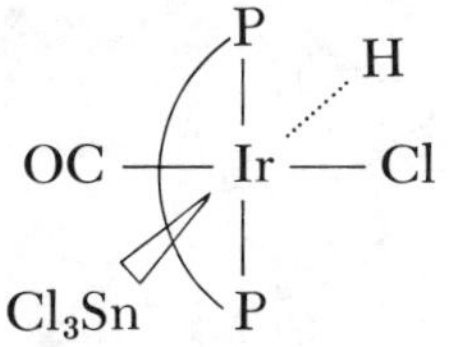

XII

PP = 2,11-bis(diphenylphosphinomethyl)-benzo[*c*]phenanthrene

A two-bond coupling constant >1.5 KHz is large; however, $^{2}J\,(^{119}Sn^{31}P)_{trans}$ in the complexes *cis*-$[PtCl(SnCl_3)P_2]$, $P_2 = 2 \times PPh_3$ or DIOP, is larger yet with values exceeding 4 KHz. This knowledge is also valuable and has assisted in distinguishing between the isomeric possibilities for the $[Pt(SnCl_3)_2P_2]$ complexes, which for monodentate phosphorus ligands are all trans: $^{2}J\,(^{119}Sn,\ ^{31}P)_{cis} \sim 200 - 250$ Hz). But 4 KHz is not the last word! We find $^{2}J\ (^{119}Sn,\ ^{117}Sn)_{trans}$ to be > 16,000 and we plan to submit these, and yet larger two-bond coupling constants, separately for publication.

Additional Physical Measurements. In all of the complexes we have encountered to date—and such molecules are reasonable models for intermediates in the hydrogenation cycle—the $SnCl_3$ group is present as a coordinated ligand and forms distinct metal–metal bonds in solution.

Further, both the hydrogenation and hydroformylation reactions are reported to require an excess of tin(II) dichloride with a Sn/Pt ratio of more than 2 (often 5 is optimal). Our NMR studies suggest that Sn/Pt > 1 affords an appreciable increase in the concentration of *trans*-$[Pt(SnCl_3)_2(PEt_3)_2]$ and higher $SnCl_2$ concentrations support a more complete formation of poly-tin species. In view of the potential relevance of these complexes it would be valuable to further characterize them. This will require not only additional complexes but also additional physicochemical information concerning the electronic structure of the ground and low-lying excited states. We have therefore measured some X-ray photoelectron ESCA spectra in the hope of obtaining a fuller picture of the charge distribution. In contrast to UV photoelectron spectroscopy, (UV–PES), where orbital interactions play a dominant role, ESCA binding energies, BEs, are recognized (*42, 43*) to depend primarily on the charge of the atom and the charges of the surrounding atoms; i.e., the higher the positive charge the more energy is needed to eject a core electron (assuming that there are similar relaxation energies).

In Table III we give BEs for some of our complexes and note the following points: (a) the $Sn(3d_{5/2})$ values of 486.7 − 487.2 eV are larger than that for the model tin(II) complex $(Et_4N)SnCl_3$ at 485.7 eV (*44*) suggesting an increased positive charge on tin due to $SnCl_3$ coordination to the platinum, and (b) the $Pt(4f_{7/2})$ values, 72.7–73.3 eV, do not deviate significantly from those of K_2PtCl_4 = 73.4 eV or other model complexes (*see* Table III).

$$\left[\gt \overset{\delta+}{Sn}: \right] \rightarrow \left[\gt \overset{\Delta+}{Sn} - Pt - \right]$$

These data may be interpreted using a classical donor–acceptor complex model in which we have a diminished positive platinum charge and an enhanced positive charge on tin. The coordination of the $SnCl_3^-$ unit decreases the Pt BE on one hand; however, the presence of a positively charged neighboring tin compensates by increasing the BE with the net result being little or no significant change in the ionization potential of the platinum core electrons. The tin core responds to the donation of the lone pair by showing an enhanced potential which is increased further by introducing a positively charged neighboring platinum. The end effect at tin is a relatively high core potential that creates the impression of a high positive charge at this metal ($(Et_4N)_2SnCl_6$ = 487.1 eV (*45*)).

The picture is at least qualitatively in agreement with earlier Sn-119 m Mössbauer measurements for a variety of transition-metal trichlorostannate complexes by Fenton and Zuckerman (*46*) who more rigorously concluded that the tin in $M–SnCl_3$ units should be considered as tin(IV).

From Table II we note that the Pt-195 NMR chemical shifts of the platinum–tin complexes fall in and even at higher field than the classical Pt(0) complexes (e.g. $[Pt(CF_3–C{\equiv}C–CF_3)(PEt_3)_2]$ = −4712 ppm, $[Pt(\textit{trans}\text{-stilbene})(PEt_3)_2]$ = −5122 (*47*)). Accordingly it is tempting to seek a relationship between the Pt-195 NMR coordination chemical shift $\Delta\delta = \delta$(complex with tin)-δ(related complex without tin) and the electron density at the metal. However, the few ESCA data for the chosen test compounds do not indicate a simple charge dependence of the NMR shielding.

A more quantitative attempt at interpreting platinum chemical shifts requires that we consider some form of the Ramsey (*48*) equation, which describes the resonance frequency, ν, in terms of the paramagnetic screening term, σ_P. Q_{AB} is a charge-density, bond-order matrix, ΔE is the average excitation energy, and r represents a distance from the nucleus for, in this case, a given d electron.

$$\nu \propto \delta H_0(1 - \sigma_P) \,|\, ^2\pi \qquad (8)$$
$$\sigma_P \propto (1/\Delta E)\langle 1/r^3 \rangle \Sigma Q_{AB}$$

Although there is some precedence for believing that ^{195}Pt chemical shifts may be understood qualitatively if the ΔE term can be estimated from visible UV data (*49*), there is an experimental difficulty in our complexes introduced by the fact that minor impurities with high extinction coefficients may distort considerably the UV curves. We have found that our nearly colorless hydrido-tin, and chloro-tin derivatives also can be isolated as yellow- or orange-colored complexes that, from ^{31}P measurements, are >95% pure. It seems that caution will be required in interpreting the UV spectra. Nevertheless, a future study

Table III. ESCA Data[a]

	$Pt(4f_{7/2})$		$Sn(3d_{5/2})$	$P(2p)$	
trans-$[PtCl(SnCl_3)(PEt_3)_2]$	73.3	$(73.2)_{cis}$[b]	487.2	131.6	
trans-$[PtH(SnCl_3)(PPh_3)_2]$	73.3	(73.0)[b]	487.0	132.0	(131.7)[b]
cis-$[PtCl(SnCl_3)(PPh_3)_2]$[c]	—	(73.2)[b]	486.8	132.5	(131.9)[b]
cis-$[Bu_4N]_2[PtCl_2(SnCl_3)_2]$	72.7	—	486.7	—	

[a] Ref. 51. Data are in electron volts relative to a carbon 1s BE of 285.0 eV.
[b] Values for the complexes where $SnCl_3$ is replaced by chlorine.
[c] Ref. 52.

Table IV. IR Data[a]

	$\tilde{\nu}_{PtH}$ $[cm^{-1}]$	$(\Delta\tilde{\nu})^b$ $[cm^{-1}]$	$\tilde{\nu}_{SnCl_3}$ $[cm^{-1}]$
trans-$[PtCl(SnCl_3)(Et_4PP)]$	—	—	353, 335, 320
trans-$[PtCl(SnCl_3)(PEt_3)_2]$	—	—	352, 329, 316
trans-$[PtH(SnCl_3)(Et_4PP)]$	2112	(− 93)	342, 320
trans-$[PtH(SnCl_3)(PEt_3)_2]$	2120	(−108)	332, 306
trans-$[PtH(SnCl_3)(PBzPh_2)_2]$	2154	(− 66)	337, 317, 308
trans-$[PtH(SnCl_3)(PPh_3)_2]$	2200	(− 45)	332, 315, 306
trans-$[PtC^1(SnCl_3)(PEt_3)_2]^c$	—	—	329, 305
trans-$[PtC^2(SnCl_3)(PBzPh_2)_2]^c$	—	—	340, 320

[a] Measured as KBr discs.
[b] $\Delta\nu$ = (ν Pt–H in the tin complex) − (ν Pt–H in the chloro complex).
[c] *See* Table I for abbreviations.

in this direction could prove helpful, although the knowledge of a single, low-energy UV transition by itself is not necessarily sufficient to claim a full understanding of the ΔE summation. The $1/r^3$ term may not be varying significantly as our ESCA data show no marked change in the charge on platinum. Since we have no satisfactory method of estimating changes in Q_{AB}, other than our IR data which is far too crude (*see* Table IV), and since an interpretation of ΔE is problematic, a deeper understanding of our δ ^{195}Pt data must await further support, perhaps in the form of molecular orbital calculations.

One could seek additional help from the values δ ^{119}Sn; however, in addition to the problems already mentioned we note that there is a considerable dependence of δ ^{119}Sn on both solvent and temperature (*50*). In view of the different sample conditions indicated by both the solubility and the dynamic characteristics of our complexes, a detailed interpretation of these values would be presumptuous.

In summation, there is reason to believe, from the ESCA and NMR studies, that the coordination of $SnCl_2$ bestows interesting properties on our complexes; however, there is as yet insufficient data to permit definitive conclusions relevant to the homogeneous hydrogenation reaction.

Acknowledgments

We thank H. Ruegger for experimental assistance and L. M. Venanzi for many helpful discussions.

Literature Cited

1. Itatani, H.; Bailar, J. C. Jr. *Ind. Eng. Chem. Prod. Res. Dev.* 1972, *11*, 146.
2. Cramer, R. D.; Jenner, E. L.; Lindsey, R. V. Jr.; Stolberg, U. G. *J. Amer. Chem. Soc.* 1963, *85*, 1691.

3. Knifton, J. F. *J. Org. Chem.* 1976, *41*, 793.
4. Schwager, I.; Knifton, J. F. *J. Catal.* 1976, *45*, 256.
5. Parshall, G. W. *J. Mol. Catal.* 1978, *4*, 243.
6. Young, J. F.; Gillard, R. D.; Wilkinson, G. *J. Chem. Soc.* 1964, 5176.
7. Baird, M. C. *J. Inorg. Nucl. Chem.* 1967, *29*, 367.
8. Cramer, R. D.; Lindsey, R. V. Jr.; Prewitt, C. T.; Stolberg, U. G. *J. Amer. Chem. Soc.* 1965, *87*, 658.
9. Lindsey, R. V.; Parshall, G. W.; Stolberg, U. G. *J. Amer. Chem. Soc.* 1965, *87*, 658.
10. Onaka, S. *Bull. Chem. Soc. Jpn.* 1975, *48*, 319.
11. Mason, R.; Robertson, G. B.; Whimp, P. O. *Chem. Comm.* 1968, 1655.
12. Clegg, W., *Acta Crystallogr.* 1978, *B34*, 278.
13. Guggenberger, L. J. *Chem. Comm.* 1968, 512.
14. Stalick, J. K.; Corfield, P. W. R.; Meek, D. W. *J. Amer. Chem. Soc.* 1972, *94*, 6194.
15. Bürgi, H. B.; Johnson, D. K.; Venanzi, L. M., unpublished results.
16. Pregosin, P. S.; Sze, S. N. *Helv. Chim. Acta* 1978, *61*, 1848.
17. Starzewski, Ostoja K. A.; Ruegger, H.; Pregosin, P. S. *Inorg. Chim. Acta* 1979, *36*, L445.
18. Jesson, J. P. In "Transition Metal Hydrides", Muetterties, E., Ed.; Marcel Dekker: New York, 1971; p. 75.
19. Baumgartner, E.; Venanzi, L. M., unpublished data.
20. Pregosin, P. S.; Sze, S. N. *Helv. Chim. Acta* 1978, *61*, 1848.
21. Starzewski, Ostoja K. A.; Pregosin, P. S., unpublished data.
22. Hsu, C. C.; Geanangel, R. A. *Inorg. Chem.* 1980, *19*, 110.
23. Hsu, C. C.; Geanangel, R. A. *Inorg. Chem.* 1977, *16*, 2529.
24. Kauffman, J. W.; Moor, D. H.; Williams, R. J. *J. Inorg. Nucl Chem.* 1977, *39*, 1165.
25. Ruegger, H.; Pregosin, P. S., unpublished data.
26. Kunz, R. W.; Pregosin, P. S. In "NMR Basic Principles and Progress"; Springer-Verlag: Heidelberg, 1979; Vol. 16.
27. Butler, G.; Eaborn, C.; Pidcock, A. *J. Organomet. Chem.* 1979, *181*, 47.
28. Eaborn, C.; Pidcock, A.; Steele, B. R. *J. Chem. Soc. Dalton* 1976, 767.
29. Eaborn, C.; Pidcock, A.; Steele, B. R. *J. Chem. Soc. Dalton* 1975, 809.
30. Starzewski, Ostoja K. A.; Pregosin, P. S. *Angew. Chem. Int. Ed.* 1980, *19*, 316.
31. Nelson, J. H.; Cooper, V.; Rudolph, R. R. *Inorg. Nucl. Chem. Lett.* 1980, *16*, 263.
32. Pidcock, A.; Richards, R. E.; Venanzi, L. M. *J. Chem. Soc. A* 1966, 1707.
33. Appleton, T. G.; Clark, H. C.; Manzer, L. E. *Coord. Chem. Rev.* 1973, *10*, 335.
34. Pople, J. A.; Santry, D. P. *Mol. Phys.* 1964, *8*, 1.
35. *Ibid*, 1965, *9*, 311.
36. Crocker, C.; Goggin, P. L.; Goodfellow, R. J. *J. Chem. Soc. Dalton* 1976, 2494.
37. Allen, F. H.; Pidcock, A.; Waterhouse, C. R. *J. Chem. Soc. A*, 1970, 2087.
38. Dingle, T. W.; Dixon, K. R. *Inorg. Chem.* 1974, *13*, 846.
39. Mather, G. G.; Pidcock, A.; Rapsey, G. J. N. *J.C.S. Dalton* 1973, 2095.
40. Verkade, J. G. *Coord. Chem. Rev.* 1972–73, *9*, 1.
41. Baumgartner, E.; Starzewski, Ostoja K. A.; Venanzi, L. M., unpublished data.
42. Riggs, W. M. *Anal. Chem.* 1972, *44*, 830.
43. Cook, C. D.; Wan, K. Y.; Gelius, U.; Hamrin, K.; Johansson, G.; Olsson, E.; Siegbahn, H.; Nordling, C.; Siegbahn, K. *J. Amer. Chem. Soc.* 1971, *93*, 1904.
44. Parshall, G. W. *Inorg. Chem.* 1972, *11*, 433.
45. Swartz, W. E.; Watts, P. H.; Lippincott, E. R.; Watts, J. C.; Huheey, J. E. *Inorg. Chem.* 1972, *11*, 2632.
46. Fenton, D. E.; Zuckerman, J. J. *Inorg. Chem.* *8*, 1771.

47. Browning, J.; Green, M.; Spencer, J. L.; Stone, F. G. A. *J. Chem. Soc. Dalton,* 1977, 278.
48. Ramsey, N. F. *Phys. Rev.* **1950,** *78,* 699.
49. Goggin, P. L.; Goodfellow, R. J.; Haddock, S. R.; Taylor, B. F.; Marshall, I. R. H. *J. Chem. Soc. Dalton* **1976,** 459.
50. Smith, P. J.; Smith, L. *Inorg. Chim. Acta Rev.* **1973,** *7,* 11.
51. Starzewski, Ostoja K. A.; Pregosin, P. S.; Sawatsky, G., unpublished data.
52. Grutsch, P. A.; Zeller, M. V.; Fehlner, T. P. *Inorg. Chem.* **1973,** *12,* 1431.

RECEIVED August 15, 1980.

3

P-31 NMR Studies of Equilibria and Ligand Exchange in Triphenylphosphine Rhodium Complex and Related Chelated Bisphosphine Rhodium Complex Hydroformylation Catalyst Systems

RODNEY V. KASTRUP, JOSEPH S. MEROLA, and ALEXIS A. OSWALD

The Analytical and Information Division and Corporate Research Science Laboratories, Exxon Research and Engineering Company, P.O. Box 45, Linden, NJ 07036

Triphenylphosphine–rhodium complex hydroformylation catalyst systems discovered by Wilkinson and developed by Union Carbide, Davy Powergas, and Johnson Matthey

$$(Ph_3P)_3Rh(CO)H + nPh_3P \rightleftharpoons (Ph_3P)_2Rh(CO)H + (n + l)Ph_3P$$

(n = 0 – 140) were studied in toluene solution by P-31 NMR in the +5°–+105°C temperature range. In general, the equilibria favored the trisphosphine rather than the bisphosphine complex. The dissociation of the tris-tertiary-phosphine–rhodium carbonyl hydride complex to the corresponding trans-bis-tertiary-phosphine–rhodium complex is postulated to provide the active intermediate for the selective terminal hydroformylation of 1-olefins.

The tris(triphenylphosphine) rhodium carbonyl hydride complex also was used via ligand exchange to obtain known chelate complexes of bisphosphines

$$Ph_2PCH_2CH_2PPh_2 \text{ and } Ph_2PCH_2CH_2CH_2PPh_2$$

0065–2393/82/0196–0043$05.50/0

for similar studies. These stabilized complexes exhibited sharply reduced ligand exchange under comparative conditions.

P-31 NMR was a powerful tool in studies correlating the structure of tertiary-phosphine–rhodium chloride complexes with their behavior as olefin hydrogenation catalysts. Triphenylphosphine–rhodium complex hydrogenation catalyst species (*1*) were studied by Tolman et al. at du Pont and Company (*2*). They found that tris(triphenylphosphine)rhodium(I) chloride (**A**) dissociates to triphenylphosphine and a highly reactive intermediate (**B**). The latter is dimerized to tetrakis(triphenylphosphine)dirhodium(I) dichloride (**C**).

$$\underset{\mathbf{A}}{2(Ph_3P)_3RhCl} \rightleftharpoons \underset{\mathbf{B}}{2(Ph_3P)_2RhCl} + 2Ph_3P \rightleftharpoons \underset{\mathbf{C}}{[(Ph_3P)_4RhCl]_2} + 2Ph_3P$$

Although **B** could not be detected spectroscopically, equilibria between **A** and **B** could be postulated from kinetic results.

The effects of added triphenylphosphine and changing temperature on ligand dissociation and equilibria were studied also. The above dimer was an active hydrogenation catalyst. The equilibrium concentration of the dimer and the rate of olefin hydrogenation catalysis by the system depend inversely on the concentration of excess phosphine ligand.

Applying P-31 NMR to the field of hydroformylation catalysis by triphenylphosphine rhodium complex-based systems is the subject of this chapter. These hydroformylation catalyst systems are of high academic and technological interest. They are effective for hydroformylating 1-olefins at low pressure and temperature and exhibit a high selectivity to *n*-aldehydes:

$$RCH = CH_2 + CO + H_2 \rightarrow \underset{n\text{-}}{RCH_2CH_2CHO} + \underset{i\text{-}}{R\underset{\displaystyle CH_3}{\underset{|}{C}}HCHO}$$

These catalyst systems were discovered and studied first by G. Wilkinson and co-workers in the 1960s (*3*). Wilkinson recognized the exceptionally mild reaction conditions and high selectivities attainable in such systems. However, Pruett and Smith of the Union Carbide Corporation reported first that the selectivity of the reaction to produce a high ratio of *n*- to *i*-aldehyde products is directly proportional to the excess triphenylphosphine ligand (*4*). In contrast, they found that the *n/i* aldehyde ratio was inversely proportional to the partial pressure of the CO reactant. These findings were used by Pruett and other Union Carbide researchers to develop an important commercial process for the continuous, low-pressure, homogeneous, liquid-phase hydroformylation of propylene to *n*-butyraldehyde using a high ratio of H_2/CO (*5, 6*).

The studies of Wilkinson et al. included IR and H-1 NMR spectroscopy of the intermediate species of this catalyst system (*7*). This led to recognizing tris(triphenylphosphine)rhodium(I) carbonyl hydride (**D**) as the key stable rhodium complex. The reactive *trans*-bis-(triphenylphosphine)rhodium(I) carbonyl hydride (**E**) resulting via the dissociation of this complex

$$\underset{\mathbf{D}}{(Ph_3P)_3Rh(CO)H} \rightleftharpoons \underset{\mathbf{E}}{(Ph_3P)_2Rh(CO)H} + Ph_3P$$

is a key intermediate in both the associative and dissociative mechanisms proposed by Wilkinson et al. (*7, 8*).

Figure 1 outlines the key intermediates of a catalytic cycle where the rate-determining step is the formation of an *n*-alkyl derivative of the trans-bisphosphine via a coordinated olefin complex. This presumed catalytic cycle appears to satisfy proposals by Cavalieri d'Oro et al. (*9*), C. V. Pittman et al. (*10*), and J. Hjortkjaer (*11*). Although no single mechanism of hydroformylation was established, the cycle is shown here to illustrate the key nature of the equilibrium between the trisphosphine (**D**) and the trans-bisphosphine (**E**).

Since there were no known definitive studies of the crucial equilibria between the tris-(triphenylphosphine)- and bis-(triphenylphosphine)rhodium(I) carbonyl hydride complexes, it was decided to study such equilibria by P-31 NMR spectroscopy. The

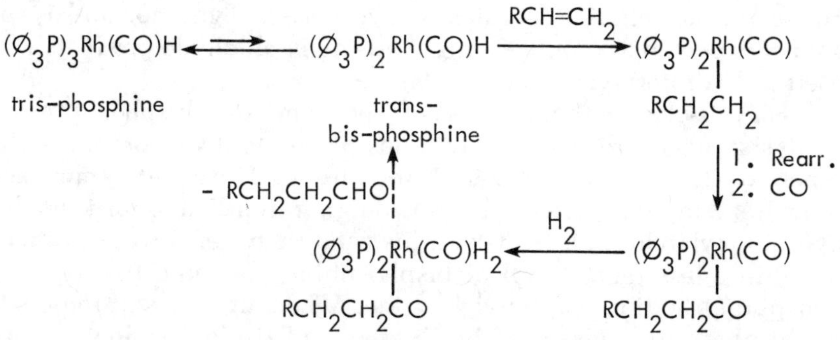

Figure 1. Ligand exchange is a proposed key step in the mechanism of phosphine–rhodium complex catalyzed hydroformylation of olefins

presently reported studies were carried out at various temperatures in varying amounts of excess triphenylphosphine. The rate of phosphine dissociation for the trisphosphine complex was determined also by P-31 NMR.

P-31 NMR studies also were carried out in a similar manner on rhodium complexes of two chelating bisphosphines—bis-1,3-diphenylphosphinopropane and bis-1,2-diphenylphosphinoethane. These complexes were generated in solution via ligand displacement from tris(triphenylphosphine)rhodium carbonyl hydride. For example, one of the possible displacement products of bis-1,3-diphenylphosphinopropane (**F**) is a cis-chelate (**G**) that can undergo dissociation to yield a chelating bisphosphine complex (**H**):

$$(\text{Ø}_3P)_3Rh(CO)H + 2\ \underset{\mathbf{F}}{\text{Ø}_2P(CH_2)_3P\text{Ø}_2} \rightleftarrows$$

$$\underset{\mathbf{G}}{CH_2\begin{matrix} \nearrow CH_2P\text{Ø}_2 \cdots \\ \searrow CH_2P\text{Ø}_2 \cdots \end{matrix} \overset{CO}{\underset{H}{Rh}} \cdots \text{Ø}_2P(CH_2)_3P\text{Ø}_2}$$

$$\Big\updownarrow$$

$$\underset{\mathbf{H}}{CH_2\begin{matrix} \nearrow CH_2P\text{Ø}_2 \cdots \\ \searrow CH_2P\text{Ø}_2 \cdots \end{matrix} \overset{CO}{\underset{H}{Rh}} + \text{Ø}_2P(CH_2)_3P\text{Ø}_2}$$

However, such chelate complexes have a cis-configuration and as such are not expected to produce a high ratio of *n*/*i* aldehyde products when used as hydroformylation catalysts.

For cis-chelate complexes of rhodium and bisphosphines as catalysts, indeed relatively low ratios of *n*/*i* aldehyde products were reported (*12*, *13*). Using a 1:1 mixture of H_2/CO at atmospheric pressure, Sanger reported *n*/*i* ratios ranging from 3 to 4 for propylene hydroformylation (*12*). However, his catalyst systems were produced by adding less than 2 mol of bisphosphine per mole tris(triphenylphosphine)rhodium carbonyl hydride. When an excess of the chelating bisphosphines was used by Pittman and Hirao (*13*), low *n*/*i* ratios close to 1 were produced from 1-pentene using a mixture of H_2/CO at 100–800 psi between 60° and 120°C.

In the P-31 NMR studies, generally a 9 : 1 mixture of the toluene and perdeuterobenzene was used as a solvent for the catalyst-complex-plus-triphenylphosphine systems. Similar solutions in toluene were used in hydroformylation studies.

Some examples of 1-butene hydroformylation studies at 90° and 120°C are shown in Table I. This olefin allowed simultaneous observation of the rate of isomerization to 2-butenes. In these studies, an approximately 5 : 1 mixture of H_2/CO gas was used at about 4 atm as an initial reactant to ensure a relatively moderate partial pressure of CO. After the reaction mixture was pressured with the initial synthesis gas mixture, an approximately 1 : 1 mixture of H_2/CO was provided as a feed gas at the same pressure. The feed gas was slightly richer in H_2 in order to provide the extra H_2 for the hydrogenation side reaction. By an appropriate adjustment of the feed gas ratio, 5 : 1 mixture of H_2 and CO was maintained throughout the reaction. This meant that the reaction was performed in the same range of CO partial pressure. Generally, the reactions were carried out to about 80% butene conversion. This resulted in a highly selective hydroformylation. Selectivities to butane hydrogenation product and 2-butene isomerization products were each less than 4% at the higher ligand concentrations of Table I. Reaction rates and conversions were estimated on the basis of the amount of feed gas required.

As shown by the data of Table I, adding increasing amounts of the three phosphines used in the present studies to tris(triphenylphosphine)rhodium(I) carbonyl hydride produces catalyst systems of reduced activity as indicated by the reduced reaction rate. As expected, an increasing excess of triphenylphosphine results in an increased 1-butene hydroformylation selectivity towards the *n*-

Table I. Effect of Ligand Excess on Catalysis[a]

Ligand Structure	*Reaction Temperature* °C	*Ligand Excess P/Rh*	*Reaction Rate* k × 10^{-3} min^{-1}	*Reaction Selectivity* n/i *Valeraldehyde Ratio*
$Ø_3P$	90	6	336	3.6
		63	140	4.0
		160	59	4.3
$Ø_2P(CH_2)_2PØ_2$	120	3	68	1.3
		15	61	1.2
		140	8	1.2
$Ø_2P(CH_2)_3PØ_2$	120	3	143	1.2
		140	21	1.2

[a] Butene hydroformylation at 90°C: [Rh] = $1 \times 10^{-3}M$, p CO = 4 atm.

valeraldehyde, i.e. an increased *n/i* aldehyde ratio. (Reduced CO partial pressure results in a further increase of the *n/i* ratio.) On the other hand, an increasing excess of the two chelating diphosphines does not result in any increase in selectivity. The same low *n/i* ratio is maintained independent of ligand excess (and CO partial pressure).

Tris(triphenylphosphine)rhodium(I) Carbonyl Hydride-Plus-Triphenylphosphine Systems

P-31 NMR studies of the triphenylphosphine rhodium complex system were carried out in the presence of six-, fifty-seven-, and one-hundred-forty fold ligand excess level (*see* Figures 3–8). Using a certain ligand excess, the spectra were determined at two sets of temperatures. Studies in the low-temperature region were carried out at 5°, 15°, and 25°C (*see* Figures 2 and 5). In the high-temperature region, the spectra were studied at 35°, 60°, 90°, and 105°C (*see* Figures 3, 6, and 7). A solution of pure tris(triphenylphosphine)rhodium carbonyl hydride was used at room temperature as a standard solution.

In the low-temperature region, the P-31 spectra of all of the solutions exhibited the doublet arising from the triphenylphosphine coordinated with rhodium. This doublet has a chemical-shift value of +39.8 ppm (relative to 85% H_3PO_4) and a coupling constant, J_{P-Rh}, of 155 Hz. As such, it arises undoubtedly from triphenylphosphine bound to rhodium in a trigonal bipyramidal complex, $(Ph_3P)_3Rh(CO)H$. None of the spectra showed any indication of the trans-bisphosphine complex that is formed by dissociation of the tris-phosphine complex. The equilibrium concentration of $(Ph_3P)_2Rh(CO)H$ was too low for detection by NMR under all of the experimental conditions used in the present studies.

In the spectra of the various solutions containing excess triphenylphosphine, the expected singlet signal at 7.2 ppm appeared. Also present in all of the solutions was a sharp singlet signal at +22.1 ppm, due to triphenylphosphine oxide, which always was formed due to oxidation by the traces of molecular oxygen present. The present rhodium complex system is a very effective catalyst for such oxidations.

With increasing temperature up to 105°C, all spectra indicated increasing exchange rates for bound and free triphenylphosphine. The linewidths of the sharp triphenylphosphine oxide signals did not change with temperature, indicating that triphenylphosphine oxide does not participate in the ligand exchange.

Exchange rates for the dissociation of triphenylphosphine from the trisphosphine complex and for its association with the bisphosphine complex were determined by lineshape analyses of the spectra. For slow exchange, the rate was determined from the linewidth, using the following relationship (*14*):

$$\frac{1}{\tau} = \frac{1}{T_{2\,obs}} - \frac{1}{T_2} = k$$

where τ is the lifetime of Ph_3P in the bound or free sites; $T_{2\,obs}$ equals the observed spin–spin relaxation time measured from the linewidth; T_2 is the spin–spin relaxation time in the absence of exchange; and k equals the transition probability for leaving the bound or free site, i.e. the measured NMR rate constant, $seconds^{-1}$.

In the case of fast exchange, the NMR rate constants were determined using a computer program that calculated rates, populations, and chemical shifts by nonlinear least-squares fitting of NMR lineshape data to the GMS equation (*15*).

In the case of sixfold excess triphenylphosphine, the low-temperature spectra (at 2.5m*M* rhodium concentration) show the expected increasing linewidths for both the free and bound ligand (*see* Figure 2). The high-temperature spectra also were studied using solutions having the same ligand-to-complex ratio but higher concentrations (7m*M* rhodium, (*see* Figure 3)). As the temperatures were increased from 35° to 105°C, a progression from slow to intermediate and then to fast ligand exchange was observed. Lineshape analyses of both mixtures indicated increasing rates of trisphosphine complex dissociation with increasing temperatures.

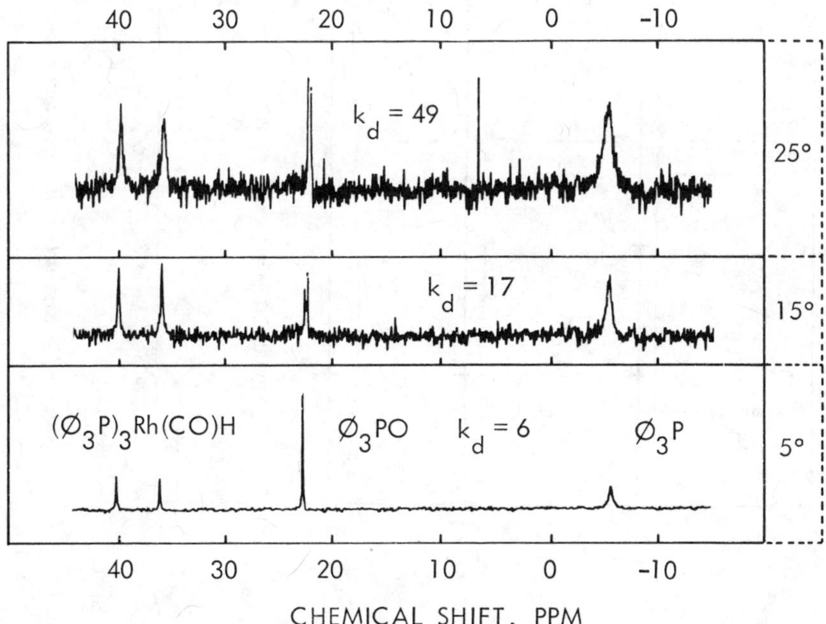

Figure 2. Ligand exchange at various temperatures— $6Ø_3P + (Ø_3P)_3Rh(CO)H$

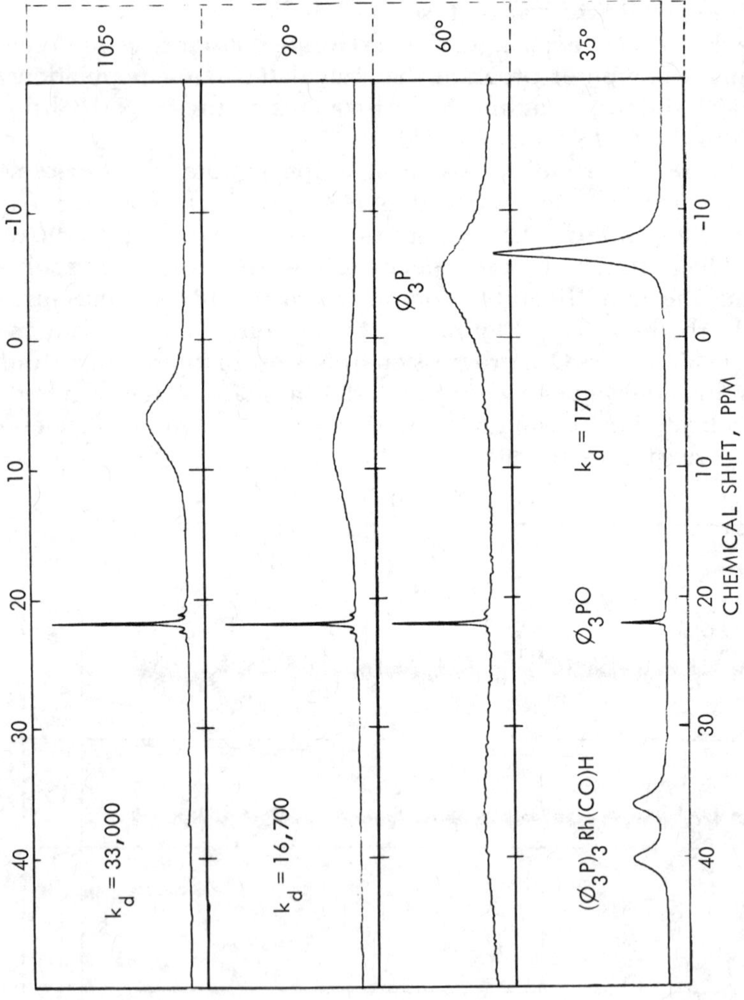

Figure 3. Ligand exchange at various temperatures—$6\emptyset_3P + (\emptyset_3P)_3Rh(CO)H$

Since NMR measures a rate constant per molecule, a first-order process will show an invariant rate constant with increasing concentration at a given temperature. In order to determine whether the triphenylphosphine dissociation and association were first-order processes, we measured the rate constants for this system at increased ligand-to-complex ratios and decreased rhodium concentration.

When a fifty-sevenfold excess of the triphenylphosphine ligand was used at a rhodium concentration of 2.5m*M*, a similar dependence of the complex's dissociation rate on the temperature was observed between 5° and 90°C (*see* Figures 4 and 5). A comparison of the spectra with those of the mixtures having a sixfold ligand excess showed that the dissociation rate is independent of the ligand-to-complex ratio. However, the association rate constants were smaller than those observed at the lower ligand-to-complex ratio. A further increase of the ligand-to-rhodium ratio to 140 (1.5m*M* rhodium) resulted in a further decrease of the association rate constant (*see* Figure 6). On the other hand, the value of the dissociation constant remained independent of the ligand excess.

From these and similar spectral analyses, it is concluded that the dissociation of triphenylphosphine from the trigonal bipyramidal tris-

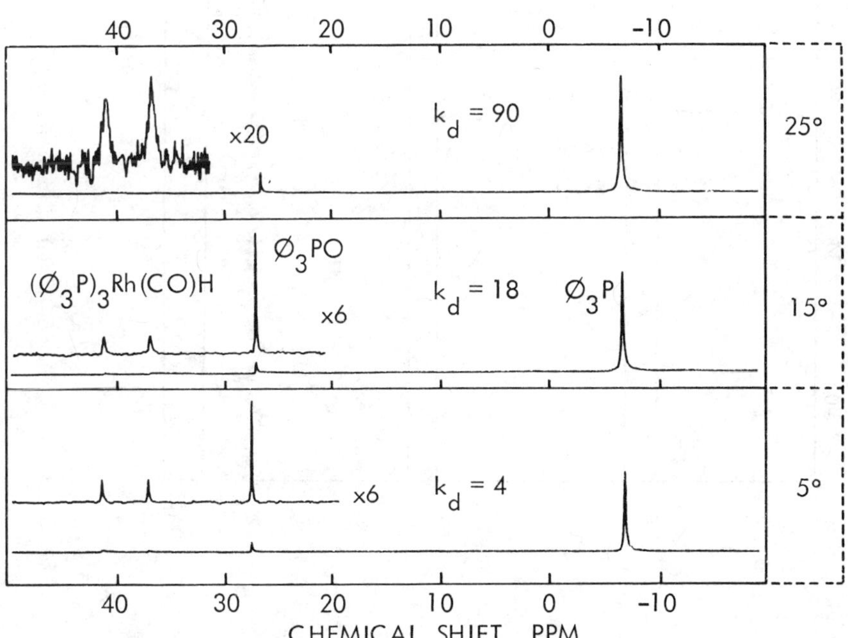

Figure 4. Ligand exchange at various temperatures— *$57Ø_3P + (Ø_3P)_3Rh(CO)H$*

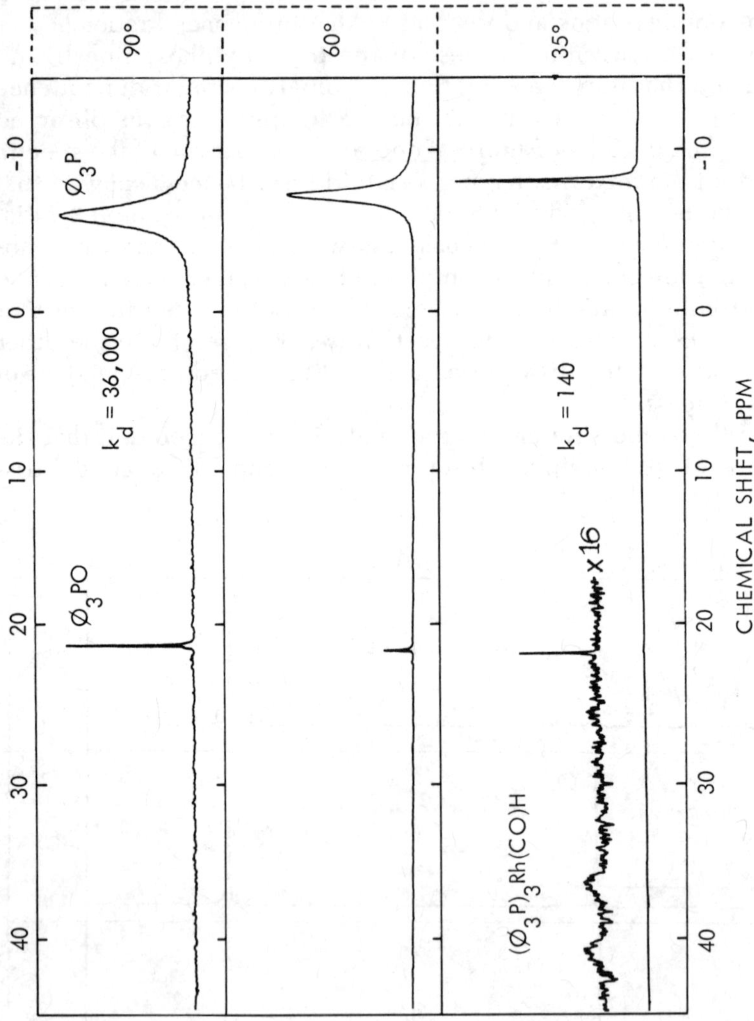

Figure 5. Ligand exchange at various temperatures—57 $Ø_3$ + $(Ø_3P)_3Rh(CO)H$

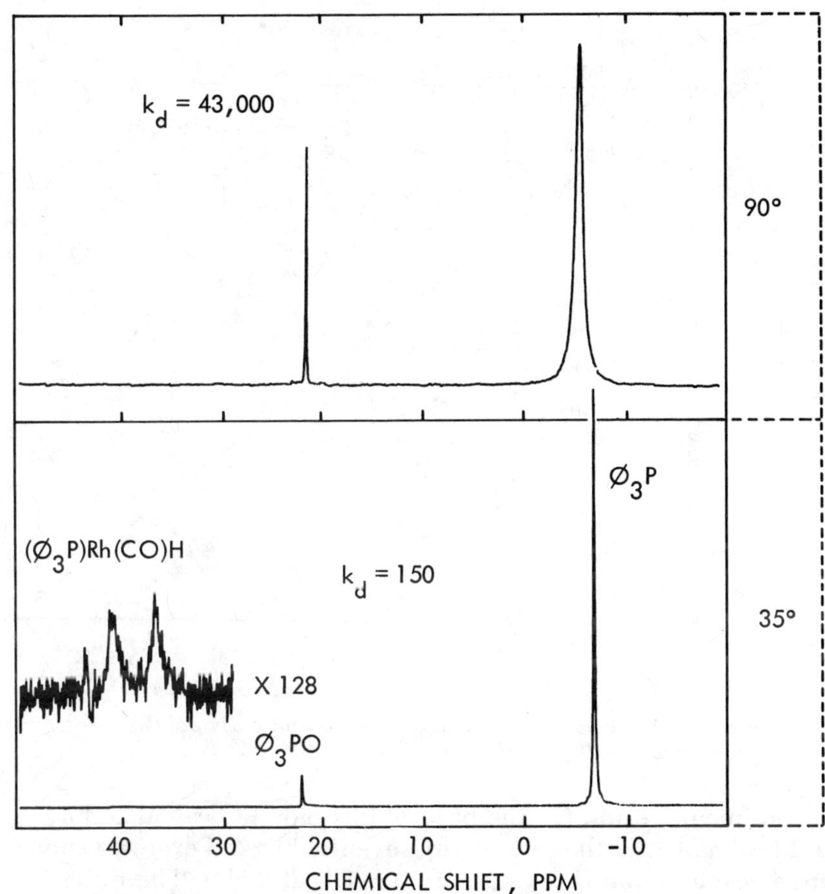

Figure 6. Ligand exchange at various temperatures— $140\O_3P + (\O_3P)_3Rh(CO)H$

phosphine complex is a first-order reaction. Obviously the reverse, i.e., association, reaction of the free triphenylphosphine with the coordinatively unsaturated bisphosphine is not of the first order.

The dissociation constant did not change with the free ligand-to-complex ratio at the concentrations of the components over the whole range studied (3.6–20.4 molal-bound and 41–170 molal-free triphenylphosphine). Accordingly, the NMR data were used for preparing Arrhenius plots correlating the dissociation constants, k_{dissoc}, i.e., the number of transitions per second, with the reciprocal absolute temperature. In this manner, the activation energy for dissociation, E_{act}, was determined.

In the case of the complex solution having a sixfold excess of triphenylphosphine, Figure 7 shows the Arrhenius plot for dissociation. The data points of the plot indicate a linear relationship within

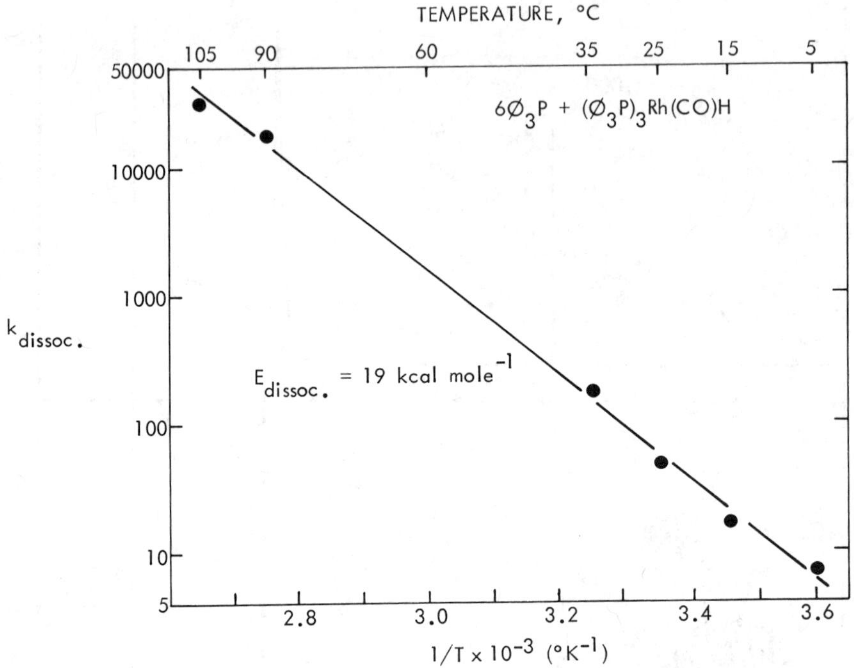

Figure 7. Arrhenius plot for $Ø_3P$–Rh complex dissociation— *$(Ø_3P)_3Rh(CO)H \rightleftarrows (Ø_3P)_2Rh(CO)H + Ø_3P$*

the experimental error. On the basis of this plot, the calculated E_{act} is 19 ± 1 kcal mol^{-1}. In the case of fifty-sevenfold ligand excess, a similar plotting results in an E_{act} value of 20 ± 1 kcal mol^{-1}. The difference between these two values indicates the magnitude of the experimental error when using this method.

It was interesting to compare the activation energy for the dissociation of the trisphosphine complex with the apparent activation energy of hydroformylation catalyzed by the trisphosphine complex-plus-excess ligand system. The hydroformylation rate had a first-order dependence on the olefin concentration and was independent of the partial pressures of CO and H_2 under the experimental conditions. Rate constants were obtained for the hydroformylation runs by measuring the amount of synthesis gas uptake as a function of time and converting gas uptake to percent conversion of 1-butene using the $n = PV \times (RT)^{-1}$ relationship. Values of $\log(1 - x)$ when x is 1-butene conversion were plotted against reaction time to yield a plot that is linear at least to the 80% conversion level to which the runs were taken. The slope of this line yields the first-order rate constant with respect to 1-butene and is expressed in units of minutes^{-1}. All of the

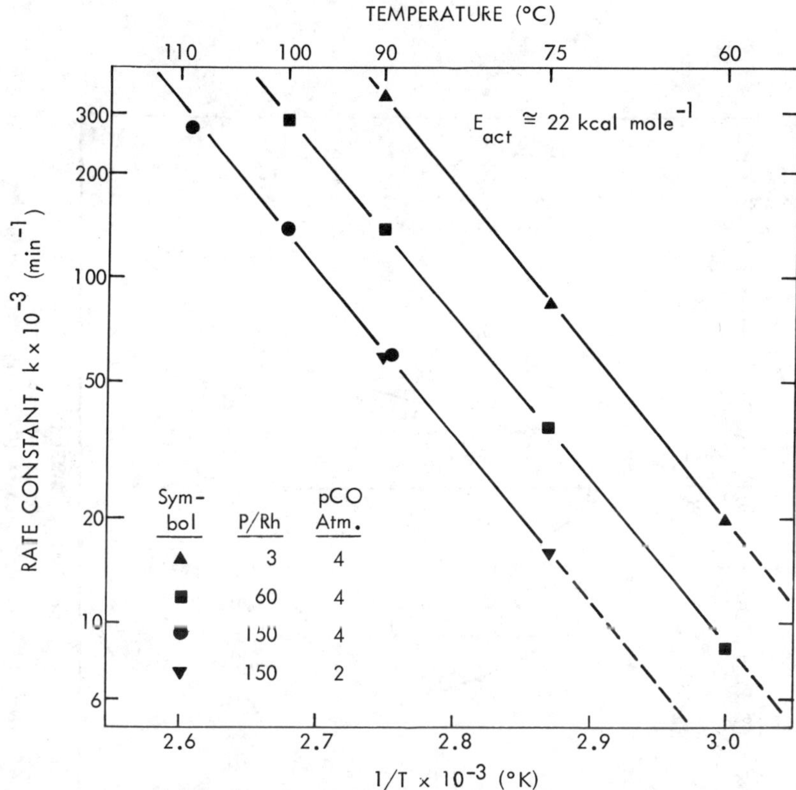

Figure 8. Arrhenius plots for $\emptyset_3P$–Rh-catalyzed 1-butene hydroformylation

rate determinations were carried out using reaction mixtures containing 0.001*M* rhodium complex. Arrhenius plots then were made for the catalyst systems having a varying excess of the triphenylphosphine ligand.

Figure 8 shows the Arrhenius plots for the 1-butene hydroformylations described in Table I. The data indicate that, as expected, the reaction rates decrease with increasing excess phosphine concentrations at all reaction temperatures. This is explained by the increasing rates of association for the active trans-bisphosphine complex with increasing phosphine excess. Thus, a reduced equilibrium concentration of the catalytically active trans-bisphosphine species results.

The slopes of the lines laid across the data points obtained at different temperatures for the three systems having different ligand-to-complex ratios were approximately the same. In other words, the activation energy for hydroformylation is about 22 kcal mol^{-1} in each of

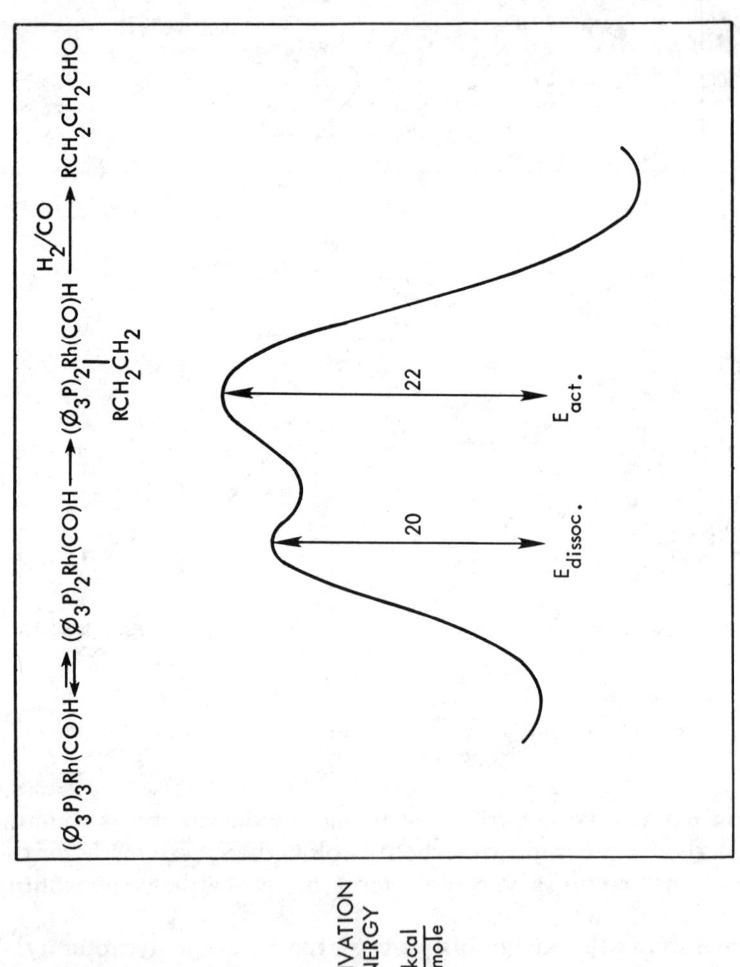

Figure 9. Energetics of catalytic intermediates in olefin hydroformylation

these systems. That means that the reaction of the catalyst complex species with the olefin has a slightly higher activation energy at all of the phosphine excess levels than that of the dissociation of the trisphosphine complex.

In accord with the data on the activation energies of complex dissociation and olefin hydroformylation, some features of the energetics of the catalytic intermediates can be recognized. These features are indicated by the simplified reaction profile shown by Figure 9. The figure shows that the dissociation of the trisphosphine complex is proposed as a key, primary step that initiates the catalytic cycle. This step has a significant activation energy of dissociation. However, reaction of the 1-olefin reactant with the trans-bisphosphine complex product of dissociation or its derivative appears to require a somewhat higher energy of activation. The transition state having the highest activation energy leads to the formation of the *n*-alkyl rhodium complex. Therefore, such a complex is indicated in the reaction scheme of the figure as a key intermediate on the way to forming the *n*-aldehyde product. However, other high-energy transition states are likely to complicate the detailed profile of the hydroformylation reaction.

In conclusion, on the basis of the two activation-energy determinations, the rate-limiting step of terminal hydroformylations is not the dissociation of the trisphosphine complex, but rather a subsequent association of a bisphosphine derivative with the 1-olefin reactant.

Chelate Complexes Derived from Tris(triphenylphosphine)rhodium(I) Carbonyl Hydride and Bisphosphines

As indicated in the introduction, bis-1,3-diphenylphosphinopropane (dppp) and bis-1,2-diphenylphosphinoethane (dppe) were reacted with tris(triphenylphosphine)rhodium(II) carbonyl hydride in toluene–deuterobenzene solution to derive cis-chelate complex hydroformylation catalysts. These complexes were expectedly nonselective terminal hydroformylation catalysts for 1-butene hydroformylation (*see* Table I) because of their cis-stereochemistry. They were also somewhat less active due to their specific structural features. The structure of these complexes in solution was studied in detail by P-31 NMR spectroscopy.

Figure 10 shows the spectra of the reaction mixtures of both chelating ligands. They show that when used in a twofold excess, both of these ligands completely displaced triphenylphosphine from the complex. Dppe has formed almost exclusively a single complex in which all of the phosphorus environments are equal. The spectrum of the complexed dppp was much more complex.

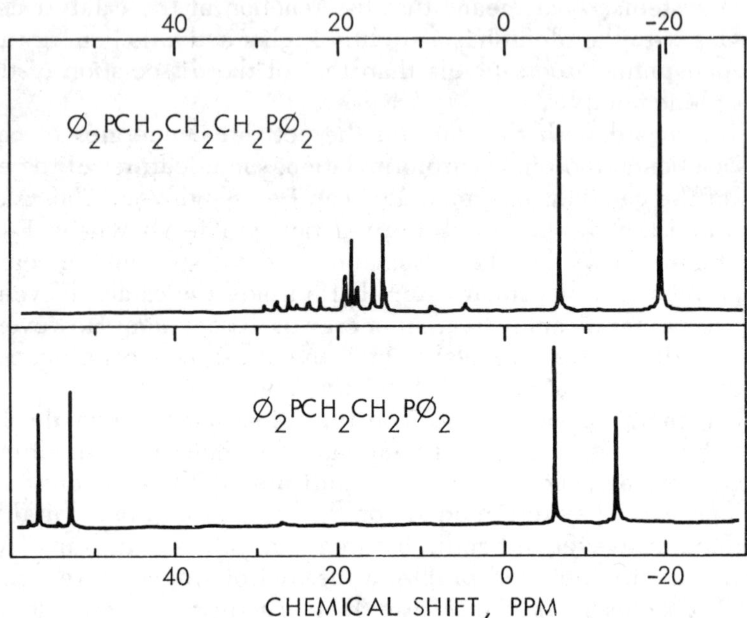

Figure 10. Ligand exchange of $(\varnothing_3P)_3Rh(CO)H$ with chelating bisphosphines—$3\varnothing_2P(CH_2)_nP\varnothing_2 + (\varnothing_3P)_3Rh(CO)H$

Figure 11 shows the spectrum of the dppe system in more detail at 35° and 90°C. The chemical shift (±54.3 ppm) and the coupling constant (P–Rh = 144 Hz) of the complexed phosphorus atoms are identical with those reported by James and Mahajan (*16*) for the known dppe rhodium hydride.

```
CH₂—PØ₂      Ø₂P—CH₂
 |       \  /       |
 |        Rh        |
 |       / | \      |
CH₂—PØ₂   |   Ø₂P—CH₂
           |
           H
```

This complex shows no ligand exchange either with triphenylphosphine or with excess dppe. The lack of affinity for CO and the strong binding preventing dppe dissociation may explain the sharply reduced activity of this catalyst at very low CO partial pressure (<1 atm) in a large excess of the dppe ligand.

Figure 12 shows the spectrum of the dppp complex in detail. The spectrum exhibits an intense doublet signal at 16.1 ppm (J_{P-Rh} = 142

Hz) and a complicated signal having a 16-line ABCX pattern. (Also present is a weak doublet signal at 6.3 ppm (J_{P-Rh} = 154 Hz) that was not assigned.)

The large doublet was assigned as the signal due to the four equivalent phosphorus nuclei in dppp rhodium hydride. The spectrum of this bicyclic complex also was described previously by James and Mahajan (*16*).

$CH_2—PØ_2$ $Ø_2P—CH_2$

CH_2 Rh CH_2

$CH_2—PØ_2$ $Ø_2P—CH_2$

H

A study of the expanded plot of the dppp complex resulted in deconvoluting the ABCX coupling pattern in terms of P–Rh and P–P couplings as indicated by Figure 12. The numerical values for the assigned chemical shifts and coupling constants are given in Table II.

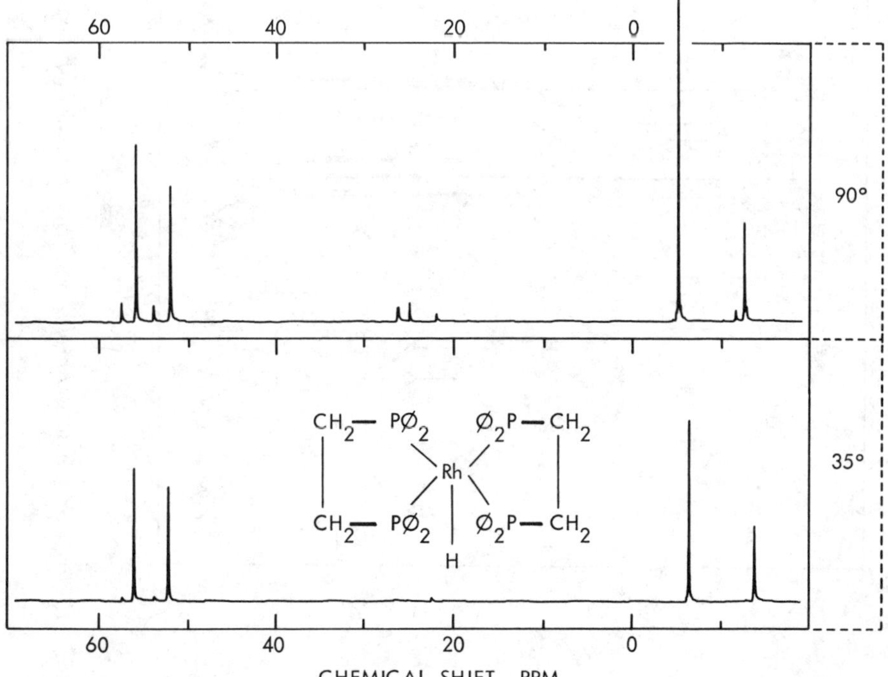

Figure 11. Ligand exchange of $Ø_3P$–Rh complex with ethylene bisphosphine—$3Ø_2PCH_2CH_2PØ_2 + (Ø_3P)_3Rh(CO)H$

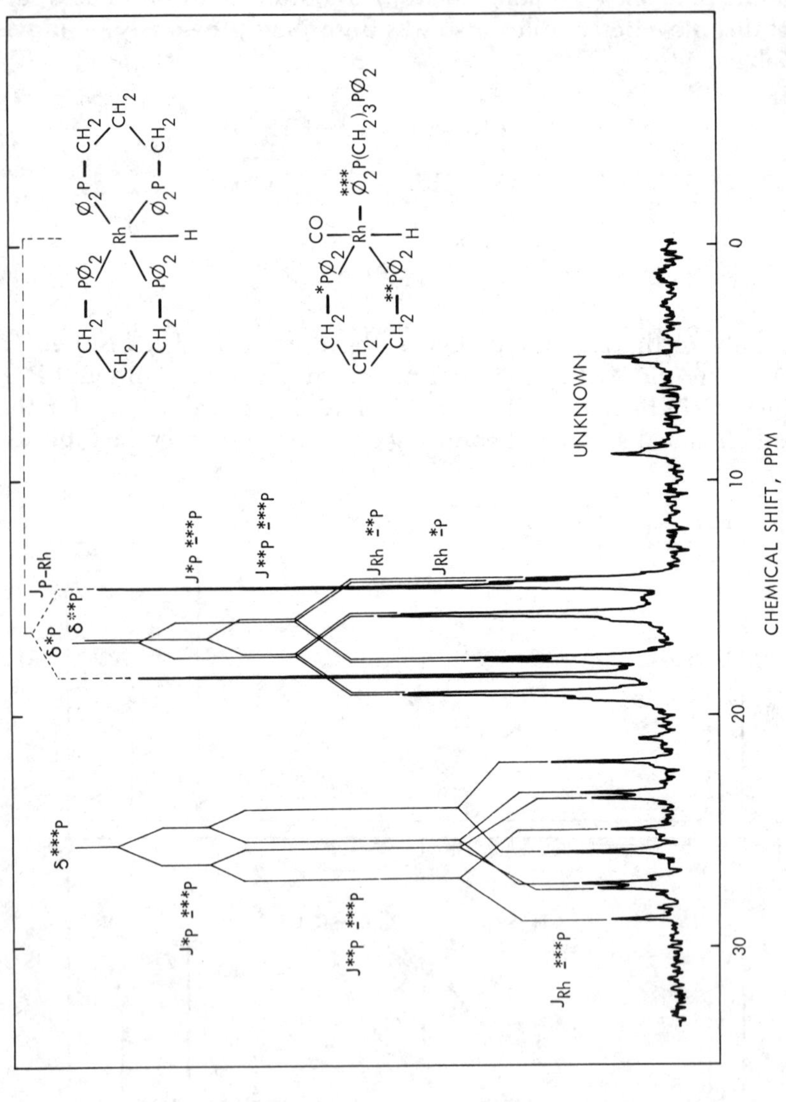

Figure 12. Chelate complexes of trimethylene bisphosphine—$3\text{Ø}_2PCH_2CH_2CH_2P\text{Ø}_2 + (\text{Ø}_3P)_3Rh(CO)H$

On the basis of the deconvolution, the ABCX pattern was assigned to originate from the three complexed nonequivalent phosphorus atoms of monocyclic 1,3-diphenylphosphinopropane 3-(diphenylphosphino)-propyldiphenylphosphinerhodium(I) carbonyl hydride.

```
            CO
 CH2—*PØ2    |
 /        \  |
CH2         Rh—Ø2***P(CH2)3PØ2
 \        /  |
 CH2—**PØ2   |
             H
```

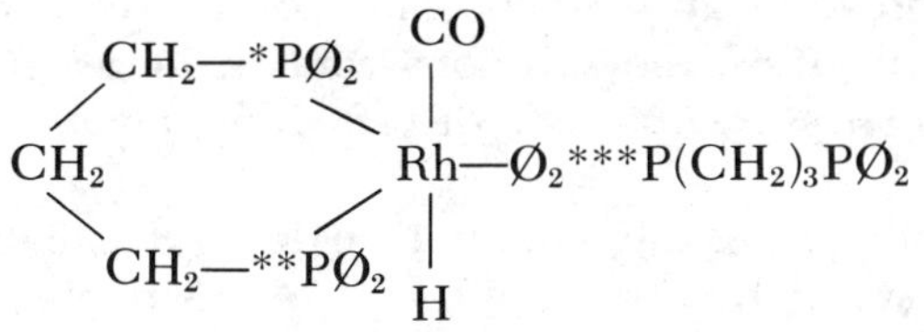

A study of the parameters of the above monocyclic phosphine (shown in Table II) indicates that the magnetic environments of the chelated phosphorus atoms in this complex and in the bicyclic complex of dppp are very similar. This is suggested due to their essentially identical chemical-shift values of about 16 ppm (and identical chemical-shift change of 36 ppm on complexation). On the other hand, the nonchelated, but complexed phosphorus atom of the monocyclic complex exhibits practically the same chemical-shift change (45 ppm) on complexation as that found for triphenylphosphine in the tris-(triphenylphosphine)rhodium(I) carbonyl hydride-plus-excess ligand system. Table II also shows that in the monocyclic complex, large P–P coupling constants are observed (about 51 and 56 Hz, respectively). These couplings are between the nonchelated bound phosphorus and

Table II. Parameters of Chelate Complexes of Trimethylene Bisphosphine

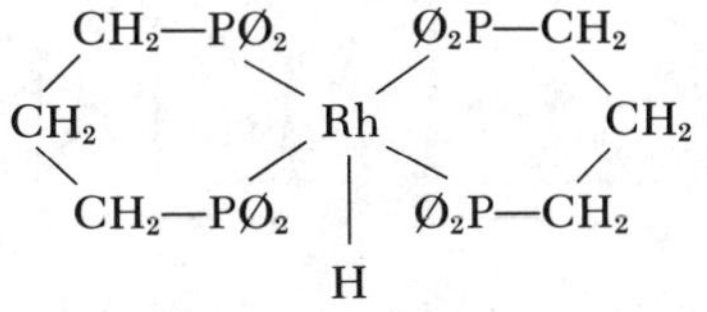

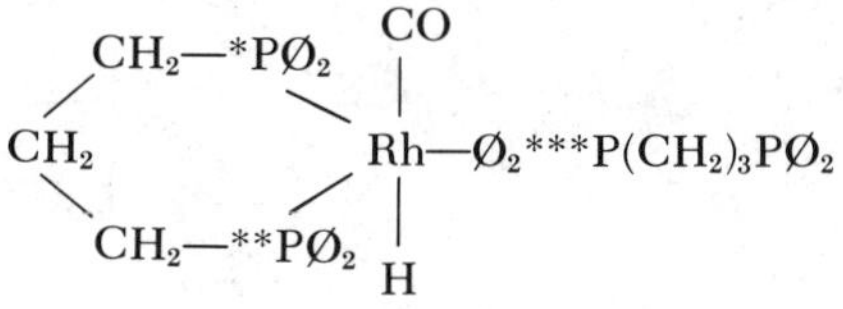

```
 CH2—PØ2      Ø2P—CH2                          CO
 /       \   /       \          CH2—*PØ2    |
CH2        Rh         CH2       /        \  |
 \       /  | \      /        CH2         Rh—Ø2***P(CH2)3PØ2
 CH2—PØ2    |  Ø2P—CH2          \        /  |
            H                    CH2—**PØ2  |
                                            H
```

Type of P	Chemical Shift, ppm: Value for Bound δ_b	Chemical Shift, ppm: Difference from Free δ_{b-f}	Coupling Constant, Hz: J_{P-Rh}	$J_{*P-***P}$	$J_{**P-***P}$
P	16.1	35.8	142	—	—
*P	16.4	36.1	144	~ 51	
**P	16.3	36.0	144		~ 56
***P	25.4	45.1	126	~ 51	~ 56

the two chelated phosphorus atoms. It is noted that coupling between the two, apparently nonequivalent, chelated phosphorus atoms is too small to observe.

Dppp solutions having increased dppp-to-rhodium ratios (15 : 140) also were studied by NMR for changes in a potential equilibrium between the bicyclic and monocyclic complexes. With increasing dppp-to-rhodium ratios, the solutions contain increasing percentages of the bicyclic complex. At a ratio of 140 : 1, no monocyclic complex is detectable.

Figure 13 illustrates the temperature dependence of the spectra of the solution derived by reacting 1 mol of tris(triphenylphosphine)-rhodium(I) carbonyl hydride with 6 mol of dppp. The maintenance of the narrow linewidth of the intense doublet signal at 16.1 ppm shows that the bicyclic complex does not dissociate up to 90°C. On the other hand, broadening signals of the complex spectrum of the monocyclic complex and that of the singlet signal of free dppp indi-

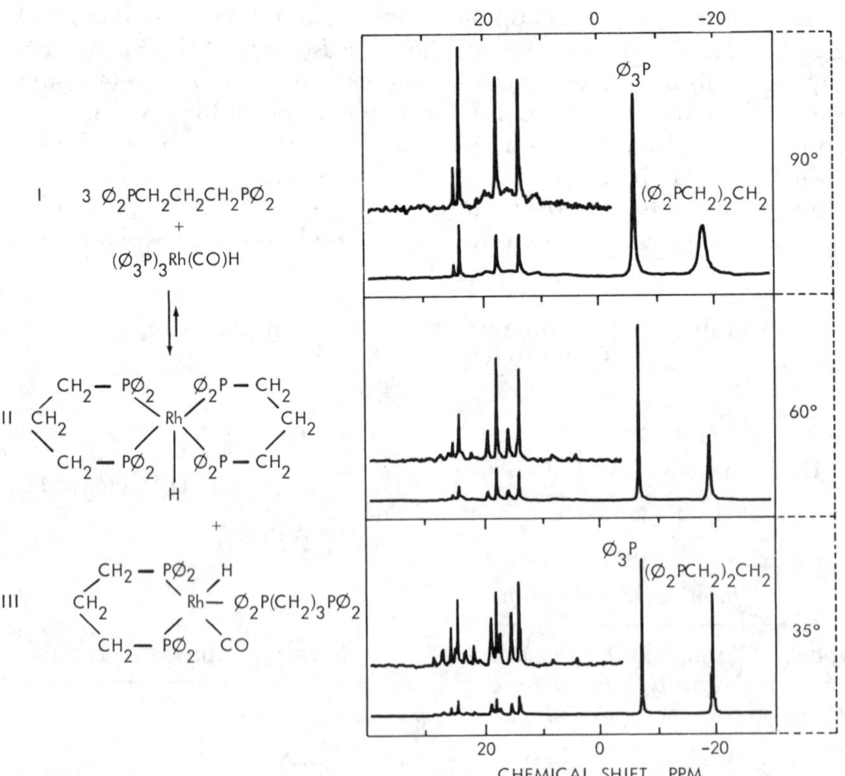

Figure 13. Ligand exchange of $\emptyset_3P$–Rh complex with trimethylene bisphosphine at different temperatures

cates that this component dissociates and exchanges with the free dppp at a faster rate with increasing temperatures. This dissociation and the associated ligand exchange are apparently due to the dissociation of the mono-coordinated bisphosphine moiety in the manner proposed in the introduction. A comparison of the spectra taken at different temperatures indicates little if any change in the relative quantities of the monocyclic and bicyclic complexes.

The variable-temperature NMR spectra help to explain the catalytic properties of the dppp complex system which were outlined previously in Table I. The reduced catalytic activity compared with the tris(triphenylphosphine) complex system is apparently due to the reduced dissociation of the cyclic complexes. For example, the 90°C spectra of Figures 3 and 13, clearly show that the ligand-exchange rate is much slower in the case of dppp. However, temperature-dependent ligand exchange of the monocyclic complex occurs and leads to cis-bisphosphine species that catalyze the hydroformylation of olefins at minimal partial pressures of CO. The hydroformylation rate of the dppp system is faster at 1 atm CO pressure than that of the dppe system. Of course, such hydroformylations are nonselective due to the cis-stereochemistry.

Conclusions

P-31 NMR studies of the tris(triphenylphosphine)rhodium(I) carbonyl hydride-plus-excess triphenylphosphine catalyst systems in toluene and benzene did not indicate any significant dissociation to the corresponding trans-bisphosphine complex in terms of detectable species. However, the existence of an equilibrium between the two complexes was established by indirect methods involving ligand-exchange studies at various excess triphenylphosphine to complex ratios. A first-order dissociation reaction of the trisphosphine complex occurs with dissociation energy of 20 ± 1 kcal mol^{-1}. The association reaction between the bisphosphine and triphenylphosphine also depends on the concentration of the reactants. In general, a correlation exists between ligand exchange and hydroformylation catalysis by this system.

Since the 1-olefin concentration-dependent hydroformylation in the presence of the above catalyst system has a slightly higher activation energy of about 22 kcal mol^{-1}, it is proposed that the rate-determining step of selective terminal 1-olefin hydroformylation may involve a transition state leading to the formation of a 1-alkyl bis-(*trans*-phosphine)rhodium carbonyl hydride complex rather than the dissociation of the trisphosphine complex.

P-31 studies established that the two chelating ligands used, i.e. dppe and dppp, effect complete ligand displacement when added to solutions of tris(triphenylphosphine)rhodium(I) carbonyl hydride. This results in nonselective hydroformylation catalyst systems. Dppe forms a known carbonyl-free bicyclic rhodium hydride complex. Dppp forms an equilibrium mixture of the analogous bicyclic compound plus a new monocyclic-trisphosphine complex having a ligand-exchange behavior similar to that of the tris(triphenylphosphine) complex. A high excess of dppp results in a mixture containing mostly the bicyclic complex.

The high stability of the bicyclic complexes in such systems and their cis-stereochemistry are suggested to explain their low activity and lack of selectivity as hydroformylation catalysts when compared with similar systems based on triphenylphosphine.

Literature Cited

1. Osborn, J. A.; Jardine, F. H.; Young, J. F.; Wilkinson, G. *J. Chem. Soc. A* **1966**, 1711.
2. Tolman, C. A.; Meakin, P. Z.; Lindner, D. L.; Jesson, J. P. *J. Am. Chem. Soc.* **1974**, *96*, 2762.
3. Evans, D.; Osborn, J. A.; Wilkinson, G. *J. Chem. Soc. A* **1968**, 3133.
4. Pruett, R. L.; Smith, J. A. *J. Org. Chem.* **1969**, *34*, 327.
5. Pruett, R. L.; Smith, J. A.; U.S. Patents 3 527 809; 3 917 661; 4 148 830.
6. Pruett, R. L. In "Advances in Organometallic Chemistry"; Stone, F. G. A.; West, R., Eds.; Academic: 1979; Vol. 17, p. 1.
7. Yagupsky, G.; Brown, C. K.; Wilkinson, G. *J. Chem. Soc. A* **1970**, 1392.
8. Brown, C. K.; Wilkinson, G. *J. Chem. Soc. A* **1970**, 2753.
9. Cavalieri d'Oro, P.; Raimondi, L.; Pagani, G.; Montrasi, G.; Gregorio, G.; Oliveri del Castillo, G. F.; Andreetta, A. *Symp. Chem. Hydroformylation Relat. React., Proc., Veszprem, Hungary,* May–June, *1972.*
10. Pittman, C. F., Jr.; Hirao, A.; Jones, C.; Hanes, R. M.; Quock, Ng; *Ann. N.Y. Acad. Sci.* **1977,** *293,* 15.
11. Hjortkjaer, J. *J. Mol. Catal.* **1979,** *5,* 377.
12. Sanger, A. R. *J. Mol. Catal.* **1977,** *3,* 221.
13. Pittman, C. V., Jr.; Hirao, A. *J. Org. Chem.* **1978,** *43,* 640.
14. Martin, M. L.; Delpuech, J. J.; Martin, G. J. "Practical NMR Spectroscopy"; Heyden: Philadelphia, 1980; Chap. 8.
15. Jackman, L., personal communication of the FORTRAN program, DNRAXl.
16. James, B. R.; Mahajan, D. *Can. J. Chem.* **1979,** *57,* 180.

RECEIVED July 21, 1980.

4

The Effect of Chelating Diphosphine Ligands on Homogeneous Catalytic Decarbonylation Reactions Using Cationic Rhodium Catalysts

D. H. DOUGHTY, M. P. ANDERSON, A. L. CASALNUOVO, M. F. McGUIGGAN, C. C. TSO, H. H. WANG, and L. H. PIGNOLET
Department of Chemistry, University of Minnesota, Minneapolis, MN 55455

Catalytic decarbonylation of aldehydes has been studied using mono- and bisdiphosphine complexes of Rh(I). The catalyst $[Rh(dppp)_2]BF_4$, where dppp = 1,3-bis(diphenylphosphino)propane, decarbonylates aldehydes homogeneously with rates that are more than two orders of magnitude faster than with $Rh(PPh_3)_3Cl$. Additionally, good catalytic activities were measured under mild thermal conditions (100 turnovers hr^{-1} for benzaldehyde at 150°C) and the reactions are highly selective. The catalysts are stable for days and turnovers in excess of 100,000 have been achieved. Experiments on a variety of aldehydes have established the general usefulness of this reaction in organic synthesis. Kinetic measurements, P-31 NMR spectroscopy, and X-ray diffraction studies have permitted some general mechanistic conclusions. For the catalytic decarbonylation of benzaldehyde, a rapid pre-equilibrium is established ($Rh(dppp)_2^+ + PhCHO \rightleftarrows (PhCHO)Rh(dppp)_2^+$) prior to oxidative addition, which is the rate-determining step. A mechanistic understanding of this reaction is important since it is a model for the discreet processes in other catalytic reactions that use organometallic catalysts.

Decarbonylation of aldehydes and acid halides is an important synthetic reaction (*1*, *2*) and using various transition-metal complexes as stoichiometric or catalytic reagents for this process has

0065–2393/82/0196–0065$05.00/0

received considerable attention (*3–8*). In this regard, tris(triphenylphosphine)chlororhodium(I), $RhCl(PPh_3)_3$, has received the most study and has been proven useful as a decarbonylation reagent in organic synthesis (*1, 8, 9*). This complex decarbonylates aldehydes and acid chlorides stoichiometrically under mild conditions (<100°C) in homogeneous solutions (*4, 9, 10, 11, 12, 13*). The overall reaction is shown by Equation 1. In cases where R contains a β-hydrogen atom, β-hydride elimination competes with Reaction 1 and unsaturated products are produced. This is especially important when X = Cl but it is of minor importance when X = H (e.g. decarbonylation of heptanal yields 86% hexane and 14% 1-hexene) (*4*). The decarbonylation reaction is catalytic above 200°C (*1, 4, 14*); however, most studies have used the reaction stoichiometrically.

$$RhCl(PPh_3)_3 + RCOX \xrightarrow[X = Cl \text{ or } H]{} RX + RhCl(CO)(PPh_3)_2 \qquad (1)$$

Reaction Mechanism Using $RhCl(PPh_3)_3$

There is general agreement on the mechanism for the stoichiometric decarbonylation of acid chlorides (*9, 14, 15, 16*). The overall mechanism is shown by Equation set 2 where X = Cl. The stoichiometric decarbonylation reaction results from initial oxidative addition of the acid chloride to $RhCl(PPh_3)_2$ (Equation 2b, X = Cl). $RhCl(PPh_3)_2$ is a very reactive, low-concentration intermediate which is likely to be solvated (*see* Equation 2a) (*17*).

$$RhCl(PPh_3)_3 \overset{K \ll 1}{\rightleftharpoons} RhCl(PPh_3)_2 + PPh_3 \qquad (2a)$$

$$RhCl(PPh_3)_2 + RCOX \rightleftharpoons RCO(X)RhCl(PPh_3)_2 \qquad (2b)$$

$$RCO(X)RhCl(PPh_3)_2 \rightleftharpoons R(X)RhCl(CO)(PPh_3)_2 \qquad (2c)$$

$$R(X)RhCl(CO)(PPh_3)_2 \longrightarrow RX + RhCl(CO)(PPh_3)_2 \qquad (2d)$$

Equations 2c and 2d show the acyl–alkyl migration and reductive elimination steps, respectively. There is good evidence that this same mechanistic scheme applies to the decarbonylation of aldehydes (*see* Equation set 2, X = H), although in this case reaction intermediates have not been isolated (*3, 5, 9, 18*). Additionally, evidence exists that the rate-determining step is oxidative addition for aldehyde decarbonylation (*see* Equation 2b, X = H) (*3, 9, 18*). Several recent reports have shown that for some special aldehydes, oxidative addition of the carbonyl–hydrogen bond indeed does occur using rhodium(I) complexes (*8, 19*). In these studies a stable chelate was formed after oxidative addition that enabled isolation and characterization of the products (*8, 19*).

At temperatures above ca. 200°C, the decarbonylation reaction can be driven catalytically (*1, 4, 14, 20*). Scheme I illustrates the proposed catalytic reaction scheme (*15, 16*). This catalytic reaction is slow (activity for benzaldehyde decarbonylation at 178°C is 10 turnovers hr^{-1}) presumably because the oxidative addition of RCO*X* to $RhCl(CO)(PPh_3)_2$ is difficult (*7, 21, 22*). Consistent with this, the rate is significantly greater when $IrCl(CO)(PPh_3)_2$ is used as the catalyst (benzaldehyde, 178°C, activity is 66 turnovers hr^{-1}) (*23*). Oxidative addition to iridium complexes is well known to be more facile than with their rhodium analogues.

Scheme I.

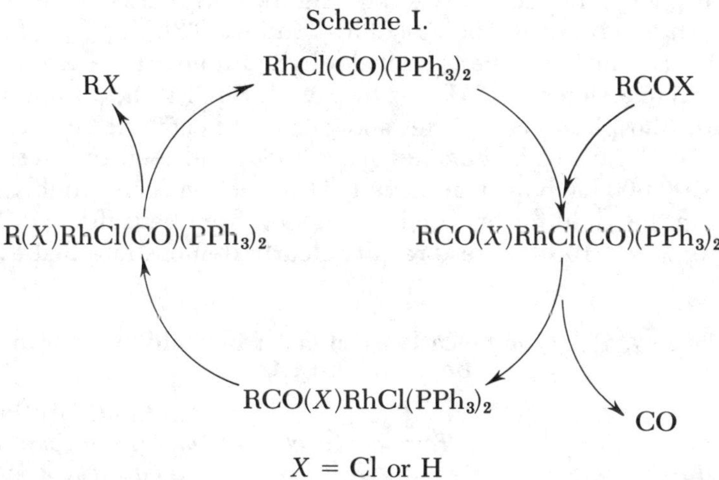

X = Cl or H

***Decarbonylation of Benzaldehyde Using Cationic Diphosphine Complexes of Rhodium* (*I*)**

In order to achieve higher catalytic activities for decarbonylation under mild thermal conditions, it is necessary to use a metal complex of sufficient basicity to promote facile oxidative addition, but also with the proper stereochemical and electronic properties so that CO loss and catalyst regeneration is rapid. With $RhCl(PPh_3)_3$, the product complex *trans*-$RhCl(CO)(PPh_3)_2$ (*see* Equation 2d) is inert to CO dissociation even at high temperature (*24*) and under UV irradiation (*25*); therefore Scheme I is required for catalytic decarbonylation. With this in mind, we synthesized a series of mono- and bis(diphosphine) complexes: [Rh(P–P)]*X* and $[Rh(P\text{–}P)_2]X$, where *X* = Cl or BF_4, and P–P = $Ph_2P(CH_2)_nPPh_2$ (hereafter named dppm, dppe, dppp, and dppb for n = 1, 2, 3, and 4, respectively) (*7, 21, 22*). These complexes are known to undergo oxidative addition reactions as well as to bind CO weakly and reversibly (*26, 27, 28, 29*). Indeed $[Rh(dppe)_2]^+$ is unreactive towards CO; however, it does function as a reasonable decarbonylation catalyst for aldehydes (vide infra).

These diphosphine complexes were allowed to react with neat benzaldehyde at several temperatures under a purified nitrogen purge in order to determine their catalytic activities for decarbonylation. The rate of benzene production was monitored by gas–liquid chromatography (GLC) as described previously (*7, 21, 22*). The experimental procedure for these reactions as well as for the syntheses of the complexes has been published (*7, 21, 22*). The results are presented in Table I. It is apparent from these results that catalytic activities using diphosphine complexes are significantly greater than with $RhCl(PPh_3)_3$ or *trans*-$RhCl(CO)(PPh_3)_2$. Indeed, $[Rh(dppp)_2]X$ (X = Cl or BF_4) catalytically converts benzaldehyde into benzene at a rate that is more than a factor of 10^2 faster than with $RhCl(PPh_3)_3$. Importantly, the bis(diphosphine) complexes show constant catalytic activity for at least several days in homogeneous solution, and total turnover numbers of 100,000 have been achieved. The reaction is also highly selective since the yield for benzene production from benzaldehyde using $[Rh(dppp)_2]^+$ is 100%. These results clearly demonstrate that the bis-

Table I. Catalytic Decarbonylation of Benzaldehyde into Benzene and CO

Catalyst[a]	*Temperature* °C	*Catalytic Activity*[b] *(moles benzene/moles catalyst/hour)*
$RhCl(PPh_3)_3$[c]	150	0.60
$[Rh(dppp)_2]^+$[d]	150	1.0×10^2
$[Rh(dppe)_2]^+$[d]	150	25
$[Rh(dppm)_2]^+$[d]	115	0.40
$[Rh(dppe)_2]^+$[d]	115	3.6
$[Rh(dppp)_2]^+$[d]	115	11
$[Rh(dppb)_2]^+$[d]	115	1.2
$[Rh(dppp)_2]^+$[d]	145	85
$[Rh(dppp)]BF_4$	145	40
RhCl(dppp)	145	30
$RhCl(PPh_3)_3$[c]	178	10
$[Rh(dppp)_2]^+$[d]	178	1.1×10^3
$[Rh(dppe)_2]^+$[d]	178	2.1×10^2
$[Rh(dppm)_2]^+$[d]	178	74

[a] Catalyst concentrations between 1×10^{-4} and $1 \times 10^{-3}M$ and neat benzaldehyde as the solvent.

[b] *See* Refs. *7* and *21* for experimental details; values averaged over a 30–40 h period and first-order dependence on [catalyst] is assumed.

[c] Same results using *trans*-$RhCl(CO)(PPh_3)_2$.

[d] Same rate for Cl^- and BF_4^- salts.

(diphosphine) complexes are excellent catalysts for the decarbonylation of benzaldehyde and for other aldehydes as well (vide infra).

The mono(diphosphine) complexes, $[Rh(dppp)]BF_4$ and RhCl(dppp), are less effective than $[Rh(dppp)_2]^+$ but are still more active than $RhCl(PPh_3)_3$. The mono(diphosphine) catalysts also decompose slowly under the reaction conditions, which renders them less useful than the bis(diphosphine) catalysts. The slower rate of decarbonylation observed with the mono(diphosphine) catalysts compared with the bis(diphosphine) catalysts presumably is due to the lower basicity of the former which retards the rate of oxidative addition (vide infra). Consistent with this is the observation that $[Rh(COD)(dppp)]BF_4$ (COD = 1,5-cyclooctadiene) shows a higher rate for catalytic decarbonylation of benzaldehyde than does $[Rh(dppp)]BF_4$ (*22*). An additional observation is that the type of anion, Cl^- or BF_4^-, has no apparent effect on decarbonylation rates for the bis(diphosphine) catalysts; however, for the mono(diphosphine) complexes the chloride salts show slightly lower rates than their tetrafluoroborate analogues.

Within the class of cationic bis(diphosphine) complexes, the catalytic activity shows a marked dependence on the chelate ring size. At all of the temperatures examined, the rate increases in the order dppm < dppc < dppp, thus reflecting the increase in chelate ring size from four to six members. However, the observed rate using the seven-membered chelate ring catalyst $[Rh(dppb)_2]^+$ is considerably slower than for the dppp and dppe analogues. An understanding of this trend is undoubtedly complex and clearly requires a knowledge of the reaction mechanism (vide infra). It is noteworthy that similar rate effects as a function of diphosphine chelate ring size have been observed by others (*30, 31, 32*) for a variety of catalytic reactions.

Synthetic Usefulness of the Reaction

In order to determine the general synthetic utility of these catalysts, and to gain mechanistic insight, a wide variety of aldehydes were decarbonylated catalytically using $[Rh(dppp)_2]^+$ (*22*). Aldehydes were chosen to illustrate the following effects: (a) steric size of substrate (*see* Table II); (b) electronic effects of substrate (*see* Table III); and (c) catalyst selectivity (*see* Table II). There is a significant rate decrease when the α-carbon atom of the aldehyde becomes substituted (*see* Table II). The decarbonylation of tertiary aldehydes occurs very slowly even at 180°C, and it is therefore of only minor synthetic utility. The effects of para substitution on benzaldehyde are interesting in that a simple relationship does not exist (*see* Table III). Electron-donating and -withdrawing groups decrease the catalytic activity compared with the unsubstituted benzaldehyde. This result is in marked contrast to that observed using $RhCl(PPh_3)_3$ (stoichiometric reaction at 100°C) where a linear Hammett plot was obtained with a positive ρ value (*18*). These observations have important mechanistic implications and will be discussed in the next section.

An important characteristic of the bis(diphosphine) catalysts is the remarkable selectivity observed in the decarbonylation products. Recall that 1-heptanal is converted into 86% hexane and 14% 1-hexene by $RhCl(PPh_3)_3$ (25°C, stoichiometric reaction) (*4*). In contrast, using $[Rh(dppp)_2]^+$ as the catalyst, the only volatile product is hexane,

Table II. Steric Effects of Aldehydes and Catalyst Selectivity[a]

Aldehyde	*Only Volatile Products*	*Temperature (°C)*	*Catalytic Activity (turnover/hour*
$C_5H_{11}CH_2CHO$ Heptanal	hexane	115 150	31 330
$C_2H_5-CH(C_2H_5)-CHO$ 2-Ethylbutanal	pentane	115	5
$Ph-C(CH_3)(C_2H_5)-CHO$ 2-Phenyl-2-Methylbutanal	2-phenylbutane	180	1.3
$Ph-CH(CH_3)-CHO$ 2-Phenylpropanal	ethylbenzene	115 150	16 67
Ph(H)C=C(H)CHO Cinnamaldehyde	Ph(H)C=CH₂ styrene	115 150	12 135
Ph(CH₃)C=C(H)CHO α-Methylcinnamaldehyde	Ph(CH₃)C=CH₂ *cis*-β-methylstyrene	150	16

[a] Catalyst is $[Rh(dppp)_2]BF_4$.

Table III. Electronic Effects of para-Substituted Benzaldehydes[a]

Aldehyde	*Product*	*Temperature (°C)*	*Catalytic Activity (turnover/hour*	σ_p
$H-C_6H_4-CHO$	benzene	115	11	0
		150	90	
$CH_3-C_6H_4-CHO$	toluene	115	3.0	−.17
		150	34	
$CH_3O-C_6H_4-CHO$	anisole	115	—	−.268
		150	10	
$Cl-C_6H_4-CHO$	chlorobenzcnc	115	0.1	+.227
		150	1.0	

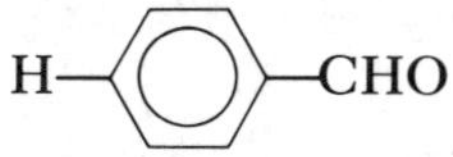

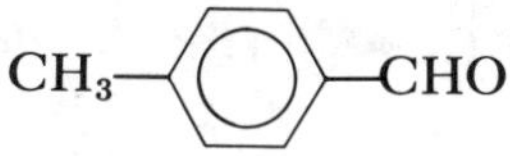

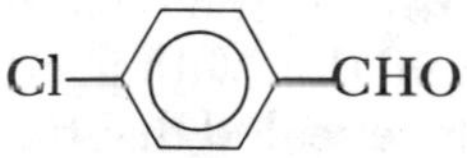

[a] Catalyst is $[Rh(dppp)_2]BF_4$.

which is produced at 150°C with a catalytic activity of 330 turnovers hr^{-1}. Additionally, only ethylbenzene is produced from 2-phenylpropanal although the possible unsaturated product, styrene, is very stable. The absence of unsaturated products using $[Rh(dppp)_2]^+$ is attributed to the lack of a vacant coordination site in appropriate reaction intermediates (vide infra) so that β-hydride elimination is unlikely to occur. In contrast, $RhCl(PPh_3)_3$ can more easily provide a vacant coordination site (*see* Equation 2) for β-hydride elimination. This argument is supported by the observation that decarbonylation of 1-heptanal using the mono(diphosphine) catalyst $[Rh(dppp)]BF_4$ produces significant amounts of 1-hexene at 150°C (the ratio of hexane/1-hexene is 2/1) (*22*). Intermediates using mono(diphosphine) catalysts are likely to be coordinatively unsaturated.

$RhCl(PPh_3)_3$ is known to isomerize double bonds during decarbonylation (*4*). For example, α-methylcinnamaldehyde is converted into *cis-* and *trans*-β-methylstyrene:

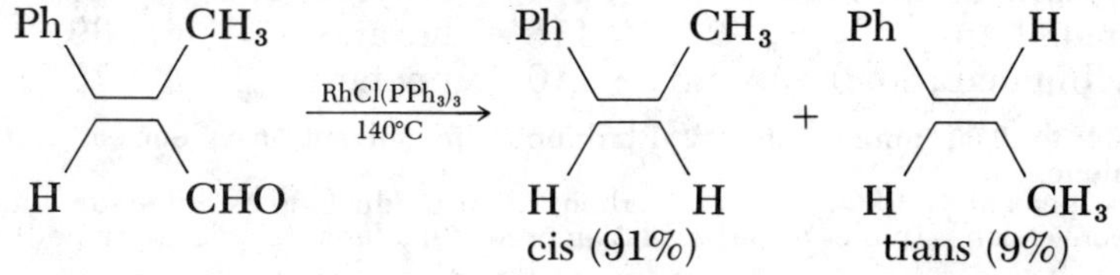

Ph CH$_3$ / H CHO $\xrightarrow[140°C]{RhCl(PPh_3)_3}$ Ph CH$_3$ / H H + Ph H / H CH$_3$

cis (91%) trans (9%)

No trans isomer is produced using $[Rh(dppp)_2]^+$ at 150°C (*see* Table II).

Walborski and Allen have shown that stiochiometric decarbonylation of an optically active aldehyde with $RhCl(PPh_3)_3$ gives complete retention of configuration (5). Using $[Rh(dppp)]BF_4$, (−) − (*R*)-2-phenyl-2-methylbutanal was decarbonylated catalytically at 165°C (0.4 turnovers hr^{-1}) into (+) − (*S*)-2-phenylbutane with 100% retention of optical activity (*21, 22*).

$$C_2H_5 \blacktriangleright \underset{\text{Ph }(-)-(R)}{\overset{CH_3}{C}} \blacktriangleleft CHO \xrightarrow[165°C]{[Rh(dppp)]^+} C_2H_5 \blacktriangleright \underset{\text{Ph }(+)-(S)}{\overset{CH_3}{C}} \blacktriangleleft H$$

For most aldehydes studied, the decarbonylation reaction occurs with high product yields (yields based on aldehyde). Typical data is shown in Table IV using $[Rh(dppp)_2]^+$ as the catalyst. However the saturated aldehydes, heptanal and 2-ethylbutanal, are decarbonylated to hexane and pentane, respectively, in rather low yield (but with good catalytic activities, *see* Table II). Further experiments revealed that the low yields observed for these aldehydes are due to thermal decomposition which also occurs in the absence of catalyst. Even lower yields (based on aldehyde) are obtained for these aldehydes using $RhCl(PPh_3)_3$ at the same temperature, since with this complex decarbonylation activities are much smaller.

The above data which are presented in Tables I–IV show that the diphosphine complexes are effective catalysts for the aldehyde decarbonylation. Although these reactions tolerate various functional groups such as carboxylic acids, ethers, ketones, olefins, and aryl chlorides,

Table IV. Yield and Conversion for Various Aldehydes[a]

Aldehyde (millimoles)	*Time (days)*	*Temperature (°C)*	*Product*	*Yield[b] (%)*	*Conversion[b] (%)*
Benzaldehyde (2.45)	2	140	benzene	79	100
Cinnamaldehyde (4.00)	2	118	styrene	95	100
2-Phenylpropanal (64.0)	1.5	175	ethylbenzene	83	93
Heptanal (1.85)	2	118	hexane	39	46
2-Ethylbutanal (60.8)	1.5	110	pentane	15	43

[a] 0.02 to 0.05 mmol catalyst, $Rh(dppp)_2^+$ in solvent *m*-xylene or α-methylnaphthalene.

[b] The percent yield = (moles decarbonylation product/moles aldehyde added) × 100; % conversion = (moles decarbonylation product/moles aldehyde consumed) × 100.

Table V. Effect of Added Reagents on Catalytic Activity[a]

Added Reagent (M)	*Temperature* (°C)	*Catalytic Activity (turnovers/hour)*
None (N_2 purge)	145	60
Acetophenone (0.8*M*)	145	50
Benzoic acid (0.2*M*)	145	50
para-Chlorotoluene (4.0*M*)	145	40
Ethylene (bubbled in place of N_2)	145	30
Benzonitrile (5.0*M*)	145	3.3
Benzoylchloride (0.8*M*)	145	0
Oxygen (bubbled in place of N_2)	145	—irreversible catalyst decomposition—
Carbon monoxide (bubbled in place of N_2)	145	0.1
$PEtPh_2$ (0.2*M*)	145	0.2
none (N_2 purge)	115	11
PPh_3 (0.05*M*)	115	7
dppp (0.03*M*)	115	0.1

[a] For conversion of benzaldehyde to benzene; $[Rh(dppp)_2]^+$ = 0.001*M*; solvent is benzaldehyde (*21, 22*).

several reagents are effective at retarding or stopping the catalytic reaction. For example, if CO is used as the purging gas, the reaction is stopped completely. Results of experiments in the presence of various reagents are presented in Table V. It is apparent from this data that reagents that can compete for a coordination site retard the decarbonylation rate. The effects of gasses such as ethylene and CO are reversible; however O_2 irreversibly decomposes the catalyst. The results shown in Table V have important mechanistic implications which will be discussed in the next section. Surprisingly, the presence of 0.8*M* benzoylchloride completely stops benzene production, and no chlorobenzene is produced. This result is unexpected since $RhCl(PPh_3)_3$ is an effective catalyst for acid chloride decarbonylation (vide supra) and illustrates one of the major reactivity changes due to the presence of chelating diphosphine ligands (*26*).

Comments on the Reaction Mechanism

A number of additional experiments have been carried out in an attempt to determine the overall mechanism for the catalytic decarbonylation reaction using $[Rh(dppp)_2]^+$. Since this reaction is a useful synthetic method and is a model for the discreet processes in other organometallic reactions, a detailed knowledge of the mechanisms and stereochemistry for the individual steps is important. The additional experiment types thus far carried out are: (a) deuterium labeling and kinetic isotope effects; (b) determination of reaction kinetics and rate law; (c) P-31 NMR studies of the catalysts and intermediates; and (d) structural characterization of the catalysts and intermediates. The results of these experiments (which are summarized here) and of those discussed already have given some mechanistic insight. However, many additional experiments are needed before details of the overall mechanism will emerge.

Deuterobenzaldehyde, C_6H_5CDO, and *p*-tolualdehyde, CH_3-C_6H_4-CHO, were mixed in equimolar amounts and decarbonylated at 140°C using $[Rh(dppp)_2]^+$ (*22*). Toluene and benzene were produced simultaneously, separated by GLC and analyzed by mass spectroscopy (MS). The isotopic purities of $C_6H_5CH_3$ and C_6H_5D were determined to be 100 ± 1% and 100 ± 4%, respectively, and therefore all possible processes that are intermolecular in aldehyde can be excluded. This experiment, along with the one that used an optically active aldehyde (vide supra), argue against a free-radical chain mechanism. These results agree with those using $RhCl(PPh_3)_3$ (*5, 9*). The deuterium isotope effect (k_H/k_D) for $C_6H_5C(H,D)O$ was 1.1 ± 0.1 using $[Rh(dppp)_2]^+$ at 140°C (*22*). Unfortunately, there have not been enough isotope effect studies on organometallic systems in order to draw meaningful conclusions. The measured (k_H/k_D) effect for phenylacetaldehyde and phenylacetaldehyde-1-*d* using $RhCl(PPh_3)_3$ is 1.86 ± 0.07 and this has been used to support oxidative addition as the rate-determining step (*9*).

The rate of benzene production from neat benzaldehyde solution was measured as a function of $[Rh(dppp)_2]^+$ concentration at 135°C. The reaction is first order in catalyst concentration as shown in Figure 1 (*21*). The rate of benzene production also was monitored as a function

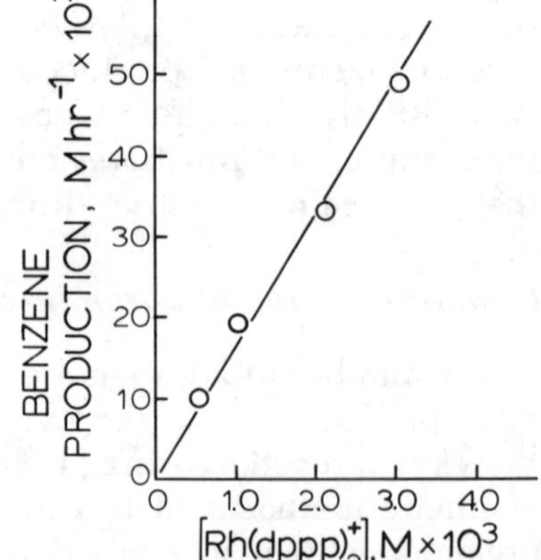

Figure 1. Rate of benzene production from neat benzaldehyde at 135°C as a function of $Rh(dppp)_2^+$ concentration

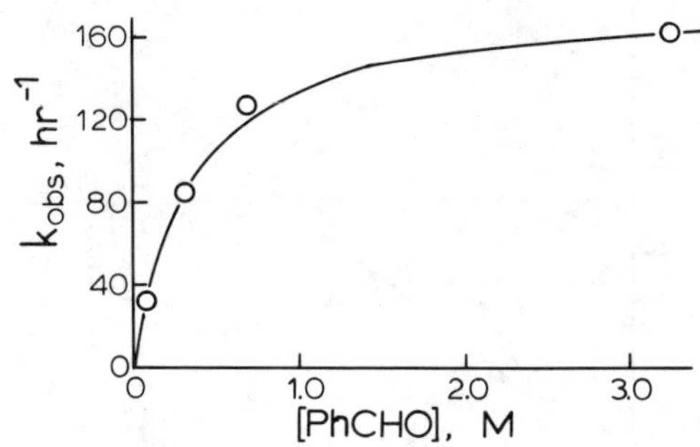

Figure 2. Rate of benzene production vs. benzaldehyde concentration at 158°C for $[Rh(dppp)_2]^+ = 6.7 \times 10^{-3}$M (22). The curved line was calculated from the plot in Figure 3 (see Equation 4).

of the benzaldehyde concentration. In this experiment the $Rh(dpp)_2^+$ concentration was held constant at $6.7 \times 10^{-3}M$ and anisole was used as the solvent. The benzaldehyde concentration was varied between 0.066 and 3.3M and the reaction temperature was maintained at 158°C. A plot of the rate constant, k_{obs}, for benzene production vs. benzaldehyde concentration is shown in Figure 2 and a saturation effect is observed. Such an effect implies that a rapid pre-equilibrium is established and is followed by the rate-determining step. Equation 3 shows steps that are consistent with the kinetic data. A standard kinetic analysis of Equation 3

$$Rh(dppp)_2^+ + PhCHO \overset{K}{\rightleftharpoons} (PhCHO)Rh(dppp)_2^+ \tag{3a}$$

$$(PhCHO)Rh(dppp)_2^+ \overset{k}{\rightleftharpoons} (PhCO)(H)Rh(dppp)_2^+ \tag{3b}$$

requires that a plot of k_{obs}^{-1} vs. $[PhCHO]^{-1}$ be linear as shown in Equation 4.

$$\text{Rate} = k_{obs}[Rh(dppp)_2^+] \tag{4a}$$

$$k_{obs} = \frac{kK\,[PhCHO]}{K\,[PhCHO] + 1} \tag{4b}$$

$$\frac{1}{k_{obs}} = \frac{1}{k} + \frac{1}{k\,K}\frac{1}{[PhCHO]} \tag{4c}$$

The plot that is presented in Figure 3 shows linear behavior and the calculated values of k and K are 170 hr^{-1} and $3.5M^{-1}$, respectively, at 158°C. Reaction Steps 3a and 3b are consistent with several additional observations.

The P-31{H-1} NMR spectrum of $[Rh(dppp)_2]BF_4$ in PhCHO solvent at 130°C shows only a doublet (δ 8.03 vs. H_3PO_4 with positive

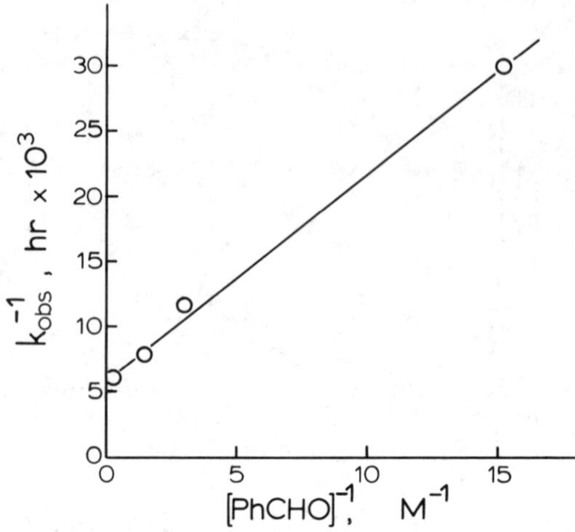

Figure 3. Plot of Equation 4c; data from Figure 2.

shifts downfield, J_{Rh-P} = 132 Hz) which is assigned to the rapidly interconverting complexes in Equation 3a. P-31{H-1} NMR data recorded at low temperature provides good evidence that a solvated complex exists. This data is presented in Figure 4 using acetone as a solvent. At temperatures below ca. −60°C an A_2B_2X pattern emerges that is consistent with the trigonal–bipyramidal Complex **I**. Above this temperature the complex becomes nonrigid on the NMR time scale. A similar observation with $[Rh(dppp)_2]Cl$ has been noted (33).

I

Since acetone is a reasonable model for benzaldehyde, the presence of Complex **I** provides support for Step 3a. (A resolved A_2B_2X pattern cannot be reached using benzaldehyde since the solvent freezes before the slow exchange limit is achieved; however, exchange broadening is evident.) Although additional experiments are needed to establish the presence of Complex **I** at higher temperatures, the fact that Complex **I** is the only species observed at low temperature and has an averaged chemical shift and Rh–P coupling constant that are close to the values observed at 130°C suggests that solvation is important at higher temperatures. Additionally, the formation of Complex **I** is a reasonable prerequisite for oxidative addition. The fact that Rh–P coupling is preserved at 130°C and that uncoordinated phosphorous resonances are not observed proves that Rh–P bond rupture does not occur in Step 3a (recall that K for equilibrium in Equation 3a is $3.5M^{-1}$).

The assignment of oxidative addition as the rate-determining step (*see* Equation 3b) is reasonable since there is no evidence for rhodium(III) intermediates in the high-temperature P-31 NMR spectrum. If any later reaction step were rate determining, except for the loss of CO from $(CO)Rh(dppp)_2^+$ (vide infra), rhodium(III) intermedi-

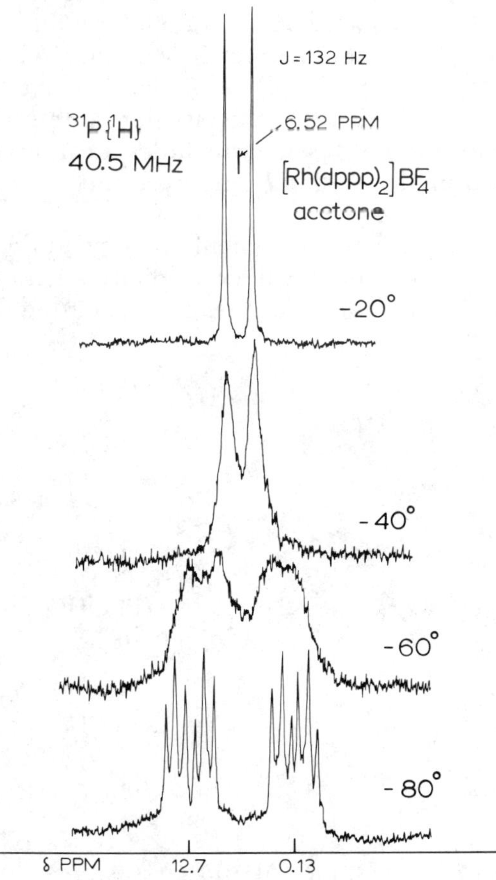

Figure 4. P-31{H-1} NMR spectra as a function of temperature for $[Rh(dppp)_2]\ BF_4$ *recorded using acetone solvent*

ates would be observable and their absence is strong evidence that Step 3b is rate determining. CO loss from $(CO)Rh(dppp)_2^+$ cannot be rate determining because (a) $(CO)Rh(dppp)_2^+$ is not observed by P-31 NMR at 130°C under catalytic reaction conditions and (b) CO is displaced readily from $(CO)Rh(dppp)_2^+$ in CH_2Cl_2 solution at room temperature by a N_2 purge as monitored by IR spectroscopy (*22*).

Reaction Steps 3a and 3b also can be used to rationalize the observed para-substituent effects presented in Table III; the more electron-releasing, para-substituted benzaldehydes retard the rate of oxidative addition (*18*) for $RhCl(PPh_3)_3$. Therefore, *p*-methyl- and *p*-methoxybenzaldehyde are expected to be decarbonylated slower than the unsubstituted benzaldehyde, as is observed in Table III. (This argument requires that Reaction 3a be saturated to the right, which is expected, in neat aldehyde solvent with electron-releasing, para-substituted benzaldehydes.) The unexpected slower rate for *p*-chloro-benzaldehyde could be accounted for if K for this aldehyde is small and saturation of equilibrium in Equation 3a is not achieved. Note that k_{obs} is a function of K and k (*see* Equation 4b) under this condition. It is also possible that the rate-determining step is different for this aldehyde. Present research includes a careful kinetic analysis using several aldehydes so that K and k can be determined independently.

In light of the previous comments it is possible to construct an overall mechanistic scheme for the decarbonylation of benzaldehyde (*see* Scheme II). Recall that various reagents retarded the decarbonyla-

Scheme II.

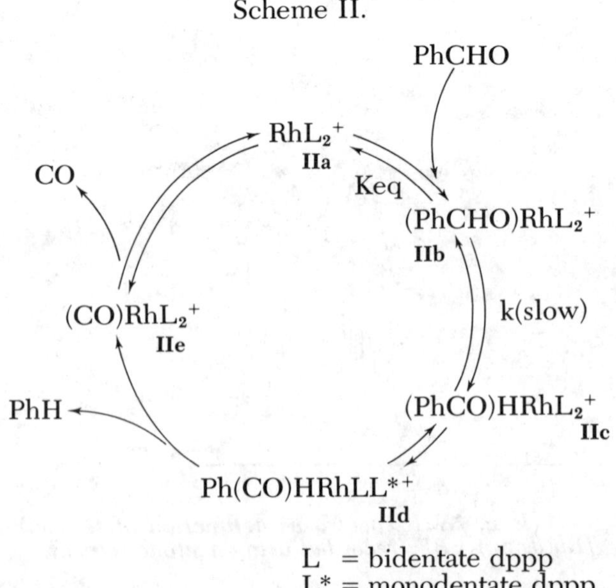

tion rate of benzaldehyde (*see* Table V). All of the effective reagents (phosphines, CO, nitriles, and ethylene) are potential ligands and presumably can compete with the pre-equilibrium step and thereby decrease k_{obs} by removing Complex **IIb**. This can be tested in principle and such experiments are in progress. In Scheme II, the only observed intermediates are **IIb** and **IIe**. Present work is designed to isolate and study other intermediates in this scheme.

Benzoylchloride presents a special case since its presence completely stops the decarbonylation of benzaldehyde and no chlorobenzene is produced. Analysis of this reaction has shown that PhCOCl reacts with $[Rh(dppp)_2]Cl$ to produce $Rh(Cl)_2(PhCO)(dppp)$. This rhodium(III) complex is inert to migration and reductive elimination and therefore no decarbonylation products are produced (*26*).

Structural Studies and Other Catalysts

In the preceding section all studies were carried out using $[Rh(dppp)_2]^+$. Although it is tempting to generalize Scheme II to other catalysts, some recent results argue against this. It is well known that the chemical properties of $[Rh(dppe)_2]BF_4$ are very different from those of $[Rh(dppp)_2]BF_4$. For example, $[Rh(dppe)_2]^+$ does not react with CO or H_2 (*29, 32, 34*) whereas both substrates react readily with $[Rh(dppp)_2]^+$ (*29, 32*). We have found that the rate of decarbonylation of benzaldehyde into benzene using $[Rh(dppe)_2]BF_4$ is not first order in catalyst concentration (*35*). (Recall that $[Rh(dppp)_2]^+$ showed first-order kinetics.) More studies are needed with this catalyst system, although a similar rate difference between $[Rh(dppe)_2]^+$ and $[Rh(dppp)_2]^+$ for catalytic hydrogenation of methylenesuccinic acid has been noted (*31*).

$[Rh(dppb)_2]BF_4$ also exhibits different reactivity towards CO. In this case the bimetallic complex $[Rh_2(CO)_4(dppb)_3](BF_4)_2$ has been isolated as the major product and characterized by single-crystal X-ray diffraction (*28*). The structure is illustrated in Figure 5. It is apparent that Rh–P bond rupture occurs more easily with dppb complexes where chelate ring strain becomes important. For example, uncoordinated dppb is observed by P-31 NMR at 25°C immediately on adding CO to a CH_2Cl_2 solution of $[Rh(dppb)_2]BF_4$ (*35*). Uncoordinated phosphorous resonances have not been observed with complexes of dppp or dppe. The recorded low-temperature P-31{H-1} NMR spectrum of $[Rh(dppb)_2]BF_4$ which used acetone as the solvent is also very complex (*see* Figure 6) and suggests solvent association and possibly the presence of dimeric species.

For this reason we very recently have determined the structure of $[Rh(dppb)_2]BF_4$ by single-crystal X-ray diffraction (*see* Figure 7) (*36*). The structure consists of discreet, monomeric molecular units with no unusually short intermolecular contacts. Importantly, the coordination

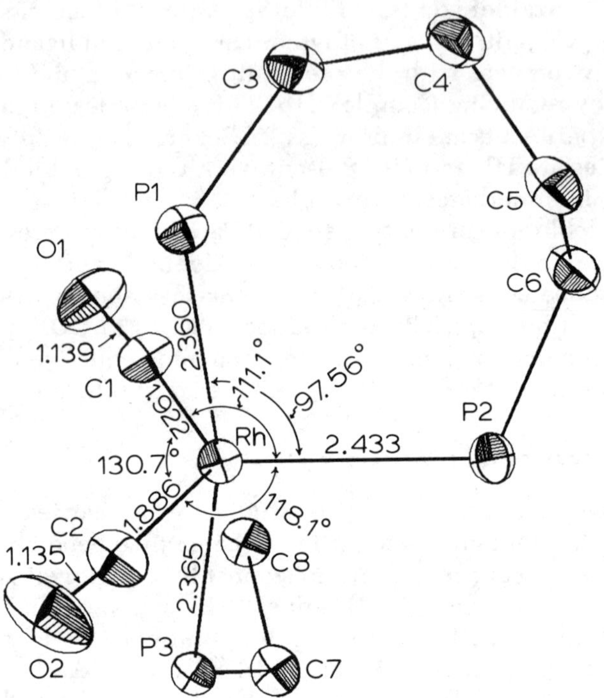

Figure 5. ORTEP drawing of the coordination core of $[Rh_2(CO)_4(dppb)_3]^{2+}$. Only the unique part of the structure is shown (an inversion center in the middle of the C8–C8 bond relates this unit to its equivalent counterpart) (28).

core geometry is distorted significantly toward the tetrahedral angle (the dihedral angle between planes containing P1, P2, Rh and P3, P4, Rh is 39°), and in this respect it is similar to that of $[Rh(PMe_3)_4]Cl$ (*37*). In addition, the Rh–P4 distance is long (2.361(1)Å) compared with the other Rh–P distances (average = 2.318(1)Å) and this may account for its reactivity. Recall that $[Rh(dppb)_2]^+$ was the poorest bis(diphosphine) catalyst examined (see Table I) for the decarbonylation of benzaldehyde. Perhaps this complex undergoes bond rupture such that Intermediate **IIb** has a monodentate dppb ligand. Although this might increase the value of K, it surely would decrease k since the Rh center would be less basic. Preliminary experiments using the triphosphine complex $[Rh(CO)(ttp)]^+$ where ttp = $PhP(CH_2CH_2CH_2PPh_2)_2$ (kindly supplied by D. W. Meek) showed a catalytic decarbonylation activity similar to that of $[Rh(dppb)_2]^+$ in support of this argument. Also, the cationic mono(diphosphine) complexes of rhodium(**I**) were less active than their bis(diphosphine) analogues (vide supra).

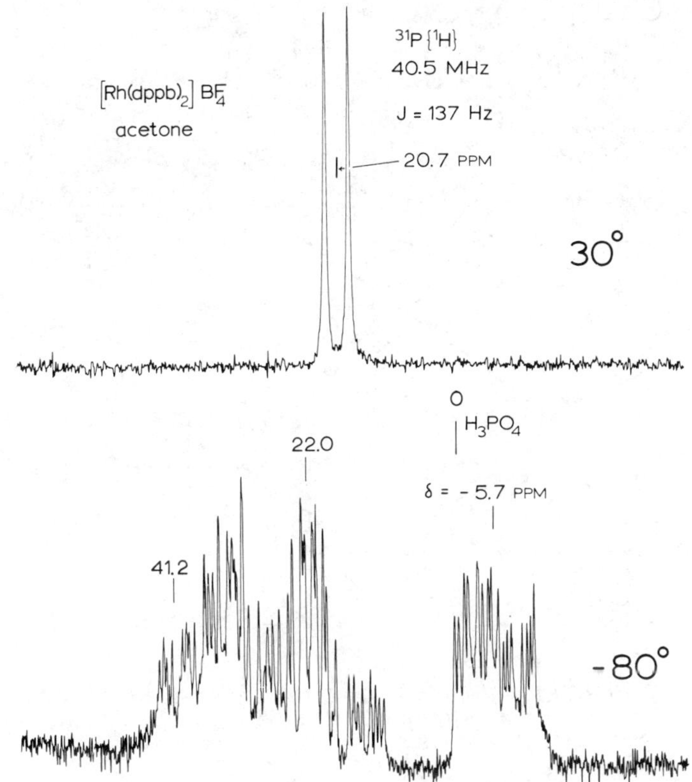

Figure 6. P-31 {H-1} NMR spectra as a function of temperature for $[Rh(dppb)_2]BF_4$ recorded using acetone solvent

These observations show that the catalytic reactions using complexes of different diphosphine ligands are likely to involve different reaction mechanisms. Also, subtle variations in metal basicity and catalyst stereochemistry can affect significantly the rate and mechanism of reaction. Therefore it is highly risky to generalize a reaction mechanism to seemingly similar catalysts. Our current research is directed towards determining these mechanistic differences.

Acknowledgment

This research was supported by the National Science Foundation (CHE78–21840 and CHE77–28505). The Johnson–Matthey Co. is acknowledged for a generous loan of rhodium chloride. M. P. Anderson acknowledges support from the Air Force Institute of Technology.

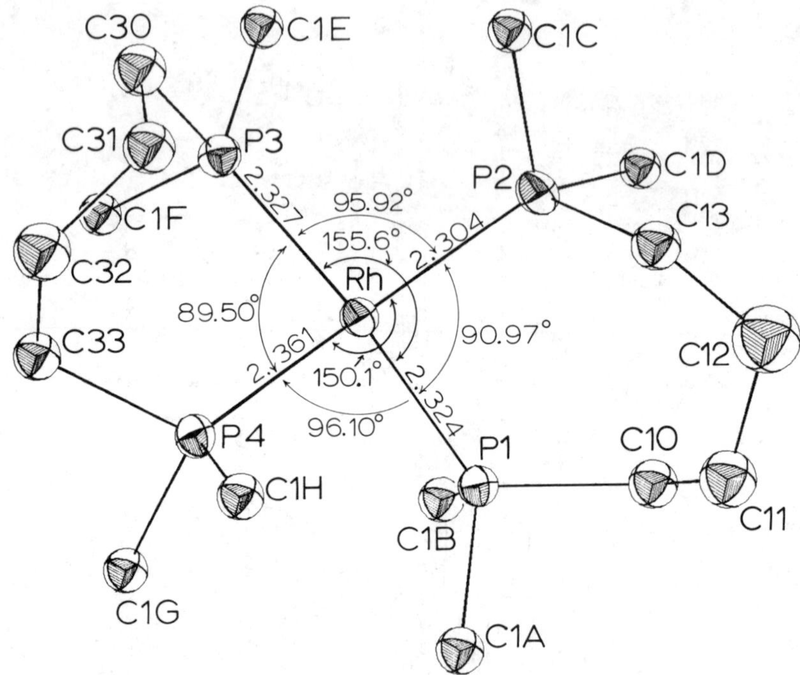

Figure 7. ORTEP drawing of the coordination core of $[Rh(dppb)_2]BF_4$ (36)

Literature Cited

1. Tsuji, J.; Ohno, K. *Synthesis* **1969,** *1,* 157.
2. Bird, C. W. "Transition Metal Intermediates in Organic Synthesis;" Logos Press: London, 1967; pp. 112, 239.
3. Baird, M. C.; Nyman, C. J.; Wilkinson, G. *J. Chem. Soc. A* **1968,** 348.
4. Ohno, K.; Tsuji, J. *J. Am. Chem. Soc.* **1968,** *90,* 99.
5. Walborsky, H. M.; Allen, L. E. *J. Am. Chem. Soc.* **1971,** *93,* 5465.
6. Schrock, R. R.; Osborn, J. A. *J. Am. Chem. Soc.* **1971,** *93,* 2397.
7. Doughty, D. H.; Pignolet, L. H. *J. Am. Chem. Soc.* **1978,** *100,* 7083.
8. Suggs, J. W. *J. Am. Chem. Soc.* **1978,** *100,* 640.
9. Kampmeier, J. A.; Harris, S. H.; Wedegaertner, D. K. *J. Org. Chem.* **1980,** *45,* 315.
10. Blum, J.; Oppenheimer, E.; Bergman, E. D. *J. Am. Chem. Soc.* **1967,** *89,* 2338.
11. Stille, J. K.; Regan, M. T. *J. Am. Chem. Soc.* **1974,** *96,* 1508.
12. Stille, J. K.; Fries, R. W. *J. Am. Chem. Soc.* **1974,** *96,* 1514.
13. Stille, J. K.; Huang, F.; Regan, M. T. *J. Am. Chem. Soc.* **1974,** *96,* 1518.
14. Stille, J. K.; Regan, M. T.; Fries, R. W.; Huang, F.; McCarley, T. In "Homogeneous Catalysis—II," **1974,** *132,* 181.
15. Lau, K. S. Y.; Becker, Y.; Huang, F.; Baenziger, N.; Stille, J. K. *J. Am. Chem. Soc.* **1977,** *99,* 5664.
16. Egglestone, D. L.; Baird, M. C.; Lock, C. J. L.; Turner, G. *J. Chem. Soc., Dalton Trans.* **1977,** 1576.
17. Baird, M. C.; Mague, J. T.; Osborn, J. A.; Wilkinson, G. *J. Chem. Soc. A* **1967,** 1347.

18. Stevens, D. J.; Nelson, D. A. "Abstracts of Papers," 177th Nat. Meet., A. C. S. *April, 1979;* INOR 47.
19. Rauchfuss, T. B. *J. Am. Chem. Soc.* **1979,** *101,* 1045.
20. Strohmeier, W.; Pfohler, P. *J. Organomet. Chem.* **1976,** *108,* 393.
21. Doughty, D. H.; McGuiggan, M. F.; Wang, H.; Pignolet, L. H. *Fund. Res. Homog. Catal.* **1979,** *3,* 909.
22. Doughty, D. H. Ph. D. Thesis, Univ. of Minnesota, Minneapolis, 1979.
23. Wang, H.; Pignolet, L. H., unpublished results.
24. Osborn, J. A.; Jardine, F. H.; Young, J. F.; Wilkinson, G. *J. Chem. Soc. A* **1966,** 1711.
25. Geoffroy, G. L.; Denton, D. A.; Kenney, M. E.; Bucks, R. R. *Inorg. Chem.* **1976,** *15,* 2382.
26. McGuiggan, M. F.; Doughty, D. H.; Pignolet, L. H. *J. Organomet. Chem.* **1980,** *185,* 241.
27. Pignolet, L. H.; Doughty, D. H.; Nowicki, S. C.; Casalnuovo, A. L. *Inorg. Chem.* **1980,** *19,* in press.
28. Pignolet, L. H.; Doughty, D. H.; Nowicki, S. C.; Anderson, M. P.; Casalnuovo, A. L., unpublished data.
29. Sanger, A. R. *J. Chem. Soc., Dalton Trans.* **1977,** 120.
30. Kawabata, Y.; Hayashi, T.; Ogata, I. *J.C.S. Chem. Commun.* **1979,** 462.
31. Poulin, J.-C.; Dang, T-P.; Kagan, H. B. *J. Organomet. Chem.* **1975,** *84,* 87.
32. James, B. R.; Mahajan, D. *Can. J. Chem.* **1979,** *57,* 180.
33. Slack, D. A.; Greveling, I.; Baird, M. C. *Inorg. Chem.* **1979,** *18,* 3125.
34. Hall, M. C.; Kilbourn, B. T.; Taylor, K. A. *J. Chem. Soc. A* **1970,** 2539.
35. Tso, C. C., unpublished results.
36. Pignolet, L. H.; Anderson, M. P., to be published.
37. Jones, R. A.; Real, F. M.; Wilkinson, G.; Galas, A. M. R.; Hursthouse, M. B.; Malik, K. M. A. *J. Chem. Soc., Dalton Trans.* **1980,** 511.

RECEIVED July 10, 1980.

5

Cobaloximes Containing Phosphorus-Donor Ligands

LUIGI G. MARZILLI, PAUL J. TOSCANO, and JAMES H. RAMSDEN[1]
Department of Chemistry, Emory University, Atlanta, GA 30322

LUCIO RANDACCIO and NEVENKA BRESCIANI–PAHOR
Istituto di Chimica, Università di Trieste, 34127 Trieste, Italy

Aspects of the chemistry of cobaloximes (particularly species containing the moiety $Co(DH)_2$≡bis*(dimethylglyoximato)cobalt) that contain P-donor ligands are discussed. Topics covered include the influence of electronic and steric effects of the P-donor ligands on NMR spectra, structure, and reactivity of the complexes. In turn, variation of the* X *group of the cobaloxime moiety,* $Co(III)(DH)_2X$*, on the reactivity, structure, and NMR spectra of the phosphorus donor ligands can be assessed for* X *groups of widely differing trans influence and trans effect. The comparative trans effect of* $(CH_3O)_2PO^-$ *and related uninegative P-donor ligands is evaluated also. Finally, the entire body of information can be compared with that available for other complexes containing P-donor ligands and to cobaloximes that do not contain P-donor ligands.*

The term cobaloximes refers to that class of cobalt complexes in which two dimethylglyoximato or other glyoximato monoanions occupy four coordination positions about cobalt and in which the four N donors are in essentially the same plane. A neutral octahedral low-spin d^6 Co(III) cobaloxime is shown in Figure 1. Cobaloximes with this oxidation state have been studied most extensively since these are more stable than the oxidation state I, II, and IV compounds. The chemistry of these latter oxidation state compounds will be reviewed

[1] Current address: Department of Chemistry, United States Military Academy, West Point, NY 10996.

0065–2393/82/0196–0085$05.00/0

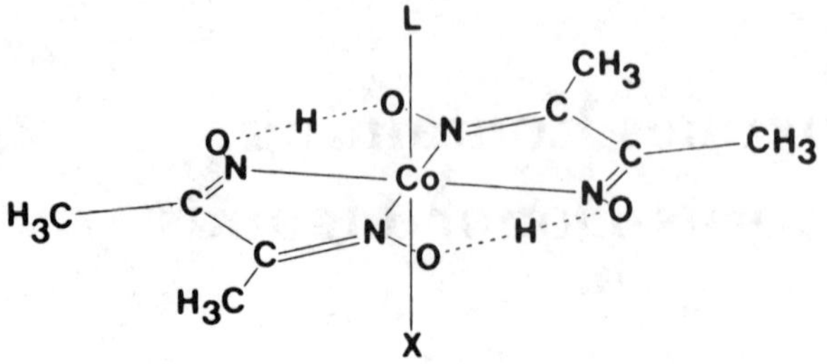

Figure 1. Schematic of an $LCo(DH)_2X$ cobaloxime

briefly first before our more detailed discussion of oxidation state III compounds. Emphasis will be placed on cobaloximes containing L = P donor ligands and only passing reference will be made to using these compounds as models for vitamin B_{12} systems (*1, 2, 3, 4, 5*).

Cobalt (I)

These compounds are probably five coordinate in basic solution and are very good nucleophiles. They are useful in preparing organocobaloximes via oxidative addition reactions. The influence of two P-donor ligands on the rate of nucleophilic attack on alkyl halides has been assessed (*6*). The reactivity of the Bu_3P was ca. five times greater than that of the ϕ_3P compound and was attributed (*5*) to the greater π bonding of the ϕ_3P ligand with the in-plane metal *d* orbitals in the 1:1 adducts (*7*). Under more acidic conditions, the Co(I) compounds form hydrides of the $LCo(DH)_2H$ type and the $Bu_3PCo(DH)_2H$ species has been isolated and studied by NMR (*8*). The compound decomposes readily to Co(II) and $1/2H_2$ (*9*). The rates and mechanism of this reaction have been examined (*10*). These hydrides can be viewed as being Co(III) compounds.

Cobalt (II)

The Co(II) compounds are usually low-spin, five-coordinate monomeric species, $LCo(DH)_2$(II) (*11*). However, $L_2Co(DH)_2$ low-spin compounds can be isolated (*11*). These latter materials are known to be five coordinate in solution and have EPR and visible spectra and magnetic moments (~2 BM) identical to those of $LCo(DH)_2$ (*12*). The X-ray structure of neither type of solid is known when L = P-donor ligand but the bis(pyridine) compound has been characterized structurally (*13*) and exhibits long Co–N(pyridine) bonds of 2.25 Å.

The Co(II) compounds react with alkyl halides according to Equation 1. The initial and rate-determining step is halogen abstraction (*14*). For L = phosphines

$$2LCo(DH)_2 + RX \rightarrow LCo(DH)_2X + LCo(DH)_2R \quad (1)$$

and RX = benzyl bromide, log k linearly decreases with increasing Tolman (*15, 16*) cone angles with the Bu_3P complex ca. 1000 times more reactive than the Cy_3P complex (*14*). It was suggested that two factors could be important in determining this trend. First, the Co–P bond length probably would be longer for bulky phosphines, decreasing their effective electron density donation to the cobalt. Second, bulky phosphines might be less readily accommodated in the octahedral Co(III) complexes than in the five-coordinate Co(II) compounds. The $(CH_3O)_3P$ complex is approximately 50 times less reactive than the $(CH_3)_3P$ complex, illustrating that electronic effects are also important. For $(XC_6H_4)_3P$ complexes, a good Hammett plot was obtained illustrating the greater reactivity of the more basic ligand of similar steric size.

Cobalt(IV)

Alkylcobaloximes, $LCo(DH)_2R^+$, are generated by the reaction of $LCo(DH)_2R$ with strong oxidizing agents such as $Ce(NO_3)_4$, K_2IrCl_6, and Br_2 (*17, 18*) or by electrooxidation (*17*). The compounds are unstable at 25°C and when L = ϕ_3P, EPR studies show no HFS with P and suggest that the Co is formally in oxidation state IV (*18*).

Cobalt(III)

Variation of X Ligand. Most of the rate, spectroscopic, and structural information available on Co(III) compounds has been obtained for neutral $LCo(DH)_2X$ compounds. In reviewing this chemistry, we will consider first variations in the X ligand and then variations in the L ligand.

Influence of X Variation on Reaction Rates. The reaction that has been studied most extensively is the ligand exchange reaction (Equation 2), most frequently with the leaving ligand L = $(CH_3O)_3P$. Using NMR

$$LCo(DH)_2X + L' \rightarrow L'Co(DH)_2X + L \quad (2)$$

techniques and noncoordinating solvents, Brown showed that this reaction was first order in [Co] and independent of entering ligand concentration (*19*). Subsequent studies using more traditional spectroscopic techniques established that the five-coordinate species formed in the rate-determining dissociative step is equally reactive towards a number of competing, entering P-donor ligands ($(CH_3)_3P$, $(CH_3O)_3P$, ϕ_3P) and N-donor ligands (pyridine, 1-methylimidazole) (*20, 21*). Ligands with very long alkyl chains (Bu_3P, n-Oct_3P) are three to eight times less reactive presumably because the alkyl chain may bend around and make the P lone pair less available for reacting with the "hot" five-coordinate species produced in the dissociative step. The variation in exchange rate for a number of X ligands in $(CH_3O)_3PCo(DH_2X$ is plotted against σ^* in Figure 2. The exchange reaction is unexpectedly rapid when X = secondary alkyl. Similar behavior is observed for bulkier phosphine ligands (ϕ_3P and ϕ_2EtP) as well as N-donor ligands. Therefore, there is no unusual feature of the rate profile which is peculiar to P-donor ligands suggesting little or no role for π bonding in these systems.

The reaction of coordinated $(CH_3O)_3P$ (*22*) shown in Equation III has been studied also as

$$(CH_3O)_3PCo(DH)_2X + Br^- \rightarrow (CH_3O)_2P(O)Co(DH)_2X^- + CH_3Br \quad (3)$$

a function of X ligand (*23*). In this case a plot of log k vs. σ^* does not exhibit a pronounced difference for secondary and primary alkyl ligands. We will return to a discussion of this reaction later in this chapter.

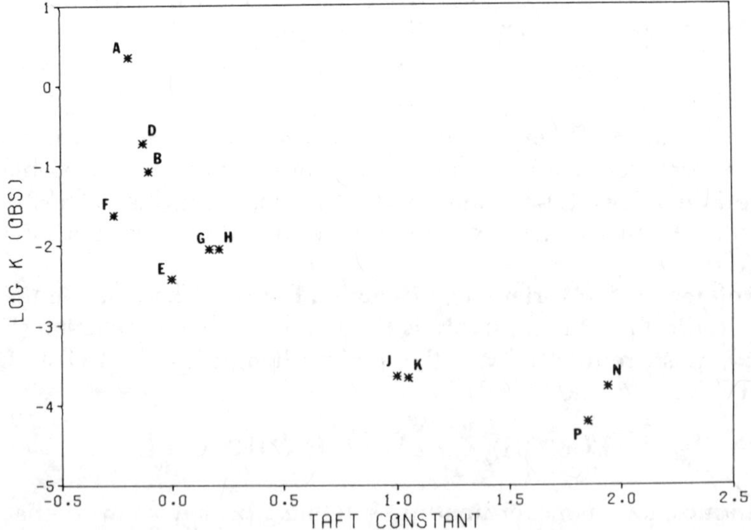

Figure 2. Plot of log k_{obs} vs. σ^* (Taft's polar substituent constant) for the ligand exchange reaction $(CH_3O)_3PCo(DH)_2X$ + 1-methylimidazole → (1-methylimidazole)$Co(DH)_2X$ + $(CH_3O)_3P$ (CH_2Cl_2, 25°C). Letters correspond to ligands as follows: A, iso-C_3H_7; B, C_2H_5; C, n-C_3H_7; D, iso-C_4H_9; E, CH_3; F, $CH_2Si(CH_3)_3$; G, $(CH_2)_3CN$; H, $CH_2C_6H_5$; I, C_6H_5; J, CH_2Br; K, CH_2Cl; L, CH_2CN; M, $P(O)(OCH_3)_2$; N, $CHCl_2$; P, $CHBr_2$; R, $SO_2C_6H_4CH_3$; S, NO_2; T, CN; U, I; V, Br; W, Cl; Y, NO_3.

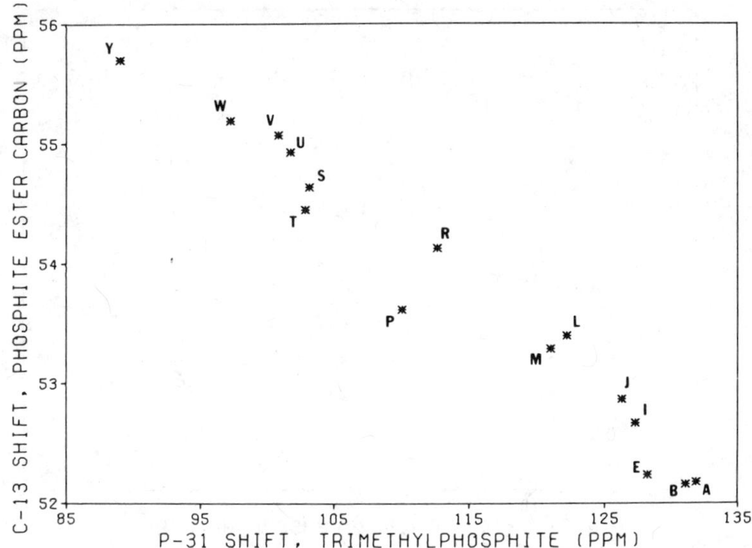

Figure 3. The C-13 shift of the ester carbon of $(CH_3O)_3PCo(DH)_2X$ vs. the P-31 shift of $(CH_3O)_3PCo(DH)_2X$ ($CDCl_3$). Letters correspond to ligands as given in the caption for Figure 2.

Influence of *X* Variation on NMR Spectra. The H-1 (*24, 25*), C-13 (*26*), and P-31 (*27*) NMR spectra of a wide range of cobaloximes with different *X* groups and L = $(CH_3O)_3P$ and Bu_3P, and less extensive data with other P-donor ligands are available. For the L = $(CH_3O)_3P$ series, the shifts of C-13 and P-31 in the coordinate phosphite are related linearly (*see* Figure 3). In turn, the C-13 shifts are correlated linearly with log *k* for Reaction 3 (*23*). Since Reaction 3 involves an S_N2 attack of the Br^- on the electron-deficient phosphite carbon, these relationships suggest that these parameters all reflect the degree to which the $Co(DH)_2X$ influences electron donation by the $(CH_3O)_3P$ ligand. Such shifts also correlate with C-13 shifts of coordinated pyridine ligands, again suggesting that π bonding is probably not having much influence on the properties of the complexes.

The P-31 nucleus is coupled to the ^1H of the oxime methyl groups. This splitting varies as *X* is varied (*24, 25*). This variation correlates linearly with changes in H-1 NMR chemical shifts in the series *t*-$BupyCo(DH)_2X$ (*24*) and it is believed that the ordering of *X* groups reflects changes in hybridization at the cobalt. Although this series closely parallels other effects of *X*, the ^{31}P–^1H coupling constants and other ^1H data do not vary significantly for secondary and primary alkyl ligands containing C and H only. In contrast ^{31}P–$^{13}C(OCH_3)$ coupling constants correlate linearly with variations in C-13 NMR shifts and

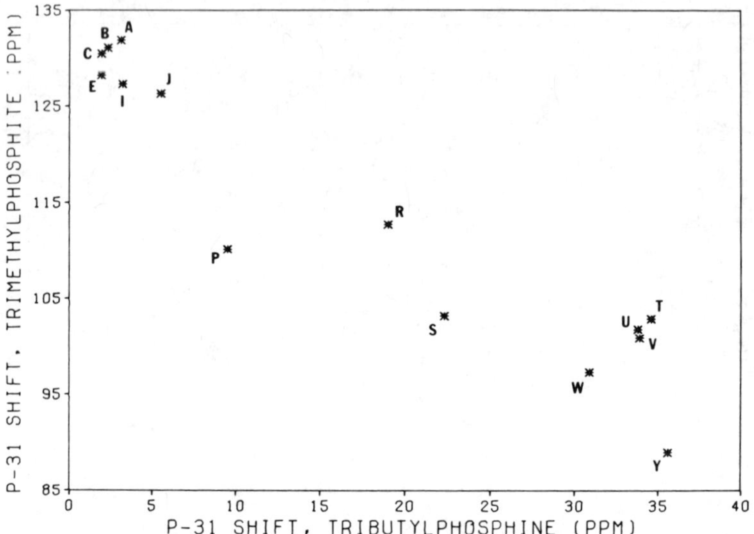

Figure 4. The P-31 shift of $(CH_3O)_3PCo(DH)_2X$ vs. the P-31 shift of $Bu_3PCo(DH)_2X(CDCl_3)$. Letters correspond to ligands as given in the caption for Figure 2.

log *k* of Reaction 3. Similarly, the ^{31}P shifts of the $(CH_3O)_3PCo(DH)_2X$ complexes correlate with the ^{13}C shifts of the phosphite ligand as noted earlier (*see* Figure 3). Therefore these shifts can be viewed as being a function of the amount of the electron donation of the phosphite ligand to the cobalt center. In this regard, it is quite interesting that the ^{31}P shifts of the $Bu_3PCo(DH)_2X$ and $(CH_3O)_3PCo(DH)_2X$ series are correlated linearly but the plot has a negative slope (*see* Figure 4). This finding suggests that the direction of the P-31 NMR shifts in metal complexes is not a good guide to the degree of electron donation since it is unlikely that the relative electron donation across the *X* series differs for Bu_3P and $(CH_3O)_3P$.

Influence of X Variation on Structure. Known Co–P bond lengths are reported in Table I. For the same ligand, the lengths increase with the increasing σ-donor power of the trans ligand *X*. Furthermore, the lengthening of the Co–P bond due to the same *X* ligand is nearly equal in all of the series (*see* Figure 5). A similar, but less regular, increase of the displacement, *d*, of the cobalt atom out of the N-donor plane towards the P ligand may be observed.

Comparison of such displacements in the Cl and CH_3 derivatives of the Cy_3P, ϕ_3P, and $(CH_3O)_3P$ ligands shows that they are significantly larger in the methyl complexes. This observation is contrary to the hypothesis (*37*) that the Co–C bond covalent character, greater than

Table I. Co–P Bond Lengths (Å) and Displacement of Co Above the Coordination Plane Towards the Phosphine (*d* (Å)) in $R_3PCo(DH)_2X$ Compounds

R_3P/X	$Co–P$	d	Ref.
Cy_3P/Cl	2.369(5)	0.10	28
CH_3	2.463(1)	0.12	29
ϕ_3P/Cl	2.327(2)	0.05	30
Br	2.331(4)	0.07	31
NO_2	2.392(3)	0.04	32
CH_3	2.418(1)	0.11	28
iso-C_3H_7	2.412(4)	0.17	29
Bu_3P/Cl	2.265(2)	0.04	33
Hypoxanthinato	2.285(2)	0.02	34
-C_6H_4N	2.342(1)	0.03	35
$(CH_3O)_3P/Cl$	2.188(4)	0.02	33
CH_3	2.256(4)	0.10	36

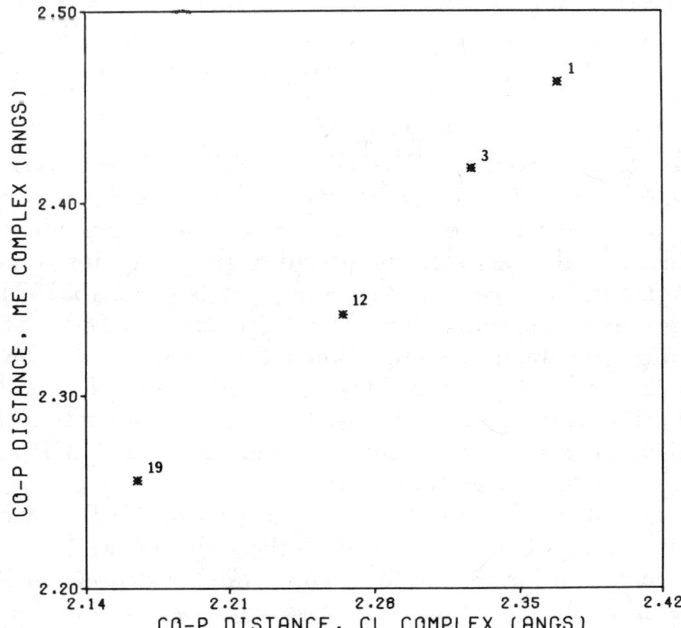

Figure 5. Comparison of the bond lengths in $LCo(DH)_2Cl$ and $LCo(DH)_2CH_3$. Numbers correspond to ligands as follows: 1, Cy_3P; 2, (iso-$Pr)_3P$; 3, ϕ_3P; 4, $Bu\ \phi_2P$; 5, $Et\ \phi_2P$; 6, $CH_3\ \phi_2P$; 7, Bz_3P; 8, $(CH_3O)\phi_2P$; 9, $(2\text{-}(CN)C_2H_4)_3P$; 10, (o-$CH_3C_6H_4O)_3P$; 12, Bu_3P; 13, $(CH_3)_2\ \phi P$; 14, $(CH_3O)_2\ \phi P$; 15, (iso-$PrO)_3P$; 16, $(\phi O)_3P$; 17, $(CH_3)_3P$; 18, $C_2H_5C(CH_2O)_3P$; 19, $(CH_3O)_3P$; 20, $(EtO)_3P$; 21, $(ClC_2H_4O)_3P$; 23, $CH_3C(CH_2O)_3P$.

would be expected for a Co(III) complex, should provoke a displacement of the metal out of the $(DH)_2$ plane towards the alkyl ligand. Without arguing about the covalency of the Co–C bond, the above differences in displacements between CH_3 and Cl couples have been attributed mainly to the difference in the bulkiness of these ligands. In fact, the more compact bulk of the Cl atom more efficiently opposes the bending and, consequently, the displacement of the cobalt induced by the phosphine ligand (*28*). Finally, it is worthwhile to note that in the more complete $\phi_3PCo(DH)_2X$ series, the lengthening of Co–P bond corresponds to the decrease of the ^{31}P chemical shifts reported recently for the analogous Bu_3P derivatives (*27*). The NO_2 compound may be an exception.

Variation in the P-Donor Ligand. INFLUENCE OF L VARIATION ON REACTION RATES AND EQUILIBRIA. The solvolytic Equation 4 in methanol (30%) water is first order (*38*). The

$$R_3PCo(DH)_2Cl + H_2O \rightarrow R_3PCo(DH)_2H_2O^+ + Cl^- \qquad (4)$$

rate increases tenfold in the series $\phi_3P < \phi_2EtP < Et_3P < Bu_3P$ and the rate for the Cy_3P complex was too rapid to measure with the techniques used. The reaction is only about one order of magnitude faster for P-donor ligands over N-donor ligands and it is likely that the bulky Cy_3P is sterically accelerating the rate.

For the two series $(RO)_3PCo(DH)_2NCS/SCN$ and $R_3PCo(DH)_2NCS/SCN$, the degree of S bonding is promoted when R is electron withdrawing within each series but S bonding is more prevalent in the R_3P series than the less-electron-donating $(RO)_3P$ series, suggesting that correlations between the series are not possible (*39*). This was the first group of compounds in which thiocyanate ambidentate equilibria could be evaluated as a function of the trans ligands. The results were confirmed later in a more extensive study (*40*). The S/N distribution does not differ greatly from that found for N-donor trans ligands (*39*). Taken together, these results suggest that neutral P-donor ligands do not have a large trans effect.

The rate at which ligands are exchanged by Bu_3P in the series $LCo(DH)_2CH_3$ and to a lesser extent in the series $LCo(DH)_2iC_3H_7$ has been evaluated preliminarily for a large number of P-donor ligands. The ease of displacement has been assessed as a function of electron-donor ability ΣX (*39*) and the Tolman cone angle (*15, 16*) of the ligand (*see* Figures 6 and 7). Again, the series $(RO)_3P$ and R_3P are not correlated readily. In the phosphite series, the rate appears to be influenced by steric bulk but increases as ΣX increases, which is consistent with the expected lower electron-donating ability of ligands with larger ΣX. For phosphines, steric bulk appears to be important, but we cannot

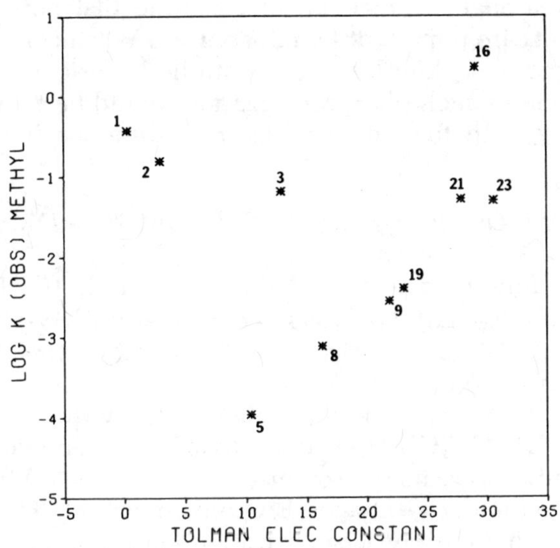

Figure 6. Plot of log k_{obs} *vs.* ΣX *for the ligand exchange reaction* $LCo(DH)_2CH_3 + Bu_3P \rightarrow Bu_3PCo(DH)_2CH_3 + L$ *(CH_2Cl_2, 25°C). Numbers correspond to ligands as given in the caption for Figure 5.*

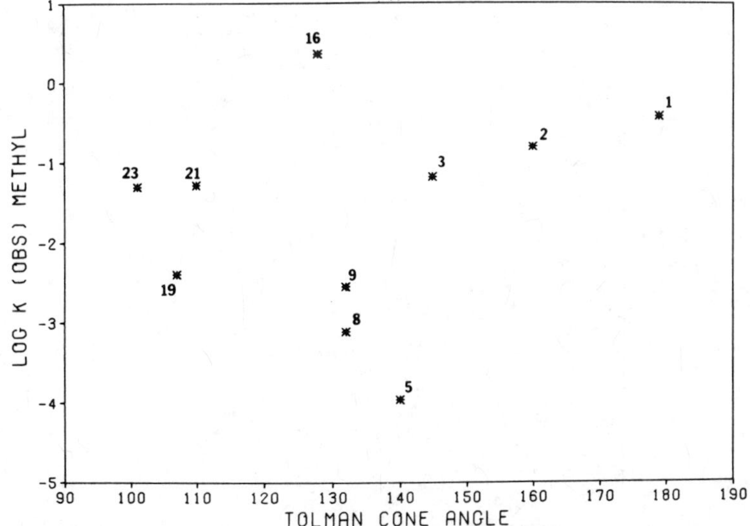

Figure 7. Plot of log k_{obs} *vs. TCA for the ligand exchange reaction* $LCo(DH)_2CH_3 + Bu_3P \rightarrow Bu_3PCo(DH)_2CH_3 + L$ *(CH_2Cl_2, 25°C). Numbers correspond to ligands as given in the caption for Figure 5.*

study phosphine ligands that are both electron donating and small since we have not yet found ligands that will displace such phosphines. It is reasonable, however, to conclude that both steric and electronic effects are important for phosphines. When complexes with sufficiently large phosphite ligands are studied, steric acceleration of the displacement of such phosphite ligands should be found.

In keeping with the rate studies, K_f values for Reaction 5 in DMSO-d_6 at

$$L + DMSOCo(DH)_2CH_3 \rightleftharpoons LCo(DH)_2CH_3 + DMSO \quad (5)$$

32°C give the following order for P-donor ligand: {$(CH_3)_3P$, Bu_3P, $(CH_3O)_2\phi P$, $(CNCH_2CH_2)_3P$ (all too large to measure)} > $(CH_3O)\phi_2P$ > (*iso*-PrO)$_3$P ~ $(CH_3O)_3P$ > $CH_3C(CH_2O)_3P$ > ϕ_3P > {$(\phi O)_3P$, $(iPr)_3P$, Cy_3P (all too small to measure)}.

Measurements of the enthalpy change for adduct formation of $(CH_3O)_3P$ and $CH_3CH_2C(CH_2O)_3P$ methylcobaloxime adducts led to similar results that were more exothermic (once corrected for solvation effects) than predicted by Dragos E/C equation (*37*). This led to the suggestion that the additional stabilization could be the result of π bonding (*37*). None of the other studies reported here give any clear indication of π bonding.

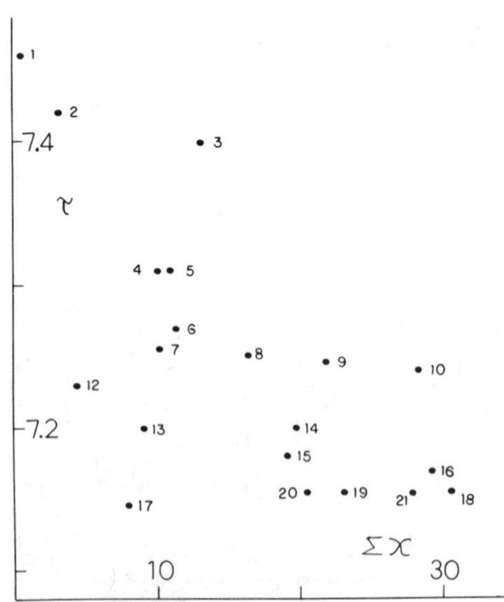

Figure 8. Plot of the chemical shift of the coordinated methanol's methyl resonance in $[LCo(DH)_2CH_3OH]^+$ *vs. Tolman's* ΣX *(cm*$^{-1}$*). Numbers correspond to ligands as given in the caption for Figure 5.*

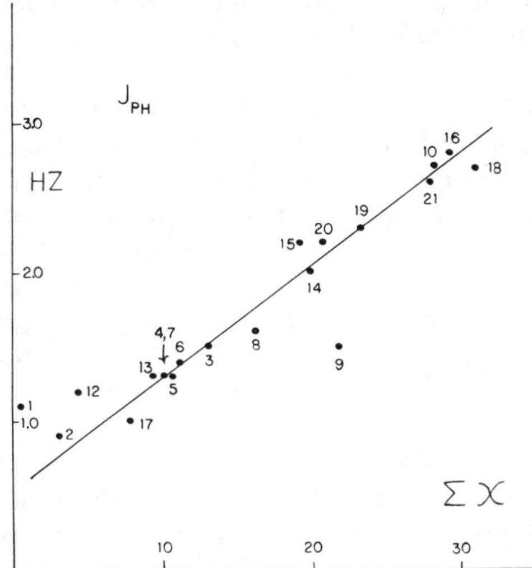

Figure 9. Plot of the P–H coupling (in hertz) of the coordinated methanol's methyl resonance in $[LCo(DH)_2CH_3OH]^+$ vs. Tolman's ΣX (centimeters^{-1}). Numbers correspond to ligands as given in the caption for Figure 5.

INFLUENCE OF L ON NMR SPECTRA. The most extensive investigation of the variation of L on NMR spectra was reported for the H-1 NMR spectra of $LCo(DH)_2(CH_3OH)^+$ complexes (in methanol) (*42, 43*). Shift and coupling constant data are plotted against ΣX in Figures 8 and 9. The coupling constants are clearly related to ΣX suggesting a close relative similarity for the interaction of L in these positional changed Co(III) compounds and in the neutral $LNi(CO)_3$ compounds. The shift data do not correlate with ΣX but do correlate quite nicely with the Tolman cone angle. This was the first evidence that possible structural distortions, (which now have been confirmed, *see* later discussion) in the cobaloxime systems could lead to changes in H-1 NMR shifts. Variations in the bulk of L also can induce changes in the H-1 NMR spectra of alkylcobaloximes (*29*). The changes in the H-1 NMR spectra of cobaloximes reflect changes in the magnetic anisotropy of cobalt (*24*). The most important property of the axial ligands that influences this shift is the nature of the ligating atom. Only in the case of P donors have we found large variations in H-1 NMR shifts. The structural basis for these shifts will be discussed later.

The effect of varying P-donor ligands on the C-13 NMR spectra of cobaloximes has not been studied in detail as yet. Limited data are available for C-59 NMR of chloro and methylcobaloximes where the order of shifts ($\phi_3P < Bu_3P < (CH_3O)_3P$) is clearly not reflective of the electron donating ability of the P-donor ligand (*44*). The NQR spectrum of the ^{14}N in some of these compounds has been analyzed also (*44*).

In much the same way that displacement rates of phosphine ligands are influenced strongly by steric bulk, the P-31 NMR coordination shifts of phosphines (Δ = complex shift − free-ligand shift) correlate with cone angle, increasing as cone angles decrease (*27*). Phosphite Δ's are negative and become less negative as ΣX increases. Although the Co–P bond length is longer for $X = CH_3$ than $X = Cl$, the relative shifts of coordinated phosphines and phosphites are the same. It was concluded (*27*) that since so many factors appear to influence P-31 shifts in cobaloximes a detailed evaluation of bonding was not possible.

INFLUENCE OF L VARIATION ON STRUCTURE. The Co–P bond lengths (*see* Table I and Figure 10) in the series $LCo(DH)_2Cl$ and $LCO(DH)_2CH_3$ reflect the bulk of the L ligand rather than any measure of P-ligand electron-donor ability (*33*). For phosphines, these lengths correlate best with ^{31}P Δ rather than the shifts of either the free or complexed ligand. From the limited phosphite data, again it appears that phosphites and phosphines are not readily correlated. However, as seen in Figure 5, the Co–P bond lengths in the $(CH_3O)_3P$ complexes fall on the line established by phosphine complexes. The significance of this correlation must await further study. However, this trend indicates that the lengthening due to the greater trans influencing power of CH_3 is not influenced by the bulk of the P ligand (*33*). The linear relationship in Figure 5 corresponds to that found between the Δ's of the CH_3 and Cl series (*27*), although in the bond length correlation the Cy_3P derivatives do not represent any exception as was found for complexes of this ligand in the P-31 NMR study (*27*).

The Co–Cl and Co–CH_3 distances in Table II do not vary greatly as the L-donor ligand is changed. The former vary from 2.277(2)(ϕ_3P) to 2.294(2)(Cy_3P, Bu_3P) and the latter from 2.014(14)(($(CH_3O)_3P$) to 2.026(6)(ϕ_3P). These values, however, appear slightly but significantly larger than those reported for $RNH_2Co(DH)_2Cl$ (R = ClC_6H_4 (*45*), $H_2NSO_2C_6H_4$ (*45*), C_6H_5 (*46*), and H (*30*)) of 2.235(3) to 2.261(4) and for $LCo(DH)_2CH_3$ (L = py (*47*), 1-methylimidazole (*47*), and H_2O (*48*)) of 1.990(5) to 2.009(6). The bond lengthening in the P donor relative to the N and O donor series may be due to a slightly greater trans influencing ability of the P donors. However, there is no correlation between the rates for Reaction 4 and the Co–Cl bond lengths.

Complexes Containing $R_2(O)P^-$ Ligands. In contrast to the relatively minor trans effect and trans influence of the wide range of neutral P-donor ligands discussed above, negative ligands based on $\phi_x(CH_3O)_yOP^-$ ($x = 0 \rightarrow 2$), $y = 0 \rightarrow 2$) exhibit an appreciable trans influence as established in the X-ray structure of (3-benzyladenine)Co-

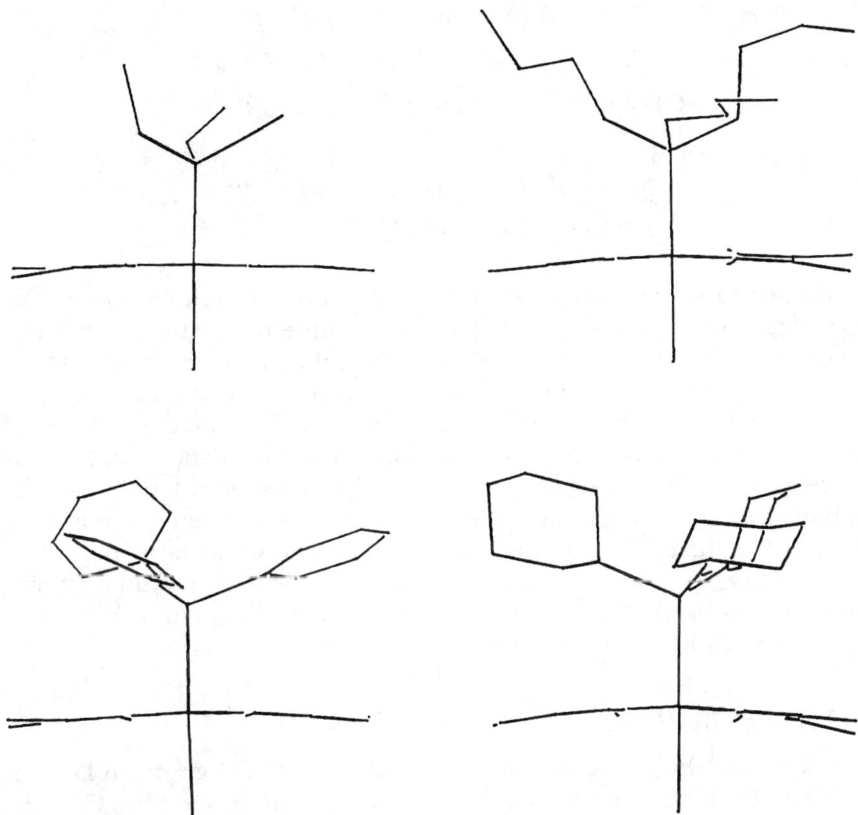

Figure 10. ORTEP drawings of chlorocobaloximes containing phosphorus ligands viewed parallel to the equatorial plane defined by the two oxime ligands. (Anisotropic thermal motion has been ignored.) Upper left, *$(CH_3O)_3PCo(DH)_2Cl$;* upper right, *$Bu_3PCo(DH)_2Cl$;* lower left, *$\phi_3PCo(DH)_2Cl$; and* lower right, *$Cy_3PCo(DH)_2Cl$.*

$(DH)_2P(O)(OCH_3)_2$ where the Co–N(7) distance is 2.101 Å and the Co–P distance is quite short (2.229) (*49*). With 4-cyanopyridine as the leaving ligand, $\phi_2(O)P^-$ and $\phi(CH_3O)(O)P^-$ have threefold greater trans effect than CH_3 and $(CH_3O)_2(O)P^-$ has a fourfold smaller trans effect than CH_3 (20). Thus, these negative phosphorus donors are quite high in the trans effect series.

Comparisons with Other Metal Systems. Although as complete a series as that reported here is not available for other metal systems, the limited data reviewed in the literature (*15, 16, 50*) suggest that both steric bulk and electronic effects will influence the M–P bond length. As the bulk of L is changed, the percentage *s* character in the M–P bond also changes (*51, 52*). The spectroscopic trends when *X* is varied

Table II. Co–X Bond Lengths in the $LCo(DH)_2X$ series[a] (X = Cl, CH_3)

L	*Co–Cl(Å)*	*Co–CH_3(Å)*
$(CH_3O)_3P$	2.280(2)	2.014(14)
Bu_3P	2.294(2)	–
ϕ_3P	2.277(2)	2.026(6)
Cy_3P	2.294(2)	2.016(5)

[a] References as in Table I.

as discussed above correlate with trends in other metal systems (*25, 53*) suggesting that our findings are not unique to cobaloximes. Similarly, the variations in coordination chemical shifts of P-donor ligands with changing L have been evaluated for other metals (*27*) and are analogous to the cobaloximes. The changes in P-31 NMR spectra with trans *X* for both a phosphite and a phosphine series showing a linear correlation with a negative slope has been observed only for the cobaloxime system. Cobaloxime complexes can accommodate a wide range of *X* ligands accounting for the wide range of rates that have been measured. The environment for the P-donor ligand is quite crowded and represents a useful contrast to studies with other metal systems with lower coordination number.

Acknowledgments

This research was supported in part by a NATO grant (to L. Marzilli and L. Randaccio), in part by a grant to L. Randaccio from C. N. R., Rome, Italy, and in part from NIH Grant No. GM 29225 (to L. Marzilli). This research was initiated at Johns Hopkins University.

Literature Cited

1. Schrauzer, G. N. *Accs. Chem. Res.* **1968,** *1,* 97.
2. Schrauzer, G. N. *Angew. Chem., Int. Ed. Eng.* **1976,** *15,* 417.
3. Schrauzer, G. N.; Windgassen, R. J. *J. Am. Chem. Soc.* **1966,** *88,* 3738.
4. Abeles, R. H.; Dolphin, D. *Accs. Chem. Res.* **1976,** *9,* 114.
5. Golding, B. T.; Kemp, T. J.; Sell, C. S.; Sellars, P. J.; Watson, W. P. *J. Chem. Soc., Perkin Trans. 2* **1978,** 839.
6. Schrauzer, G. N.; Deutsch, E. *J. Am. Chem. Soc.* **1969,** *91,* 3341.
7. Schrauzer, G. N.; Windgassen, R. J.; Kohnle, J. *Chem. Ber.* **1965,** *98,* 3324.
8. Schrauzer, G. N.; Holland, R. J. *J. Am. Chem. Soc.* **1971,** *93,* 1505.
9. Schrauzer, G. N.; Deutsch, E.; Windgassen, R. J. *J. Am. Chem. Soc.* **1968,** *90,* 2441.
10. Chao, T-H.; Espenson, J. H. *J. Am. Chem. Soc.* **1978,** *100,* 129.
11. Schrauzer, G. N. *Inorg. Synth.* **1968,** *11,* 61.
12. Schneider, P. W.; Phelan, P. F.; Halpern, J. *J. Am. Chem. Soc.* **1969,** *91,* 77.
13. Fallon, G. D.; Gatehouse, B. M. *Cryst. Struct. Commun.* **1978,** *7,* 263.
14. Halpern, J.; Phelan, P. F. *J. Am. Chem. Soc.* **1972,** *94,* 1881.
15. Tolman, C. A. *J. Am. Chem. Soc.* **1970,** *92,* 2956.
16. Tolman, C. A. *Chem. Rev.* **1977,** *77,* 313.

17. Halpern, J.; Chan, M. S.; Roche, T. S.; Tom, G. M. *Acta Chem. Scand., Ser. A* **1979**, *33*, 141.
18. Topich, J.; Halpern, J. *Inorg. Chem.* **1979**, *18*, 1339.
19. Guschl, R. J.; Stewart, R. S.; Brown, T. L. *Inorg. Chem.* **1974**, *13*, 417.
20. Trogler, W. C.; Stewart, R. C.; Marzilli, L. G. *J. Am. Chem. Soc.* **1974**, *96*, 3697.
21. Stewart, R. C.; Marzilli, L. G. *J. Am. Chem. Soc.* **1978**, *100*, 817.
22. Trogler, W. C.; Epps, L. A.; Marzilli, L. G. *Inorg. Chem.* **1975**, *14*, 2748.
23. Toscano, P. J.; Marzilli, L. G. *Inorg. Chem.* **1979**, *18*, 421.
24. Trogler, W. C.; Stewart, R. C.; Epps, L. A.; Marzilli, L. G. *Inorg. Chem.* **1974**, *13*, 1564.
25. Marzilli, L. G.; Politzer, P.; Trogler, W. C.; Stewart, R. C. *Inorg. Chem.* **1975**, *14*, 2389.
26. Stewart, R. C.; Marzilli, L. G. *Inorg. Chem.* **1977**, *16*, 424.
27. Kargol, J. A.; Crecely, R. W.; Burmeister, J. L.; Toscano, P. J.; Marzilli, L. G. *Inorg. Chim. Acta* **1980**, *40*, 79.
28. Bresciani-Pahor, N.; Calligaris, M.; Randaccio, L.; Marzilli, L. G. *Inorg. Chim. Acta* **1979**, *32*, 181.
29. Bresciani-Pahor, N.; Randaccio, L.; Toscano, P. J.; Marzilli, L. G., to be published.
30. Brückner, S.; Randaccio, L. *J. Chem. Soc., Dalton Trans.* **1974**, 1017.
31. Calligaris, M., to be published.
32. Bresciani-Pahor, N.; Calligaris, M.; Randaccio, L. *Inorg. Chim. Acta* **1978**, *27*, 47.
33. Ibid., **1980**, *39*, 173.
34. Marzilli, L. G.; Sorrell, T.; Kistenmacher, T. J. *J. Am. Chem. Soc.* **1975**, *97*, 3351.
35. Adams, W. W.; Lenhert, P. G. *Acta Crystallogr., Sect. B* **1973**, *29*, 2412.
36. Stucky, G. D.; Swanson, J. S., personal communication.
37. Courtright, R. L.; Drago, R. S.; Nusz, J. A.; Nozari, M. S. *Inorg. Chem.* **1973**, *12*, 2809.
38. Costa, G.; Tauzher, G.; Puxeddu, A. *Inorg. Chim. Acta* **1969**, *3*, 41.
39. Marzilli, L. G.; Stewart, R. C.; Epps, L. A.; Allen, J. B. *J. Am. Chem. Soc.* **1973**, *95*, 5796.
40. Kargol, J. A., Ph.D. Thesis, Univ. of Delaware, Newark, 1979.
41. Tolman, C. A. *J. Am. Chem. Soc.* **1970**, *92*, 2953.
42. Trogler, W. C.; Marzilli, L. G. *J. Am. Chem. Soc.* **1974**, *96*, 7589.
43. Trogler, W. C.; Marzilli, L. G. *Inorg. Chem.* **1975**, *14*, 2942.
44. La Rossa, R. A.; Brown, T. L. *J. Am. Chem. Soc.* **1974**, *96*, 2072.
45. Palenik, G. J.; Sullivan, D. A.; Naik, D. V. *J. Am. Chem. Soc.* **1976**, *98*, 1177.
46. Botoshanskii, M. M.; Simonov, Yu. A.; Malinovskii, T. I.; Ablov, A. V.; Bologa, O. A. *Dokl. Akad. Nauk. SSSR (Chem.), Eng. Trans.* **1975**, *225*, 625.
47. Bigotto, A.; Zangrando, E.; Randaccio, L. *J. Chem. Soc., Dalton Trans.* **1976**, 96.
48. McFadden, D. L.; McPhail, A. T. *J. Chem. Soc., Dalton Trans.* **1974**, 363.
49. Toscano, P. J.; Chiang, C. C.; Kistenmacher, T. J.; Marzilli, L. G. *Inorg. Chem.*, in press.
50. Randaccio, L. *Izvj. Jugoslav. Centr. Krist. (Zagreb)* **1977**, *12*, 35.
51. Wayland, B. B.; Abd-Elmageed, M. E. *J. Am. Chem. Soc.* **1974**, *96*, 4809.
52. Wayland, B. B.; Abd-Elmageed, M. E.; Mehne, L. F. *Inorg. Chem.* **1975**, *14*, 1456.
53. Gofman, M. M.; Nefedov, V. I. *Inorg. Chim. Acta* **1978**, *28*, 1.

RECEIVED July 10, 1980.

6

Stereopopulation Control on Cyclometallation Reactions and the Stability of Very Large Chelate Rings

B. L. SHAW
School of Chemistry, The University, Leeds LS2 9JT, England

The effects of nonbonded interactions in tertiary phosphine–transition–metal complexes on (a) cyclometallation and C–H activation and (b) the stability and conformations of large chelate rings are discussed. The effects of a bulky substituent such as a t-*butyl on a tertiary phosphine ligand are attributed to (a) favorable conformations around P–metal, P–C, and metal–halogen and bonds caused by nonbonded interactions with* t-*butyl or similar groups (stereopopulation control) and to (b) a decrease in the loss of internal rotational entropy which occurs on cyclization. Several sterically demanding, suitably positioned substituents will be necessary to stabilize large rings. The effects are illustrated with the 20- and 16-atom ring chelates* $[Pd_2Cl_4\{Bu^t_2P(CH_2)_7PBu^t_2\}_2]$, $[Rh_2H_2Cl_4\{Bu^t_2P(CH_2)_5PBu^t_2\}_2]$, *the hydrides* $[RhHCl\{Bu^t_2PCH_2CH_2CRCH_2CH_2PBu^t_2\}]R$ = *H(fluxional) =* CH_3*(not fluxional), olefinic and carbene complexes, and some new hydrides of iridium.*

This chapter is concerned with the effects of nonbonded interactions in tertiary phosphine–transition-metal complexes on (a) cyclometallation and C–H activation and (b) the stability and conformations of large chelate rings. It also deals with how nonbonded interactions might be used in other areas of chemistry.

In 1970 we showed that the tendency for cyclometallation of a tertiary phosphine ligand could be increased enormously by the presence of bulky substituents on the tertiary phosphine (*1*). Thus with Structure **I**, cyclometallation usually occurs and occurs rapidly to give Structure **II** if R = *t*-butyl(Bu^t); if R = Ph, it either occurs less rapidly

0065–2393/82/0196–0101$05.00/0

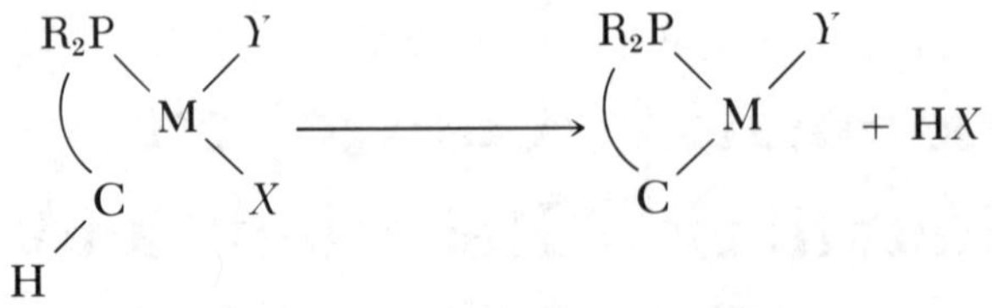

than with R = *t*-butyl or not at all, and if R = methyl (Me) it almost invariably does not occur. We have used the effect to cyclometallate various types of carbon—aromatic, benzylic, secondary and tertiary carbons, purely aliphatic, alkyl ethers, etc (*2–13*) and also to promote *O*-metallation (demethylation) i.e., Complex **III** → Complex **IV** (*5–10*).

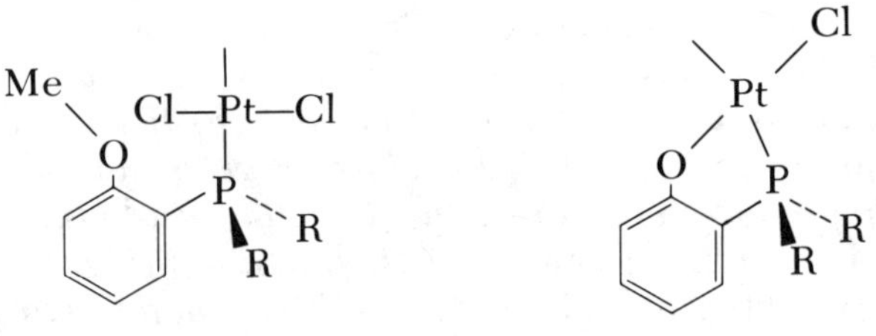

If in Complex **III** R = *t*-butyl, *O*-metallation occurs readily; if R = phenyl (Ph) it occurs less readily and if R = CH_3 then it doesn't occur at all. We explained the effect of *t*-butyl groups in terms of (a) favorable conformations around P–metal, P–C, and Pt–Cl bonds caused by nonbonded interactions with *t*-butyl groups and (b) a decrease in the loss of internal rotational entropy which occurs on cyclization. Variable-temperature NMR studies show that PBu^t_2 moieties cause restricted rotation around metal–P bonds (*14*). There is a close analogy between the effect of sterically demanding substituents such as *t*-butyls on phosphorus and the Thorpe–Ingold or *gem*-dialkyl effect in carbocyclic and related chemistry (*15*). Although the nature and causes of the effect now are well understood, they are not familiar to many chemists and therefore it is necessary to give a brief resumé of the effect before discussing how it relates to our work.

The Thorpe–Ingold or gem-Dialkyl Effect—Stereopopulation Control or Orbital Steering

Thorpe and Ingold were interested in the synthesis of small (3- and 6-membered) and aliphatic rings and found that the replacement of a CH_2 group by a CMe_2 group invariably increased (i) the rate of cyclization and (ii) the yield of cyclized product and its stability relative to an open-chain precursor or possible alternative product (*16, 17, 18, 19*). Since then many other examples have been found and there are apparently no exceptions (*18, 19, 20*). In highly sterically hindered situations the effect can be very large indeed, e.g. by converting a

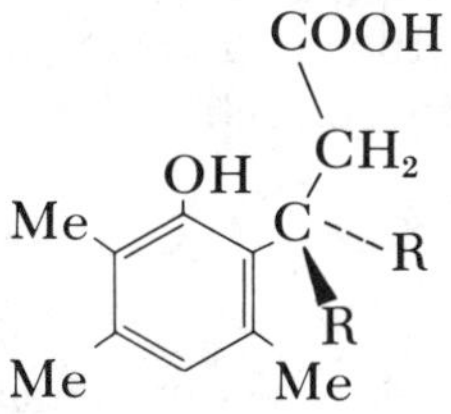

COOH, CH_2, OH, Me, C, R, R, Me, Me

V

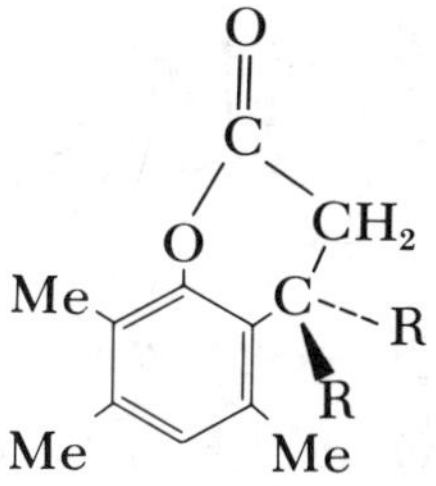

O, C, O, CH_2, Me, C, R, R, Me, Me

VI

phenolic acid of Type **V** to the lactone (Complex **VI**) the rate of lactonization with R = Me is faster than with R = H by a factor of 2×10^{10} (*21, 22*)! The equilibrium position lactone $\rightleftharpoons$ open-chain phenol also is influenced highly by the *gem*-dimethyl group in favor of the cyclic product. It now is generally considered that both entropy and enthalpy factors contribute towards the Thorpe–Ingold effect, also called stereopopulation control (*21, 22*), orbital steering (*23*), or rotamer distribution (*24*) of which I think stereopopulation control is the best term to use. Thus, as was first pointed out by Hammond (*25*), a *gem*-dialkyl group restricts the rotation around adjacent bonds in an open-chain precursor and therefore the loss of internal rotational entropy that occurs on cyclization is reduced (a $T\Delta S$ effect). Later, Allinger and Zalkow (*26*) pointed out that a *gem*-dimethyl group would decrease the extra gauche interactions which occur on cyclization (a ΔH effect). What does not seem to have been generally appreciated is that because these $T\Delta S$ and ΔH effects are due to nonbonding interactions, it should be possible to apply them widely in chemistry. Thus they are a function of bond lengths, bond angles, torsion angles, and atomic sizes. Whereas methyl or *gem*-dimethyl substituents have a pronounced effect in carbon chemistry, with larger tetrahedral atoms such as phosphorus, silicon, or metals, larger groups such as *t*-butyl would be necessary (*12, 15*). The effect should apply also to other geometries such as square planar or square pyramidal. For a more comprehensive discussion of the effect and its explanation the reader is asked to consult Refs. *12, 18–22*.

Although the Thorpe–Ingold effect has been widely used and studied with small rings, there have been very few attempts to use it to stabilize large rings. One would imagine that for large rings several sterically demanding substituents, suitably positioned, would be necessary to have a pronounced effect and that with such substituents the $T\Delta S$ and ΔH contributions would be cumulative. There are examples in nature where sterically demanding substituents probably confer stability on large rings such as the macrolides (*27*) and cyclic peptide/depsipeptide antibiotics. Thus the 18-atom ring of enniatin carries six isopropyl substituents positioned in an equatorial fashion around the ring (*28*)—the presence of alternating *R*- and *S*- centers allows this. H-1 NMR studies suggest that the conformation of free enniatin is similar to that of its complex with potassium ion.

The Stabilization of Large Chelate Rings

In addition to our own work, a variety of large-ring chelate complexes using flexible bidentate ligands has been described. There are also many examples where long-chain, flexible bidentate ligands preferentially give open-chain (polynuclear) rather than large-ring chelate complexes. Therefore it would appear that sometimes a large chelate ring is favored and sometimes open-chain polymers are preferred and that the free-energy difference between the two possibilities is not large. Some large chelate ring complexes dissociate in solution (*29, 30*) and for others there is little or no information as to what is present in their solution. We have shown, however, that long-chain diphosphines carrying sterically demanding PBu^t_2 end-groups, e.g., of types $Bu^t_2P(CH_2)_nPBu^t_2$ (n = 5–10 or 12), $Bu^t_2PC{\equiv}C(CH_2)_nC{\equiv}CPBu^t_2$ (n = 4 or 5), $Bu^t_2P(CH_2)_4C{\equiv}C(CH_2)_4PBu^t_2$, etc. form large ring chelates (12- to 72-membered) with platinum group metal halides, carbonyl halides, etc., which are particularly stable with respect to open-chain polymer formation (*12*). We have studied many of these large-ring chelates by P-31 NMR spectroscopy and shown that they do not ring-open or decompose in solution (*12, 31*). To discuss the reasons why I think these large rings are so stable I will use as examples the large chelate rings formed from palladium(II) chloride and diphosphines of the type $Bu^t_2P(CH_2)_nPBu^t_2$ (n = 5–10 or 12). The bulky PBu^t_2 groups generally will not coordinate in mutually cis positions, only in mutually trans positions. With n = 9, 10, or 12, mononuclear complexes *trans*-$[PdCl_2\{Bu^t_2P(CH_2)_nPBu^t_2\}]$ are formed but shorter methylene chains are not sufficiently long to span trans positions. Thus with $Bu^t_2P(CH_2)_7PBu^t_2$ the 20-atom ring complex *trans*-$[Pd_2Cl_4\{Bu^t_2P(CH_2)_7PBu^t_2\}_2]$ forms in good yield and appears to be stable indefinitely in solution (*12*). The crystal structure is shown in Figure 1. The $-(CH_2)_7-$ chains are in the extended (preferred) conformation and the PBu^t_2 groups occupy "corner" positions with Pd–P–C–C torsion angles very close to 60° as in the Newman projection (Complex **VII**). It is suggested that Cl–Pd–Cl is less steri-

Bu^t H — H — ---$PdCl_2$ — Bu^t CH_2CH_2

VII

R H — CH_2CH_2 — ---$PdCl_2$ — R H

VIII

cally demanding than *t*-butyl and that Complex **VII** is the preferred conformation. If the R groups on phosphorus are less sterically demanding than $PdCl_2$ (e.g., if R = ethyl) then the preferred conformation would be (Structure **VIII**). Thus I would suggest that the effect of the eight *t*-butyl groups is to eliminate torsional strain in the ring but that with a less sterically demanding group such as methyl there would be torsional strain at the four corner positions of the ring. We have shown from NMR studies that there is a large energy barrier to rotation around a metal–P bond in an MCl_2–PBu^t_2R moiety because of strong nonbonding interaction between the chlorines and the *t*-butyl groups (*14*). Thus in the structure shown in Figure 1 one chlorine is gauche with respect to two sets of $-PBu^t_2$ groups, which are eclipsed and the other chlorine lies above the two inward-pointing pseudoequatorial hydrogens of the β-methylene groups. I suggest that this arrangement of the Cl–Pd–Cl grouping is a preferred one. It is an arrangement that has been observed in many crystal structures on similar compounds.

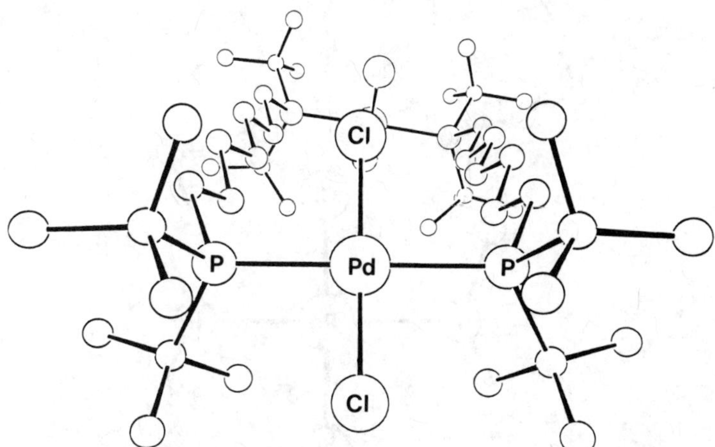

Figure 1. ORTEP drawing of the structure of the 20-atom ring complex $[Pd_2Cl_4\{Bu^t_2P(CH_2)_7PBu^t_2\}_2]$*. A small packing effect causes one P–Pd–P to be inclined 9° to the other.*

Often in cyclization reactions the formation of a ring introduces some ring strain or torsional strain which is not present in an open-chain precursor or an alternative product, i.e., an open-chain polymer. In our compounds the effect of the PBu^t_2 is to eliminate the ring strain and is thus analogous to one of the factors (i.e., the ΔH term) that promotes the *gem*-dimethyl effect in organic compounds (*see* earlier discussion). Additionally there will be an internal entropy effect. On cyclization of an open-chain precursor, internal rotational entropy is lost. However, in an open-chain precursor just prior to the formation of the 20-atom ring shown in Figure 1, viz. $Bu^t_2P(CH_2)_7PBu^t_2PdCl_2Bu^t_2P(CH_2)_7PBu^t_2PdCl_2$ (or a closely related complex), there would be restricted rotation around several of the bonds and the loss of internal rotational entropy on ring formation would be less than if the phosphorus atoms carried less sterically demanding substituents, e.g., methyls.

Because of steric interaction between the Pd–Cl and PBu^t_2 groups, which are eclipsed, the 20-atom ring adopts a barge conformation analogous to the boat conformation of cyclohexane. Similarly the sulfur-containing, 16-atom ring chelate $[Pd_2Cl_4\{Bu^tS(CH_2)_5SBu^t\}_2]$ adopts a barge conformation (32). In contrast, the 16- and 20-atom ring

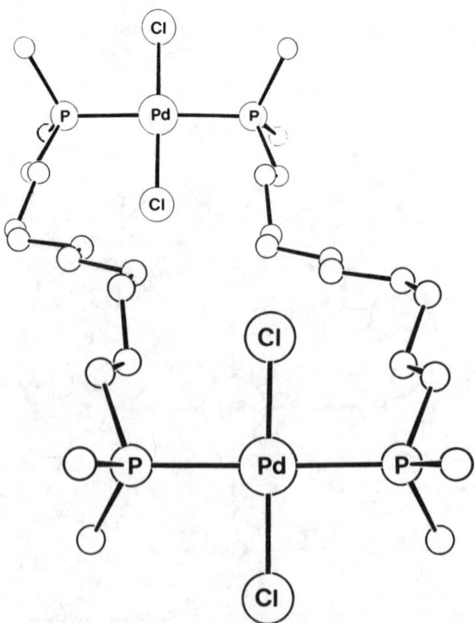

Figure 2. ORTEP drawing of the structure of the 26-atom ring complex $[Pd_2Cl_4\{Bu^t_2P(CH_2)_{10}PBu^t_2\}_2]$*. For clarity neither the t-butyl methyls nor the hydrogens are shown.*

chelates containing a *N,N,N,N*-tetramethylpolymethylene diamine, of type *trans*-[Pd_2Cl_4{$Me_2N(CH_2)_n$-NMe_2}$_2$] (n = 5 or 7) adopt a sofa conformation (*33*) analogous to the chair conformation of cyclohexane. In these complexes the NMe_2 groups occupy corner positions: with sp^3-nitrogen, which is even smaller than carbon, *gem*-dimethyl groups are sufficiently sterically demanding to have the desired stabilizing effect (*33*).

With a polymethylene diphosphine, $Bu^t_2P(CH_2)_nPBu^t_2$, with an even number of methylene groups, a chelate with a strain-free conformation cannot form. The crystal structure of the 26-atom ring chelate [Pd_2Cl_4{$Bu^t_2P(CH_2)_{10}PBu^t_2$}$_2$] is shown in Figure 2. There is a twist in the middle of the polymethylene chain although at either end the arrangement of atoms is very similar to that found in the heptamethylene complex (*see* Figure 1). A similar explanation can be given for the stabilizing effect of the bulky end-groups in other large chelate rings which we have synthesized.

Cyclometallation Reactions

In some of our systems large chelate ring formation and cyclometallation occur together since large nonbonding interactions influence both of them (*see* e.g., some palladium(II) and platinum(II) systems (*12, 13*)). What is remarkable is that under mild conditions the central carbon of a pentamethylene chain is metallated, a reaction that is possibly unique in chemistry. Treatment of rhodium trichloride with the pentamethylene diphosphine $Bu^t_2P(CH_2)_5PBu^t_2$ gives a binuclear square pyramidal rhodium(III) hydride [$Rh_2H_2Cl_4${$Bu^t_2P(CH_2)_5PBu^t_2$}$_2$] with a 16-atom ring, a cyclometallated rhodium(III) hydride, [RhHCl{$Bu^t_2PCH_2CH_2CHCH_2CH_2PBu^t_2$}], and a small amount of an olefinic diphosphine–rhodium(I) complex (vide infra). The 16-atom chelate exists in solution as predominantly one rotamer, with about 10% of a second rotamer (P-31 and H-1 NMR evidence) (*34*). The probable structures of these two rotamers are Complexes **IX** and **X** or vice versa. The cyclometallated hydride complex [RhHCl{$Bu^t_2PCH_2$-$CH_2CHCH_2Ch_2PBu^t_2$}] has the configuration of Structure **XI** by X-ray crystallography, i.e., with a transoid H—C—Rh—H arrangement. This is the arrangement that one would expect to get by the concerted oxidative addition of a CH_2 group to rhodium(I).

We find that on treating the 16-atom ring binuclear dihydride tetrachloride complex with a base such as 2-methylpyridine to remove the elements of hydrogen chloride from the rhodium, smooth conversion to the cyclometallated hydride complex occurs. A possible sequence of reactions leading to this overall conversion is shown in Scheme I.

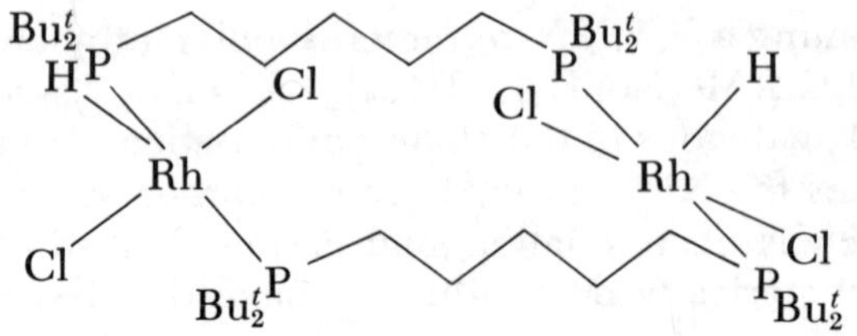

IX

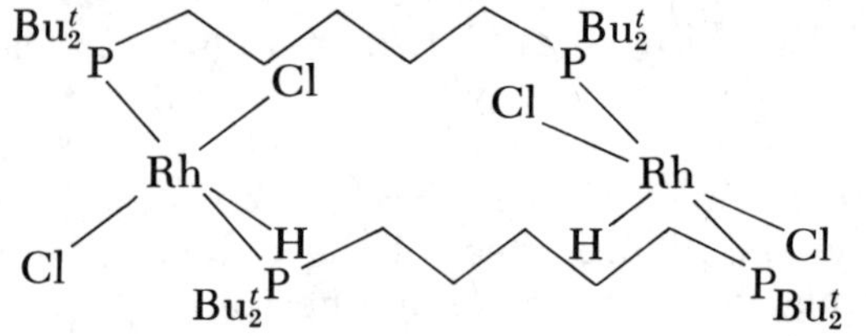

X

The H-1 NMR spectrum of the cyclometallated hydride (Structure **XI**) shows no hydride resonance at room temperature, but on cooling a broad resonance starts to appear, and at $-62°C$ it shows resolved coupling to ^{31}P (19 Hz) and ^{103}Rh (55 Hz). We ascribe this temperature-dependent behavior to the occurrence of a rapid, reversible C–H/Rh–H fission such as shown in Scheme II. Alternatively there may be rapid, reversible oxidative additions, Structure **XII** $\rightleftharpoons$ Structure **XI**, as in Scheme I.

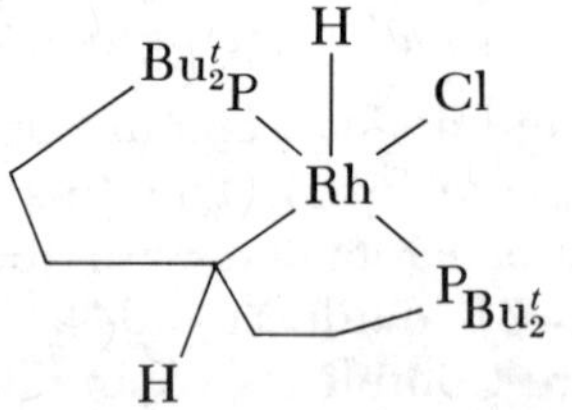

XI

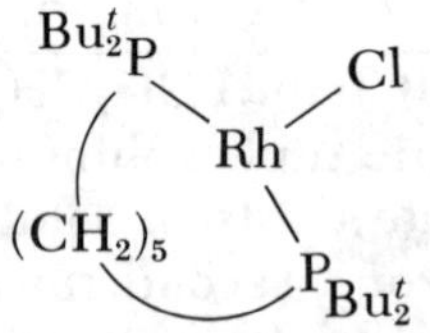

XII

In support of this interpretation the ^{103}Rh resonances obtained by ^{1}H-{^{31}P, ^{103}Rh} INDOR experiments at room temperature show an equal coupling to two protons with an averaged coupling constant of 24 Hz (*34*). Since $^{1}J(RhH)$ is 55 Hz and positive and $^{2}J(RhCH)$ is known to be a few Hertz and negative (*35*), then the observed value of the coupling constant is as would be expected. As mentioned above a third complex was formed in the reaction between $RhCl_3$ and $Bu^t_2P(CH_2)_5PBu^t_2$. This was not obtained in the pure state but was characterized by an AMX P-31 NMR pattern with an extremely large

Scheme I. Showing a possible sequence of reactions leading to the fluxional hydride (Structure ***XI****)*

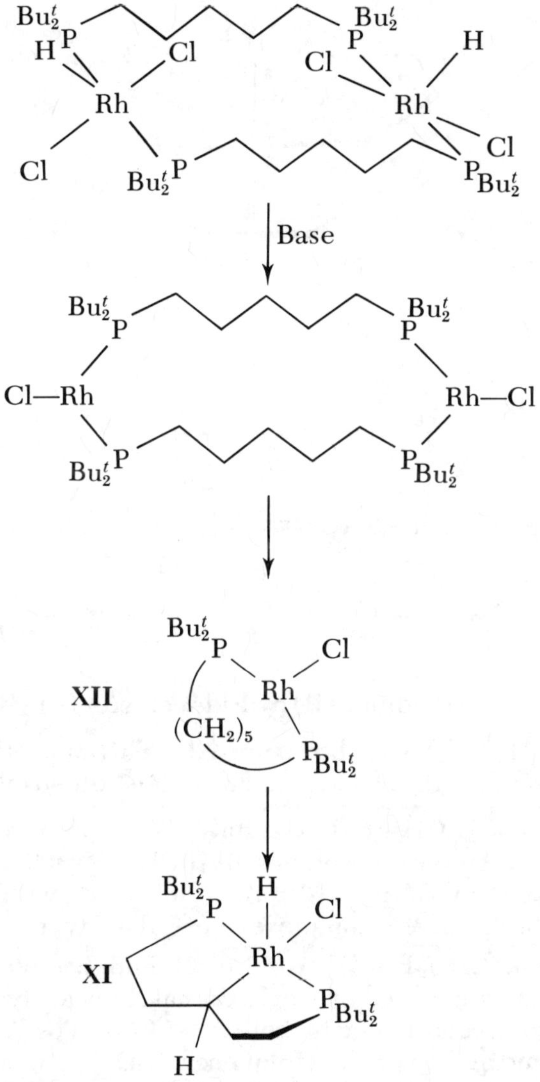

Base

XII

XI

(115 ppm) difference in chemical shift between the two phosphorus nuclei. The resonance at lower frequency ($\delta = -35.9$ ppm) is characteristic of a strained, 4-membered ring (shifts to high frequency are positive).

In the H-1 NMR spectrum a broad resonance (at $\delta = 4.2$ ppm) assigned to olefinic protons is observed. The complex is volatile and

*Scheme II. Possible processes to explain the fluxionality of the hydride (Structure **XI**): (A) concerted hydrogen shift mechanism; (B) stepwise process involving a carbene intermediate.*

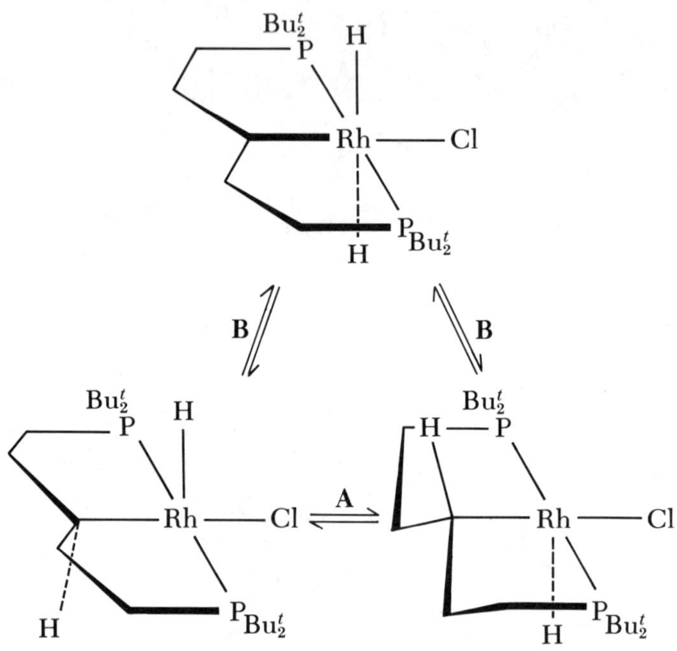

cosublimes with the rhodium(III) hydride complex [$RhHCl\{Bu^t_2PCH_2$-$CH_2CHCH_2CH_2PBu^t_2\}$ (*see* Structure **XI**). This and other evidence leads us to formulate the complex as an olefinic–diphosphine complex [$RhCl\{Bu^t_2PCH_2CH_2CH{=}CHCH_2PBu^t_2\}$]. We also have synthesized the two new olefinic diphosphines $Bu^t_2PCH_2CH{=}CHCH_2PBu^t_2$ and $Bu^t_2PCH_2CH_2CH{=}CHCH_2CH_2PBu^t_2$; these with rhodium trichloride both give complexes of the type [$RhCl\{Bu^t_2P$-$(CH_2)_xCH{=}CH(CH_2)_xPBu^t_2\}$] ($x = 1$ or 2). Both complexes have had their structures determined by single-crystal X-ray analysis and have a *trans*-C=C bond coordinated to rhodium (*34, 36*). The ^{31}P shifts for the $x = 1$ compound are to very low frequency (−42.5 ppm) consequent on the strained environment around the phosphorus nuclei. In contrast we find that the Rh-103 NMR shift for [$RhCl\{Bu^t_2PCH_2CH{=}CHCH_2PBu^t_2\}$] (+ 983 ppm relative to Ξ (^{103}Rh) = 3.16 MHz) is at a much higher frequency than for [$RhCl\{Bu^t_2PCH_2CH_2CH{=}CHCH_2CH_2PBu^t_2\}$] (243 ppm).

We also have studied the action of $Bu^t_2PCH_2CH_2CHMeCH_2$-$CH_2PBu^t_2$ on $RhCl_3$ in refluxing 2-propanol. This gives the 16-atom ring, binuclear chelate $[Rh_2H_2Cl_4\{Bu^t_2PCH_2CH_2CHMeCH_2CH_2PBu^t_2\}_2]$ which P-31 and H-1 NMR spectroscopies show to be a mixture of six very similar species (isomers or rotamers) in solution. They are probably analogues of Structures **IX** and **X** with the presence of the methyl group in the center of each chain increasing the number of isomers (rotamers). This six-component mixture when heated with 2-methylpyridine gave a single, cyclometallated hydrido complex $[RhHCl\{Bu^t_2PCH_2CH_2CMeCH_2CH_2PBu^t_2\}]$ in 75% yield. The structure of this complex has been determined by single-crystal X-ray analysis and, as expected, has the transoid arrangement of Me–C–Rh–H as in Projection **XIII**.

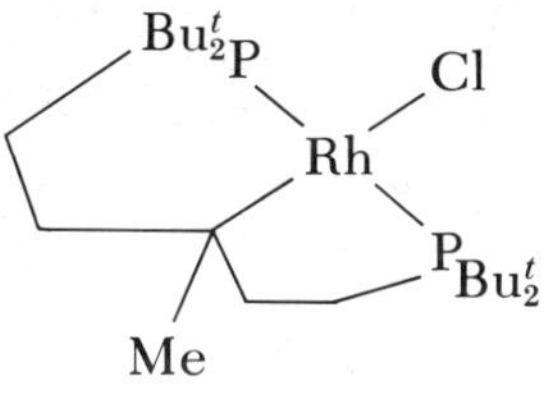

XIII

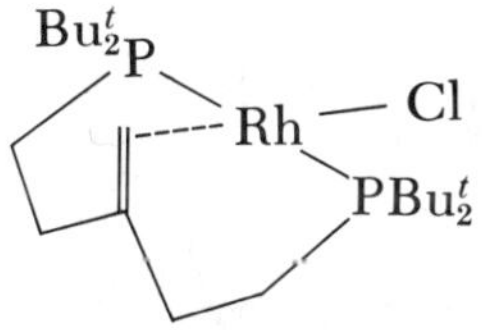

XIV

This hydride is nonfluxional and gives the expected H-1, C-13, and P-31 NM spectra (*34*). However, on heating its solution in 2-propanol, especially in the presence of a few percent of water, it dehydrogenates irreversibly to give the olefin–rhodium(I) complex $[RhCl\{Bu^t_2PCH_2CH_2C(=CH_2)CH_2CH_2PBu^t_2\}]$ (*see* Structure **XIV**) (crystal structure). We also have found that the hexamethylene diphosphine $Bu^t_2P(CH_2)_6PBu^t_2$ dehydrogenates when treated with rhodium trichloride to give the olefin diphosphine–rhodium(I) complex $[RhCl\{Bu^t_2PCH_2CH_2CH{=}CHCH_2CH_2PBu^t_2\}]$ reported above.

Some Iridium–Diphosphine Chemistry

Treatment of iridium trichloride with $Bu^t_2P(CH_2)_5PBu^t_2$ gives a mixture of binuclear five-coordinate iridium(III) hydrides, $[Ir_2H_2Cl_4\{Bu^t_2P(CH_2)_5PBu^t_2\}_2]$ with 16-atom rings and a red, volatile, five-coordinate hydride $[IrHCl\{Bu^t_2PCH_2CH_2CHCH_2CH_2PBu^t_2\}]$ which is not fluxional (this is in contrast to the rhodium species discussed earlier) (*37, 38*). It shows the expected H-1, C-13, and P-31 NMR patterns, etc. Crystals suitable for X-ray analysis have not been obtained yet. One would expect by analogy with rhodium that the stereochemistry of this hydride would be Structure **XI**. However, it loses dihydrogen reversibly on heating to give the very unusual

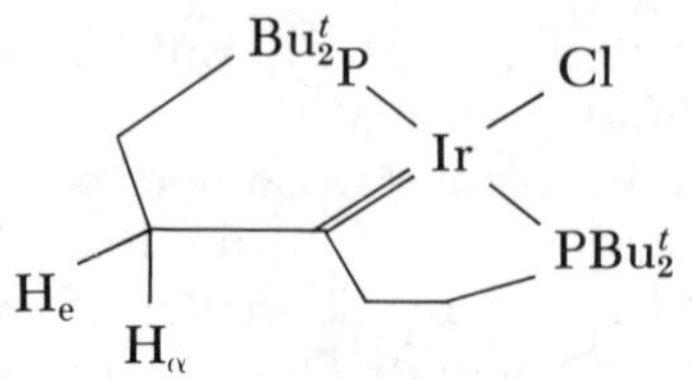

Bu^t_2P Cl
Ir
PBu^t_2
H_e
H_α

XV

carbene or ylide Complex **XV**, the structure that was established very accurately by X-ray diffraction (37). This carbene rapidly takes up dihydrogen at room temperature and 1 atm to give back $[IrHCl(Bu^t_2PCH_2CH_2CHCH_2\text{-}CH_2PBu^t_2]$. Dihydrogen usually adds to double bonds or metal atoms in a cis fashion and this suggests that the stereochemistry might be that of Structure **XVI**. The stereochemistry of

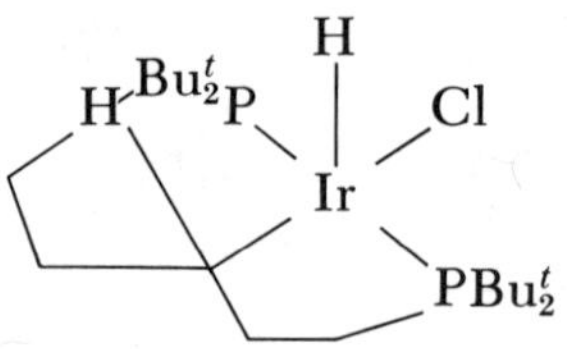

H
H Bu^t_2P Cl
Ir
PBu^t_2

XVI

this iridium hydride still has to be resolved. An unusual feature of the carbene complex (**XV**) is the presence of a quintet H-1 NMR pattern at -2.98τ due to two hydrogens. This collapses to an approximate 1:2:1 triplet in an H-1{P-31} experiment. It seems most likely that the two hydrogens with the very unusual chemical shift are adjacent to the unique (carbene or ylide) carbon atom either in pseudo-equatorial (H_e) or pseudoaxial (H_a) positions (*see* Structure **XV**). In an attempt to resolve this question we have substituted a methyl group in the 2-position of the polymethylene chain viz. $Bu^t_2PCH_2CHMeCH_2CH_2CH_2PBu^t_2$. This together with $[Ir_2Cl_2(C_8H_{14})_4]$ (C_8H_{14} = cyclooctene) rapidly gives a mixture of two cyclometallated hydrides $[IrHCl\{Bu^t_2PCH_2CHMeCHCH_2CH_2PBu^t_2\}]$, probably corresponding to the single methyl group being either in an axial-**XVII** or an equatorial-**XVIII** position. In solution, conversion of the axial to the equatorial isomer slowly takes place and is virtually complete after ca. 15 h at room temperature. The crystal structure of the equatorial isomer (*see* Structure **XVIII**) has been determined by X-ray diffraction, except that the hydrogen attached to iridium was not located (38). However, we have not been able to convert this hydride (Structure **XVIII**) into the corresponding carbene, e.g., under conditions in

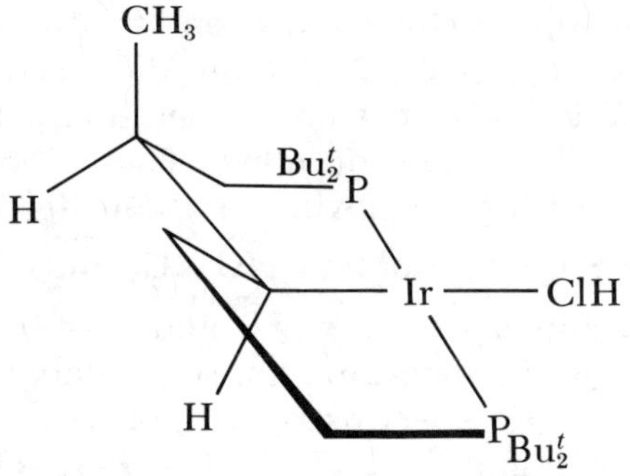

XVII

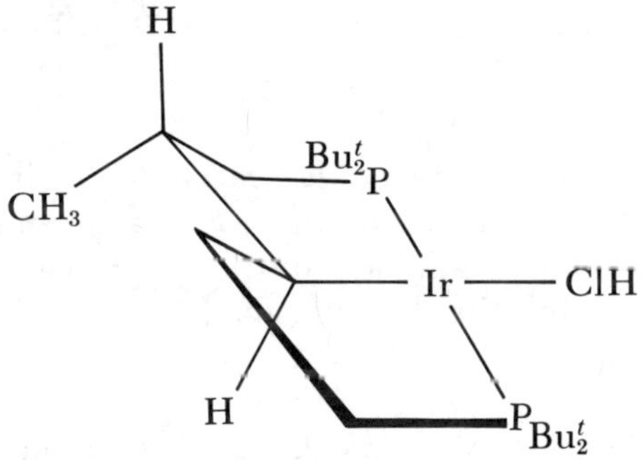

XVIII

which $[IrHCl\{Bu^t_2PCH_2CH_2CHCH_2PBu^t_2\}]$ gives the carbene (Structure **XV**) its methyl analogue (Structure **XVIII**) gives a complex mixture which we have not separated. We find that $[IrHCl\{Bu^t_2PCH_2$-$CH_2CHCH_2CH_2PBu^t_2\}]$ under dihydrogen (1 atm) in the presence of sodium propan-2-oxide rapidly gives a tetrahydride, $[IrH_4\{Bu^t_2PCH_2$-$CH_2CHCH_2CH_2PBu^t_2\}]$ (*38*). Pentahydrides of iridium, $[IrH_5(PR_3)_2]$, are fluxional even at low temperatures, showing magnetically equivalent hydrogens (*39, 40*). We hoped that the rigidity of a $Bu^t_2PCH_2$-$CH_2CHCH_2CH_2PBu^t_2$ ligand would render the hydridic hydrogens less prone to be fluxional. However, we find that $[IrH_4\{Bu^t_2PCH_2CH_2$-$CHCH_2CH_2PBu^t_2\}]$ is fluxional on the NMR time scale, even at −100°C with all four hydrogens equally coupled to both phosphorus nuclei as established by a P-31–H-1 experiment in which only the organic hydrogens were decoupled from the phosphorus resonance. Similarly the equatorial 2-methyl-analogue (Structure **XVIII**) with base and hydrogen gives $[IrH_4\{Bu^t_2PCH_2CHMeCHCH_2CH_2PBu^t_2\}]$. In this molecule the four hydrogens attached to iridium must be chemically nonequivalent. However, even at $-100°C$ the hydrogens are apparently magnetically equivalent on the NMR time scale, e.g., the P-31–H-1 (organic hydrogens only) spectrum is an ABX_4 pattern consisting of four 1:4:6:4:1 quintets. These tetrahydrides react rapidly with either carbon monoxide or isonitriles (L) with loss of dihydrogen to give complexes of type $[\overline{IrH_xL_y\{Bu^t_2PCH_2CH_2CHCH_2\text{-}}$ $\overline{CH_2PBu^t_2}\}]$ ($x = 2$ and $y = 1$ or $x = 0$ and $y = 2$). However, with Bu^tNC one major intermediate product, so far only detected in solution by P-31-{H-1} NMR spectroscopy, appears to be a *bis*-iridium complex with four bridging hydrogens. The P-31–{H-1 (organic hydrogens only)} pattern is very complex but symmetrical about a midpoint (*38*). Thus the iridium chemistry with these medium-chain diphosphines is unusual but some aspects of it still need clarification.

Acknowledgments

I am grateful to the Science Research Council for support, to Johnson Matthey, Ltd. for the generous loan of rare-metal salts, to my co-workers whose names appear in the references, and to W. S. McDonald for supplying the drawing for Figure 1.

Literature Cited

1. Cheney, A. J.; Mann, B. E.; Shaw, B. L.; Slade, R. M. *Chem. Commun.* **1970**, 1176.
2. Cheney, A. J.; Shaw, B. L. *J. Chem. Soc. Dalton* **1972**, 754.
3. Gill, D. F.; Mann, B. E.; Shaw, B. L. *J. Chem. Soc. Dalton* 1973, 270.
4. Clerici, M. G.; Shaw, B. L.; Weeks, B. *Chem. Commun.* **1973**, 516.
5. Jones, C. E.; Shaw, B. L.; Turtle, B. L. *J. Chem. Soc. Dalton,* **1974**, 992.
6. Shaw, B. L.; Truelock, M. M. *J. Organomet. Chem.* **1975**, *102*, 517.
7. Moulton, C. J.; Shaw, B. L. *J. Chem. Soc. Dalton* **1976**, 1020.
8. Mason, R.; Textor, M.; Al-Salem, N. A.; Shaw, B. L. *Chem. Commun.* **1976**, 292.
9. Empsall, H. D.; Hyde, E. M.; Pawson, D.; Shaw, B. L. *J. Chem. Soc. Dalton* **1977**, 1292.
10. Empsall, H. D.; Heys, P. N.; Shaw, B. L. *J. Chem. Soc. Dalton* **1978**, 257.
11. Empsall, H. D.; Heys, P. N.; McDonald, W. S.; Norton, M. C.; Shaw, B. L. *J. Chem. Soc. Dalton* **1978**, 1119.
12. Al-Salem, N. A.; Empsall, H. D.; Markham, R.; Shaw, B. L.; Weeks, B. *J. Chem. Soc. Dalton* **1979**, 1972.
13. Al-Salem, N. A.; McDonald, W. S.; Markham, R.; Norton, M. C.; Shaw, B. L. *J. Chem. Soc. Dalton* **1980**, 59.
14. Mann, B. E.; Masters, C.; Shaw, B. L.; Stainbank, R. E. *Chem. Commun.* **1971**, 1103.
15. Shaw, B. L. *J. Am. Chem. Soc.* **1975**, 97, 3856.
16. Beesley, R. M.; Ingold, C. K.; Thorpe, J. F. *J. Chem. Soc.* **1915**, *107*, 1080.
17. Ingold, C. K. *J. Chem. Soc.* **1921**, *119*, 305, 951.
18. Eliel, E. L. "Stereochemistry of Carbon Compounds;" McGraw–Hill: New York, 1962.
19. Eliel, E. L.; Allinger, N. L.; Angyll, S. J.; Morrison, G. A. "Conformational Analysis;" Interscience: New York, 1965.
20. Capon, B.; McManus, S. P. "Neighboring Group Participation," Plenum: New York, 1976; Vol. 1, p. 58–70.

21. Milstien, S.; Cohen, L. A. *Proc. Nat. Acad. Sci. U.S.A.* **1970,** 67, 1143.
22. Milstien, S.; Cohen, L. A. *J. Am. Chem. Soc.* **1972,** *94,* 9158.
23. Storm, D. R.; Koshland, D. E. *Proc. Nat. Acad. Sci. U.S.A.* **1970,** *66,* 445.
24. Bruce, T. C.; Pandit, U. K. *J. Am. Chem. Soc.* **1960,** *82,* 5258.
25. Hammond, G. S. "Steric Effects in Organic Chemistry"; Newman, M. S., Ed.; Wiley: New York, 1956; p. 468.
26. Allinger, N. L.; Zalkow, V. *J. Org. Chem.* **1960,** *25,* 701.
27. Masamune, S.; Bates, G. S.; Corcoran, J. W. *Angew. Chem. Int. Ed.* **1977,** *16,* 585.
28. Dobler, M.; Dunitz, J. D.; Krajewski, J. *J. Mol. Biol.* **1969,** *42,* 603.
29. Alcock, N. A.; Brown, J. M.; Jeffery, J. C. *J. Chem. Soc. Dalton* **1977,** 888.
30. Chottard, J. C.; Mulliez, E.; Girault, J. P.; Munsuy, D. *Chem. Commun.* **1974,** 780.
31. Shaw, B. L.; Shepherd, I., unpublished data.
32. Errington, J.; McDonald, W. S.; Shaw, B. L. *J. Chem. Soc. Dalton* **1980,** 2309.
33. Constable, A.; McDonald, W. S.; Shaw, B. L. *J. Chem. Soc. Dalton* **1979,** 496.
34. Crocker, C.; Errington, R. J.; Markham, R.; Moulton, C. J.; Odell, K. J.; Shaw, B. L. *J. Am. Chem. Soc.* **1980,** *102,* 4373.
35. Hyde, E. M.; Kennedy, J. D.; Shaw, B. L.; McFarlane, W. *J. Chem. Soc. Dalton* **1977,** 1571.
36. Mason, R.; Scollary, G.; Moyle, B.; Hardcastle, K. I.; Shaw, B. L.; Moulton, C. J. *J. Organomet. Chem.* **1976,** *113,* C49.
37. Empsall, H. D.; Hyde, E. M.; Markham, R.; McDonald, W. S.; Norton, M. C.; Shaw, B. L.; Weeks, B. *Chem. Commun.* **1977,** 589.
38. Errington, J.; McDonald, W. S.; Shaw, B. L., unpublished data.
39. Mann, B. E.; Masters, C.; Shaw, B. L. *Chem. Commun.* **1970,** 703.
40. Mann, B. E.; Masters, C.; Shaw, B. L. *J. Inorg. Nucl. Chem.* **1971,** 2195.

RECEIVED July 10, 1980.

7

Preparation, Structure, and Reactions of $Rh_2H_4(P(isopropyl)_3)_4$

DAVID L. THORN and JAMES A. IBERS

Department of Chemistry, Northwestern University, Evanston, IL 60201

A new hydrido-bridged rhodium dimer, $Rh_2H_4(P(iso\text{-}Pr)_3)_4$, has been prepared by reacting $RhClH_2(P(isoPr)_3)_2$ with sodium hydride in tetrahydrofuran (THF) and by reacting hydrogen with a THF solution of $Rh(\eta^3\text{-}C_3H_5)\text{-}(P(iso\text{-}Pr)_3)_2$. The compound crystallizes in space group $C^5_{2h}\text{-}P2_1/n$ of the monoclinic system with four formula units and approximately 0.6 THF molecules in a cell of dimensions at −150°C of a = 11.463(10) Å, b = 16.288(15) Å, c = 23.891(23) Å, β = 93.88(6)°. The structure has been refined to an R index on F_o^2 of 0.153, based on 207 variable parameters and 4997 F_o^2 values obtained at −150°C. The compound consists of two nonequivalent RhP_2 units with a Rh–Rh separation of 2.618(3) Å and an interplanar dihedral angle of 72.6°. Chemical means were used to establish that the material was a tetrahydride. Possible positions for the four hydrido ligands have been established from residual electron-density peaks and from bond distances and angles around the rhodium and phosphorus atoms. The most probable positions yield a Rh(III)–Rh(I) mixed-valence compound, $RhHP_2\text{–}RhP_2$, held together by three hydrido bridges. As such the compound may be an intermediate in the hydrogenation of $RhH(P(iso\text{-}Pr)_3)_{2,3}$ and the dehydrogenation of $RhH_3(P(iso\text{-}Pr)_3)_2$.

As a continuation of studies of the reactivities of rhodium complexes containing bulky phosphine ligands (*1–6*), we have been investigating alternative routes for synthesizing halogen-free hydrido rhodium complexes with these bulky phosphine ligands. In this context we happened to react $RhClH_2(P(iso\text{-}Pr)_3)_2$ with solid NaH (*7*) and obtained an unexpected dimeric product, $Rh_2H_4(P(iso\text{-}Pr)_3)_4$. This

0065–2393/82/0196–0117$05.00/0

complex crystallized as small black polyhedra that resembled the earlier reported compound $RhH(P(t\text{-Bu})_3)_2$ (*2, 4*) both in appearance and in the IR spectrum, and we initially tentatively formulated this new compound as $[RhH(P(\text{iso-Pr})_3)_2]_x$ ($x = 1$ or 2). A low-temperature, single-crystal X-ray structure determination revealed the dimeric nature of the complex and also structural details that suggested the presence of a greater number of hydrido ligands. In this publication we report the low-temperature crystal structure of $Rh_2H_4(P(\text{iso-Pr})_3)_4$ and its reactions with dinitrogen and hexene.

Experimental

All manipulations were carried out under an inert atmosphere (argon or nitrogen). Solvents were dried by refluxing over sodium/benzophenone under nitrogen and were degassed further by vacuum transfer or freeze–pump–thaw cycles before use. IR spectra were recorded on a Perkin–Elmer 283 spectrometer; P-31 and H-1 NMR spectra were recorded on a JEOL FX90Q spectrometer. Proton chemical shifts are reported in parts per million downfield from tetramethylsilane(TMS). Dry crystalline sodium hydride and $P(iso\text{-Pr})_3$ were obtained from Alpha Products and from Strem Chemical Co., respectively. Gas chromatographic (GC) analyses were performed on a Hewlett–Packard 5750 chromatograph instrument equipped with a flame ionization detector and a 6-ft, $\frac{1}{8}$-in. diameter column packed with 0.1% Supelco SP100 on 80/100 carbopack C.

$RhClH_2(P(\text{iso-Pr})_3)_2$, (*5, 8*). $(RhCl(C_8H_{14})_2)_2$ (*9*) (0.69 g, 1.92 mmol rhodium) and $P(\text{iso-Pr})_3$ (0.81 g, 5.06 mmol) were added to 10 mL THF under nitrogen. The resulting purple–brown solution was stirred overnight at room temperature under a hydrogen atmosphere. The solution then was dried in vacuum and the residue was recrystallized from hexane to yield yellow crystals (0.435 g, 0.94 mmol, 49%); IR (Nujol) 2140 cm^{-1}. H-1 NMR (THF-d_8): δ-22.8 (d of t, $J_{H\text{-}Rh}$ 27 Hz, $J_{H\text{-}P}$ 13 Hz); δ 1.26 (q, $J_{H\text{-}H+H\text{-}P}$ 7 Hz); δ 2.32 m; carbon and hydrogen analysis.

$Rh_2H_4(P(\text{iso-Pr})_3)_4$: Method A. $RhClH_2(P(\text{iso-Pr})_3)_2$ (0.21 g, 0.45 mmol) and sodium hydride (~0.1 g) were added to one chamber of a two-chambered flask containing a glass wool plug between the two chambers. About 8 mL of THF were distilled into the reaction chamber. The mixture then was put under 1 atm of argon and stirred at room temperature. After 2 d the resulting dark mixture was filtered through the glass wool into the second chamber, concentrated in vacuum to ca. 1 mL, and cooled to −20°C. Black crystals began to form after ~24 h. After 3 weeks the crystals were collected and dried briefly in vacuum. Yield: 0.04 g, 0.047 mmol dimer, 21%.

$Rh_2H_4(P(\text{iso-Pr})_3)_4$: Method B. A method analogous to that used by Sivak and Muetterties (*10*) for making allyl rhodium bisphosphite complexes was used to prepare $Rh(\eta^3\text{-}C_3H_5)(P(\text{iso-Pr})_3)_2$. H-1 NMR ($C_6D_6$) of $Rh(\eta^3\text{-}C_3H_5)(P(\text{iso-Pr})_3)_2$: δ 1.19 (d of d, $J_{H\text{-}H}$ 7 Hz, $J_{H\text{-}P}$ 11 Hz); δ 2.05 m; δ 3.10 (d,$J_{Hsyn\text{-}H}$ 7 Hz); δ 4.75 m; carbon and hydrogen analysis. Bubbling hydrogen gas through a THF solution of $Rh(\eta^3\text{-}C_3H_5)(P(\text{iso-Pr})_3)_2$ for 1 h at room temperature resulted in its complete conversion to $Rh_2H_4(P(\text{iso-Pr})_3)_4$, as judged by NMR spectrum of the solution.

X-Ray Data Collection. For preliminary room-temperature X-ray photographic examination, a well-formed polyhedral single crystal of Rh_2H_4(P(iso-

$Pr)_3)_4$ was sealed in a glass capillary under an argon atmosphere. Weissenberg and precession photographs revealed monoclinic symmetry and systematic absences consistent with the space group C_{2h}^5-$P2_1/n$. The density calculated from the room-temperature unit-cell constants is 1.22 g/cm^3 for 4 molecules of $Rh_2H_4(P(iso\text{-}Pr)_3)_4$ in the cell, which agrees with the density of 1.17(2) g/cm^3 measured by flotation in aqueous $ZnCl_2$.

For data collection this same crystal was removed from its capillary and mounted directly on a computer-controlled, four-circle Picker diffractometer, where it was bathed continuously in a stream of cold (~−150°C) dry nitrogen gas. (The diffractometer was run under the disk-oriented Vanderbilt system (*11*) and the design of the low-temperature apparatus is that of Huffman (*12*).) At this temperature the cell constants, derived from the setting angles of 15 hand-centered reflections in the range $22.2° < 2\theta < 27.0°$, (*13*), are $a = 11.463(10)$, $b = 16.288(15)$, $c = 23.891(23)$ Å, $\beta = 93.88(6)°$, and $V = 4450$ Å^3. Owing to the high mosaicity of the crystal (typical peaks were 0.8° to 1.0° wide, half-height to half-height in ω) data were collected in an omega step scan mode. Each peak was scanned in 41 steps, counting for 1 s at each step, across a total width in ω of 3.0°. Background counts were collected for 4 s at the low- and high-angle extremes. Weak reflections ($F_o^2 \leq 3\sigma(F_o^2)$) were rescanned and background counts were collected for a total of 14 s at the low- and high-angle extremes. A total of 5371 data were collected in the range $3.0 < 2\theta < 42.5°$, of which the 4997 unique data were used in the refinement. The intensities of six standard reflections were remeasured every 100 reflections, and they did not vary significantly during the course of data collection. Information about the crystal and data collection is summarized in Table I.

Solution and Refinement of the Structure. Scattering factors for the hydrogen (*14*) and nonhydrogen (*15*) atoms are those used previously. Anomalous dispersion terms (*16*) were included in F_c for rhodium and phosphorus atoms. For the processing of the data and solution and refinement of the structure procedures and computer programs standard in this laboratory were used. (*See*, e.g., Ref. *17*).) Trial absorption corrections calculated for a random selection of reflections gave transmission factors in the range 0.71 to 0.73; therefore a full absorption correction was considered to be unnecessary.

The locations of the rhodium atoms were obtained using direct methods, and all remaining nonhydrogen atoms (except for solvent atoms, vide infra) were located unambiguously in a subsequent Fourier map. After two cycles of isotropic refinement a difference Fourier map revealed locations of many of the alkyl hydrogen atoms and also some peaks clustered about the inversion center $(0, \frac{1}{2}, \frac{1}{2})$ which strongly suggested a disordered solvent molecule (THF) of partial occupancy. The isopropyl hydrogen atoms were placed in idealized locations (C–H: 0.95 Å, tetrahedral angles), assigned isotropic thermal parameters 1.0 Å^2 greater than those of their respective attached carbon atoms, and kept fixed in all subsequent refinement. The peaks around $(0, \frac{1}{2}, \frac{1}{2})$ were modeled by a THF molecule which was refined subsequently as a rigid group (*18*) with a single isotropic thermal parameter and a single, refined occupancy number. In the final cycles the rhodium and phosphorus atoms were refined anisotropically and all of the carbon atoms and the rigid solvent molecule were refined isotropically for a total of 207 variables. Final refinement was on F^2 with all of the unique data included. The refinement converged to values of R and Rw on F^2 of 0.153 and 0.172 and to an error in an observation of unit weight of 1.12 electrons^2. The refined occupancy number of the THF molecule is 0.60(4) molecules per unit cell. A final difference Fourier map revealed no peaks above 0.8 e/Å^3 except for peaks around the Rh centers, some

Table I. Summary of Crystal Data and Intensity Collection for $Rh_2H_4(P(iso\text{-}Pr)_3)_4$

Compound	$Rh_2H_4(P(iso\text{-}Pr)_3)_4$
Formula	$C_{36}H_{88}P_4Rh_2$
Temperature	−150°C
Formula weight	850.81 amu
Space group	C_{2h}^5–$P2_1/n$
a	11.463(10) Å
b	16.288(15) Å
c	23.891(23) Å
β	93.88(6)°
V	4450 $Å^3$
Z	4
Density (calculated)	1.266 g/cm^3 (−150°C)
Density (observed)	1.17(2) g/cm^3 (20°C)
Crystal volume	0.052 mm^3
Radiation	MoKα (λ(MoKα_1) = 0.709300 Å) from monochromator
Linear absorption coefficient	8.93 cm^{-1}
Transmission factors	0.71 to 0.73
Take-off angle	3.9°
Scan method	ω-step scan
Scan range	3.0° in ω in 41 steps
Scan speed	1 s/step (41 s to scan reflection)
Background at each end	4 s for $F_o^2 > 3\sigma(F_o^2)$, 14 s for $F_o^2 < 3\sigma(F_o^2)$
2θ limits	$3.0° < 2\theta < 42.5$
Final no. of variables	207
Total data collected	5371
Unique data used in final refinement[a]	4997
Unique data, $F_o^2 > 3\sigma(F_o^2)$	2508
R (on F_o) for $F_o^2 > 3\sigma(F_o^2)$	0.070
Rw (on F_o) for $F_o^2 > 3\sigma(F_o^2)$	0.070
R (on F_o^2), all data[a]	0.153
Rw (on F_o^2), all data[a]	0.172
Error in observation of unit weight	1.12 $electrons^2$

[a] This includes reflections with $F_o^2 < 0$.

of which may be hydrido hydrogen atoms (vide infra). Table II contains the final positional and thermal parameters of the nonhydrogen atoms, excluding the solvent; Table III contains parameters for the rigidly refined (*18*) THF solvent molecule. (A table of 10 $|F_0|$ vs. 10 $|F_c|$ for the reflections used in the refinement and a table of alkyl hydrogen atom positions has been deposited as NAPS Document No. 03802 with NAPS, c/o Microfiche Publications, P.O. Box 3513, Grand Central Station, New York, N.Y. 10017.)

Results and Discussion

Synthesis and Spectra of $Rh_2H_4(P(iso\text{-}Pr)_3)_4$. This complex was the totally unexpected product of the reaction of $RhClH_2(P(iso\text{-}Pr)_3)_2$ with solid sodium hydride in THF under an argon atmosphere. Once when the reaction was performed under a hydrogen atmosphere a yellow solid was obtained, believed to be $RhH_3(P(iso\text{-}Pr)_3)_2$ (*4*); repeating this reaction resulted either in the formation of the dimer $Rh_2H_4(P(iso\text{-}Pr)_3)_4$ or of no tractable products. The alternative synthetic method, reacting $Rh(\eta^3\text{-}C_3H_5)(P(iso\text{-}Pr)_3)_2$ with hydrogen (10), gives the dimeric product $Rh_2H_4(P(iso\text{-}Pr)_3)_4$ in excellent yield and reasonably free from by-products. In both reactions we suspect the formation of the dimer $Rh_2H_4(P(iso\text{-}Pr)_3)_4$ proceeds via the initial formation of $RhH_3(P(iso\text{-}Pr)_3)_2$, with subsequent loss of hydrogen and dimerization. (Loss of hydrogen from a dimeric intermediate is related closely to the bimolecular loss of hydrogen proposed by Norton (*19*).) The IR spectrum of $Rh_2H_4(P(iso\text{-}Pr)_3)_4$ in the solid state (Nujol mull) shows only a weak terminal Rh–H stretching band at 2040 cm^{-1}. Bridging Rh–H absorption bands have not been observed.

The hydrido-hydrogen region of the room-temperature H-1 NMR spectrum of $Rh_2H_4(P(iso\text{-}Pr)_3)_4$ (in THF-d_8) shows only two very broad and featureless peaks at δ −13.3 and δ −11.6 ppm. No significant change was observed in the spectrum when the sample was cooled to −20°C; precipitation occurred at lower temperatures. The alkyl–hydrogen region of the H-1 NMR spectrum is also uninformative, consisting of broad multiplets at δ 1.3 and δ 2.0 ppm. At room temperature the P-31{H-1} NMR spectrum (THF-d_8) is complicated. The most prominent peaks are a doublet of doublets; $J_{P\text{-}Rh}$ 108.5 Hz, $J_{P\text{-}Rh'}$ 3.2 Hz, and a second doublet ($J_{P\text{-}Rh}$ 152.6 Hz) 19.3 ppm upfield from the first, all of approximately equal intensity. This pattern is consistent with a structure in which each rhodium center is bound to two mutually equivalent phosphine ligands, yet the two rhodium centers remain nonequivalent (vide infra). There are numerous additional peaks in the P-31{H-1} NMR spectrum that are less intense including one that is assignable to free phosphine; this may result from partial fragmentation of the dimer (vide infra). Fluxional processes analogous to those discovered for $Rh_2H_4(P(O\text{–}iso\text{-}Pr)_3)_4$ (*10, 20*) may be occurring in the present system as well.

Reactions of $Rh_2H_4(P(iso\text{-}Pr)_3)_4$. As might be expected from previous work (*4*), this complex rapidly reacts with nitrogen. The IR spectrum of a Nujol mull of the solid, prepared under argon, shows only the Rh–H stretching band at 2040 cm^{-1}. After brief exposure of the mull to nitrogen a second weak band at 2140 cm^{-1} may be discerned which is assignable to the N–N stretch of $RhH(N_2)(P(iso\text{-}Pr)_3)_2$ (*4*). Bubbling

Table II. Positional and Thermal Parameters for the Nongroup Atoms of $Rh_2H_4(P(iso\text{-}Pr)_3)_4$

Atom	x[a]	y	z	B*11*[b] *or* B,Å^2	B22	B33	B*12*	B*13*	B23
Rh(1)	−0.126289(98)	0.287908(76)	0.106579(52)	187.(10)	161.1(54)	92.2(27)	25.0(68)	14.1(40)	7.7(36)
Rh(2)	−0.065488(99)	0.343030(78)	0.207304(53)	153.2(99)	184.1(58)	99.8(27)	5.7(67)	4.7(41)	3.5(36)
P(1)	−0.27302(34)	0.34986(28)	0.05535(18)	260.(34)	208.(20)	120.4(98)	60.(23)	0.(14)	24.(12)
P(2)	−0.03761(34)	0.20285(27)	0.04866(17)	280.(34)	229.(20)	93.0(90)	28.(23)	26.(14)	−10.(12)
P(3)	−0.17199(33)	0.27882(26)	0.27552(17)	235.(33)	196.(19)	84.7(88)	5.(22)	−4.(14)	1.(11)
P(4)	0.10592(34)	0.39596(27)	0.24405(19)	240.(35)	181.(20)	150.(10)	−23.(22)	−6.(15)	−8.(12)
C(11)	−0.3179(13)	0.32324(96)	−0.01865(64)	2.39(33)					
C(12)	−0.4095(13)	0.3454(10)	0.09446(64)	2.66(34)					
C(13)	−0.2508(13)	0.46455(97)	0.05100(65)	2.44(34)					
C(14)	−0.3843(14)	0.3855(10)	−0.05545(68)	2.77(36)					
C(15)	−0.3769(15)	0.2398(11)	−0.02361(70)	3.48(40)					
C(16)	−0.4441(14)	0.2570(10)	0.10945(67)	2.83(36)					
C(17)	−0.5181(14)	0.3925(10)	0.06990(67)	2.91(37)					
C(18)	−0.1404(15)	0.4800(11)	0.02260(71)	3.43(39)					
C(19)	−0.2500(15)	0.5067(11)	0.10778(71)	3.41(39)					
C(21)	−0.0660(12)	0.20471(93)	−0.02824(59)	1.92(30)					
C(22)	0.1238(13)	0.2103(10)	0.06386(64)	2.64(34)					
C(23)	−0.0600(13)	0.09059(96)	0.06533(64)	2.26(33)					
C(24)	−0.0223(12)	0.28392(93)	−0.05461(59)	1.81(30)					
C(25)	−0.0316(14)	0.1311(10)	−0.06358(68)	2.99(38)					
C(26)	0.1630(14)	0.2990(10)	0.05682(66)	2.93(36)					
C(27)	0.2018(13)	0.1540(11)	0.03274(65)	2.94(35)					

Table II. Positional and Thermal Parameters for the Nongroup Atoms of $Rh_2H_4(P(iso\text{-}Pr)_3)_4$

Atom	x[a]	y	z	B*11*[b] *or* B,A^2	B22	B33	B*12*	B*13*	B23
C(28)	−0.1886(14)	0.06823(98)	0.05439(66)	2.64(35)					
C(29)	−0.0135(14)	0.0682(10)	0.12417(66)	2.80(36)					
C(31)	−0.1823(12)	0.16410(95)	0.27080(60)	1.86(31)					
C(32)	−0.3328(12)	0.30433(86)	0.26123(57)	1.30(29)					
C(33)	−0.1349(12)	0.29866(94)	0.35178(60)	2.01(31)					
C(34)	−0.0581(13)	0.12690(96)	0.27855(64)	2.52(35)					
C(35)	−0.2427(13)	0.13427(93)	0.21718(62)	2.19(33)					
C(36)	−0.3584(14)	0.3966(10)	0.25560(66)	2.86(36)					
C(37)	−0.4153(14)	0.26453(97)	0.30072(67)	2.90(36)					
C(38)	−0.1732(13)	0.23435(94)	0.39388(63)	2.36(34)					
C(39)	−0.1657(14)	0.3849(10)	0.36863(68)	2.98(37)					
C(41)	0.2261(13)	0.31950(93)	0.26212(63)	2.25(33)					
C(42)	0.1012(15)	0.4522(11)	0.30971(71)	3.57(40)					
C(43)	0.1737(15)	0.4615(11)	0.19317(70)	3.35(39)					
C(44)	0.2535(15)	0.2696(11)	0.21056(72)	3.71(41)					
C(45)	0.1988(14)	0.2668(11)	0.31049(70)	3.33(39)					
C(46)	0.0317(17)	0.5316(13)	0.30529(81)	4.90(43)					
C(47)	0.2175(18)	0.4762(13)	0.34360(86)	5.61(52)					
C(48)	0.2983(19)	0.4968(13)	0.20701(87)	5.89(54)					
C(49)	0.0990(16)	0.5328(12)	0.16973(76)	4.33(44)					

[a] Estimated standard deviations in the least significant figure(s) are given in parentheses in this and all subsequent tables.

[b] The form of the anisotropic thermal ellipsoid is: $\exp[-(B11H^2 + B22K^2 + B33L^2 + 2B12HK + 2B13HL + 2B23KL)]$. The quantities given in the table are the thermal coefficients × 10^5.

Table III. Derived Parameters for the Rigid Group Atoms of $Rh_2H_4(P(iso\text{-}Pr)_3)_4$

Atom	x	y	z	B,A^2	*Atom*	x	y	z	B,A^2
O(1)	0.0002(89)	0.5480(50)	0.4849(38)	6.4(18)	C(3)	−0.032(10)	0.4409(68)	0.5509(40)	6.4(18)
C(1)	0.0596(89)	0.4751(63)	0.4643(39)	6.4(18)	C(4)	−0.0565(91)	0.5312(64)	0.5371(39)	6.4(18)
C(2)	0.039(10)	0.4066(50)	0.5064(50)	6.4(18)					

Rigid Group Parameters

Group	X_C[a]	Y_C	Z_C	δ[b]	ϵ	η
THF	0.0022(76)	0.4804(49)	0.5087(33)	−3.121(62)	−2.667(58)	−2.138(62)

[a] X_C, Y_C, and Z_C are the fractional coordinates of the origin of the rigid group.
[b] The rigid group orientation angles δ, ϵ, and η (radians) have been defined previously (*18*).

nitrogen through a toluene solution of $Rh_2H_4(P(iso\text{-}Pr)_3)_4$, followed by evaporation of the toluene and recrystallization of the yellow–orange residue from hexane under nitrogen, results in the formation of $[RhH(P(iso\text{-}Pr)_3)_2]_2(\mu\text{-}N_2)$ (*4*). Crystals of the latter product were identified positively by their measured density (1.24(1) g/cm^3; compared with 1.260 g/cm^3 (*4*)) and the lattice constants obtained from X-ray film data.

The observed reactivity towards nitrogen indicates a possible dissociation of the dimeric molecule into monomeric fragments, with one conceivable fragmentation mode indicated in Equation 1.

$$Rh_2H_4(P(iso\text{-}Pr)_3)_4 \rightleftharpoons RhH(P(iso\text{-}Pr)_3)_2 + RhH_3(P(iso\text{-}Pr)_3)_2 \quad (1)$$

However, fragmentation is not occurring fast enough to average the signals in the P-31 NMR spectrum. Both of the monomeric complexes resulting from this fragmentation mode (*see* Equation 1) are known to react rapidly with nitrogen to give $RhH(N_2)(P(iso\text{-}Pr)_3)_2$, which then dimerizes to give $[RhH(P(iso\text{-}Pr)_3)_2]_2(\mu\text{-}N_2)$ (*4*). Hydrogen appears to react only slowly or not at all with $Rh_2H_4(P(iso\text{-}Pr)_3)_4$ at room temperature. We have not observed a thermal loss of hydrogen from room-temperature solutions of the dimer and the solid appears to be very stable under argon at room temperature.

Room-temperature solutions of $Rh_2H_4(P(iso\text{-}Pr)_3)_4$ in deuterobenzene undergo rapid H–D exchange with the solvent, as indicated by the H-1 NMR spectrum; the signal of $C_6H_nD_{6-n}$ rapidly gains intensity and the hydrido signals disappear. The reaction is complicated by an apparent concurrent H–D exchange with the methyl hydrogen atoms of the triisopropylphosphine ligands. The active species for the exchange reaction(s) might be the dimer or monomeric complexes derived from its fragmentation. Mononuclear bis(triisopropylphosphine)rhodium complexes are active catalysts for H–D exchange reactions (*21*).

None of the experiments or spectral data presented above provides good evidence for our formulation of the dimer as a tetrahydride. In fact we initially had believed the compound to be a dihydride, $Rh_2H_2(P(iso\text{-}Pr)_3)_4$, or possibly a monomeric monohydride, $RhH(P(iso\text{-}Pr)_3)_2$ (*2, 4*). The dimeric nature is proved by the crystal structure. We began to suspect the presence of more hydrido ligands after completing the structure and learning of the analogous triisopropylphosphite complexes studied by Sivak and Muetterties (*10, 20*).

Our present evidence for the tetrahydride is derived from an experiment in which a large excess of trimethylphosphite was vacuum-transferred into a degassed, frozen solution of a known amount of the dimeric complex. The solution was warmed to room temperature and stirred for 1 h, during which time the color changed from dark green to red and finally to yellow, and hydrogen gas was evolved. The solution was then refrozen and the amount of gas present was measured with a Toepler pump. One mol of hydrogen gas was evolved per mole of dimer. The final solution was dried in vacuum and the residue consisted entirely of $RhH(P(OMe)_3)_4$ (H-1 NMR analysis (*10*)). This sequence of reactions, summarized in Scheme I, is consistent only with a formulation of the dimer as a tetrahydride.

Scheme I.

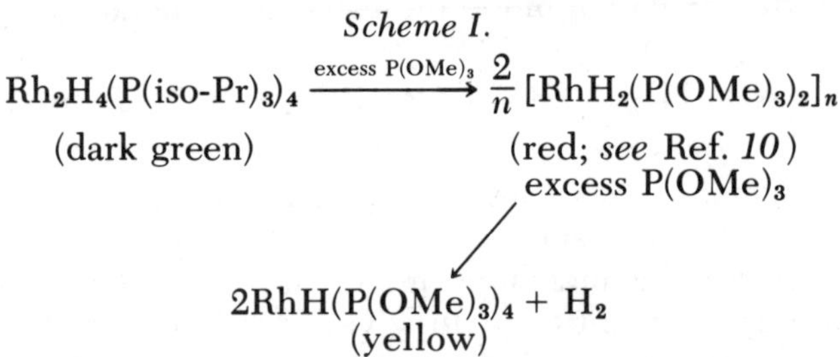

Further evidence for a tetrahydride comes from the reaction of a known amount of the dimer with an excess of 1-hexene. After several hours the amount of hexane produced was 1.4 ± 0.2 mmol of hexane per millimole of dimer (gas chromatography (GC) analysis). This suggests a rather complicated reaction but requires more than one hydrido ligand per rhodium center. The nature of the reaction and the resulting residual compound(s) are still under investigation.

Description of the Structure. The molecular structure of the present complex closely resembles that of the $[Pt_2H_3((t\text{-}Bu)_2P(CH_2)_3\text{-}P(t\text{-}Bu)_2)_2]^+$ cation (*22*). A formal electronic analogy can be made between the two complexes; $Pt_2H_3(PR_3)_4^+$ is isoelectronic with a hypothetical compound $Rh_2H_3(PR_3)_4^-$ which can be protonated (conceptually) to give the complex $Rh_2H_4(PR_3)_4$. The structural resemblance undoubtedly results from this close electronic similarity. Another species to which the present complex is related, both by structural similarity and electronic analogy, is $[Ir_2H_5(PPh_3)_4]^+$ (*23, 24*). Deprotonation of a hypothetical rhodium dimer of this formulation, $Rh_2H_5(PR_3)_4^+$, would give the present complex.

A drawing of the present molecule is presented in Figure 1. Each rhodium center is coordinated by two triisopropylphosphine ligands, the other rhodium atom, and several hydrido hydrogen ligands (vide infra). Atom Rh(2) lies 0.46 Å above the plane defined by atoms P(1), P(2), and Rh(1); atom Rh(1) lies 0.94 Å above the plane defined by atoms Rh(2), P(3), and P(4). The dihedral angle between the planes defined by Rh(1), P(1), P(2), and Rh(2), P(3), P(4) is 72.6°. (Hydrido-bridged, transition-metal dimers of the formula $M_2H_xL_4$ exhibit a

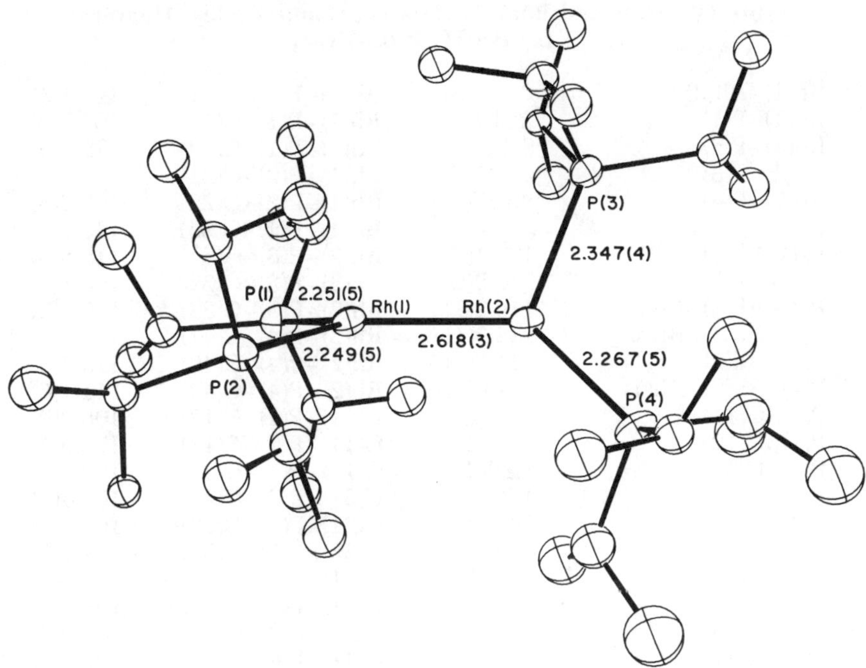

Figure 1. Sketch of the $Rh_2H_4(P(iso\text{-}Pr)_3)_4$ molecule. Hydrogen atoms are omitted. The other atoms are drawn as 50% probability ellipsoids.

range of dihedral angles that has been probed theoretically (*25*).) Selected bond distances and angles are listed in Table IV. The Rh–Rh separation of 2.618(3) Å is slightly shorter than the distance of 2.650(1) Å in $(RhH(P(O\text{-}iso\text{-}Pr)_3)_2)_2$ and considerably shorter than the Rh–Rh separations reported for $(RhH(P(OMe)_3)_2)_3$ (the average is 2.81 Å) (*20, 26, 27*). The separation is perhaps below the range expected for a formal Rh–Rh single bond, e.g., 2.88(2) Å in $Rh_2(PF_3)_8$ (*28*), but need not be construed as indicating a formal single or multiple bond in the present case (*20, 23, 24, 29, 30, 31*).

The unit-cell packing is illustrated in Figure 2. Packing of the dimeric molecules appears to be determined by intermolecular van der Waals and steric interactions among the alkyl hydrogen atoms. The closest intermolecular H · · · H contact is 2.31 Å between atom H(2) on C(25) and atom H(1) on C(37).

The two rhodium centers are distinctly nonequivalent in the solid state. The Rh(1)–P(1) and Rh(1)–P(2) bond lengths (*see* Table IV) are insignificantly different and are both shorter than the Rh(2)–P(3) and Rh(2)–P(4) bond lengths. The Rh(1)–P bond lengths (averaging 2.250(6) Å) are the shortest ever observed in a triisopropylphosphine–rhodium structure. Values found previously in low-temperature struc-

Table IV. Selected Bond Distances (Å) and Angles (Degrees) for $Rh_2H_4(P(iso\text{-}Pr)_3)_4$[a]

Rh(1)–Rh(2)	2.618(3)	Rh(1)–P(1)–C(11)	124.1(5)
Rh(1)–P(1)	2.251(5)	Rh(1)–P(1)–C(12)	109.2(5)
Rh(1)–P(2)	2.249(5)	Rh(1)–P(1)–C(13)	112.0(5)
Rh(2)–P(3)	2.347(4)	Rh(1)–P(2)–C(21)	123.1(5)
Rh(2)–P(4)	2.267(5)	Rh(1)–P(2)–C(22)	108.9(5)
		Rh(1)–P(2)–C(23)	113.1(5)
P(1)–Rh(1)–P(2)	107.1(2)	Rh(2)–P(3)–C(31)	115.8(5)
P(3)–Rh(2)–P(4)	112.2(2)	Rh(2)–P(3)–C(32)	108.8(5)
P(1)–Rh(1)–Rh(2)	119.2(1)	Rh(2)–P(3)–C(33)	120.3(5)
P(2)–Rh(1)–Rh(2)	132.3(1)	Rh(2)–P(4)–C(41)	116.0(5)
P(3)–Rh(2)–Rh(1)	111.5(1)	Rh(2)–P(4)–C(42)	116.7(6)
P(4)–Rh(2)–Rh(1)	131.0(1)	Rh(2)–P(4)–C(43)	111.1(6)
P(1)–C(11)	1.86(2)	C(11)–P(1)–C(12)	106.2(7)
P(1)–C(12)	1.88(2)	C(11)–P(1)–C(13)	102.0(7)
P(1)–C(13)	1.89(2)	C(12)–P(1)–C(13)	100.7(7)
P(2)–C(21)	1.84(2)	C(21)–P(2)–C(22)	107.4(7)
P(2)–C(22)	1.87(2)	C(21)–P(2)–C(23)	102.0(7)
P(2)–C(23)	1.89(2)	C(22)–P(2)–C(23)	99.7(7)
P(3)–C(31)	1.88(2)	C(31)–P(3)–C(32)	98.7(6)
P(3)–C(32)	1.90(1)	C(31)–P(3)–C(33)	103.9(7)
P(3)–C(33)	1.87(1)	C(32)–P(3)–C(33)	106.9(6)
P(4)–C(41)	1.89(2)	C(41)–P(4)–C(42)	101.5(7)
P(4)–C(42)	1.82(2)	C(41)–P(4)–C(43)	101.6(7)
P(4)–C(43)	1.83(2)	C(42)–P(4)–C(43)	108.5(8)
C(11)–C(14)	1.51(2)		
C(11)–C(15)	1.52(2)		
C(12)–C(16)	1.54(2)		
C(12)–C(17)	1.54(2)		
C(13)–C(18)	1.50(2)		
C(13)–C(19)	1.52(2)		
C(21)–C(24)	1.53(2)		
C(21)–C(25)	1.53(2)		
C(22)–C(26)	1.53(2)		
C(22)–C(27)	1.51(2)		
C(23)–C(28)	1.52(2)		
C(23)–C(29)	1.51(2)		
C(31)–C(34)	1.55(2)		
C(31)–C(35)	1.50(2)		
C(32)–C(36)	1.54(2)		
C(32)–C(37)	1.52(2)		
C(33)–C(38)	1.54(2)		
C(33)–C(39)	1.51(2)		
C(41)–C(44)	1.53(2)		
C(41)–C(45)	1.49(2)		
C(42)–C(46)	1.52(2)		
C(42)–C(47)	1.57(2)		
C(43)–C(48)	1.55(2)		
C(43)–C(49)	1.53(2)		

[a] The numbering scheme for carbon atoms is identical with that used in Ref. *4*.

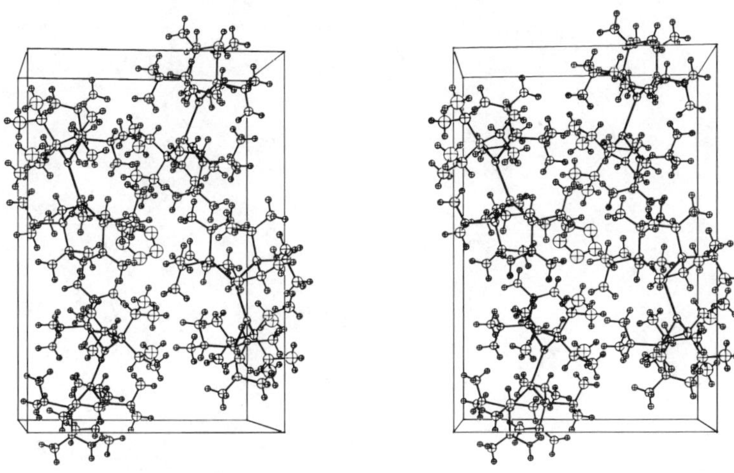

Figure 2. Packing diagram of $Rh_2H_4(P(iso\text{-}Pr)_3)_4$ with hydride ligands omitted. The 50% probability ellipsoids are shown, except for alkyl hydrogen atoms which are drawn artifically small. The partially occupied positions of the THF molecules are shown.

ture determinations are 2.273(3) Å (ave) (*1*), 2.348(1) Å (*32, 33*) in Rh(I)–dinitrogen complexes, 2.31(1) Å (ave) in a Rh(III)–bicarbonato complex (*3*), and 2.294(3) (ave) and 2.330(1) Å in $RhH(P(iso\text{-}Pr)_3)_3$ (*6*). The P(1)–Rh–P(2) bond angle, 107.1(2)°, is the smallest such angle thus far found for triisopropylphosphine ligands bonded to rhodium, smaller even than the angle 109.2(8)° found between cis-phosphine ligands in $RhH(P(iso\text{-}Pr)_3)_3$ (*6*). As in $RhH(P(iso\text{-}Pr)_3)_3$ the small P–Rh–P angle causes significant distortions within the triisopropylphosphine ligands (*see* Table IV) (*6*). The small angle and accompanying steric strain probably would not prevail in the present molecule unless there were other ligands (hydrido hydrogen atoms, vide infra) around the rhodium centers.

Possible Hydrido Hydrogen Atom Locations. It is possible to obtain reasonably unambiguous hydrido hydrogen atom positions from low- or even room-temperature X-ray diffraction data (*34*), but in the present structure the high mosaicity of the crystal—possibly a result of partial solvent loss—with resulting diminution in the quality and quantity of the diffraction data has prevented this. However, based on the angles and bond distances around the rhodium and phosphorus atoms, and the locations of residual electron-density peaks in the final difference Fourier map, reasonable sites of some of the hydrido hydrogen atoms can be postulated. Around the Rh(2) center the most likely site of a hydrogen atom is trans to atom P(3) (*see* Diagram **I** below), suggested by the large Rh(1)–Rh(2)–P(4) angle and the long Rh(2)–

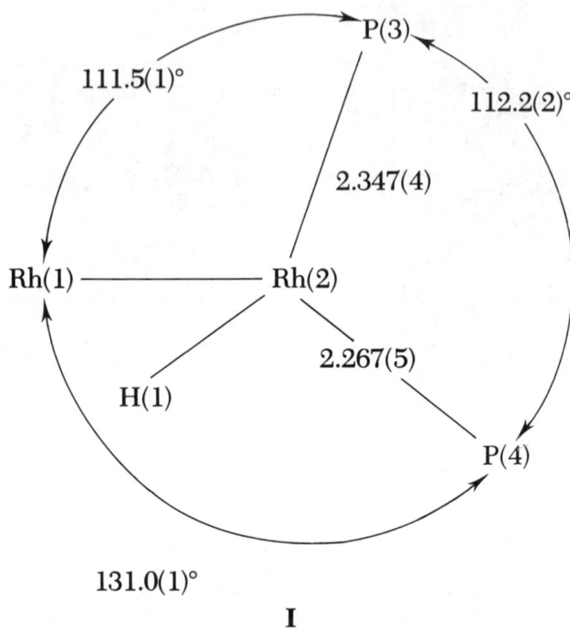

I

P(3) bond distance. The largest residual electron-density peak in the final difference Fourier map is at that position and a hydrogen atom H(1) can be placed there with some confidence.

The placement of hydrogen atoms about the Rh(1) center is less certain. Angles and Rh(1)–P distances (*see* Diagram **II** below) suggest

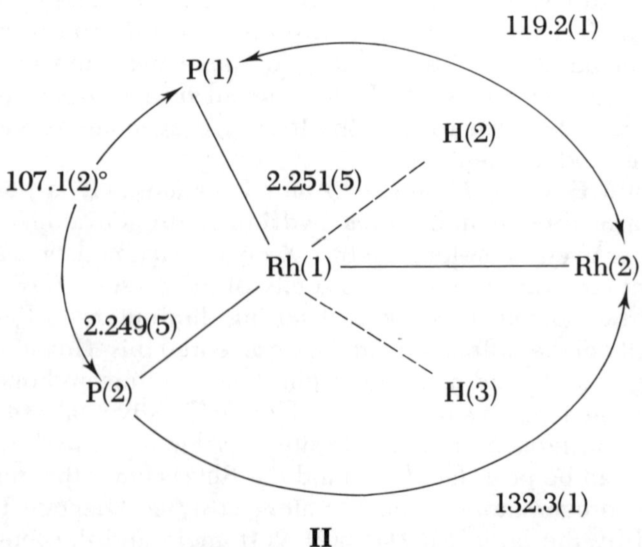

II

the possible presence of two hydrido hydrogen atoms, H(2) and H(3), in or approximately in the local coordination plane. Again there are residual electron-density peaks at about these locations.

These three hydrogen atoms, H(1), H(2), and H(3), are properly considered to be bridging hydrido ligands. The position of the fourth hydrido hydrogen atom is not at all apparent from the structural data. From the IR spectrum it is reasonable to assume at least one terminal hydrido ligand, but whether it is bonded to atom Rh(1) or Rh(2) is uncertain. One likely location is on atom Rh(1), above the Rh(1), P(1), P(2) plane, approximately trans to the bridging atom H(1). The resulting structure is sketched below in Diagram **III**. It is the structure Sivak and Muetterties (*10*) have postulated for the analogous complex $Rh_2H_4(P(O\text{-iso-Pr})_3)_4$, based on their NMR evidence.

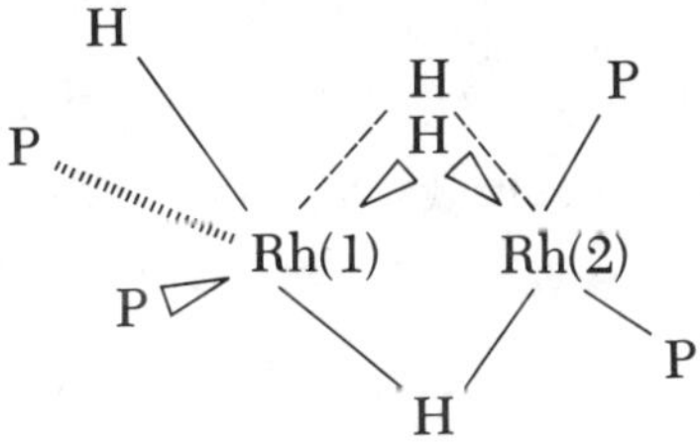

IIIa

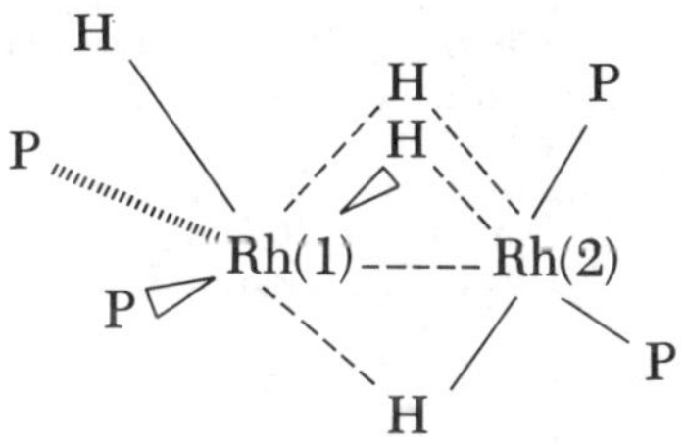

IIIb

An attractive feature of this structure is that if it is drawn as in Diagram **IIIb** it can be envisioned as a Rh(III)–Rh(I) mixed-valence compound held together by hydrido bridges. In this picture the Rh(1) center is in the +3 oxidation state, and would be expected to have a smaller covalent radius and shorter Rh–P bond lengths than those of the Rh(2) center, which is ostensibly in the +1 oxidation state. It is possible, however, to envision a terminal hydrido hydrogen ligand bonded to Rh(2). There are, in fact, several residual electron-density peaks around the Rh(2) center which are probably noise but which might be ascribable to a terminally bound hydrido hydrogen ligand.

Summary and Conclusions

The present complex, $Rh_2H_4(P(\text{iso-Pr})_3)_4$, is yet another member of the large family of RhH_xP_y complexes that recently have been prepared and studied (*1–6, 10*). It can be viewed conceptually and perhaps chemically as well as a combination of $RhH(P(\text{iso-Pr})_3)_2$ with $RhH_3(P(\text{iso-Pr})_3)_2$, and it may be an intermediate in the hydrogenation of $RhH(P(\text{iso-Pr})_3)_{2,3}$ and the dehydrogenation of $RhH_3(P(\text{iso-Pr})_3)_2$ (*4*). Its chemistry has only begun to be explored; however it—or the species resulting from its fragmentation—is capable of undergoing H–D exchange with C_6D_6 and may be catalytically active in numerous other reactions.

Acknowledgments

We thank Johnson Matthey, Inc., Malvern, Pennsylvania for the loan of rhodium used in this study. We wish to thank E. L. Muetterties for helpful suggestions and for communicating results prior to publication. This work was supported by the U.S. National Science Foundation (CHE 76–10335).

Literature Cited

1. Hoffmann, P. R.; Yoshida, T.; Okano, T.; Otsuka, S.; Ibers, J. A. *Inorg. Chem.* **1976,** *15,* 2462–2466.
2. Yoshida, T.; Okano, T.; Otsuka, S. *J. Chem. Soc. Chem. Commun.* **1978,** 855–856.
3. Yoshida, T.; Thorn, D. L.; Okano, T.; Ibers, J. A.; Otsuka, S. *J. Am. Chem. Soc.* **1979,** *101,* 4212–4221.
4. Yoshida, T.; Okano, T.; Thorn, D. L.; Tulip, T. H.; Otsuka, S.; Ibers, J. A. *J. Organometal. Chem.* **1979,** *181,* 183–201.
5. Yoshida, T.; Otsuka, S.; Matsumoto, M.; Nakatsu, K. *Inorg. Chim. Acta* **1978,** *29,* L257–L259.
6. Yoshida, T.; Thorn, D. L.; Okano, T.; Otsuka, S.; Ibers, J. A. *J. Am. Chem. Soc.* **1980,** *102,* 6451–6957.
7. Werner, H.; Feser, R.; Buchner, W. *Chem. Ber.* **1979,** 112, 834–843.
8. Van Gaal, H. L. M., Ph.D. Thesis, Katholieke Universiteit te Nijmegen, 1977.
9. van der Ent, A.; Onderdelinden, A. L.; *Inorg. Synth.* **1973,** *14,* 92–93.
10. Sivak, A. J.; Muetterties, E. L. *J. Am. Chem. Soc.* **1979,** *101,* 4878–4887.
11. Lenhert, P. G. *J. Appl. Crystallogr.* **1975,** *8,* 568–570.
12. Huffman, J. C., Ph.D. Thesis, Indiana Univ., 1974.
13. Corfield, P. W. R.; Doedens, R. J.; Ibers, J. A. *Inorg. Chem.* **1967,** *6,* 197–204.
14. Stewart, R. F.; Davidson, E. R.; Simpson, W. T. *J. Chem. Phys.* **1965,** *42,* 3175–3187.
15. Cromer, D. T.; Waber, J. T. "International Tables for X-ray Crystallography"; Kynoch: Birmingham, England, 1974; Vol. 4, Table 2.2A.
16. Cromer, D. T. "International Tables for X-ray Crystallography;" Kynoch: Birmingham, England, 1974; Vol. 4, Table 2.3.1.
17. Waters, J. M.; Ibers, J. A. *Inorg. Chem.* **1977,** *16,* 3273–3277.
18. La Placa, S. J.; Ibers. J. A. *Acta Crystallogr.* **1965,** *18,* 511–519.
19. Norton, J. R. *Acc. Chem. Res.* **1979,** *12,* 139–145.
20. Brown, R. K.; Williams, J. M.; Fredrich, M. F.; Day, V. W.; Sivak, A. J.; Muetterties, E. L. *Proc. Natl. Acad. Sci. USA* **1979,** *76,* 2099–2102.
21. Yoshida, T.; Okano, T.; Saito, K.; Otsuka, S. *Inorg. Chim. Acta* **1980,** *44,* L135–L136.
22. Tulip, T. H.; Yamagata, T.; Yoshida, T.; Wilson, R. D.; Ibers, J. A.; Otsuka, S. *Inorg. Chem.* **1979,** *18,* 2239–2250.
23. Crabtree, R. H.; Felkin, H.; Morris, G. E.; King, T. J.; Richards, J. A. *J. Organometal. Chem.* **1976,** *113,* C7–C9.
24. Crabtree, R. H.; Felkin, H.; Morris, G. E. *J. Organometal. Chem.* **1977,** *141,* 205–215.

25. Barnett, B. L.; Krüger, C.; Tsay, Y.-H.; Summerville, R. H.; Hoffmann, R. *Chem. Ber.* **1977,** *110,* 3900–3909.
26. Day, V. W.; Fredrich, M. F.; Reddy, G. S.; Sivak, A. J.; Pretzer, W. R.; Muetterties, E. L. *J. Am. Chem. Soc.* **1977,** *99,* 8091–8093.
27. Brown, R. K.; Williams, J. M.; Sivak, A. J.; Muetterties, E. L. *Inorg. Chem.* **1980,** *19,* 370–374.
28. Bennett, M. A.; Johnson, R. N.; Turney, T. W. *Inorg. Chem.* **1976,** *15,* 2938–2941.
29. Bau, R.; Teller, R. G.; Kirtley, S. W.; Koetzle, T. F. *Acc. Chem. Res.* **1979,** *12,* 176–183.
30. Dedieu, A.; Albright, T. A.; Hoffmann, R. *J. Am. Chem. Soc.* **1979,** *101,* 3141–3151.
31. Summerville, R. H.; Hoffmann, R. *J. Am. Chem. Soc.* **1979,** *101,* 3821–3831.
32. Thorn, D. L.; Tulip, T. H.; Ibers, J. A. *J. Chem. Soc. Dalton* **1979,** 2022–2025.
33. Busetto, C.; D'Alfonso, A.; Maspero, F.; Perego, G.; Zazzetta, A. *J. Chem. Soc. Dalton* **1977,** 1828–1834.
34. Ibers, J. A. In "Transition Metal Hydrides," *Adv. Chem. Ser.* **1978,** *167,* 26–35.

RECEIVED July 10, 1980.

8

Splitting of a Water Molecule by Trialkylphosphine Complexes of Platinum and Rhodium and Its Chemical Applications

T. YOSHIDA and SEI OTSUKA

Department of Chemistry, Faculty of Engineering Science, Osaka University, Toyonaka, Osaka, 560 Japan

Oxidative addition of water to PtL_n *(*$n = 3$*,* $L = PEt_3$*;* $n = 2, 3$*,* $L = P(iso\text{-}Pr)_3$*) produces PtH(OH) species. Similarly, addition to* $RhHL_3$ *(*$L = PEt_3$*,* $P(iso\text{-}Pr)_3$*) or* $(RhHL_2)_2(\mu\text{-}N_2)$ *(*$L = P(cyclo\text{-}C_6H_{11})_3$*) gave* $RhH_2(OH)$ *species. The systems* H_2O/PtL_n *and* $H_2O/RhHL_n$ *in coordinating solvents like pyridine exhibit strong basicity comparable with that of NaOH. These systems serve as catalysts for hydrating nitriles and activated olefins. They also catalyze H–D exchange reactions between* D_2O *and activated hydrocarbons. Stoichiometric* H_2 *evolution from the dihydrido rhodium(III) species* $RhH_2(OH)$ *can be effected by adding electron-withdrawing CO or RNC. Facile reduction of* CO_2 *with the water adduct of* $RhHL_n$ *was observed also.*

Current studies on solar energy conversion and storage have focused interest on water splitting into H_2 and O_2 (*1*). Certain transition metal compounds, e.g. $Ru(bipy)_3^{n+}$ (*2*) and $Rh_2(1,3\text{-diisocyanopropane})_4^{2+}$ (*3*), have been proposed as a low-energy system for the catalytic photodissociation of water. However their efficiency still must be improved. We have been interested in the oxidative addition of water to low-valence transition metals which yields species M(H)OH. This reaction can be viewed as a two-electron transfer process from the metal ion to a water molecule resulting in the H^- and OH^- ligands. A monohydrido complex is expected to give dihydrido species $M(H)_2OH$, a potential candidate for H_2 evolution. This chapter

0065–2393/82/0196–0135$05.00/0

is a brief account of our recent studies on the reaction of water with PtL_3 and $RhHL_3$ (L = trialkylphosphines) and some chemical applications of the water adducts.

Oxidative Addition of Water

The addition of water to low-valence transition metal compounds has been reported for a few cases, e.g., formation of $Os_3H(OH)(CO)_{10}$ from $Os_3(CO)_{12}$ (*4*) and $[RhH(OH)en_2]^+$ from $[Rhen_2]^+$ (*5*). Prior to our studies, however, the chemistry of hydridohydroxo metal species MH(OH) has remained virtually unexplored. The adduct formation was studied in depth for PtL_2 and PtL_3 (L = trialkylphosphines) (*6*). Using PtL_2 (L = P(iso-Pr)$_3$), the adduct, *trans*-PtH(OH)L_2, could be isolated. It is thermally unstable, reflecting the weak bonding between the soft platinum(II) ion and hard OH^- ion. This also is manifested in the ion-pair complex formation (*see* Equation 1) from $Pt(PEt_3)_3$ in aqueous THF or pyridine and its extensive dissociation (Equation 2).

$$PtL_3 + H_2O \overset{K_o}{\rightleftharpoons} PtHL_3{}^+OH^-, \ L = PEt_3 \tag{1}$$

$$PtHL_3{}^+OH^- \overset{K_d}{\rightleftharpoons} PtHL_3{}^+ + OH^- \tag{2}$$

The conductivity of system $Pt(PEt_3)/H_2O$ in pyridine was measured for a range of water concentration. The dependence of the equivalent conductance on H_2O concentration (at 0.5°C) was analyzed by Fuoss treatment to quantitatively assess the equilibria (*see* Equations 1 and 2); $K_o = 0.6(0.3)M^{-1}$, $K_d = 4.2(0.2) \times 10^{-2} \exp(-11.9(0.1)/[H_2O])\ M$. The value ($1.3 \times 10^{-2}$ for $[H_2O] = 20M$) of product $K_o \cdot K_d$ suggests an extensive dissociation of OH^-. Consistent with this is the apparent pH of $Pt(PEt_3)_3/H_2O$ (*see* Table I) comparable with that of NaOH (~ 0.01*M*) in the same medium.

Table I. Apparent pH Values in Aqueous Pyridine[a] at 20°C

Compound	*Apparent pH*	*Compound*	*Apparent pH*
$RhH(PEt_3)_3$	14.5	$Pt(PEt_3)_3$	14.3
RhH[P(iso-Pr)$_3$]$_3$	14.2	Pt[P(iso-Pr)$_3$]$_3$	14.1
		NaOH	13.5

[a] [Compound] = $9.8 \times 10^{-3}M$. Volume ratio of H_2O/pyridine = 2/3.

Pt[P(iso-Pr)$_3$]$_3$ tends to release one of the ligands even in the solid state due to the ligand bulk. The dissociation of its dilute solution (< 0.01*M*) is nearly complete in coordinating solvents like pyridine. The solution behavior of Pt[P(iso-Pr)$_3$]$_3$/H_2O is more complex than that of Pt–$(PEt_3)_3/H_2O$. Following essentially the Fuoss treatment, the conductivity data were analyzed satisfactorily in terms of Equations 3–6 (6). In this case each equilibrium constant cannot be determined. Instead composite constants $(1 + K_s)K_o$ and $K_sK_d^o/(1 + K_s)$ are obtained, where $K_d^o = \exp(-P/[H_2O])/K_d$ (P = proportionality constant).

$$PtL_3 \rightleftharpoons PtL_2 + L, \; L = P(iso\text{-}Pr)_3 \quad (3)$$

$$PtL_2 + H_2O \overset{K_o}{\rightleftharpoons} PtH(OH)L_2 \quad (4)$$

$$PtH(OH)L_2 + S \overset{K_s}{\rightleftharpoons} PtH(S)L_2{}^+OH^- \quad (5)$$

$$PtH(S)L_2{}^+OH^- \overset{K_d}{\rightleftharpoons} PtH(S)L_2{}^+ + OH^- \quad (6)$$

Water adds to $RhHL_3$ at an ambient temperature. The adduct formed from $RhH(PEt_3)_3$ in pyridine is the pentacoordinate species $[RhH_2(PEt_3)_3]OH$ (Complex 1a), which is isolatable as $[RhH_2(PEt_3)_3]BPh_4$ (Complex **1b**). The trigonal bipyramidal structure of Complex **1b** is inferred readily from H-1 NMR (δ −11.7 ppm, double double triplet, J_{Rh-h} – 11.0, $J_{H-P(equatorial)}$ = 124.5, $J_{H-P(axial)}$ = 16.5 Hz).

$$\left[RhH_2L_3 \right]^+ X \qquad (L = PEt_3)$$

Ia: X = OH

Ib: X = BPh_4

$RhH[P(iso\text{-}Pr)_3]_3$ with water in pyridine gives the corresponding adduct, $\{RhH_2(py)_2L_2\}OH$ (Complex **2a**), which is isolated again as its BPh_4 salt (Complex **2b**), (IR: ν(Rh–H) 2112, 2076 cm^{-1}. H-1 NMR: Rh–H, δ −19.9(q), $J_{Rh-H} = J_{P-H}$ = 18.4 Hz). The adduct undergoes facile reductive elimination of water in dry solvents. The formation of the water adduct, $\{RhH_2(py)_2[P(cyclo\text{-}C_6H_{11})_3]_2\}BPh_4$ was observed also for the reaction of $\{RhH[P(cyclo\text{-}C_6H_{11})_3]_2\}_2(\mu\text{-}N_2)$ with water in the presence of $BPh_4{}^-$.

These ion-pair hydroxo rhodium(III) compounds are also strong bases (*see* Table I). It may be relevant to note here that $Pt(PPh_3)_3$ does not give any indication of water–adduct formation.

$$RhHL_3 + H_2O \xrightarrow{-L} \left[RhH_2(py)_2L_2 \right]^+ X \qquad (L = P(iso\text{-}Pr)_3) \quad (7)$$

2a: X = OH

2b: X = BPh_4

Hydration of Unsaturated Compounds

The PtL_n (n = 2,3)/H_2O system in THF can serve as catalysts for hydrating nitriles and activated olefins like acrylonitrile (*6*). Pt[P(iso-Pr)$_3$]$_3$ is perhaps one of the most active catalysts among PtL_n. With this catalyst (0.1 mmol), $CH_2 = CHCN$ (5 mL) is hydrated (1 mL H_2O, 80°C, 20 h) to give $CH_2 = CHCONH_2$ (36 mol/mol catalyst), $HOCH_2CH_2CN$ (49), and $(NCCH_2CH_2)_2O$ (418) together with the dimer, $CH_2C(CN)CH_2CH_2CN$ (52.5).

The rate-determining step is likely to be the attack of an outer-sphere OH^- on the coordinated olefin (Equation 8) or the nitrile group (Equation 9). Reductive elimination regenerating PtL_2 or hydrolysis giving the water adduct $PtH(OH)L_2$ completes the catalytic cycle.

$$[(NC)CH{=}CH(R)\cdots PtHL_2]OH \longrightarrow (HOCHR)(NC)CH{-}PtHL_2 \qquad (8)$$

$$[RCN{-}PtHL_2]OH \longrightarrow R{-}\underset{\underset{O}{\|}}{C}{-}NH{-}PtHL_2 \qquad (9)$$

H–D Exchange Reaction of Hydrocarbons

The C–H bonds adjacent to a carbonyl, sulfoxy, or nitro group undergo facile H–D exchange with D_2O in the presence of a catalytic amount of PtL_3 (L = PEt_3, P(iso-Pr)$_3$). Typically the $Pt(PEt_3)_3$ (0.09 mmol)-catalyzed exchange reaction of $PhCOCH_3$ (8.3 mmol) with D_2O (43 mmol) in THF (3 mL) occurs at 80°C resulting in 78% D_2O incorporation in the methyl group after 18 h (6). The rate-determining step was the condensation reaction yielding the σ-alkyl complex (Equation 10). Here again reductive elimination or hydrolysis completes the catalytic cycle.

$$PhCOCH_3 + PtDL_3{}^+OD^- \rightarrow PtD(CH_2COPh)L_3 + DHO \qquad (10)$$

An unexpected feature in the PtL_3-catalyzed reaction is deuterium incorporation at the α-olefinic carbon of $PhCH = CHCOCH_3$, which is not achieved with aqueous NaOH. H–D exchange occurs not only at the α-olefinic carbon but also at the methyl carbon of $PhCOCH = CHCH_3$, a result also unobservable in catalysis by NaOH.

For aromatic hydrocarbons, the PtL_3/H_2O system fails to catalyze the exchange reaction. $RhHL_3$ (L = P(iso-Pr)$_3$) or $[RhHL_2]_2(\mu\text{-}N_2)$ (L = P(cyclo-C_6H_{11})$_3$), a precursor of $RhHL_2$, proved to be an efficient catalyst for this reaction (*7*). Thus a mixture of RhH[P(iso-Pr)$_3$]$_3$ (0.1 mmol), D_2O (27 mmol), and pyridine (5.4 mmol) gave (80°C, 20 h) deuterated pyridine without positional preference; i.e., 62%, 59%, and 58% at 2,6-, 3,5-, and 4-positions, respectively.

For monosubstituted benzenes, faster deuteration was observed for PhF than for $PhOCH_3$ [relative rate: *o,p*-H(PhF) 9.8, *m*-H(PhF) 5.6, *o,p*-H($PhOCH_3$) 1.9, *m*-H($PhOCH_3$) 5.0, CH_3O 1.0]. This is consistent with a mechanism involving oxidative addition of ArH as the rate-determining step. However, the relative rates of H–D exchange of Ar*X* (*X* = Me_2N, MeO, CH_3, CH_3CO, or F) do not appear to follow the simple Hammett rule, probably due to coordination of the substituent heteroatom. This effect is manifested in the deuteration of methyl protons of anisole and toluene. In view of the minor aromatic site preference observed for the exchange of C_6H_5F and $C_6H_5CH_3$, the possibility of C–H bond activation through η^6-coordination of the aromatic ring cannot be excluded.

H_2–D_2O Exchange Reaction

$RhHL_3$ (L = P(iso-Pr)$_3$) or $(RhHL_2)_2(\mu\text{-}N_2)$ (L = P(cyclo-C_6H_{11})$_3$) catalyzes effectively the exchange reaction. Thus a solution of D_2O (0.11 mol) and RhH[P(iso-Pr)$_3$]$_3$ (0.2 mmol) in THF (5 mL) was heated with H_2 (100 atm, 0.45 mol) at 80°C. The reaction reached an equilibrium after 20 h where the reaction mixture contained 75% H_2O, 22% DHO, and 3% D_2O. A similar result was obtained with the μ-N_2 complex (7).

The H–D exchange probably involves RhH_2DL_2 and the water adduct $[RhHD(S)_2L_2]OD$ (Complex **2a**). The isolation and spectral characterization of the latter complex has been given already in the previous section. The former RhH_3L_2 was prepared by treating $RhHL_n$ (L = P(*t*-Bu)$_3$, P(cyclo-C_6H_{11})$_3$, and P(iso-Pr)$_3$) with H_2 and was characterized well (8). The catalytic reaction sequence then may be described with the following scheme.

It may be noted that PtL_3 (L = PEt_3, P(iso-Pr)$_3$) is less effective for the exchange.

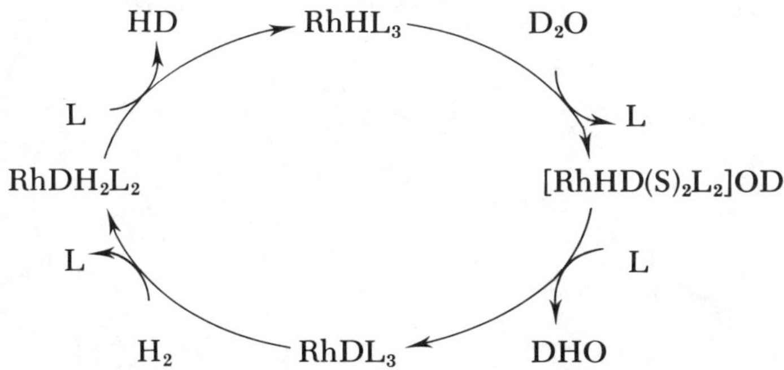

H_2 Evolution from Water by Rhodium Complexes

Since the transition metal compounds proposed for the catalytic photocleavage of water (*2*, *3*) appear to be inefficient, factors influencing the dihydrogen evolution from cis-dihydrido compounds deserve scrutiny. We know that dihydrogen is readily dissociated thermally from $[RhH_2(S)_2(PPh_3)_3]^+$ (S = acetone) (*9*, *10*). The complex cannot be formed from water. $RhH(PPh)_3$ lacks a sufficient nucleophilic character to form a cis-dihydro complex by adding water. The question then is how we could release H_2 from Complex **2a**. When heated to 90°C in dioxane, Complex **2a** merely decomposes into an intractable oil; irradiation (low-pressure Hg lamp) is also ineffective.

The two pyridine ligands of Complex **2b** then was displaced with bipyridine(bipy) to give $[RhH_2L_2(bipy)]^+$ (Complex **3**). The molecular structure of Complex 3 involving cis-dihydrido ligation is deduced readily from its IR (υ(Rh–H): 2080, 2135 cm^{-1}) and H-1 NMR (Rh–H, − 17.21(q), $J_{H-P} = J_{H-P}$ = 15.6 Hz; CH_3, 1.02(q), $^3J_{H-P} + {^5J_{H-P}}$ = 12.0 Hz, J_{H-H} = 60 Hz). Complex 3 was somewhat more stable than Complex **2b**.

On adding *t*-BuNC (0.3 mmol) to a THF solution of Complex **2b** (0.1 mmol), a stoichiometric amount of H_2 evolved with effervescence within a few minutes. The compound *trans*-{Rh(*t*-BuNC)$_2$[P(iso-Pr)$_3$]$_2$}BPh_4(4) was isolated (80%) from the reaction mixture (*see* Scheme I).

Scheme I.

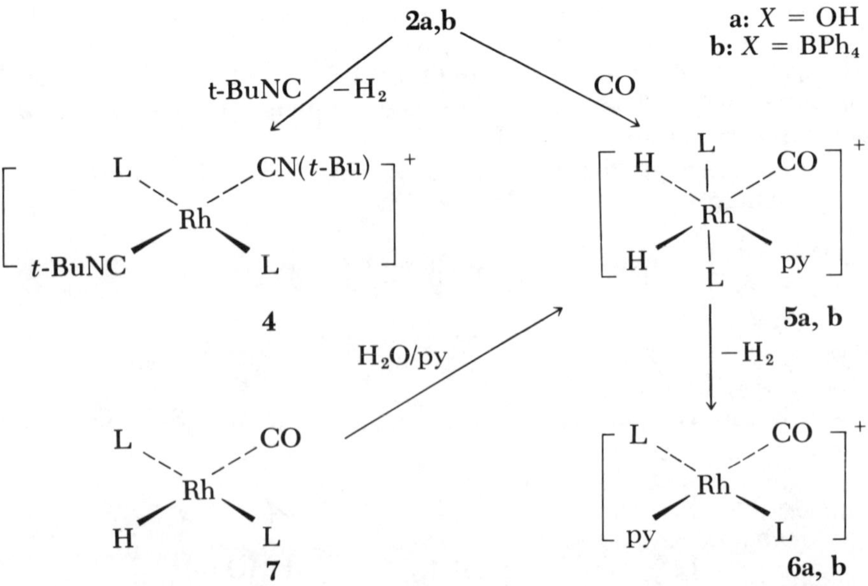

2a,b

a: *X* = OH
b: *X* = BPh_4

t-BuNC / $-H_2$

CO

$[Rh(L)_2(CN(t\text{-}Bu))(t\text{-}BuNC)]^+$ — L, CN(*t*-Bu), Rh, *t*-BuNC, L

4

$[RhH_2(L)_2(CO)(py)]^+$ — H, L, CO, Rh, H, L, py

5a, b

H_2O/py

$-H_2$

L, CO, Rh, H, L

7

$[Rh(L)_2(CO)(py)]^+$ — L, CO, Rh, py, L

6a, b

A reaction of CO with $[RhH_2(S)_2(PPh_3)_2]^+$ (S = py) reportedly gave $[Rh(CO)(S)(PPh_3)_2]^+$ (*9, 10*). A similar displacement of two hydrido ligands with CO was observed for Complex **2b** resulting in Structure **6**. Thus, brisk H_2 evolution occurs when CO is introduced into a THF solution of Complex **2b** yielding *trans*-$\{Rh(CO)(py)[P(iso\text{-}Pr)_3]_2\}BPh_4$ (Complex **6b**); IR v(CO), 1985 cm^{-1}. The reaction probably involves intermediate Complex **5** since an analogous compound, $[RhH_2(CO)(PEt_3)_3]^+$, is formed by treating $[RhH_2(PEt_3)_3]OH$, a water adduct of $RhH(PEt_3)_3$ with CO (*see* Scheme I).

From these results it is now clear that electron-donating ligands like pyridine or dipyridine, coplanar with the two hydride ligands, stabilize the Rh–H bonding in Complex **2** while electron-withdrawing ligands like RNC or CO destabilize it. The stability of $[RhH_2(CO)(PEt_3)_3]BPh_4$ deserves comment. It releases H_2 only when the THF solution is heated at 80°C. Due to experimental difficulties, we were unable to obtain unambiguous structural information for this compound. We tentatively assign the structure shown below.

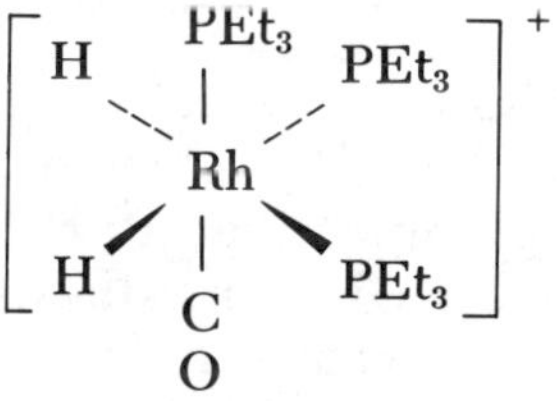

$$\left[\begin{array}{ccc} H & PEt_3 & PEt_3 \\ & Rh & \\ H & CO & PEt_3 \end{array}\right]^+$$

This complex shows v(Rh–H) at 2030, 2005 cm^{-1} which may be compared with those (2015, 2003 cm^{-1}) of $[RhH_2(PEt_3)_4]^+$ and those (2038, 1972 cm^{-1}) of $[RhH_2(PEt_3)_3]^+$. The postulated structure, involving the axial ligation of CO (cis to the hydride ligands), is consistent with the observed stability of the Rh–H bonds. The alternative structure with a coplanar CO ligand should destabilize the Rh–H bonds.

Cis-dihydrido-bicarbonato and -formato complexes, RhH_2BL_2 (B = HCO_3 and O_2CH, L = $P(iso\text{-}Pr)_3$) were prepared (*11*). The ν(Rh–H) bands of the former appear at 2120 and 2140 cm^{-1} and the latter at 2130 cm^{-1}. Consistent with these relatively higher ν(Rh–H) frequencies, neither compound releases H_2 at an ambient temperature. Dissociation of H_2 from these compounds occurs at a higher temperature (> 90°C).

A hydridocarbonyl compound, *trans*-$[RhH(CO)L_2]$ (Complex **7**, L = $P(iso\text{-}Pr)_3$), prepared from $RhHL_3$ and methanol, reacts with water in pyridine producing H_2 (> 70%) and Complex **6a** which could be confirmed by its crystalline BPh_4 salt. The formation of Complex **6a** from Complex **7** probably involves the intermediate Complex **5a** (*see* Scheme I).

The reaction of $RhHL_3$ (L = P(iso-Pr)$_3$) with CO was thought to give Complex **7**. Instead, the product formed from the reaction in dry hexane under a CO atmosphere was a binuclear Rh(0) carbonyl compound, $Rh_2(CO)_3L_3$ (Complex **8**, IR ν(CO): 1732, 1768, and 1957 cm^{-1}) (*12*). The similar reaction of $RhHL_3$ containing bulkier phosphines produced $Rh_2(CO)_4L_2$ (L = P(cyclo-C_6H_{11})$_3$, PPh(*t*-Bu)$_2$, P(*t*-Bu)$_3$). These dinuclear carbonyl compounds can be prepared by treating *trans*-RhH(CO)L_2 with CO at room temperature. The formation of $Rh_2(CO)_3L_3$ requires several steps, and a plausible sequence is proposed below.

$$RhHL_3 \xrightarrow{CO} trans\text{-}RhH(CO)L_2 \xrightarrow{CO} [RhH(CO)_2L_2]$$

$$\rightarrow \tfrac{1}{2}[L_2(CO)_2Rh(\mu\text{-}H)_2Rh(CO)_2L_2] \xrightarrow{-\frac{1}{2}H_2} \tfrac{1}{2}[Rh_2(CO)_4L_4]$$

$$Rh_2(CO)_4L_4 \xrightarrow{-CO,\ -L} Rh_2(CO)_3L_3$$

$Rh_2(CO)_3L_3$ (L = P(iso-Pr)$_3$) reacts with water producing H_2 and Complex **6a** which with CO regenerates Complex **7** producing CO_2. These observations imply that $RhHL_3$, $[RhH_2(py)_2L_2]^+OH^-$, RhH(CO)L_2, $[Rh(CO)(py)L_2]^+OH^-$, and $Rh_2(CO)_3L_3$ should be able to catalyze the water gas shift reaction. Indeed, these chloride-free systems proved to be the active catalysts, as we describe elsewhere (*12*).

Reduction of Carbon Dioxide

The reaction of $Rh_2H_2(\mu\text{-}N_2)$[P(cyclo-C_6H_{11})$_3$]$_4$ with a stoichiometric amount of CO_2 in wet THF gave a dihydridobicarbonato complex, *trans*-Rh(H_2)(O_2COH)[P(cyclo-C_6H_{11})$_3$]$_2$ (Complex **IXa**). Unexpectedly, the reaction of $Rh_2H_2(\mu\text{-}N_2)$-[P(cyclo-C_6H_{11})$_3$]$_4$ with an excess of CO_2 in wet THF at room temperature produced a carbonylbicarbonato compound, *trans*-Rh(CO)(O_2COH)[P(cyclo-C_6H_{11})$_3$]$_2$ (Complex **10a**). Similarly $RhH_2(O_2COH)$[P(iso-Pr)$_3$]$_2$ (Complex **9b**)

$$H_2RhL_2(\kappa^2\text{-}O_2C\text{—}OH) + CO_2 \longrightarrow H_2O + OC\text{—}RhL_2\text{—}O_2COH$$

9a, b **10a, b**

a: L = P(c − C_6H_{11})$_3$
b: L = P(i − Pr)$_3$

was obtained from $RhH[P(iso\text{-}Pr)_3]_3$. Facile reduction of CO_2 to CO was observed also when Complexes **9a** and **9b** was stirred with CO_2 (1 atm) in dry pyridine at room temperature producing H_2O and Complexes **10a** and **10b** (75%). The reaction of Complex **9b** with CO in toluene gave a μ-carbonato dimer (Complex **11**) in addition to Complex **10b**. Complex **11** is formed presumably by the intermolecular oxidative addition of Complex **10b**. Interestingly, the formation of Complex **11** was not observed for the reaction of Complex **9a** with CO_2 in pyridine.

$$2OC{-}\underset{L}{\overset{L}{Rh}}{-}O_2COH \xrightarrow{-H_2O} OC{-}\underset{L}{\overset{L}{Rh}}{-}O_2CO{-}\underset{L}{\overset{L\ \cdot\cdot CO}{Rh}}(H){-}O_2COH$$

10b

$$\rightarrow [(OC)RhL_2]_2(\mu{-}CO_3) + H_2CO_3$$

11

The reduction of CO_2 also occurred in the reaction with a cis-dihydridoformato complex, $RhH_2(O_2CH)L_2$, in pyridine. The major product here is $Rh(CO)(O_2CH)L_2$ and a small amount of Complex **11**.

$$H_2\underset{L}{\overset{L}{Rh}}(O_2CH) + CO_2 \xrightarrow{-H_2O} OC{-}\underset{L}{\overset{L}{Rh}}{-}O_2CH$$

Conclusions

The strong nucleophilicity of PtL_n and $RhHL_n$ endowed by trialkylphosphine ligands L allows facile oxidative addition of a water molecule to give the PtH(OH) and $RhH_2(OH)$ species, respectively. Their M–OH bonds tend to dissociate extensively in coordinating solvents. Dihydrogen can be released from the $RhH_2(OH)$ species by adding an electron-withdrawing ligand such as CO or RNC. The versatility of these water adducts has been demonstrated; e.g., they serve as catalysts for hydrating nitriles and activated olefins and for H–D exchange reactions between D_2O and H_2, activated C–H bonds, aromatic hydrocarbons, etc. Of particular interest may be the astonishingly facile reduction of CO_2 with the $H_2O/RhHL_n$ system.

Literature Cited

1. Balzani, V.; Moggi, L.; Manfrin, M. F.; Bolletta, F.; Gleria, M. *Science* **1975,** *189,* 852.
2. Brugger, P.-A.; Grätzel, M. *J. Am. Chem. Soc.* **1980,** *102,* 2461.
3. Miskowski, V. M.; Sigal, I. S.; Mann, K. R.; Gray, H. B.; Milder, S. J.; Hammond, G. S.; Rayson, P. R. *J. Am. Chem. Soc.* **1979,** *101,* 2027.
4. Eady, C. R.; Johnson, B. F. G.; Lewis, J. *J. Chem. Soc. Dalton* **1977,** 838.
5. Gillard, R. D.; Heaton, B. T.; Vaughan, D. H. *J. Chem. Soc. A* **1970,** 3126.
6. Yoshida, T.; Matsuda, T.; Okano, T.; Kitani, T.; Otsuka, S. *J. Am. Chem. Soc.* **1979,** *101,* 2027.
7. Yoshida, T.; Okano, T.; Saito, K.; Otsuka, S. *Inorg. Chim. Acta* **1980,** *44,* L135.
8. Yoshida, T.; Okano, T.; Thorn, D. L.; Tulip, T. H.; Otsuka, S.; Ibers, J. A. *J. Organometal. Chem.* **1979,** *181,* 183.
9. Schrock, R. R.; Osbourn, J. A. *J. Am. Chem. Soc.* **1971,** *93,* 2397.
10. Ibid., **1976,** *98,* 2134.
11. Yoshida, T.; Thorn, D. L.; Okano, T.; Ibers, J. A.; Otsuka, S. *J. Am. Chem. Soc.* **1979,** *101,* 4212.
12. Yoshida, T.; Otsuka, S. *J. Am. Chem. Soc.,* in press.

RECEIVED July 10, 1980.

9

Tricyclohexylphosphine Complexes of Ruthenium, Rhodium, and Iridium and Their Reactivity Toward Gas Molecules

BRIAN R. JAMES and MICHAEL PREECE

Department of Chemistry, University of British Columbia, British Columbia, Vancouver V6T 1Y6, Canada

STEPHEN D. ROBINSON

Department of Chemistry, King's College, Strand, London WC2R 2LS, England

Single-stage syntheses from commercially available chlorides and spectroscopic characterization are given for several tricyclohexylphosphine (PCy_3) complexes of ruthenium [$HRuCl(CO)(PCy_3)_2$, $Ru(CO)_3(PCy_3)_2$], rhodium [$HRhCl_2(PCy_3)_2$, trans-*$RhCl(CO)(PCy_3)_2$], and iridium [$HIrCl_2(PCy_3)_2$,* trans-*$IrCl(CO)(PCy_3)_2$]. The iridium(I) precursor [$IrCl(C_8H_{14})_2$]$_2$ has been used to yield $HIrCl_2(PCy_3)_2(dma)$ (dma = oxygen-bonded N,N'-dimethylacetamide), $H_2IrCl(PCy_3)_2$, and deuterated derivatives, $HIr(OH)Cl(CH_3CN)(PCy_3)_2$, $IrCl(C_6H_5CN)(PCy_3)_2$, and $IrCl(PCy_3)_2O_2$. Further data are presented on previously reported complexes such as $H_2IrCl[P(C_6H_9)Cy_2](PCy_3)$ containing a coordinated cyclohexenyl ring and $M(COD)Cl(PCy_3)$(M = Rh and Ir; COD = 1,5-cyclooctadiene). The carbonyls $H_2IrCl(CO)(PCy_3)_2$ and two isomers of $HIrCl_2(CO)(PCy_3)_2$ are characterized also. Attempts to isolate PCy_3 complexes containing CO_2 were unsuccessful.*

The work reported in this chapter has resulted from studies initiated by a NATO grant awarded for a project concerning activation of CO_2 by transition-metal complexes (*1*). The first structural

0065–2393/82/0196–0145$05.00/0

report of a transition-metal complex containing coordinated CO_2 described the bis(tricyclohexylphosphine)nickel(O) species, $Ni(CO_2)(PCy_3)_2$(*2*), and this led us to investigate the tricyclohexylphosphine complexes of ruthenium, rhodium, and iridium. This basic and bulky phosphine has been used widely (*3*) to stabilize species not usually isolable with less bulky phosphines such as triphenylphosphine, and is also a good ancillary ligand for a coordinatively unsaturated complex likely to react with small gas molecules. This had been well documented for rhodium (*4, 5*), and in situ solutions of $RhCl(PCy_3)_2$, for example, gave isolable adducts with O_2, N_2, C_2H_4, CO, and H_2 (*5*). At the time our studies were initiated little had been reported about ruthenium and iridium species, although carbonyls such as $RuCl_2(CO)(PCy_3)_2$, $HRuCl(CO)(PCy_3)_2$ (*6, 7, 8*), $Ru(CO)_3(PCy_3)_2$ (*9*), and $IrCl(CO)(PCy_3)_2$ (*4*) were known.

Since 1977 several papers on rhodium and iridium systems relevant to the present work have appeared (*10–16*). One of these by Vrieze's group (*15*), which reported partial dehydrogenation of tricyclohexylphosphine coordinated to iridium(I) and rhodium(I), overlapped with some of our studies which were reported almost simultaneously at a conference (*17*).

This chapter describes simple single-stage syntheses of a range of hydridocarbonyl and hydridocarbonyl tricyclohexylphosphine complexes of ruthenium, rhodium, and iridium, attempts at forming CO_2 complexes from rhodium(I) and iridium(I) precursors, and dehydrogenation of the cyclohexyl ring of the phosphine. It also reports preliminary data on the formation of a hydrido(hydroxo) species by oxidative addition of water to an iridium(I) complex and the characterization and chemistry of a dioxygen complex, $IrCl(O_2)(PCy_3)_2$.

We have published a brief report (*18*) on some of the studies described in this chapter.

Results and Discussion

The spectroscopic data for some of the synthesized complexes are given in Table I. Several of the syntheses were of the single-stage, small-scale type developed by one of us (*19, 20*) for preparing tertiary-phosphine-type complexes of Group VIII platinum metals. The method involves rapid mixing of refluxing ethanolic solutions of a commercially available halide and the phosphine, sometimes in the presence of KOH, $NaBH_4$, or HCHO.

Ruthenium Complexes. The $HRuCl(CO)(PCy_3)_2$ Complex **1**, like the $HRuCl(CO)P_2$ analogues ($P = PBu^t_2Me$, PBu^t_2Et) (*21*), is air-sensitive, but unlike the latter it is light stable. The high-field H-1 NMR is a triplet indicating the equivalence of the two phosphines, which is confirmed by the singlet in the P-31 NMR. The high ν(Ru–

Table I. Spectroscopic Data for Some of the Synthesized Complexes

Complex[a]	$\nu(CO)$[b]	$\nu(M-H(D))$[b]	$\nu(M-Cl)$[b]	$\tau(M-H)$[c]	$\delta(P\text{-}31)$[c]	$J_{(P-H)}$[c]	Ref.[d]
1. $HRuCl(CO)P_2$	1900	2107	295	34.5 (t)	+ 46.7 (s)	18 Hz	TW
	1905	2030	295	34.6 (t)	—	17.5 Hz	*6, 8, 23*
2. $Ru(CO)_3P_2$	1870				+ 63.5 (s)		TW
	1873				—		*9*
3. $HIrCl_2P_2$		not found	317	42.3 (t)	+ 20.9 (s)	11 Hz	TW
			317	—	—	—	*4*[e]
4. $HIrCl_2P_2(dma)$	1628	2300	300				TW
5. $HIrCl_2(CO)P_2$	1980	2105	310	18.2 (t)	+ 2.3 (s)	15 Hz	TW
6. $HIrCl_2(CO)P_2$	2010	2275	310, 260	26.1 (t)	− 0.7 (s)	12 Hz	TW
	2015	2276	310, 255	—	—	—	*4*
7. H_2IrClP_2		2230	287	42.3 (t)	+ 41.4 (s)	12 Hz	TW
		2240	285	43	+ 41.9	12.5 Hz	*15*
8. D_2IrClP_2		1605	287		+ 41.6 (s)		TW
9. $HDIrClP_2$		1605, 2230	287	42.4 (t)	—	12 Hz	TW
10. $H_2IrCl(CO)P_2$	1970	2200, 2100	255	30.1 (d. of t)[f]	+ 24.2 (s)	13 Hz	TW
				18.4 (d. of t)		18 Hz	
12. H_2IrClP^*P[g]		2210, 2125	255	33.2, 20.5	+ 32.2, + 12.3[h]		TW
				35.3, 17.2	+ 33.6, + 11.5[h]		
		2210, 2120	250	—	+ 30.6, + 14.3[i]		*15*
					+ 33.3, + 13.6[j]		
16. $HRhCl_2P_2$		1943	345	40.5 (d. of t)[k]		12 Hz	TW
		—	342	—		—	*13*
17. $RhCl(COD)P$			275		+ 26.1 (d)[l]		TW

Note: Data for $HDIrCl(CO)P_2$ (**11**), $IrCl(C_6H_5CN)P$ (**13**), $HIr(OH)Cl(CH_3CN)P_2$ (**14**), $IrCl(O_2)P_2$ (**15**), and *trans*-$IrCl(CO)P_2$, $IrCl(COD)P$, and *trans*-$RhCl(CO)P_2$ are discussed in the text. [a] P = tricyclohexylphosphine. [b] IR recorded as Nujol mulls on Perkin Elmer Model 457 (centimeters^{-1}). [c] NMR data recorded in $CDCl_3$ or C_6D_6 solution; H-1 (100 MHz) relative to TMS; P-31–{H-1} (40.5 MHz) relative to 85% H_3PO_4, downfield shifts being positive; s = singlet; d = doublet; t = triplet. [d] TW = this work. [e] Data quoted for $IrCl_2P_2$ (*4*). [f] $^2J_{(H_A-H_B)} = 5$ Hz. [g] $P^* = P(C_6H_9)Cy_2$ [h] AB quartet; $J_{(P-P)} = 336$ Hz. [i] AB quartet; $J_{(P-P)} = 335$ Hz. [j] AB quartet; $J_{(P-P)} = 337$ Hz. [k] $J_{(Rh-H)} = 30$ Hz. [l] $J_{(Rh-P)} = 141.9$ Hz; olefinic protons detected at τ values of 4.77 and 6.46, trans to phosphorus and chlorine, respectively.

Cl), 295 cm^{-1}, is indicative of chloride trans to CO (*22*), and thus the stereochemistry shown in Structure 1 is favored, although a trigonal bypyramidal structure with axial phosphines cannot be ruled out (*21*). The $HRuCl(CO)(PCy_3)_2$ complex prepared earlier (*6, 8, 23*) appears to be the same isomer judging from the high-field H-1 NMR (*23*) and the ν(CO) and ν(Ru–Cl) values (*6, 8*), although our ν(Ru–H) at 2107 cm^{-1} is much higher than the 2030-cm^{-1} value which was reported (*6*).

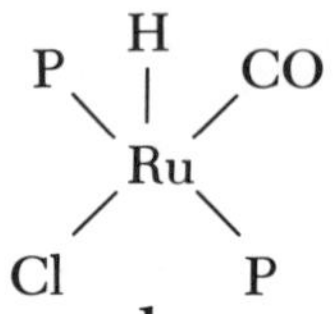

```
     H
P    |    CO
  \  |  /
    Ru
  /    \
Cl       P
     1
```

On standing in air, the carbonyl band shifts to 1940 cm^{-1} (cf. 1935 cm^{-1} in Ref. *6*) and the hydride band to 2040 cm^{-1}, while a broad band appears centered around 1150 cm^{-1}, together with a weak band at 1220 cm^{-1}, which are almost certainly due to phosphine oxide (*24*). The brown oxidation product has not been isolated in a pure state but a hydridocarbonyl containing coordinated phosphine oxide is indicated strongly. We have not been able to detect a possible dioxygen intermediate even at low temperatures in solution, although the osmium analogue $HOsCl(CO)(PCy_3)_2O_2$ is characterized well (*8*).

Complex **1** is known (*8*) to form complexes with CO, SO_2, and CS_2 (which inserts to give a dithioformate), but we found no reactivity toward CO_2 in toluene solution.

Complex **1** can be synthesized also in somewhat lower yields by simply refluxing $RuCl_3 \cdot 3H_2O$ and PCy_3 in methanol. In contrast to the PPh_3 reaction which yields $RuCl_2(PPh_3)_3$ (*25*), the PCy_3 appears basic enough to promote the base-catalyzed carbonyl abstraction reaction.

The $Ru(CO)_3(PCy_3)_2$ complex, **2**, is characterized by a single, strong ν(CO) centered at 1870 cm^{-1}, and the singlet in the P-31 NMR shows equivalent phosphines; like the PPh_3 analogue (*26, 27*), a trigonal bipyramidal structure with equatorial CO ligands is indicated.

Iridium Complexes. The air-stable, rose-colored monohydride $HIrCl_2(PCy_3)_2$, Complex **3**, may be prepared directly from a commercially available chloride, or by adding HCl to a toluene solution containing the cyclooctene dimer $[IrCl(COT)_2]_2$ and PCy_3. The six-coordinate, yellow Complex **4** containing oxygen-bonded dma, ν(CO) 1628 cm^{-1} (*28*), also is isolated readily.

The ν(Ir–H) of Complex **3** is not observed; however Shaw's group (*29*) has noted also that the metal hydride stretch is weak or unobservable in five-coordinate, square pyramidal iridium and rhodium hydrides containing bulky phosphines. Based on the high-field H-1 NMR triplet, the P-31 singlet, and the single ν(Ir–Cl), Complex **3** is assigned the square pyramidal structure shown in Reaction 1; a trigonal bipyramidal structure with axial chlorides satisfies the spectral data, but it is not favored on steric grounds or by the findings of Shaw's group (29).

Complex **4** shows the ν(Ir–H) at 2300 cm^{-1} which is relatively high and presumably reflects the weak trans-influence of the amide ligand; Complex **4** in benzene or CH_2Cl_2 readily dissociates to Complex **3** and free dma. The deep red solutions of Complex **3** react further with gaseous HCl to give yellow solutions. This reaction is reversed readily when HCl is removed, but the nature of the forward reaction (possibly oxidative addition or addition of chloride to the vacant coordination site) has not been elucidated.

A diamagnetic rose complex formulated as the iridium(II) species $IrCl_2(PCy_3)_2$, ν(Ir–Cl) 317 cm^{-1}, prepared from aqueous isopropyl alcohol solutions containing $IrCl_6^{3-}$ and PCy_3 (*4*), must be the hydride complex, **3**.

Adding CO to Complex **3** gives the white, hydridocarbonyl $HIrCl_2(CO)(PCy_3)_2$. The equivalent phosphines (P-31 NMR), single ν(Ir–Cl),

```
      H                  H                  H
  P   |   Cl         P   |   Cl         P   |   Cl                 P       Cl
    \ | /              \ | /              \ | /                      \   /
     Ir     --CO-->     Ir      -->        Ir      <==-HCl/+HCl==>    Ir        (1)
    /   \              / | \              / | \                      /   \
  Cl     P          Cl   |   P         CO   |   P                 CO       P
                         CO                 Cl
      3                  5                  6
```

and dramatic lowering of the τ value to 18.2, indicating the introduction of a high trans influence ligand trans to the hydride, are all consistent with Structure **5**. On standing in the solid state or in solution Complex **5** isomerizes to Complex **6**, and there is also loss of HCl to yield the known compound *trans*-$IrCl(CO)(PCy_3)_2$ (*4, 15, 30*). During the isomerization, the P-31 chemical shift remains essentially unchanged indicating retention of mutually trans phosphine ligands, ν(Ir–Cl) splits into two bands, and the ν(Ir–H) and ν(CO) shifts to higher wave numbers. The data for Complex **6** are consistent with the geometry shown, which is that postulated on IR evidence alone for the identical compound made via addition of HCl to *trans*-$IrCl(CO)(PCy_3)_2$ (*4*). Our synthesis of this Vaska-type complex from $[IrCl(COT)_2]_2$ offers an alternative to that given in the literature (*4, 30*).

The 1,5-cyclooctadiene complex $IrCl(COD)PCy_3$ is formed when phosphine is added to the easily made and handled $[HIrCl_2(COD)]_2$

dimer (*31*). The phosphine is sufficiently basic to remove the HCl and accomplish the familiar bridge-splitting reaction. The complex has been made previously via the somewhat more elusive $[IrCl(COD)]_2$ complex and was characterized by proton NMR (*14*). The phosphine appears as a singlet, δ(P-31) + 14.1.

The chemistry of in situ $Ir(I)/PCy_3$ systems is complicated by an interesting dehydrogenation of a cyclohexyl ring (vide infra), but under an H_2 atmosphere $[IrCl(COT)_2]_2/PCy_3$ solutions readily yield the dihydride complex $H_2IrCl(PCy_3)_2$, **7**, as discovered independently by Vrieze's group (*15*). The spectroscopic data indicate a trigonal bypyramidal structure with axial phosphines (*15*). We have monitored formation of the dihydride at 20°C by following an irreversible absorption of 1.0 mol H_2/iridium, and the noted hydrogenation of the cyclooctene (*15*) subsequently takes place. Catalytic hydrogenation of monoenes using Complex **7** under ambient conditions occurs but very slowly.

An attempted synthesis of the deuterated analogue of Complex **7** using a D_2 instead of a H_2 atmosphere leads to the expected dideuteride, Complex **8**, together with appreciable amounts of the dihydride and HD complex, 9. The IR ν(Ir–H and Ir–D) clearly reveals a hydride and a deuteride, while the high-field H-1 NMR shows two distinct triplets at τ42.3 and τ42.4, assignable to $H_2IrCl(PCy_3)_2$ and $HDIrCl(PC_3)_2$, respectively.

Further evidence for the presence of the HD complex comes from carbonylation studies. $H_2IrCl(PCy_3)_2$ reacts with 1 atm of CO to give $H_2IrCl(CO)(PCy_3)_2$ defined unambiguously by its spectroscopic parameters as Complex **10**. The P-31 singlet indicates equivalent phosphines, while the two doublets of triplets at τ30.1 and τ18.4 are assigned to H_A(trans to chlorine) and H_A(trans to CO), respectively. Carbonylation of the product obtained from the deuteration reaction gives Species **11** in which the $J_{(H-H)}$ coupling is lost in the two high-field resonances and each six-line pattern simplifies to a triplet (*see* Figure 1). Species **11** must be mainly the two isomers of $HDIrCl(CO)(PCy_3)_2$. Since H–D couplings are generally about one-seventh of H–H couplings, such interactions are not usually resolvable.

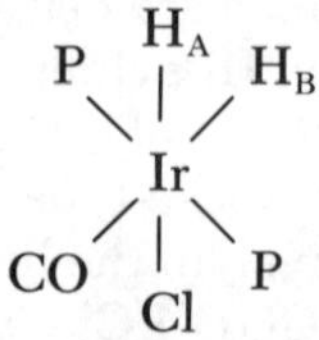

```
       HA                    H(D)
  P    |    HB          P    |    D(H)
    \  |  /               \  |  /
       Ir                    Ir
    /  |  \               /  |  \
 CO    |    P          CO    |    P
       Cl                    Cl

       10                    11
```

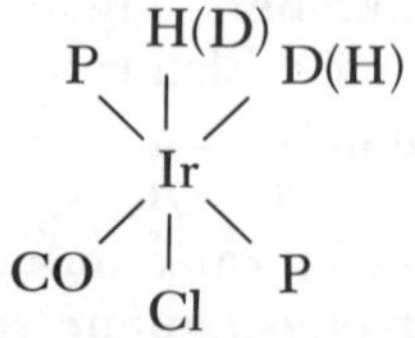

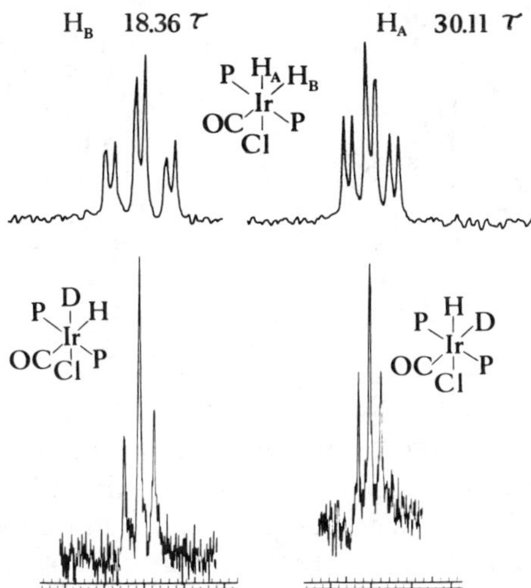

Figure 1. High-field H-1 NMR for $H_2IrCl(CO)(PCy_3)_2$ (Complex **10***) and isomers of $HDIrCl(CO)(PCy_3)_2$ (Complex* **11***)*

The mechanism of the HD exchange must involve the ligand cyclohexyl ring, presumably via a C–H cleavage (oxidative addition; *see* Refs. *32* and *33*) within the initially formed five-coordinate $D_2IrCl(PCy_3)_2$ complex. The IR of the deuterated product shows several bands of $H_2IrCl(PCy_3)_2$ shifted by about $\sqrt{2}$ (e.g. $837 \rightarrow 600\ cm^{-1}$ and $795 \rightarrow 566\ cm^{-1}$). Treatment of the dihydride with deuterium leads to the dideuteride (and vice versa), although dihydride formation is essentially irreversible under the same conditions. This again shows that the exchange process must involve the phosphine. There is no direct evidence for an ortho-metallated species since the P-31 NMR of $H_2IrCl(PCy_3)_2$ remains as a singlet even at −85°C (*15*). The coordinatively saturated complex $H_2IrCl(CO)(PCy_3)_2$ shows no exchange with D_2.

$IrCl(PCy_3)_2$ Chemistry and Dehydrogenation of a Cyclohexyl Ring. An investigation into the chemistry of $IrCl(PCy_3)_2$ species seemed warranted in view of the reactivity of the rhodium analogue (*5*, *12*) and the $IrCl(PPh_3)_2$ species (*34*) toward small molecules.

Treatment of $[IrCl(COT)_2]_2$ with PCy_3 in toluene at 20°C under argon yields $H_2IrCl[P(C_6H_9)Cy_2](PCy_3)$, Complex **12**, in which one cyclohexyl ring has been dehydrogenated to a cyclohexene, the double bond of which is coordinated to the iridium. This interesting reaction was discovered independently by Vrieze's group (*15*). However these workers had used refluxing conditions and this had led to loss of H_2 and formation of four-coordinate iridium(I) species $IrCl[P(C_6H_9)Cy_2](PCy_3)$ as a mixture of isomers with 4.5- and 5.5-

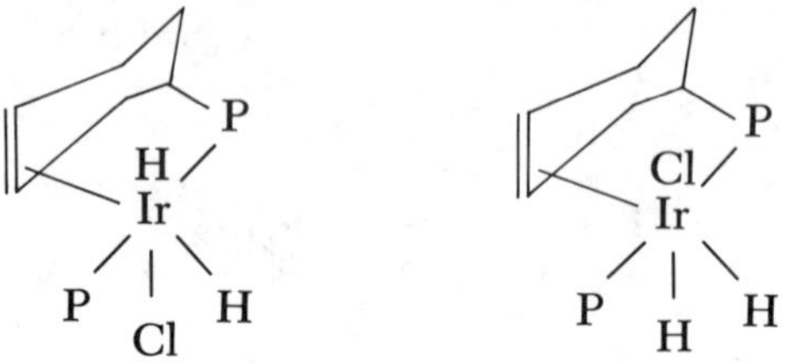

12

membered rings. Treatment of this mixture with H_2 gave the two 5.5-membered-ring cis-dihydrides shown in Complex **12**, as well as a dihydride of a 4.5-membered ring. Our spectroscopic data for Complex **12** are in excellent agreement with theirs; the P-31 is well-resolved at 20°C, whereas the Dutch workers obtained a more complex reaction mixture which required lower temperatures for resolution. We also were able to obtain the high-field but poorly resolved τ values at 20.5 and 33.2 for the dominant isomer of Complex **12** (80%), and at 17.2 and 35.3 for the other isomer (20%), the higher τ value in each case probably referring to the hydride which is trans to chloride. Crabtree et al. (*35*) have reported recently on dehydrogenation of cycloalkanes using cationic iridium complexes containing triphenylphosphine.

There is no hydrogenation of the double bond in Complex **12** (the reverse of its synthesis) using an H_2 atmosphere up to 80°C, although such hydrogenation is well documented for unsaturated phosphines such as *o*-vinylphenyldiphenylphosphine when coordinated at rhodium (*36*). Dihydrogen transfer within species such as $[H_2Ir(diene)(PPh_3)_2]^+$ has been demonstrated (*37*), but here both hydrogens are cis to a coordinated olefinic bond, whereas in Complex **12**, one hydrogen is trans. For an overall cis-addition of hydrogens to an olefin within an octahedral olefin–dihydride intermediate, it is not necessary to have both hydrogens cis to the olefin. The hydrogens are transferred successively and it is clear in some cases that an isomerization within an alkyl–hydride intermediate is necessary to yield a species with cis-disposed alkyl and hydride ligands prior to reductive elimination of saturated product (*38*). It is possible that any such isomerization pathway following an initially formed hydridoalkyl from Complex **12** is difficult.

The $IrCl(PCy_3)_2$ moiety could be stabilized in the solid state by isolation as the benzonitrile complex $IrCl(C_6H_5CN)(PCy_3)_2$, Complex **13**, which was characterized by microanalysis and IR, ν(coordinated CN) 2250 cm^{-1}. No intramolecular dehydrogenation is apparent, although the nitrile readily dissociates in solution and Complex **12** is detected by NMR.

Reversible Splitting of Water: $HIr(OH)Cl(CH_3CN)(PCy_3)_2$ and $IrCl(O_2)(PCy_3)_2$. While attempting to synthesize $IrCl(CH_3CN)(PCy_3)_2$ from the cyclooctene precursor, we inadvertently isolated a white solid that appears to be $HIr(OH)Cl(CH_3CN)(PCy_3)_2$, Complex **14**. The ν(CN) is found at 2282 cm^{-1} and a strong, sharp peak at 3490 cm^{-1} with no accompanying band around 1630 cm^{-1} is attributed to coordinated hydroxide rather than to water. A 290-cm^{-1} band is attributed to ν(Ir–Cl). A hydride band is not observed in the IR because it is either too weak or obscured by the ν(CN). However high-field H-1 NMR measured immediately on dissolution in C_6D_6 sometimes shows a resonance at τ34.3 that rapidly decays with the appearance of others at τ values of 19.8 and 31.5, attributable to Complex **12**, whose formation also was confirmed by P-31 NMR. On dissolution in CH_2Cl_2, Complex **14** also yields stoichiometric amounts of CH_3CN and H_2O, as evidenced by gas chromatography. Reaction 2 summarizes the reversible formation of Complex **14**, which is thought to be produced by oxidative addition of trace water to an in situ $IrCl(PCy_3)_2$ species.

$$\underset{\mathbf{14}}{HIr(OH)Cl(CH_3CN)(PCy_3)_2} \rightleftharpoons H_2O + CH_3CN + IrCl(PCy_3)_2 \downarrow \underset{\mathbf{12}}{H_2IrCl[P(C_6H_9)Cy_2](PCy_3)} \qquad (2)$$

Such oxidative addition of water to platinum(O) and rhodium(I) phosphine complexes, including cyclohexylphosphine derivatives, has been demonstrated by Otsuka and co-workers (*16*, *39*, *40*) and some of the systems have involved trace water in carefully dried solvents (*16*).

One disturbing result concerning our formulation of Complex **14**, however, is that the use of CH_3CN doped with D_2O again gives the hydroxo product without any incorporation of deuterium in the ligand which is thought to be derived from the H_2O splitting! There could be exchange with the PCy_3 ligands, although it is difficult to discern any isotope shifts in the IR. Whether an Ir–H or Ir–D bond is present is more equivocal because of the nonappearance of ν(Ir–H) in the IR and the uncertainty of its detection by NMR. Exchange via cationic species such as $[HIrCl(CH_3CN)(PCy_3)_2]^+OH^-$, involving C–H cleavage in the acetonitrile (*41*), is a possibility. Further work is needed to clarify this problem. The formation of H_2O and CH_3CN on dissolution of Complex **14** argues strongly for the suggested formulation.

Further indirect evidence for formation of Complex **14** via $IrCl(PCy_3)_2$ and Reaction 2 is the isolation of the purple dioxygen adduct $IrCl(O_2)(PCy_3)_2$, Complex **15**, in high yield by treatment of Complex **14** with O_2. On dissolution in CH_2Cl_2, the white solid Complex **14** yields an orange solution (in situ $IrCl(PCy_3)_2$) that under O_2 rapidly adopts the purple color of the dioxygen complex (1.0 mol O_2 is absorbed per mole of iridium.) The ν(O–O) at 940 cm^{-1} is unusually high for a Group VIII metal peroxo species (*42*), but the rhodium analogue $RhCl(O_2)(PCy_3)_2$ shows this band at the even higher frequency of 993 cm^{-1}. This was rationalized on the basis of very weak π-back-bonding (Rh–O) within a monomeric, square-planar structure containing side-on coordinated dioxygen (*12*). The single ν(Ir–Cl) at 340 cm^{-1} (cf. at 328 cm^{-1} for the rhodium complex) is consistent with the monomeric formulation, although we cannot rule out entirely a dimeric formulation containing bridging dioxygen moieties, as found for $[RhCl(O_2)(PPh_3)_2]_2$ (*43*). We have not been able to get NMR data for Complex **15** because of limited solubility in solvents and a rapid decomposition of such solutions to give phosphine oxide, Species **12** (via $IrCl(PCy_3)_2$), and other unidentified products.

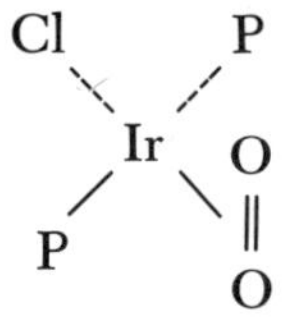

Cl P Ir O P O

15

The dioxygen complex shows oxygen transfer reactions characteristic of species containing coordinated peroxide (*42*), and, for example, we have isolated a sulfato complex from the reaction with SO_2 and a carbonato complex from a reaction with CO_2 in the presence of excess phosphine, and we have detected a peroxycarbonato species from the reaction with CO_2 (*44*). A reaction with H_2 simply replaces the O_2 to yield $H_2IrCl(PCy_3)_2$, Complex **7** (*44*).

The dioxygen complex is light sensitive, but it is otherwise stable in air. The O_2 is not removed on pumping and this adds further indirect support (*45*) against a widely accepted correlation between O–O bond strength (and length) and O_2 lability (*46*). The high ν(O–O) value could be attributed to a weak metal–oxygen interaction and a labile O_2 moiety (*46*), but such reasoning is based on the incorrect assumption that ν(O–O) is a pure mode (*45*). The solid state, light-sensitive reaction (sunlight, 2–3 days), which results in a color change from purple to brown, is accompanied by a decrease in intensity of the 940-cm^{-1} band and the appearance of bands with increasing intensity at 1135 cm^{-1}, almost certainly due to phosphine oxide (*24*) and 780 cm^{-1}. The latter could possibly be an Ir = O stretch (*47*). Reactions shown in

Equation 3 are indicated and are analogous to the one reported recently by Ledon et al. for a molybdenum porphyrin system (*see* Reaction 4) (*48, 49*).

$$IrCl(O_2)(PCy_3)_2 \xrightarrow{h\nu} O = IrCl(PCy_3)(OPCy_3) \text{ or } IrCl(OPCy_3)_2 \quad (3)$$

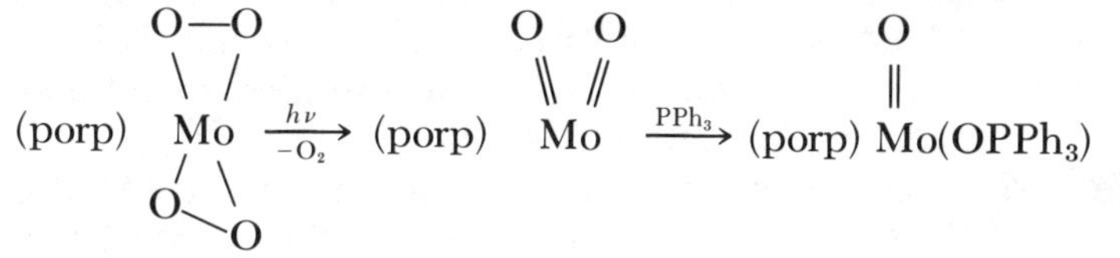

$$(porp)\,Mo(O_2)_2 \xrightarrow[-O_2]{h\nu} (porp)\,Mo(=O)_2 \xrightarrow{PPh_3} (porp)\,Mo(=O)(OPPh_3) \quad (4)$$

Such reactions are interesting as models for oxygen atom transfer in mechanisms of dioxygen activation by cytochrome P450 enzyme systems (*49, 50, 51*).

Attempts to synthesize Complex **15** by reacting O_2 with either in situ $[IrCl(COT)_2]_2/PCy_3$ systems or $IrCl(C_6H_5CN)(PCy_3)_2$ gave very low yields.

Rhodium Complexes. The complexes $HRhCl_2(PCy_3)_2$, Complex **16**, and *trans*-$RhCl(CO)(PCy_3)_2$ are made readily via single-stage syntheses involving solutions of $RhCl_3 \cdot 3H_2O$ and PCy_3. The hydride was obtained previously by adding HCl to the isolated $RhCl(PCy_3)_2$ complex (*13*). The presence of the hydride had been inferred, although we detected this in the IR and in the NMR. The doublet of triplets observed in the NMR together with the single ν(Rh–Cl) is consistent with the square pyramidal stereochemistry (cf. Complex **3**) originally formulated (*13*). The hydride is photosensitive, being converted to the well-characterized, square-planar, green rhodium(II) species, *trans*-$RhCl_2(PCy_3)_2$ (*13*). The synthesis of $RhCl(CO)(PCy_3)_2$ is more convenient than that described in the literature (*4*), and our spectroscopic data are the same as those previously reported (*4, 15*). A second convenient synthesis from CH_2Cl_2/methanol solutions of $[RhCl(COD)]_2$ and PCy_3 must involve a base (phosphine)-promoted carbonyl abstraction reaction, since in the absence of the methanol a simple bridge-splitting reaction results in the formation of $RhCl(COD)(PCy_3)$, Complex **17**. This COD complex has been synthesized previously but spectroscopic data were not reported (*52*).

Forming $H_2RhCl(PCy_3)_2$ from $Rh(I)/PCy_3$ solutions under H_2 is well documented (*5, 13, 15*), and Vrieze et al. (*15*) also have noted its formation in the absence of H_2 together with the dehydrogenated species $RhCl[P(C_6H_9)Cy_2](PCy_3)$ containing a 4.5-membered ring (cf. Structure **12**). We find that toluene solutions of $[RhCl(COT)_2]_2$ with PCy_3 at 20°C under argon also yield some of the monohydride $HRhCl_2(PCy_3)_2$.

Formation of Carbon Dioxide Complexes. As mentioned in the introduction, our initial interest in synthesizing the PCy_3 complexes was in their potential for binding CO_2. However, except for the formation of peroxycarbonate and carbonato complexes from $IrCl(O_2)(PCy_3)_2$ (*44*), which is well-established chemistry for some platinum metal peroxide complexes (*42*) (but, to our knowledge, not with PCy_3 systems), we have not been able to isolate any CO_2 complexes or even carbonate or bicarbonate species which are formed sometimes in the presence of adventitious water (*16*).

We were hopeful when toluene solutions of the $[MCl(COT)_2]_2$ complexes (M = Rh and Ir), in the presence of PCy_3 at 25°C under 1 atm CO_2 (phosphine: dimer = 4) absorbed 1 mol CO_2 per metal atom. The rhodium solution, for example, changed gradually from yellow to red and, in the constant pressure apparatus used, the measured gas uptake analyzed excellently for the pseudo-first-order rate law $k[Rh]$, with $k = 9.3 \times 10^{-4} s^{-1}$. However, such CO_2 uptake was observed in only five experiments; twenty others revealed no reactivity whatsoever! The reasons for the irreproducibility have not been traced; water content and visible light are not problems. The systems certainly are complicated by the "blank" dehydrogenation reactions (vide supra).

Spectroscopic analysis of the products obtained from the rhodium/CO_2 solutions showed the presence of $HRhCl_2(PCy_3)_2$ [ν(Rh–H) 1943 cm^{-1}, τ40.5], its photodecomposition product, $RhCl_2(PCy_3)_2$, and $RhCl(CO)(PCy_3)_2$ [ν(CO) 1942 cm^{-1}]. The hydride is formed in the absence of CO_2 (vide supra), while the carbonyl could be formed by Reactions 5 (*53, 54*) or 6 (*16*).

$$M(PCy_3) + CO_2 \rightarrow M(CO) + OPCy_3 \tag{5}$$

$$H_2M + CO_2 \rightarrow M(CO) + H_2O \tag{6}$$

(Reaction 6 has been documented for $H_2Rh(O_2COH)(PCy_3)_2$ (*16*).) Other IR bands at 1580(s) and 1235(m) cm^{-1} that result from the solution CO_2 reactions could arise from a CO_2 complex (*54*), a formate complex formed via CO_2 insertion into Rh–H (*16*), or even a bicarbonate formed via Reaction 7, which has been documented again for a rhodium/PCy_3 system (*16*).

$$HM + CO_2 \xrightarrow{H_2O} H_2M(O_2COH) \tag{7}$$

The iridium/CO_2 reactions gave products including the dihydride Complex **12** [ν(Ir–H) 2210, 2125, τ17, τ35] and $IrCl(CO)(PCy_3)_2$ [ν(CO) 1930 cm^{-1}]. Bands at 1755(s) and 1235(w) cm^{-1} are assigned more readily to coordinated CO_2 (*54, 55*) than formate or bicarbonate (*16*).

During our frustrating studies on the in situ rhodium/PCy_3 systems, a note appeared describing isolation of $RhCl(CO_2)(PBu_3^n)_2$ from the reaction of CO_2 with in situ $[RhCl(C_2H_4)_2]_2/PBu_3^n$ solutions (*54*). Few details were reported but we found no evidence for any reaction of $[RhCl(olefin)_2]_2/PBu_3^n$ solutions with 1 atm CO_2 at 20°C (olefin = C_2H_4, COT); for example, with $[RhCl(C_2H_4)_2]_2/4PBu_3^n$ under CO_2, 2.0 mol C_2H_4 are evolved rapidly per mole of rhodium and there is no subsequent CO_2 absorption over several hours.

The study of CO_2 complexes of platinum metal complexes remains an enigma, at least in our hands.

Experimental

General. Dried, degassed reagent-grade solvents were used throughout, and dma was purified as described previously (*56*). Complexes were made under inert atmospheres using Schlenk-tube techniques and were washed with $C_2H_5OH/H_2O/C_2H_5OH/$ hexane before drying unless stated otherwise. Microanalyses were done by P. Borda of the chemistry department of the University of British Columbia. The constant-pressure apparatus used for measuring rates and stoichiometries of gas absorption or evolution has been described elsewhere (*56*).

Tricyclohexylphosphine was obtained from Strem Chemicals. Ruthenium, rhodium, and iridium trichlorides were obtained as trihydrates from Johnson, Matthey Limited. Iridium tetrachloride was obtained from Platinum Chemicals. The precursor complexes $[RhCl(COD)]_2$ (*57*), $[RhCl(COT)_2]_2$ (*58*), $[RhCl(C_2H_4)_2]_2$ (*59*), $[IrCl(COD)]_2$ (*14*), and $[HIrCl_2(COD)]_2$ (*31*) were made according to the literature procedures.

Ruthenium Complexes. CARBONYLCHLOROHYDRIDOBIS(TRICYCLOHEXYLPHOSPHINE)RUTHENIUM(II), $HRuCl(CO)(PCy_3)_2$, COMPLEX 1. Ethanol solutions of $RuCl_3 \cdot 3H_2O$ (0.2 g in 7 mL) and $NaBH_4$ (0.15 g in 7 mL) were added quickly and successively to a stirring, boiling ethanol solution of PCy_3 (1.1 g in 40 mL). Refluxing for 10 min precipitated a mustard-colored powder (75%). We found: carbon, 60.8; hydrogen, 9.0. $C_{37}H_{67}OClP_2Ru$ requires carbon, 61.2; hydrogen, 9.3.

TRICARBONYLBIS(TRICYCLOHEXYLPHOSPHINE)RUTHENIUM(O), $Ru(CO)_3(PCy_3)_2$, COMPLEX 2. Solutions of $RuCl_3 \cdot 3H_2O$ (0.07 g in 5 mL 2-methoxyethanol), HCHO (5 mL, 40% aqueous), and $NaBH_4$ (0.05 g in 5 mL 2-methoxyethanol) were added quickly and successively to a stirring, boiling 2-methoxyethanol solution of PCy_3 (0.4 g in 15 mL). Refluxing for 30 min yielded pale yellow microcrystals (30%). We found: carbon, 62.2; hydrogen, 8.4. $C_{39}H_{66}O_3P_2Ru$ requires carbon, 62.8; hydrogen, 8.8.

Iridium Complexes. DICHLOROHYDRIDOBIS(TRICYCLOHEXYLPHOSPHINE)IRIDIUM(III), $HIrCl_2(PCy_3)_2$, COMPLEX 3. (a) A solution of $IrCl_4$ in 2-methoxyethanol (0.1 g in 5 mL) was added quickly to a stirring, boiling 2-methoxyethanol solution of PCy_3 (0.4 g in 15 mL) to yield rose-colored crystals (80%). We found: carbon, 52.3; hydrogen, 8.3. $C_{36}H_{67}Cl_2P_2Ir$ requires carbon, 52.4; hydrogen, 8.2. (b) Tricyclohexylphosphine (0.3 g) and $[IrCl(COT)_2]_2$ (0.22 g) were dissolved in 5 mL toluene to give an orange solution through which HCl gas was bubbled; the solution immediately turned red and then yellow, but on standing the red color returned. We concentrated the solution and added hexane to induced crystallization of the product (65%).

DICHLORO(*N,N'*-DIMETHYLACETAMIDE)HYDRIDOBIS(TRICYCLOHEXYLPHOSPHINE)IRIDIUM(III), $HIrCl_2(PCy_3)_2(dma)$, COMPLEX 4. The dma · HCl adduct was prepared as a white solid by bubbling HCl gas through dma. This adduct (0.1 g) was added to a solution of PCy_3 (0.3 g) and $[IrCl(COT)_2]_2$ (0.15 g) in 6 mL toluene/dma (2 : 1 v/v). The yellow product precipitated for more than 16 h; this was washed with toluene and hexane (95%). We found: carbon, 54.5; hydrogen, 8.4; nitrogen, 1.9. $C_{40}H_{76}ONCl_2P_2Ir$ requires carbon, 54.9; hydrogen, 8.7; nitrogen, 1.6. Dissolution of this hydride in CH_2Cl_2 or benzene yields the rose color of Complex 3, which is isolated readily by adding hexane.

CARBONYLDICHLOROHYDRIDOBIS(TRICYCLOHEXYLPHOSPHINE)IRIDIUM (III), $HIrCl_2(CO)(PCy_3)_2$, COMPLEX 5. Carbon monoxide was bubbled through a CH_2Cl_2 solution of Complex 3 or 4 (0.1 g in 3 mL). The resulting pale yellow solution was concentrated to an oil to which hexane (3 mL) then was added. The pale yellow precipitate was washed liberally with hexane (80%). We found: carbon, 52.3; hydrogen, 8.0. $C_{37}H_{67}OCl_2P_2Ir$ requires carbon, 52.1; hydrogen, 7.8.

Trans-CARBONYLCHLOROBIS(TRICYCLOHEXYLPHOSPHINE)IRIDIUM (I), $IrCl(CO)(PCy_3)_2$. Solid PCy_3 (0.2 g) and $[IrCl(COT)_2]_2$ (0.1 g) were placed under a CO atmosphere. The mixture turned blackish green, but when 10 mL toluene was added the stirred mixture became lighter and after 2 hr it precipitated a yellow solid that was collected and washed with ethanol (80%). We found: carbon, 54.2; hydrogen, 8.2. $C_{37}H_{66}OClP_2Ir$ requires carbon, 54.4; hydrogen, 8.2.

CHLORO(1,5-CYCLOOCTADIENE)(TRICYCLOHEXYLPHOSPHINE)IRIDIUM(I), $IrCl(COD)PCy_3$. The dimer $[HIrCl_2(COD)]_2$ (0.1 g) and PCy_3 (0.2 g) were stirred in 5 mL benzene for 2 h. The bright yellow solution was concentrated and diethyl ether was added to induce crystallization (65%). The same synthesis can be performed using $[IrCl(COD)]_2$. We found: carbon, 50.8; hydrogen, 7.5. $C_{26}H_{45}ClPIr$ requires carbon, 50.7; hydrogen, 7.4.

CHLORODIHYDRIDOBIS(TRICYCLOHEXYLPHOSPHINE)IRIDIUM(III), $H_2IrCl(PCy_3)_2$, COMPLEX 7. This orange complex was prepared according to a literature procedure (*15*). We found: carbon, 54.9; hydrogen, 8.9. $C_{36}H_{68}ClP_2Ir$ requires carbon, 54.7; hydrogen, 8.6.

CARBONYLCHLORODIHYDRIDOBIS(TRICYCLOHEXYLPHOSPHINE)IRIDIUM-(III), $H_2IrCl(CO)(PCy_3)_2$, COMPLEX 10. This complex was prepared from Complex 7 by the same procedure that 5 was prepared from 3. We found: carbon, 54.6; hydrogen, 8.3. $C_{37}H_{68}OClP_2Ir$ requires carbon, 54.3; hydrogen, 8.4.

CHLORODIHYDRIDO[DICYCLOHEXYL(CYCLOHEX-3-ENYL)PHOSPHINE] (TRICYCLOHEXYLPHOSPHINE)IRIDIUM(III), $H_2IrCl[P(C_6H_9)Cy_2](PCy_3)$, COMPLEX 12. The dimer $[IrCl(COT)_2]_2$ (0.15 g) and PCy_3 (0.3 g) were stirred together in 7 mL toluene for 12 h during which time the solution turned from orange to yellow. The solvent was removed and the resultant oil was dissolved in 10 mL hexane and refrigerated. The product which precipitated slowly as a pale yellow crystalline solid, was filtered and washed with hexane (45%). We found: carbon, 55.0; hydrogen, 8.6. $C_{36}H_{66}ClP_2Ir$ requires carbon, 54.8; hydrogen, 8.4.

BENZONITRILECHLOROBIS(TRICYCLOHEXYLPHOSPHINE)IRIDIUM(I), $IrCl(C_6H_5CN)(PCy_3)_2$, COMPLEX 13. Benzonitrile (0.5 mL) was added to 5 mL toluene containing PCy_3 (0.28 g) and $[IrCl(COT)_2]_2$ (0.15 g). There was an immediate color change from orange to yellow. The product precipitated as large pale yellow crystals (70%) on reducing the volume and adding 5 mL hexane. We found: carbon, 60.2; hydrogen, 8.1; nitrogen, 1.6. The material contains toluene that was not removed on pumping. The formula $C_{43}H_{71}NClP_2Ir$ plus 1 mol toluene requires carbon, 61.1; hydrogen, 8.0; nitrogen, 1.4.

ACETONITRILECHLOROHYDRIDOHYDROXOBIS(TRICYCLOHEXYLPHOSPHINE)IRIDIUM(III), $HIr(OH)Cl(CH_3CN)(PCy_3)_2$, COMPLEX **14**. Acetonitrile (0.5 mL) was added to 5 mL toluene containing PCy_3 (0.28 g) and $[IrCl(COT)_2]_2$ (0.15 g). The resulting yellow solution was reduced to an oil and 5 mL hexane was added. Stirring the orange solution for 15 min yields a white solid which was filtered and washed sparingly with hexane (80%). We found: carbon, 54.1; hydrogen, 8.5; nitrogen, 1.5; chlorine, 4.2. $C_{38}H_{71}ONClP_2Ir$ requires carbon, 53.8; hydrogen, 8.5; nitrogen, 1.7; chlorine, 4.2.

CHLORO(PEROXO)BIS(TRICYCLOHEXYLPHOSPHINE)IRIDIUM(III), $IrCl(O_2)(PCy_3)_2$, COMPLEX **15**. Complex **14** (0.3 g) was dissolved in 7 mL CH_2Cl_2 under argon; admission of an O_2 atmosphere resulted in a color change from orange to purple. Hexane (5 mL) was added and the total volume was reduced to about 3 mL. The purple, light-sensitive compound that precipitated was washed with hexane and dried (80%). We found: carbon, 52.4; hydrogen, 8.1. $C_{36}H_{66}O_2ClP_2Ir$ requires carbon, 52.4; hydrogen, 8.1.

Rhodium Complexes. DICHLOROHYDRIDOBIS(TRICYCLOHEXYLPHOSPHINE)RHODIUM(III), $HRhCl_2(PCy_3)_2$, COMPLEX **16**. An ethanol (or H_2O/acetone, 1:4 v/v) solution of $RhCl_3 \cdot 3H_2O$ (0.07 g in 5 mL) was added to a stirring, boiling ethanol (or acetone) solution of PCy_3 (0.7 g in 20 mL); light-sensitive, red microcrystals are formed (80%). We found: carbon, 58.7; hydrogen, 9.1. $C_{36}H_{67}Cl_2P_2Rh$ requires carbon, 58.8; hydrogen, 9.1.

Trans-CARBONYLCHLOROBIS(TRICYCLOHEXYLPHOSPHINE)RHODIUM(I), $RhCl(CO)(PCy_3)_2$. (a) Solutions of $RhCl_3 \cdot 3H_2O$ (0.3 g in 10 mL ethanol), HCHO (5 mL, 40% aqueous), and $NaBH_4$ (0.1 g in 10 mL ethanol) were added rapidly and successively to a stirring, boiling ethanol solution of PCy_3 (1.4 g in 40 mL) to give the yellow crystalline product (70%). We found: carbon, 60.5; hydrogen, 9.1. $C_{37}H_{66}OClP_2Rh$ requires carbon, 61.1; hydrogen, 9.2. (b) Tricyclohexylphosphine (1.6 g) and $[RhCl(COD)]_2$ (0.2 g) were stirred in 4 mL CH_2Cl_2/10 mL MeOH and the solution then was reduced in volume. The product (85%) precipitated on adding diethyl ether.

CHLORO(1,5 - CYCLOOCTADIENE)(TRICYCLOHEXYLPHOSPHINE)RHODIUM(I), $RhCl(COD)(PCy_3)$, COMPLEX **17**. The dimer $[RhCl(COD)]_2$ (0.2 g) and PCy_3 (1.6 g) were stirred in 10 mL CH_2Cl_2 for 10 min. The volume was reduced and diethyl ether was added to induce crystallization of the yellow product. We found: carbon, 59.5; hydrogen, 8.4. $C_{26}H_{45}ClPRh$ requires carbon, 59.3; hydrogen, 8.6.

Concluding Remarks

Although our original objective to discover platinum metal systems that bind and activate CO_2 has not been attained, we feel that some significant chemistry has been realized. The dehydrogenation of the cyclohexyl group is interesting in terms of C–H activation. Oxidative addition of water at a transition-metal center is likely to be an important step in a range of catalytic processes involving gas molecules, particularly CO_2 and CO, and the isolation of the dioxygen complex coupled with its photoactivity could be relevant for catalytic oxygenations.

Acknowledgment

We thank NATO for a grant which allowed us to initiate this work and the Natural Sciences and Engineering Research Council of Canada for subsequent financial support. Johnson, Matthey, Ltd. generously loaned us the ruthenium, rhodium, and iridium trichlorides.

Literature Cited

1. Eisenberg, R.; Hendrikson, D. E. *Adv. Catal.* **1979,** *28,* 79.
2. Aresta, M.; Nobile, C. F.; Albano, V. G. Forni, E.; Manassero, M.; *J. Chem. Soc. Chem. Commun.* **1975,** 636.
3. Tolman, C. A. *Chem. Rev.* **1977,** *77,* 313.
4. Moers, F. G.; de Jong, J. A. M.; Beaumont, P. M. H. *J. Inorg. Nucl. Chem.* **1973,** *35,* 1915.
5. Van Gaal, H. L. M.; Moers, F. G.; Steggerda, J. J. *J. Organomet. Chem.* **1974,** *65,* C43.
6. Moers, F. G.; Langhout, J. P. *Rec. Trav. Chim.* **1972,** *91,* 591.
7. Moers, F. G.; Ten Hoedt, R. W. M.; Langhout, J. P. *J. Organomet. Chem.* **1974,** *65,* 93.
8. Moers, F. G.; Ten Hoedt, R. W. M.; Langhout, J. P. *J. Inorg. Nucl. Chem.* **1974,** *36,* 2279.
9. Hieber, W.; Frey, V.; John, P. *Chem. Ber.* **1967,** *100,* 1961.
10. van Gaal, H. L. M.; Verlaan, J. P. J. *J. Organomet. Chem.* **1977,** *133,* 93.
11. Yoshida, T.; Okano, T.; Otsuka, S. *J. Chem. Soc. Chem. Commun.* **1978,** 855.
12. van Gaal, H. L. M.; van den Bekerom, F. L. A. *J. Organomet. Chem.* **1977,** *134,* 237.
13. van Gaal, H. L. M.; Verlaak, J. M. J.; Posno, T. *Inorg. Chim. Acta* **1977,** *23,* 43.
14. Crabtree, R. H.; Morris, G. E. *J. Organomet. Chem.* **1977,** *135,* 395.
15. Hietkamp, S.; Stufkens, D. J.; Vrieze, K. *J. Organomet. Chem.* **1978,** *152,* 347.
16. Yoshida, T.; Thorn, D. L.; Okano, T.; Ibers, J. A.; Otsuka, S. *J. Am. Chem. Soc.* **1979,** *101,* 4212.
17. James, B. R.; Preece, M.; Robinson, S. D. *61st Can. Chem. Conf., Winnipeg, 1978;* Abstract IN 31.
18. James, B. R.; Preece, M.; Robinson, S. D. *Inorg. Chim. Acta* **1979,** *34,* L219.
19. Ahmed, N.; Levison, J. J.; Robinson, S. D.; Uttley, M. F. *J. Chem. Soc., Dalton Trans.* **1975,** 370.
20. Ahmed, N.; Levison, J. J.; Robinson, S. D.; Uttley, M. F. *Inorg. Synth.* **1974,** *15,* 45.
21. Gill, D. F.; Shaw, B. L. *Inorg. Chim. Acta* **1979,** *32,* 19.
22. Jesson, J. P. In "Transition Metal Hydrides"; Muetterties, E. L., Ed.; Marcel Dekker: New York, 1971; p. 75.
23. Johnson, B. F. G.; Segal, J. A. *J. Chem. Soc., Dalton Trans.* **1973,** 478.
24. Cotton, F. A.; Barnes, R. D.; Bannister, E. *J. Chem. Soc.* **1960,** 2199.
25. Stephenson, T. A.; Wilkinson, G. *J. Inorg. Nucl. Chem.* **1966,** *28,* 945.
26. Levison, J. J.; Robinson, S. D. *J. Chem. Soc.* (A) **1970,** 2947.
27. Collman, J. P.; Roper, W. R. *J. Am. Chem. Soc.* **1965,** *87,* 4008.
28. Carty, A. *Can. J. Chem.* **1966,** *44,* 1881.
29. Empsall, H. D.; Hyde, E. M.; Mentzer, E.; Shaw, B. L. *J. Chem. Soc., Dalton Trans.* **1977,** 2285.
30. Strohmeier, W.; Onoda, T. Z. *Naturforsch.* **1969,** *23B,* 1377.
31. Robinson, S. D.; Shaw, B. L. *J. Chem. Soc.* **1965,** 4998.

32. Parshall, G. W. *Acc. Chem. Res.* **1975,** *8,* 113.
33. Webster, D. E. *Adv. Organomet. Chem.* **1977,** *15,* 147.
34. van der Ent, A.; Onderdelinden, A. L. *Inorg. Chim. Acta* **1973,** *7,* 203.
35. Crabtree, R. H.; Mihelcic, J. M.; Quirk, J. M. *J. Am. Chem. Soc.* **1979,** *101,* 7738.
36. Clark, P. W.; Hartwell, G. E. *J. Organomet. Chem.* **1977,** *139,* 385.
37. Crabtree, R. H. *Acc. Chem. Res.* **1979,** *12,* 331.
38. James, B. R. *Adv. Organomet. Chem.* **1979,** *17,* 319.
39. Yoshida, T.; Ueda, Y.; Otsuka, S. *J. Am. Chem. Soc.* **1978,** *100,* 3941.
40. Yoshida, T.; Otsuka, S., Chapter 8 in this book.
41. English, A. D.; Herskovitz, T. *J. Am. Chem. Soc.* **1977,** *99,* 1648.
42. Lyons, J. E. In "Aspects of Homogeneous Catalysis"; Ugo, R., Ed.; Reidel: Dordrecht, 1977; Vol. 3, p. 1.
43. Bennett, M. J.; Donaldson, P. B. *J. Am. Chem. Soc.* **1971,** *93,* 3307.
44. James, B. R.; Preece, M., unpublished data.
45. Laing, M.; Nolte, M. J.; Singleton, E. *J. Chem. Soc. Chem. Commun.* **1975,** 660.
46. Cotton, F. A.; Wilkinson, G. "Advanced Inorganic Chemistry", 3rd ed.; Wiley: New York, 1972; pp. 635, 783.
47. Adams, D. M. "Metal-Ligand and Related Vibrations"; Edward Arnold: London, 1967; Chap. 5.
48. Ledon, H.; Bonnet, M.; Lallemand, J.-Y. *J. Chem. Soc. Chem. Commun.* **1979,** 702.
49. Ledon, H.; Bonnet, M. *J. Mol. Catal.* **1980,** *7,* 309.
50. Ullrich, V. *J. Mol. Catal.* **1980,** *7,* 159.
51. Groves, J.; Kruper, W. J.; Nemo, T. E.; Myers, R. S. *J. Mol. Catal.* **1980,** *7,* 169.
52. de Waal, D. J. A., Robb, W. *Inorg. Chim. Acta* **1978,** *26,* 91.
53. Nicholas, K. M. *J. Organomet. Chem.* **1980,** *188,* C10.
54. Aresta, M.; Nobile, C. F. *Inorg. Chim Acta,* **1977,** *24,* L49.
55. Aresta, M.; Nobile, C. F. *J. Chem. Soc., Dalton Trans.* **1977,** 708.
56. James, B. R.; Rempel, G. L. *Discuss. Faraday Soc.* **1968,** *46,* 48.
57. Chatt, J.; Venanzi, L. M. *J. Chem. Soc.* **1957,** 4735.
58. van der Ent, A.; Onderdelinden, A. L. *Inorg. Syn.* **1973,** *14,* 92.
59. Cramer, R. *Inorg. Chem.* **1962,** *1,* 722.

RECEIVED July 10, 1980.

10

Phosphido-Bridged Iron Group Clusters

ARTHUR J. CARTY
Guelph–Waterloo Centre for Graduate Work in Chemistry, Waterloo Campus, University of Waterloo, Waterloo, Ontario, N2L 3G1 Canada

The synthesis, structural chemistry, and reactivity of some iron group phosphido-bridged carbonyl clusters containing μ_2- and μ_3-bound acetylides are described. Phosphinoacetylenes $Ph_2PC{\equiv}CR$ ($R = Ph$, Bu^t, Pr^i), in reactions with $Fe_2(CO)_9$ or $M_3(CO)_{12}$ (M = Ru and Os) are useful precursors for clusters with bridging PPh_2 and $-C{\equiv}CR$ groups. Open $M_3(CO)_9(C{\equiv}CR)(PPh_2)$ (M = Ru, Os; $R = Pr^i$, Bu^t) and closed $Ru_3(CO)_8(C{\equiv}CR)(PPh_2)$ clusters with μ_3-$C{\equiv}CR$ groups and phosphido bridges have been characterized. Dimers $Ru_2(CO)_6(C{\equiv}CR)$-(PPh_2) ($R = Ph$, Bu^t and Pr^i) and $Ru_2(CO)_5(C{\equiv}CPr^i)$-$(PPh_2)(Ph_2PC{\equiv}CPr^i)$ with μ_2-η^2-acetylides as well as trimeric $Ru_3(CO)_6(C{\equiv}CR)_2(PPh_2)_2(Ph_2PC{\equiv}CR)$ ($R = Bu^t$ and Pr^i) having symmetrical μ_2- and edge-bridging μ_2-η^2-acetylides are described. Tetranuclear $Ru_4(CO)_{10}(C=CHPr^i)(OH)(PPh_2)$ has μ_4-vinylidene and μ_3-hydroxo groups while $Os_3(CO)_9(COOEt)(OH)(Ph_2$-$PC{\equiv}CPr^i)$ contains μ_2-hydroxo and μ_2-ester groups. Trends in P-31 NMR shifts for bridging PPh_2 groups are discussed. The unusual reactivity of μ_2-η^2-acetylides towards uncharged carbon nucleophiles (carbenes and isonitriles) is described.

The phosphido group PR_2 long has been recognized as an excellent bridging ligand and a large number of phosphido-bridged transition metal complexes are now known. The early synthetic work of Hayter (*1*) and the structural studies of Dahl and co-workers (*2*) are perhaps particularly noteworthy in the context of phosphido-group coordination chemistry. Recently attention has begun to focus on the

0065–2393/82/0196–0163$07.50/0

potential of the PR_2 ligand as a building block for metal clusters (*3*) and on the utility of phosphido-bridged complexes in homogeneous catalysis.

Some Useful Characteristics of the Phosphido Bridge

As illustrated in Table I, the PR_2 (and AsR_2) group is a one-electron ligand when bonded terminally and a 3-electron donor when bridging. These electronic characteristics have been advantageous in synthesizing mixed-homonuclear and -heteronuclear complexes earlier by Haines (*4*) and more recently in a wider context by Vahrenkamp and co-workers (*5*). The PR_2 bridge is also a flexible yet stable entity capable of supporting a wide range of bonding and nonbonding metal–metal distances. Thus in the binuclear molecule $Fe_2(CO)_6(PPh_2)_2$ the Fe–Fe bond of length 2.623(2) Å is supported by phosphido bridges with Fe–P–Fe angles of 72.0° ave. (*6*). However in the dianion $[Fe_2(CO)_6(PPh_2)_2]^{2-}$ the Fe—Fe distance (3.630(3) Å) is nonbonding and the Fe–P–Fe bridge angle has opened to 105.5(1)°. As another example of phosphido-bridge versatility, the trinuclear cluster

Table I. Characteristics of the Phosphido Bridge

1. :PR_2 one-electron donor when terminal
   three-electron donor when bridging

```
R     R
 \   /
   P
 /   \
M     M
```

2. For M(PR₂)M the stereochemistry at phosphorus is a distorted tetrahedral.
3. The bridge is extremely flexible.

```
   P
 /   \
M     M
```

   M–P–M angles can vary from ~70° to ~106° for first-row transition metals. Values above the tetrahedral value are possible, especially when only a single PR_2 bridge is present.

   Examples.

   $Fe_2(CO)_6(PPh_2)_2$ (*6*); Fe(1)–Fe(2) 2.623(3) Å; ∠Fe(1)–P–Fe(2) 72.0(1)°.

   $[Fe_2(CO)_6(PPh_2)_2]^{2-}$ (*6*); Fe(1)---Fe(2) 3.630(3) Å; ∠Fe(1)–P–Fe(2) 105.5(1)°.

   $Fe_3(CO)_8[Ph_2PC_4(CF_3)_2](PPh_2)$ (*7*); Fe(1)---Fe(2) 3.59 Å; ∠Fe(1)–P(1)–Fe(2) 98.0(1)°.

   $Fe_3(CO)_7[Ph_2PC_4(CF_3)_2](PPh_2)$ (*8*); Fe(1)–Fe(2) 2.665(8) Å; Fe(1)–P(1)–Fe(2) 73.3(2)°.

4. The PR_2 bridges are stable. There is preferential reactivity at other sites.

Scheme I. Phosphido-bridged complexes: reactivity at other sites.

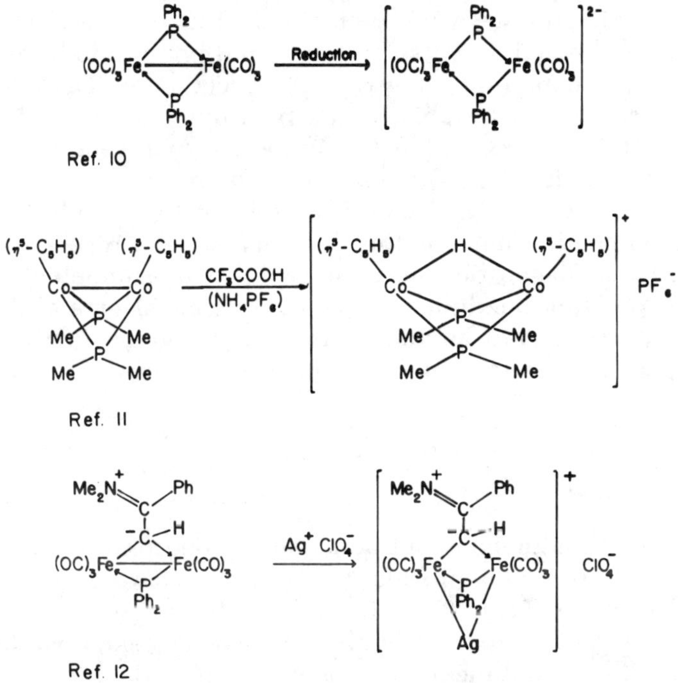

Ref. 10

Ref. 11

Ref. 12

$Fe_3(CO)_8[Ph_2PC_4(CF_3)_2](PPh_2)$, which contains one phosphido group bridging two nonbonding iron atoms (Fe(1)—Fe(2) = 3.59 Å) at a 98.0(1)° angle (*7*), is converted via loss of CO and bond formation between Fe(1) and Fe(2) to $Fe_3(CO)_7[Ph_2PC_4(CF_3)_2](PPh_2)$ where Fe(1)–Fe(2) is 2.665(8) Å and Fe(1)–P–Fe(2) is 73.3(2)° (*8*). As an example of an intermediate value for the bridge angle at phosphorus, the metal atoms in $[Ph_4As][Mo_2(C_5H_5)_2(CO)_4(PMe_2)]$ at a distance of 3.157(2) Å subtend an 83.6(1)° angle at the bridging phosphorus atom (*9*). Such bridging characteristics are potentially useful in designing and exploiting bi- and polynuclear metal derivatives as catalysts for elaborating unsaturated molecules such as CO, olefins, and acetylenes. Facile metal–metal bond formation and cleavage may play a key role in such processes. An additional and attractive feature of the phosphido bridge is its relative inertness in chemical reactions. This chemical stability allows the investigation of reactivity associated with metal–metal bonds or unsaturated groups bound in multisite fashion to two or more metal atoms. Scheme I lists a few examples where two-electron reduction (*10*), protonation (*11*), or attack by a metal electrophile (silver ion) (*12*) has occurred at a metal–metal bond resulting either in cleavage or lengthening of the bond with retention of the phosphido bridges. One of the long-term interests of our group has been to examine patterns of reactivity for multisite bound ligands.

In this chapter we wish to focus on the synthesis, structural chemistry, and reactivity of some iron-group phosphido-bridged carbonyl clusters containing sideways-bound acetylides. The role of the PR_2 group here is as a stable, unreactive, but flexible bridging coligand which facilitates exploitation of the unusual reactivity patterns for these μ_2-η^2- and μ_3-η^2-bound hydrocarbyl groups.

Examples of open and closed clusters with phosphido groups bridging nonbonding and bonding dimetal fragments respectively will be described. Diagnostic trends in P-31 NMR parameters of PPh_2 bridges supporting bonding and nonbonding metal—metal distances will be discussed and an indication will be given of the hyperreactivity resulting from multisite bonding of acetylides in metal clusters.

Synthetic Methods

For the simultaneous introduction of phosphido groups and acetylides into bi- or polynuclear complexes we have used the rather facile cleavage of the P–C(sp) bond of a phosphinoalkyne that frequently occurs when reacting with a low-valent metal complex. The principle is illustrated nicely in Scheme II where the binuclear complex $Fe_2(CO)_6$ ($C{\equiv}CBu^t$)(PPh_2) is synthesized from $Fe_2(CO)_9$ and $Ph_2PC{\equiv}CBu^t$. The intermediates $Fe(CO)_4(Ph_2PC{\equiv}CBu^t)$ and $Fe_2(CO)_8(Ph_2PC{\equiv}CBu^t)$ in this sequence have been characterized fully, the latter by X-ray crystallography (*13*). The final step in this reaction is presumed to be oxidative insertion into the P–C bond of the coordinated ligand.

The reaction of the trinuclear clusters $M_3(CO)_{12}$ (M = Ru and Os) with $Ph_2PC{\equiv}CR$ (R = Bu^t, Pr^i, Ph) in a hydrocarbon solvent in the presence of anhydrous Me_3NO afforded good yields of the red (M = Ru) or yellow (M = Os) monosubstituted derivatives $M_3(CO)_{11}(Ph_2PC{\equiv}CR)$ (*see* Scheme III). An X-ray analysis of $Ru_3(CO)_{11}(Ph_2PC{\equiv}CPh)$ (*see* Figure 1) showed that a phosphorus-coordinated phosphinoacetylene has replaced an equatorial CO group of $Ru_3(CO)_{12}$ (*14*). The dangling acetylene, held proximate to a neighboring metal is notable. Careful treatment of $Ru_3(CO)_{11}(Ph_2PC{\equiv}CR)$ with Me_3NO at −10°C for 24 h gave bright yellow $Ru_3(CO)_9(C{\equiv}CR)(PPh_2)$ (*see* Scheme II) via P–C bond cleavage. The osmium analogues were obtained on refluxing $Os_3(CO)_{11}(Ph_2PC{\equiv}CR)$ (R = Bu^t and Pr^i) in decalin for 5 h or on treatment with Me_3NO in heptane for 5 hr. The role of Me_3NO in these reactions is likely to generate a vacant coordination site on an adjacent metal atom, thus

Scheme II.

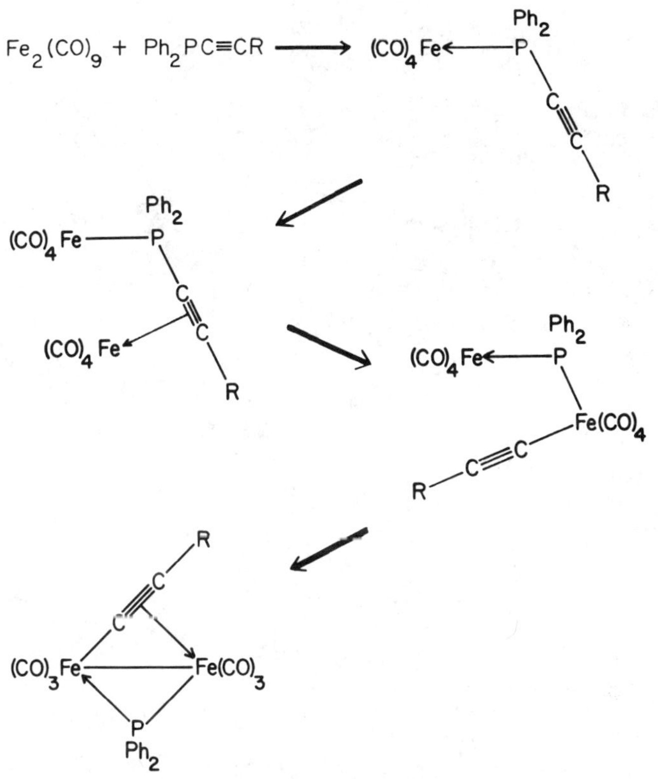

$$Fe_2(CO)_9 + Ph_2PC{\equiv}CR \longrightarrow (CO)_4Fe{\leftarrow}P(Ph_2)C{\equiv}CR$$

facilitating interaction of the dangling acetylene and P–C bond cleavage. Although products and yields in these reactions (vide infra) are quite sensitive to conditions and reagents, particularly to solvent and the choice of anhydrous or hydrated Me_3NO, it is clear that thermal or Me_3NO-promoted P–C bond cleavage of a phosphinoacetylene can give usable yields of trinuclear clusters with phosphido bridges.

Other products (*see* Scheme III) from the fragmentation of $Ru_3(CO)_{11}$ $(Ph_2PC{\equiv}CR)$ include the rather stable dimers $Ru_2(CO)_6$-$(C{\equiv}CR)(PPh_2)$ and another trimeric molecule $Ru_3(CO)_8(C{\equiv}CR)$-(PPh_2) which, however, is obtained more readily by the decomposition of $Ru_3(CO)_9(C{\equiv}CR)(PPh_2)$, which is metastable in solution. The structural relationship of these molecules will be described in the next section. As might be expected because of the greater stability of Os_3 than Ru_3 clusters, none of the binuclear complexes $Os_2(CO)_6$-$(C{\equiv}CR)(PPh_2)$ are observed when $Os_3(CO)_{11}(Ph_2PC{\equiv}CR)$ is thermolized. Indeed the synthesis of $Os_3(CO)_9(C{\equiv}CR)(PPh_2)$ from $Os_3(CO)_{11}$-$(Ph_2PC{\equiv}CR)$ is a remarkably clean reaction.

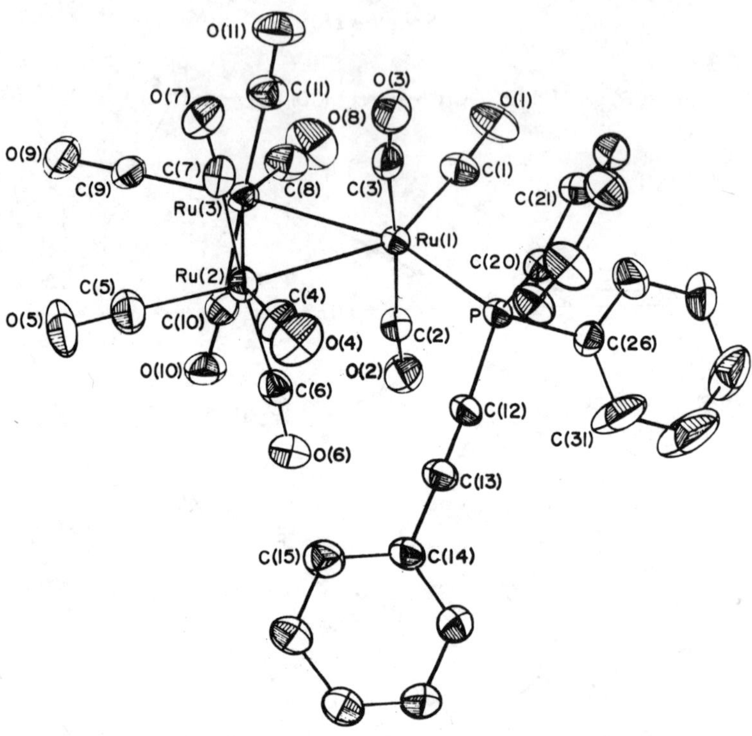

Figure 1. X-ray structure of $Ru_3(CO)_{11}(Ph_2PC{\equiv}CPh)$

Scheme III.

1) $Ru_3(CO)_{12} + Ph_2PC{\equiv}CR + Me_3NO \longrightarrow Me_3N + CO_2$

$+ Ru_3(CO)_{11}(Ph_2PC{\equiv}CR)$ (75%)

$+ Ru_3(CO)_{10}(Ph_2PC{\equiv}CR)_2$ (10%)

($R = Bu^t, Pr^i, Ph$)

2) $Ru_3(CO)_{11}(Ph_2PC{\equiv}CR) \xrightarrow[\text{R. T.}]{\text{14 DAYS}} Ru_2(CO)_6(Ph_2P)(C{\equiv}CR)$ (20%)

$+ Ru_3(CO)_9(Ph_2P)(C{\equiv}CR)$ (10%)

+ Other Products ($R = Pr^i$)

3) $Ru_3(CO)_{11}(Ph_2PC{\equiv}CR) + Me_3NO \xrightarrow{0^\circ}$

$Ru_3(CO)_9(Ph_2P)(C{\equiv}CR)$ (50%)

+ Other Products ($R = Pr^i$)

Using $Me_3NO \cdot 2H_2O$ in place of anhydrous Me_3NO in the reaction with $Ru_3(CO)_{11}(Ph_2PC{\equiv}CPr^i)$ generates a fourth tetranuclear cluster, $Ru_4(CO)_{10}(C = CHPr^i)(OH)(PPh_2)$ albeit in relatively poor yield (*see* Scheme IV) as well as somewhat different ratios of the two trinuclear clusters and $Ru_2(CO)_6(C{\equiv}CPr^i)(PPh_2)$. The structure of this highly unusual Ru_4 cluster will be discussed later.

Using anhydrous Me_3NO in the decomposition of $Os_3(CO)_{11}$-$(Ph_2PC{\equiv}CPr^i)$ in the presence of ethanol (*see* Scheme V) leads to excellent yields of another interesting cluster type $Os_3(CO)_9(OH)$-$(COOEt)(Ph_2PC{\equiv}CPr^i)$ containing a sideways-bound ester group and a bridging hydroxide. The stoichiometry of the initial reaction of $Os_3(CO)_{12}$ with ligand in the presence of ethanol and Me_3NO has not been established completely as yet, but there is evidence for an active species $Os_3(CO)_{10}(Ph_2C{\equiv}CPr^i)(L)$, where L may be a metastable ligand resulting from attack by Me_3NO on a carbonyl group. Adding

Scheme IV. $Ru_4(CO)_{10}(OH)(PPh_2)(C = CHPr^i)$

$Ru_3(CO)_{11}(PPh_2C{\equiv}CPr^i)$ + 2 $Me_3NO \cdot 2H_2O$

heptane, −78 °C

Warm to RT, 24 hrs.

$Ru_3(CO)_9(C{\equiv}CPr^i)(PPh_2)$ 20%

$Ru_2(CO)_6(C{\equiv}CPr^i)(PPh_2)$ 30%

$Ru_3(CO)_8(C{\equiv}CPr^i)(PPh_2)$ <10%

$Ru_4(CO)_{10}(OH)(PPh_2)(C{=}CHPr^i)$ 10%

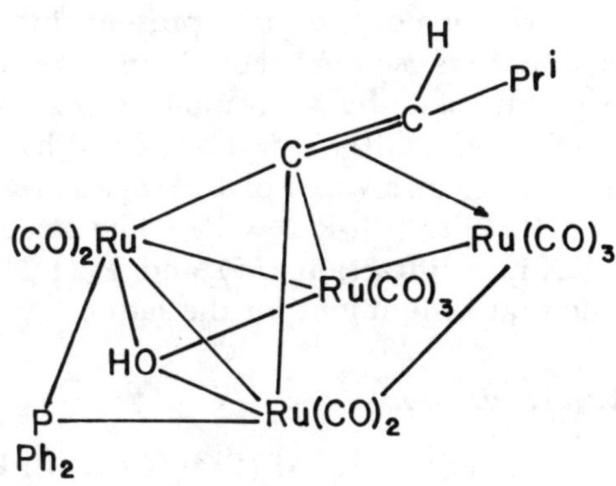

Scheme V. $Os_3(CO)_9(CO_2Et)(OH)(Ph_2PC{\equiv}CPr^i)$

$Os_3(CO)_{12}$ + $Ph_2PC{\equiv}CPr^i$
(L)

2 Me_3NO
EtOH
75° heptane

$Os_3(CO)_9(CO_2Et)(OH)L$ + $Os_3(CO)_{11}L$
(x-ray)

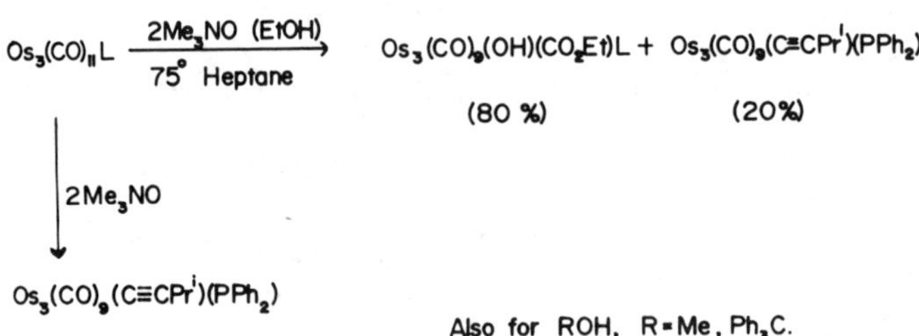

ethanol to this species and eliminating Me_3N then might generate the observed complex. Further work on this intriguing sequence is underway.

The reaction of $Ru_3(CO)_{12}$ with 3 mol of phosphinoalkyne in heptane at 75°C is illustrated in Scheme VI. The red trisubstitution product $Ru_3(CO)_9(Ph_2PC{\equiv}CPr^i)_3$, readily identified by its characteristic ν(CO) spectrum (ν(CO) $CHCl_3$ 2049m, 2009m, 1991s, 1978s, 1952m cm^{-1}), fragments on heating to give $Ru_2(CO)_5(C{\equiv}CR)(PPh_2)$-$(Ph_2PC{\equiv}CR)$ as the major product present in solution as a 5:1 mixture of trans and cis isomers. For R = Pr^i, only the trans species was found in the solid state by X-ray analysis (*14*). A second trinuclear complex $Ru_3(CO)_6(\mu_2\text{-}C{\equiv}CR)$ $(\mu^2\text{-}\eta^2\text{-}C{\equiv}CR)(PPh_2)_2(Ph_2PC{\equiv}CR)$ with two phosphido bridges also was isolated together with small amounts of $Ru_3(CO_7(C{\equiv}CR)(PPh_2)(Ph_2PC{\equiv}CR)_2$. The structure of the former was solved by X-ray diffraction (*15*) and P-31 NMR data led to a reasonable structural assignment for the latter.

Structural Characterization

$M_3(CO)_9(C{\equiv}CR)PPh_2$ and $M_3(CO)_8(C{\equiv}CR)PPh_2$: Clusters with $\mu_3\text{-}\eta^2$-Acetylides. Structural data are available for both of the trimeric

Scheme VI. Reaction of $Ru_3(CO)_{12}$ with $Ph_2PC{\equiv}CR$ ($R = Pr^i$, Bu^t)

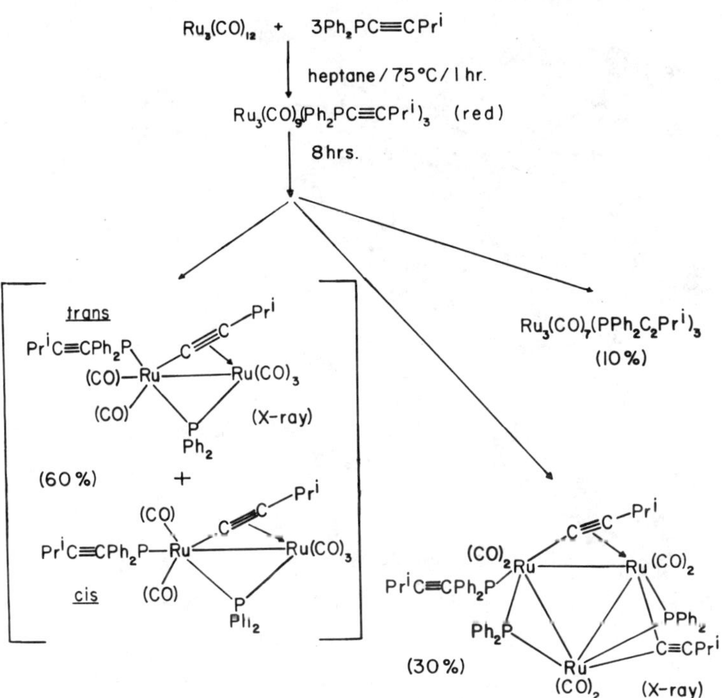

clusters $M_3(CO)_9(C{\equiv}CPr^i)(PPh_2)$ (M = Ru and Os). An ORTEP II plot of the structure of the ruthenium derivative is shown in Figure 2. Pertinent bond distances and angles are compared and contrasted with structural data for the binuclear species $Fe_2(CO)_6(C{\equiv}CBu^t)(PPh_2)$, $Fe_2(CO)_4(C{\equiv}CBu^t)(PPh_2)(CNBu^t)_2$, and $Ru_2(CO)_6(C{\equiv}CBu^t)(PPh_2)$ in Figure 3. In each molecule of $M_3(CO)_9(C{\equiv}CPr^i)(PPh_2)$ the metal atoms form an isosceles triangle with two strong metal–metal bonds (ave. Ru–Ru 2.839 Å and ave. Os–Os = 2.879 Å) and one open edge (Ru(2)—Ru(3) 3.466(1) Å and Os(2)—Os(3) 3.508(1) Å). Since the metal(2)—metal(3) distances are considerably longer than twice the covalent radii of ruthenium (1.42 Å) and osmium (1.44 Å) these interactions must be very weak at best. The open edge of each triangle is bridged by a PPh_2 group with 92.8(0)° angles at the bridging atom (M = Ru) and 93.2(0)° (M = Os). The acetylide is σ-bonded to M(1) and η^2-bonded to M(2) and M(3). Thus the μ_3-η^2-alkynyl groups contribute five-electrons to the clusters. As illustrated in Figure 3, μ_3 bonding of the acetylide produces a significantly larger distortion of the acetylide than μ_2 bonding as in $Ru_2(CO)_6(C{\equiv}CBu^t)(PPh_2)$. In particular the bend-back angles at the acetylenic carbon atoms are larger and the acetylenic triple bonds are significantly longer in the μ_3

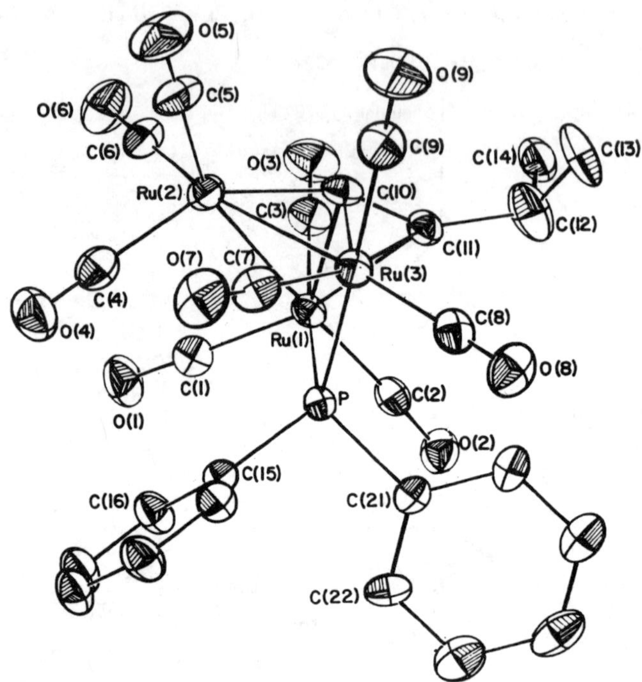

Figure 2. An ORTEP II plot of the molecular structure of $Ru_3(CO)_9$-$(C{\equiv}CPr^i)(PPh_2)$

species. The severely trans bent geometry of the acetylide and the open environment at the α-acetylenic carbon atom are also notable. All of these structural features could be interesting in the context of reactivity associated with the hydrocarbyl unit. By analogy, with the unique chemistry of the iron dimers $Fe_2(CO)_6(C{\equiv}CR)PPh_2$ (*16, 17, 18, 19*), the acetylide in $M_3(CO)_9(C{\equiv}CR)(PPh_2)$ should be extremely electrophilic and hence susceptible to attack by weak nucleophiles.

There is an interesting contrast between $M_3(CO)_9(C{\equiv}CR)(PPh_2)$ and $Ru_3(CO)_9(C{\equiv}CBu^t)(H)$ (*20, 21*). In the latter a strong Ru–Ru bond and a bridging hydride replace the PPh_2 group in $M_3(CO)_9$-$(C{\equiv}CR)(PPh_2)$. Thus the electronic characteristics of the PPh_2 group exert a remarkable influence on cluster geometry even to the extent of opening a relatively stable Ru_3 triangle.

Although $M_3(CO)_9(C{\equiv}CR)(PPh_2)$ (M = Ru and Os) have closed-shell configurations in the absence of a third metal–metal bond and are stable in the solid, the ruthenium complex is metastable in solution, losing CO on standing to generate $Ru_3(CO)_8(C{\equiv}CR)(PPh_2)$ (R = Pr^i and Bu^t). The structure of the *t*-butyl derivative is shown in Figure 4. Within the closed triangle of metal atoms there are three strong Ru–Ru bonds (Ru(1)–Ru(2) of 2.8151(4), Ru(1)–Ru(3) of 2.7084(5), and Ru(2)–Ru(3) of 2.8257(5) Å). The phosphido group, previously bridging the open edge of $Ru_3(CO)_9(C{\equiv}CR)(PPh_2)$ now links ruthenium(1) and ruthenium(3) with the μ_3-acetylide σ-bonded to ruthenium(1). The remaining edges of the Ru_3 triangle are bridged by carbonyl groups. There are marked differences between the phosphido bridges in $M_3(CO)_9(C{\equiv}CPr^i)(PPh_2)$ (M = Ru and Os) and $Ru_3(CO)_8(C{\equiv}CBu^t)(PPh_2)$ with the Ru–P–Ru angle of 92.8(0)° in the former contrasting sharply with the acute angle Ru(1)–P–Ru(2) of 74.4(0)° in $Ru_3(CO)_8(C{\equiv}CBu^t)(PPh_2)$. These changes are quite evident in the P-31 NMR

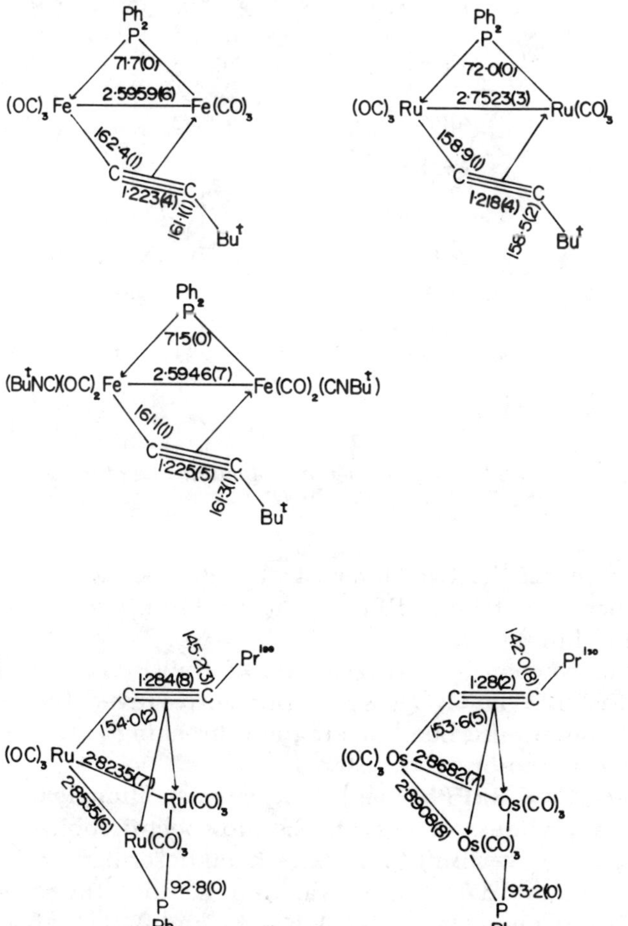

Figure 3. A comparison of structural parameters for $Fe_2(CO)_6(C{\equiv}CBu^t)(PPh_2)$, $Ru_2(CO)_6(C{\equiv}CBu^t)(PPh_2)$, $Fe_2(CO)_4(C{\equiv}CBu^t)(PPh_2)(CNBu^t)_2$, $Ru_3(CO)_9(C{\equiv}CPr^i)(PPh_2)$, and $Os_3(CO)_9(C{\equiv}CPr^i)(PPh_2)$

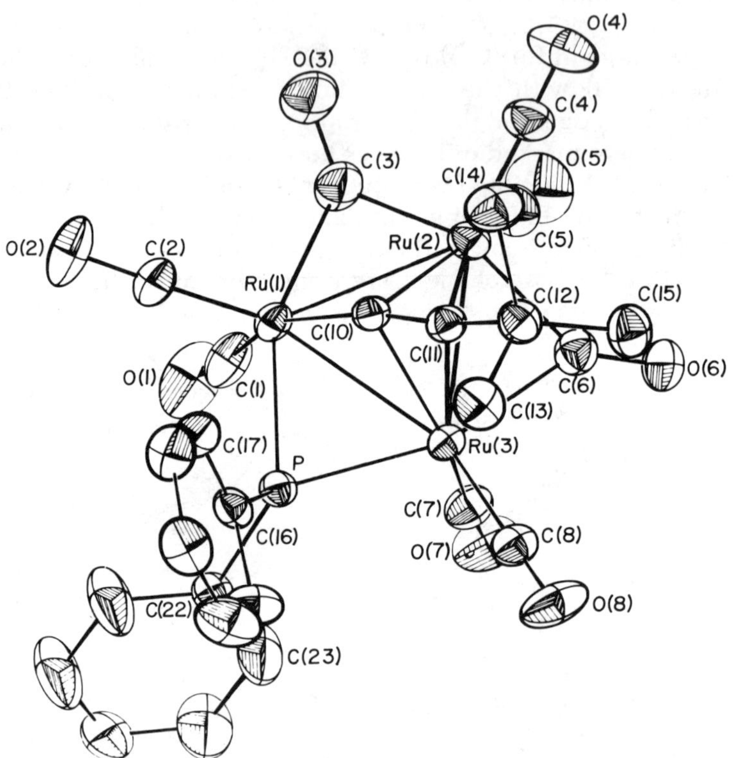

Figure 4. An ORTEP II plot of the molecular structure of $Ru_3(CO)_8$-($C{\equiv}CBu^t$)(PPh_2)

chemical shifts of the two different cluster types with δ (P-31) at a much higher field for the PPh_2 group bridging two, noninteracting metals (vide infra).

The conversion of $Ru_3(CO)_9(C{\equiv}CBu^t)(PPh_2)$ to $Ru_3(CO)_8$-($C{\equiv}CBu^t$)(PPh_2) via CO loss is apparently reversible under CO pressure. Thus Ru–Ru bond making and breaking is relatively facile in these phosphido-bridged systems.

$M_2(CO)_6(C{\equiv}CR)(PPh_2)$ and Derivatives: Binuclear Molecules with μ_2-η^2-Acetylides. The synthesis and spectroscopic characterization of $Fe_2(CO)_6(C{\equiv}CBu^t)(PPh_2)$ have been described previously (22) but no structural data were available at that time. Ruthenium analogues $Ru_2(CO)_6(C{\equiv}CR)(PPh_2)$ (R = Pr^i and Bu^t) obtained reacting $Ru_3(CO)_{11}(Ph_2PC{\equiv}CR)$ with Me_3NO (*see* Scheme III) or Me_3NO $2H_2O$ (*see* Scheme IV) have similar structures. A stereoview of $Ru_2(CO)_6(C{\equiv}CBu^t)PPh_2$ is shown in Figure 5 and structural data for $M_2(CO)_6(C{\equiv}CBu^t)(PPh_2)$(M = Fe and Ru) are compared in Figure 3.

It is interesting that despite the differences in metal–metal bond lengths (Fe(1)–Fe(2) 2.5959(6) Å and Ru(1)–Ru(2) 2.7523(3) Å), the metal–P–metal angles (71.7(0)°, M = Fe and 72.0(0)°, M = Ru) are essentially identical due to a corresponding increase in metal–P bond lengths. The acetylide is μ_2-η^2-bound in both complexes and while the acetylenic triple bond lengths (1.223(4) Å, M = Fe and 1.218(4) Å, M = Ru) are not significantly different, the bend-back angles at the acetylene (17.6° and 18.9°, M = Fe and 21.1° and 21.5°, M = Ru) may suggest a slightly stronger acetylene–metal interaction in the ruthenium complex. The minimal effect of CO substitution by Bu^tNC on the nature of the central core of the iron complex is nicely illustrated by the comparison of Figure 3.

A major product from the reaction of $Ru_3(CO)_{12}$ with 3 mol of phosphinoalkyne (*see* Scheme VI) is $Ru_2(CO)_5(C{\equiv}CR)(PPh_2)(Ph_2PC{\equiv}CR)$, a phosphine substitution product of the parent $Ru_2(CO)_6(C{\equiv}CR)(PPh_2)$. An X-ray analysis of $Ru_2(CO)_5(C{\equiv}CPr^i)(PPh_2)(Ph_2PC{\equiv}CPr^i)$ revealed the structure shown in Figure 6. The unfragmented phosphine occupies a site approximately trans to the phosphido bridge, with the acetylide μ_2-η^2-coordinated as in the parent. In solution

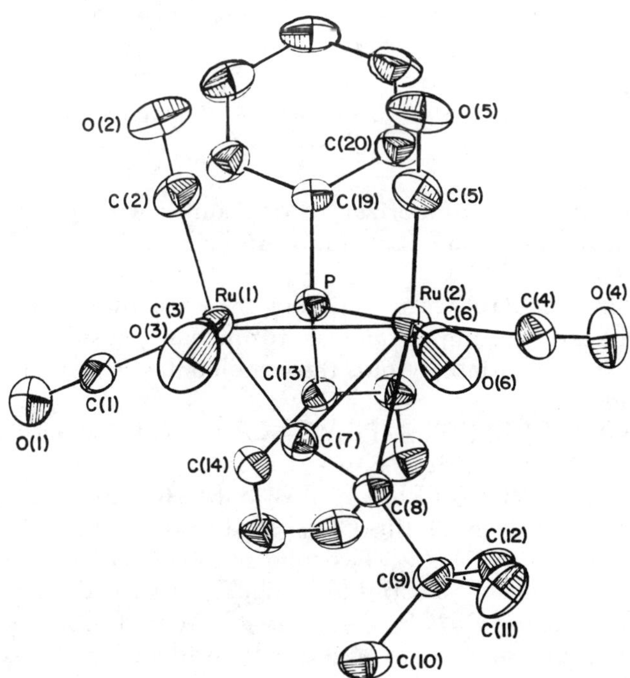

Figure 5. A perspective view of the molecular structure of $Ru_2(CO)_6$-$(C{\equiv}CBu^t)(PPh_2)$

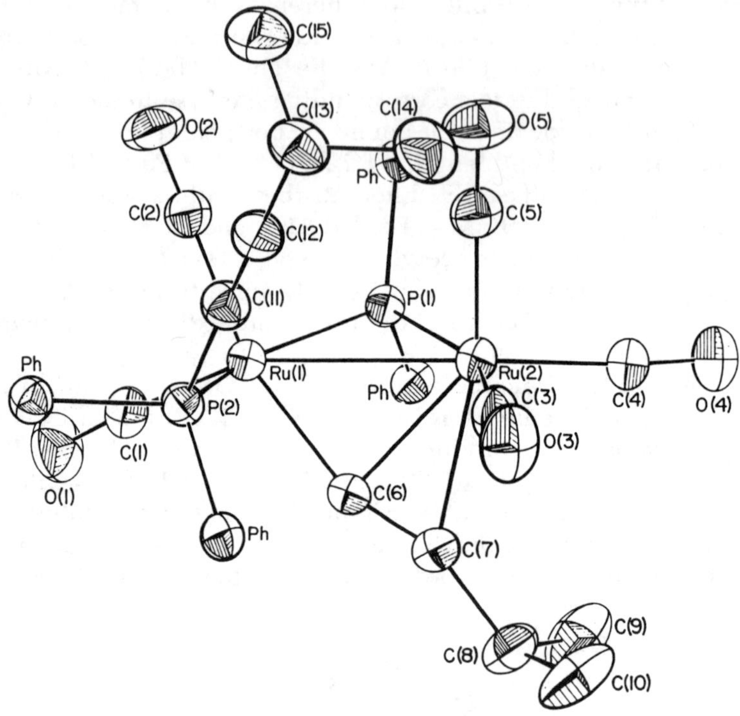

Figure 6. The molecular structure of $Ru_2(CO)_5(C{\equiv}CPr^i)(PPh_2)$-$(Ph_2PC{\equiv}CPr^i)$

this product partially isomerizes to an isomer with the phosphine occupying a site cis to the phosphido bridge as judged by P-31 NMR (vide infra).

These iron and ruthenium dimers are exceedingly robust compounds. Indeed decomposition of trinuclear phosphidoacetylide-bridged clusters invariably yields these molecules as major decomposition products.

$Ru_3(CO)_6(C{\equiv}CR)_2(PPh_2)_2(Ph_2PC{\equiv}CR)$: A Cluster with Symmetrical μ_2- and Unsymmetrical μ_2-η^2-Acetylides. The trinuclear molecules $Ru_3(CO)_6(C{\equiv}CR)_2(PPh_2)_2(Ph_2PC{\equiv}CR)$ (R = Pr^i and Bu^t) produced during pyrolysis of $Ru_3(CO)_9$ $(Ph_2PC{\equiv}CR)_3$ have fascinating structures (*see* Figure 7) (*15*). Two phosphinoalkyne units have fragmented generating two phosphido bridges and two acetylides. However, whereas one $-C{\equiv}CR$ group is μ_2-η^2-coordinated across the Ru(1)–Ru(2) edge, the other symmetrically bridges Ru(2) and Ru(3) as a one-electron ligand. Although a few examples of symmetrically bridging acetylides are known (*23, 24*), this is the first example of a carbonyl cluster with electron-deficient μ_2-acetylide bridging. Since both phosphido groups bridge metal–metal bonds, the Ru–P–Ru angles (Ru(1)–P(1)–Ru(3) 79.9(1)° and Ru(2)–P(2)–Ru(3) 74.4(1)°) are acute. Clearly, with the ability of acetylides to bind as terminal, μ_2, μ_2-η^2, or μ_3-η^2 ligands, the possibilities for structural isomerism and fluxionality in such systems are great. One of the most intriguing features of $Ru_3(CO)_6(C{\equiv}CBu^t)_2(PPh_2)_2(Ph_2PC{\equiv}CBu^t)$ is the occurence of three types of acetylide in the same molecule, thus offering the possibility of examining the influence of the bonding mode on reactivity.

Finally, a minor product from $Ru_3(CO)_{12}$ + $Ph_2PC{\equiv}CPr^i$ has the formula $Ru_3(CO)_7(Ph_2PC{\equiv}CPr^i)_3$. P-31 NMR (*see* Table II) unequivocally establishes that this complex contains only one phosphido group and two unfragmented phosphines. Coupling constants indicate that one phosphine is trans to the PPh_2 bridge (J_{P-P} = 234 Hz) and the other one is cis to PPh_2 (J = 22 Hz). Thus the two phosphines are attached to the same two ruthenium atoms as the PPh_2 group. This structure suggests a smooth conversion to $Ru_3(CO)_6(C{\equiv}CPr^i)_2(PPh_2)_2$-$(Ph_2PC{\equiv}CPr^i)$ via fragmentation of a second phosphinoalkyne, which is observed.

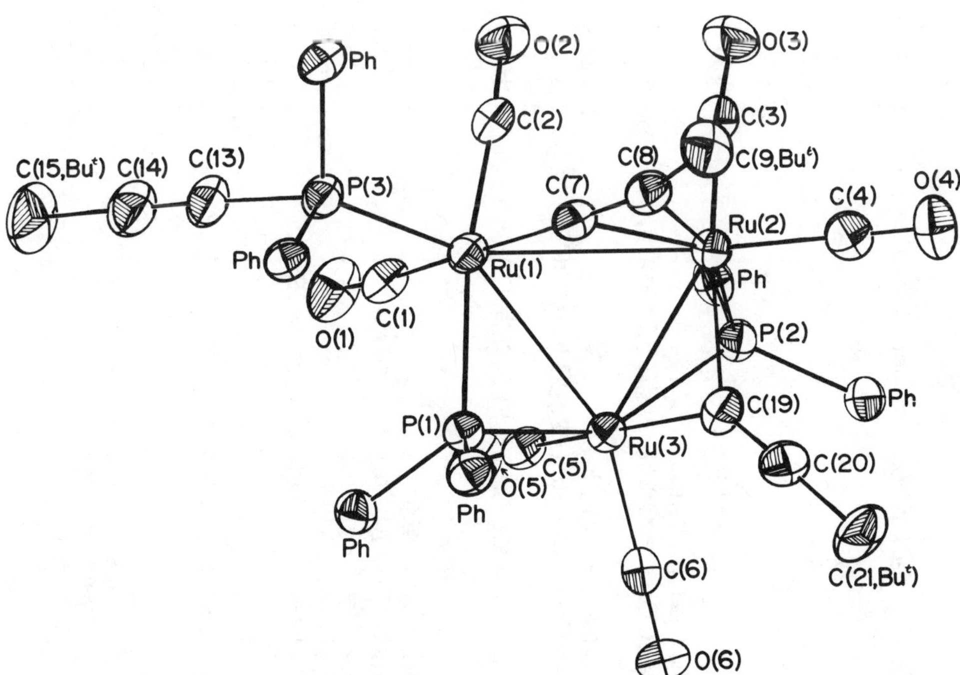

Figure 7. The molecular structure of $Ru_3(CO)_6(C{\equiv}CBu^t)_2(PPh_2)_2$-$(Ph_2PC{\equiv}CBu^t)$

Table II. P-31 NMR Parameters for Phosphinoacetylene and Phosphido-Bridged Iron-Group Complexes

Compound	$\delta(P\text{-}31)$[a]	J[b]	μ-PPh_2 M–P–M *Angle (degrees)*
$Fe_2(CO)_6(C{\equiv}CBu^t)(PPh_2)$	s 148.3	—	71.7(0)
trans-$Fe_2(CO)_5(C{\equiv}CPr^i)(PPh_2)(Ph_2PC{\equiv}CPr^i)$[c]	d 153.8	95 (PPh_2)	—
	d 36.6	95 ($Ph_2PC{\equiv}CPr^i$)	
$Ru_2(CO)_6(C{\equiv}CBu^t)(PPh_2)$	s 123.9	—	72.0(0)
trans-$Ru_2(CO)_5(C{\equiv}CPr^i)(PPh_2)(Ph_2PC{\equiv}CPr^i)$[c]	d 127.0	215 (PPh_2)	73.4(0)
	d 16.3	215 ($Ph_2PC{\equiv}CPr^i$)	
cis-$Ru_2(CO)_5(C{\equiv}CPr^i)(PPh_2)(Ph_2PC{\equiv}CPr^i)$[c]	d 108.0	17 (PPh_2)	—
	d 17.1	17 ($Ph_2PC{\equiv}CPr^i$)	
$Ru_3(CO)_{11}(Ph_2PC{\equiv}CBu^t)$	s 3.1	—	—
$Ru_3(CO)_{10}(Ph_2PC{\equiv}CPr^i)_2$	s 3.5	—	—
$Ru_3(CO)_9(C{\equiv}CPr^i)(PPh_2)$	s −51.8	—	92.8(0)
$Ru_3(CO)_8(C{\equiv}CPr^i)(PPh_2)$	s 113.0	—	74.4(0)
$Os_3(CO)_9(C{\equiv}CPr^i)(PPh_2)$	s −65.4	—	93.2(0)
$Ru_3(CO)_6(C{\equiv}CBu^t)_2(PPh_2)_2(Ph_2PC{\equiv}CBu^t)$	q 11.9	$J' = 1, J'' = 12$ ($Ph_2PC{\equiv}CBu^t$)	—
	q 152.3	$J = 144, J' = 1$ (PPh_2)	79.9
	q 121.8	$J = 144, J'' = 12$ (PPh_2)	75.9
$Ru_3(CO)_7(C{\equiv}CPr^i)(PPh_2)(Ph_2PC{\equiv}CPr^i)_2$	q 14.0	$J' = 234, J'' < 1$	—
	q 14.7	$J = 22.0, J'' < 1$	—
	q 111.9	$J = 22.0, J' = 234$ (PPh_2)	—

[a] Chemical shifts in parts per million from 85% H_3PO_4. Downfield values are positive.
[b] Coupling constants in hertz.
[c] In the trans compound the $Ph_2PC{\equiv}CR$ ligand is approximately trans to the phosphido bridge. The probable configuration for the cis species is with the phosphine cis to the phosphido bridge and trans to the metal–metal bond.

$Ru_4(CO)_{10}(C = CHPr^i)(OH)(PPh_2)$: A Tetranuclear Cluster with Bridging Vinylidene and Hydroxo Groups. The reaction of $Ru_3(CO)_{11}$-($Ph_2PC{\equiv}CPr^i$) in the presence of $Me_3NO \cdot 2H_2O$ gave 10% of an unusual cluster $Ru_4(CO)_{10}(C = CHPr^i)(OH)(PPh_2)$ whose X-ray structure is shown in Figure 8. In the tetranuclear complex the four ruthenium atoms adopt a butterfly arrangement with four strong Ru–Ru bonds (Ru(1)–Ru(2) 2.758(1), Ru(1)–Ru(3) 2.703(1), Ru(2)–Ru(4) 2.800(1), Ru(3)–Ru(4) 2.803(1) Å) and two open edges (Ru(1)–Ru(4) 4.1244(6) Å and Ru(2)–Ru(3) 3.4559(6) Å). The Ru(1)–Ru(3) bond is bridged by a phosphido group (Ru(1)–P–(Ru(3) 71.4(0)°) and a remarkable μ_4-η^2-vinylidene group is σ-bonded to ruthenium(1), ruthenium(2), and ruthenium(3) and η^2-coordinated to ruthenium(4). Although there are numerous examples of vinylidene complexes (*25–30*), the multisite interaction with four metal atoms in $Ru_4(CO)_{10}$-($C = CHPr^i$)(OH)(PPh_2) is unusual. A further novel feature is the presence of a triply bridging hydroxo ligand on the Ru(1)–Ru(2)–Ru(3) face.

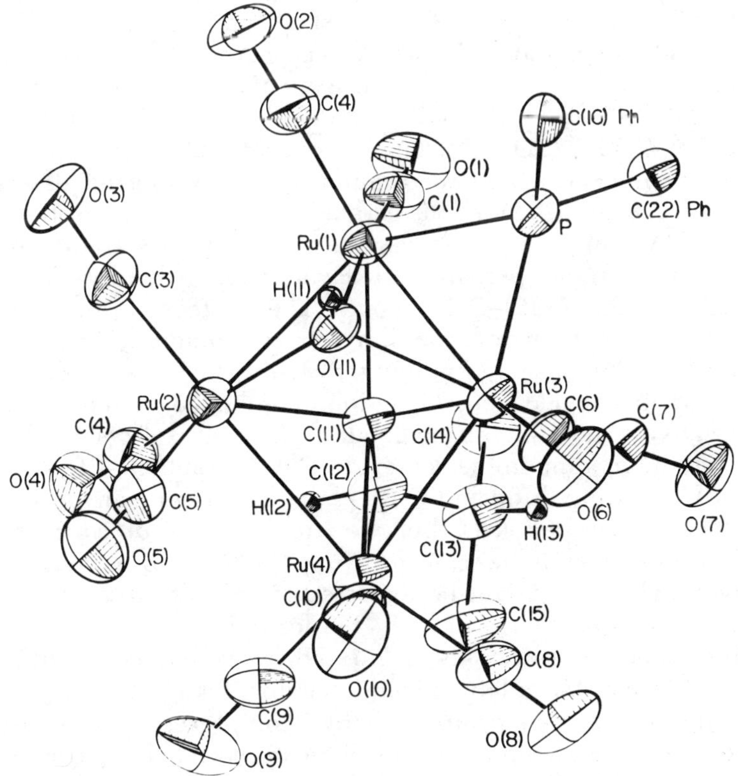

Figure 8. A perspective view of the molecular structure of $Ru_4(CO)_{10}$-($C = CHPr^i$)(OH)(PPh_2)

The hydrogen atom on the hydroxyl oxygen was located clearly in a difference Fourier synthesis and confirmed further by the presence of a ν(OH) band at 3510 cm^{-1} in the IR spectrum. Although the origin of the vinylidene hydrogen atom and the hydroxo ligand has not been established conclusively as yet, it seems likely that a water molecule from $Me_3NO \cdot 2H_2O$ is the precursor of both. While the catalytic properties of this remarkable molecule are being examined only now, it is worth mentioning that there is considerable current interest in clusters with hydroxo ligands owing to the possible relevance of surface hydroxyl groups in the catalytic reduction of CO to CH_4 (*31*).

$Os_3(CO)_9(COOEt)(OH)(Ph_2PC{\equiv}CPr^i)$: An Open Triangular Cluster with a Sideways-Bound Ester Group and a Bridging Hydroxo Ligand. Excellent yields of this complex were obtained from $Os_3(CO)_{12}$ as indicated in Scheme V. The corresponding $Os_3(CO)_9$-(COOR)(OH)($Ph_2PC{\equiv}CPr^i$) (R = Me or CPh_3) compounds also have been characterized. The structure of the ethyl ester has been determined by X-ray analysis (*see* Figure 9) and consists of an open triangle of osmium atoms with two strongly bonding Os–Os distances (Os(1)–Os(2) 2.885 Å and Os(2)–Os(3) 2.883 Å) and an essentially nonbonding Os(1)—Os(3) separation of 3.385 Å. The open edge is bridged by a sideways-bound ester group and the oxygen atom of a hydroxo ligand. The Os(1)–C(12) (2.11 Å), Os(3)—C(12) (2.91 Å), Os(3)–O(11) (2.16 Å), and C(12)–O(11) (1.29 Å) distances suggest that oxygen(11) but not carbon(12) is coordinated to osmium(3). The preferred description then is of an ester σ-bonded via carbon to osmium(1) and σ-bonded to osmium(3) via an oxygen lone pair. An alternative description in terms of an ethoxy-carbene perhaps is disfavored somewhat by the relative shortness of the C(12)–O(11) bond. The molecule obeys the 18-electron rule if both ester and hydroxyl ligands contribute three electrons to the cluster, implying the absence of an Os(1)—Os(3) bond. Bridging acyl (*32, 33, 34*) and formamido (*35, 36*) complexes are still relatively rare in organometallic chemistry although they are obviously interesting in the context of CO elaboration. A few osmium derivatives with hydride and C(R)O ligands bridging an Os–Os bond have been described (*34*), but we are unaware of corresponding bridging esters. As is the case for $Ru_4(CO)_{10}(C = CHPr^i)(OH)(PPh_2)$, the hydroxyl ligand might be expected to exhibit protonic character and be a potential source of hydrogen for reductions.

Trends and Implications of P-31 NMR Parameters. With X-ray structural data now available in our laboratory for a large number of bi- and polynuclear iron-group carbonyl complexes with phosphido bridges, we have attempted to correlate structural changes with P-31 NMR chemical shifts and coupling constants in an effort to establish

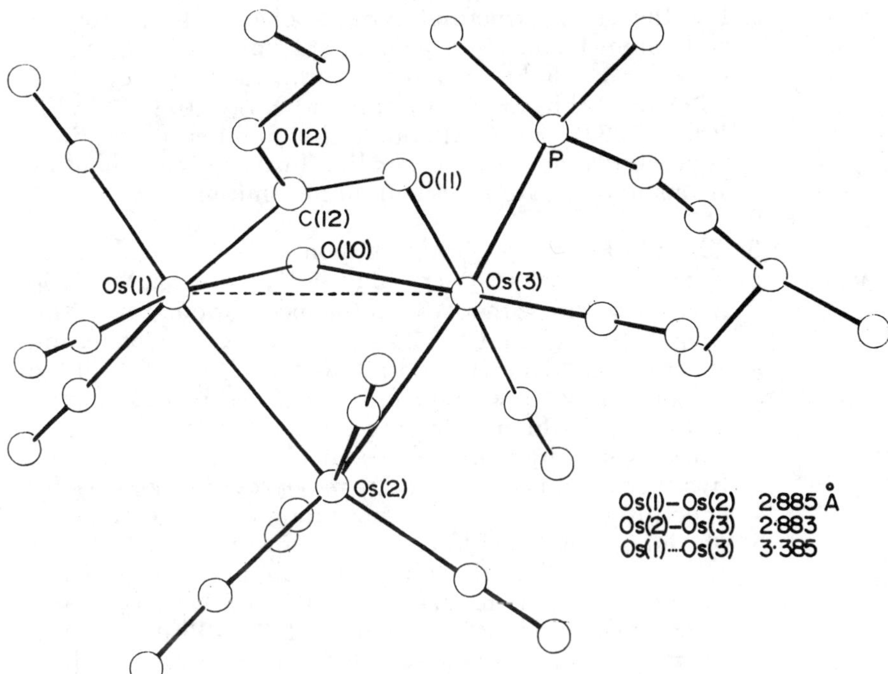

Figure 9. The crystalline structure of $Os_3(CO)_9(COOEt)(OH)$-($Ph_2PC{\equiv}CPr^i$) as shown by X-ray diffraction. The hydrogen atom on the bridging hydroxo ligand is not shown.

empirical relationships of potential diagnostic significance. A selection of $\delta(^{31}P)$ and J values for these clusters is listed in Table II together with pertinent metal–P–metal angles where available.

General Observations

1. The δ (P-31) values for PPh_2 groups bridging an Fe–Fe or Ru–Ru bond are shifted considerably downfield of δ values for the phosphorus-coordinated, intact phosphinoacetylenes. The chemical shifts for $Ru_3(CO)_{11}(Ph_2PC{\equiv}CBu^t)$ (+3.1 ppm), $Ru_2(CO)_6(C{\equiv}CBu^t)(PPh_2)$ (+123.9 ppm), and *trans*-$Ru_2(CO)_5(C{\equiv}CPr^i)(PPh_2)(Ph_2PC{\equiv}CPr^i)$ (+127.0 ppm, PPh_2 and +16.3 ppm ($Ph_2PC{\equiv}CPr^i$) provide a nice illustration of this trend. Thus this type of phosphido bridge is easily distinguishable by P-31 NMR. In the complex $Ru_3(CO)_7(Ph_2PC{\equiv}CPr^i)_3$, whose structure has not been determined yet, the P-31 spectrum suggests one phosphido bridge and two intact phosphines.

2. For the same structural type, a change from iron to ruthenium results in a general shift of P-31 resonances to a higher field. A typical comparison is $Fe_2(CO)_6$-($C{\equiv}CBu^t$)PPh_2 (+148.3 ppm) and $Ru_2(CO)_6$($C{\equiv}CBu^t$)-PPh_2 (123.9 ppm). Although the number of osmium compounds measured is small, a further high-field shift seems to occur from ruthenium to osmium.

3. Phosphido groups bridging nonbonding metals have P-31 resonances at a very high field. Thus for $M_3(CO)_9$-($C{\equiv}CPr^i$)(PPh_2) the P-31 resonances appear at −51.8 ppm (M = Ru) and −65.4 ppm (M = Os). Converting $Ru_3(CO)_9$($C{\equiv}CPr^i$)(PPh_2) to $Ru_3(CO)_8$($C{\equiv}CPr^i$)(PPh_2) results in a large downfield shift of δ (P-31) to +113.0 ppm as the Ru–P–Ru angle changes from 92.8(0)° to 74.4(0)°. These very large differences in chemical shift for the two types of phosphido bridge may provide a very useful spectroscopic probe to diagnose the presence or absence of metal–metal bonding. It is interesting that the P-31 chemical shift of the phosphido groups bridging the Pd–Pd bonds in the cation $[Pd_3Cl(PPh_2)_2(PPh_3)_3]^+$ lies very substantially downfield of the P-31 (PPh_2) resonance in the nonmetal–metal-bonded dimer $Pd_2Cl_2(PPh_2)_2$-$(PEt_3)_2$ (*37*). The PPh_2 resonances in the related platinum dimers $Pt_2Cl_2(PPh_2)_2L_2$ (L = tertiary phosphine) that lack a significant Pt–Pt bond also lie at a very high field (*38*). More recent data for $[Ir(COD)(PPh_2)_2]_2$ (δ = 166 ppm) which contains an Ir–Ir bond and for $[Rh(COD)(PPh_2)_2]_2$ (δ = −71 ppm) which lacks a Rh–Rh bond (*39*) confirm that the relationship of δ (P-31) to metal–P–metal angle (or M—M separation) may be a more general one. Even the relatively small change in Mo–Mo distance (0.11 Å) or Mo–P–Mo angle (1.2°) from $Mo_2(C_5H_5)_2(CO)_4(H)$-(PMe_2) to $Ph_4As[Mo_2(C_5H_5)_2(CO)_4(PMe_2)]$ may contribute to the large P-31 chemical shift difference (146.4 vs. 61.6 ppm) for the PMe_2 bridges in these two compounds (*9*).

Although there are indications that for PR_2 groups bridging noninteracting metals δ (P-31) becomes more negative as the angle M–P–M opens, considerably more data will be required to establish this relationship. For binuclear iron complexes with Fe–Fe bonds a definite correlation of δ (P-31) with ∠Fe–P–Fe is evident, with larger, positive shifts corresponding to larger angles at the phosphido bridge (vide infra).

4. P–P coupling constants provide additional useful information (*see* Table II). For phosphines occupying sites trans to phosphido bridges on the same metal atom, $^2J_{P-P}$

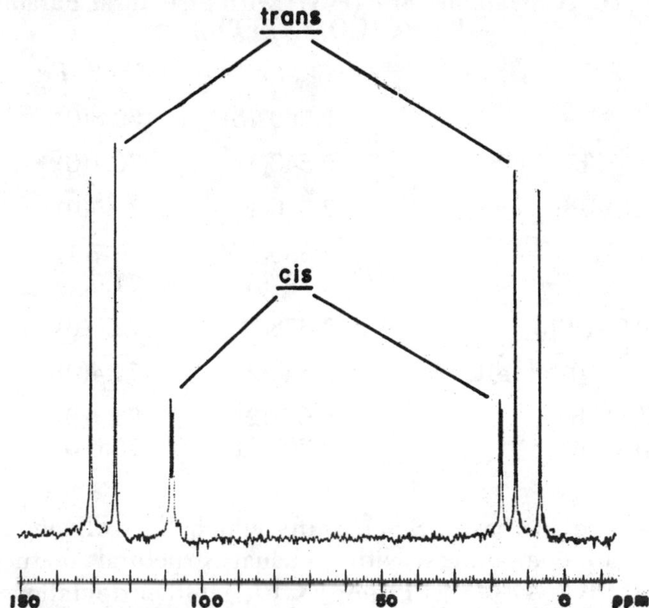

Figure 10. P-31 NMR spectrum of a mixture of isomers of $Ru_2(CO)_5$-($C{\equiv}CPr^i$)(PPh_2)($Ph_2PC{\equiv}CPr^i$). The isomer ratio is ~5:1 in favor of the trans isomer.

is quite large (e.g., ^2J = 215 Hz in *trans*-$Ru_2(CO)_5$-($C{\equiv}CPr^i$)(PPh_2)($Ph_2PC{\equiv}CPr^i$)). Cis couplings are much smaller (e.g., ^2J = 17 Hz in *cis*-$Ru_2(CO)_5$($C{\equiv}CPr^i$)(PPh_2)-($Ph_2PC{\equiv}CPr^i$)). Three bond couplings through two metal atoms between phosphines and phosphido bridges are generally small (e.g., $^3J_{P-P}$ = 1.0 Hz in $Ru_3(CO)_6$-$(C{\equiv}CPr^i)_2(PPh_2)_2$($Ph_2PC{\equiv}CPr^i$).

A P-31 NMR spectrum of a mixture of trans and cis isomers of $Ru_2(CO)_5$($C{\equiv}CPr^i$)(PPh_2)($Ph_2PC{\equiv}CPr^i$) illustrating chemical shift differences between phosphine and the phosphido bridge and between cis and trans couplings is shown in Figure 10.

P-31 Chemical Shift–Bond Angle Correlations in Binuclear Iron Carbonyls

As alluded to earlier, we have amassed a large body of X-ray data for binuclear iron carbonyl complexes of the $Fe_2(CO)_6(X)(PPh_2)$ type where X represents a three-electron donor ligand. The P-31 chemical shifts for these compounds are tabulated with Fe–P–Fe angles in Table III and a plot of δ vs. angle is shown in Figure 11. Although the correlation is by no means perfect there is an unmistakable increase in

Table III. Correlation of δ (P-31) with Structural Parameters for $Fe_2(CO)_6(X)(PPh_2)$

X	*l(Fe–Fe)(Å)*	*∠Fe–P–Fe*	*δ(ppm)*
Chlorine	2.5607(5)	69.8(0)	142.0
$\overset{\ominus}{C}HC(\overset{\oplus}{N}Et_2)Ph$	2.548(1)	70.1(0)	153.9
$\overset{\ominus}{C}HC(\overset{\oplus}{N}HCy)Ph$	2.576(1)	70.5(0)	154.0
$\overset{\ominus}{C}C(Cy_2\overset{\oplus}{P}H)Ph$	2.550(2)	70.7(1)	123.0
C≡CPh	2.597(3)	71.6(1)	148.4
$C(\overset{\oplus}{N}HCy)\overset{\ominus}{C}H(Ph)$	2.628(1)	72.5(0)	183.5
$C(\overset{\oplus}{C}NMe(CH_2)_2NMe)\overset{\ominus}{C}Ph$	2.644(2)	73.4(0)	190.3
$C(\overset{\oplus}{C}\overset{\ominus}{N}Bu^t)CPh$	2.671(2)	74.1(0)	194.2
$CH_2C(Ph)NMe$	2.707(1)	75.6(0)	198.5

δ as ∠Fe–P–Fe increases. Some shifts which lie well off the curve correspond to compounds with unusual structural features (e.g., $[Fe_2(CO)_6[CHC(NEt_2)Ph](PPh_2)Ag]^+C10_4^-$). Other deviations may be due to changes in ligand back-bonding ability (e.g., $Fe_2(CO)_5$-(C≡CPh)(PPh_2)PPh_3). The range of bond angles is only ~6° yet δ P-31) varies from ~ 120–200 ppm. This sensitivity of the P-31 shift to bond angle has been used with a considerable degree of success to predict the ligating characteristics of zwitterionic hydrocarbyl ligands (*40*) and to estimate the Fe–P–Fe angle in $Fe_2(CO)_6(Cl)PPh_2$ (*41*).

With the dependence of δ (P-31) on the metal–P–metal angle reasonably established for this admittedly restricted series of iron complexes and with some indications that δ becomes increasingly negative (i.e., upfield) as the bond angle increases for PR_2 groups bridging non-bonded metals, it is tempting to suggest that there may be a continuous curve relating δ and metal–P–metal (or *l*(M—M)). Similar relationships have been found for δ (P-31) vs. ∠O–P–O in phosphate esters (*42*). In this regard, data for complexes with metal–P–metal angles in the 80°–90° range would be most interesting.

The Reactivity of μ_2-η^2 and μ_3-η^2-Acetylides. Our long-term goal in this work is to examine patterns of reactivity for multisite-bound unsaturated ligands as a basis for understanding likely sequences of reactions in reactions catalyzed by metal clusters. The availability and structural characterization of species such as those illustrated in Figure 3 presents a rather unique opportunity to explore and contrast reactivities for different metals and ligand-bonding modes. Previous work (*16, 17, 18*) has shown abundantly that the μ_2-η^2-acetylide in $Fe_2(CO)_6$(C≡CR)PPh_2 (R = Ph) is hyper-reactive to nucleophiles

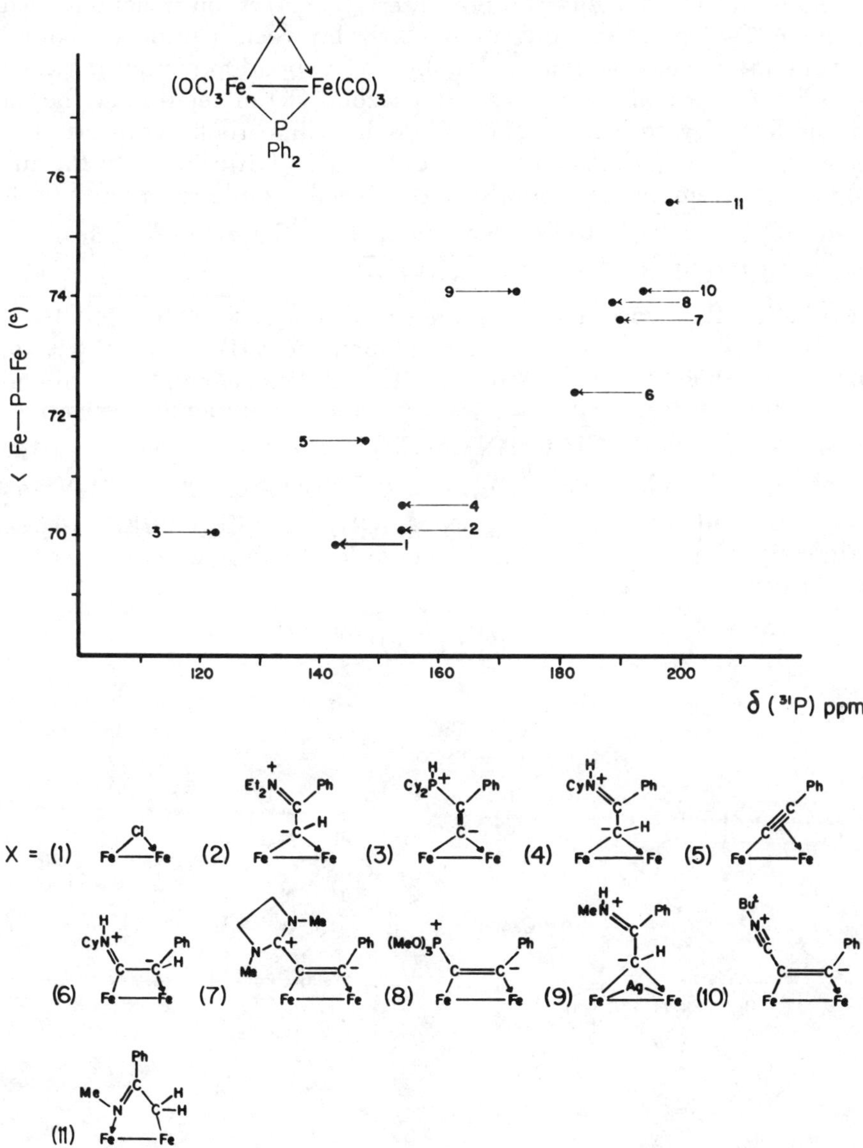

Figure 11. A plot of δ (P-31) vs. bond angle at phosphorus for phosphido-bridged complexes $Fe_2(CO)_6(X)(PPh_2)$ where X is a three-electron donor.

including primary and secondary amines (*16*), secondary and tertiary phosphines (*17*), and phosphites (*18*). In the context of homogenous catalysis, however, more interest centers on reactions that involve C–C bond making, particularly between a multisite-bound ligand and a weak carbon nucleophile. C–C bond formation between coordinated or elaborated CO and a second CO molecule may be for example a key step in Fischer–Tropsch hydrocarbon synthesis (*43*). The acetylidic carbon atoms in $Fe_2(CO)_6(C{\equiv}CR)PPh_2$ (R = Ph and Bu^t) are extremely electrophilic and in a remarkable reaction the carbene $:\overline{CN(Me)CH_2CH_2N}Me$, generated from the electron-rich olefin $Me\overline{NCH_2CH_2N(Me)C}{=}\overline{CN(Me)CH_2CH_2N}Me$, attacks the α-carbon generating the zwitterionic complex $Fe_2(CO)_6[C\{\overline{CN(Me)CH_2CH_2N}Me\}CR](PPh_2)$ via C–C coupling (*see* Scheme VII). While this C–C bond formation reaction competes with CO displacement by carbene especially at elevated temperatures, the zwitterionic derivatives $Fe_2(CO)_6[C\{\overline{CN(Me)CH_2CH_2N}Me\{CR](PPh_2)$ do not convert to the substitution product $Fe_2(CO)_5(C{\equiv}CR)(PPh_2)(\overline{CN(Me)CH_2CH_2N}Me)$. An X-ray study of $Fe_2(CO)_6[C\{\overline{CN(Me)CH_2CH_2N}Me\}CPh](PPh_2)$ (*see* Figure 12) has confirmed that C(acetylide)–C(carbene) coupling has occurred.

Scheme VII.

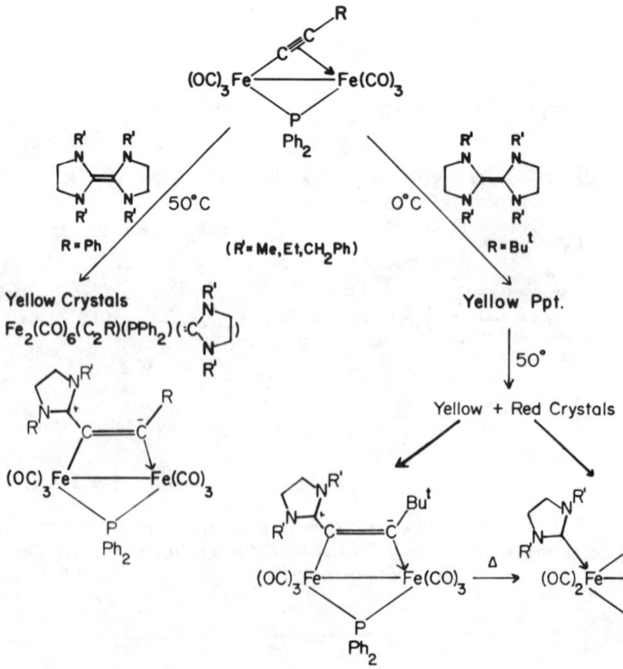

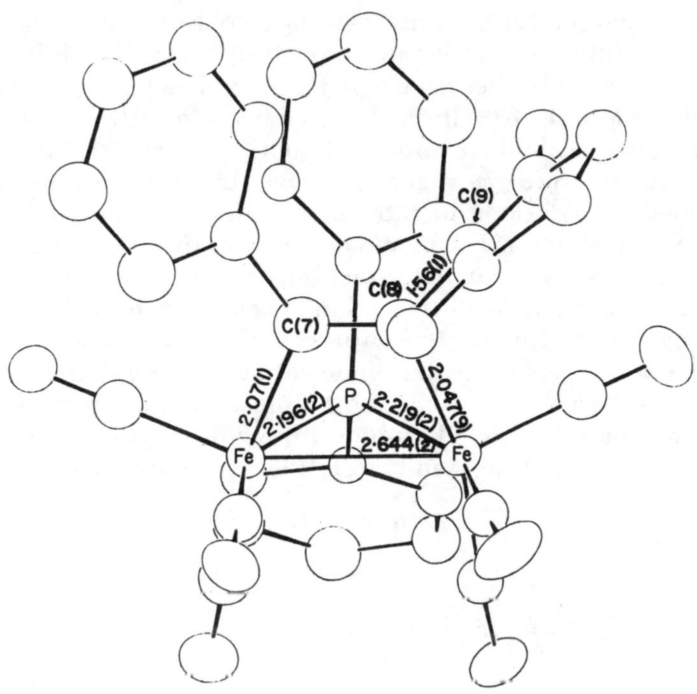

Figure 12. An ORTEP II plot of the molecular structure of $Fe_2(CO)_6$-{$C\overline{CN(Me)CH_2CH_2N}Me$}CPh}($PPh_2$) showing some pertinent bond lengths

Equally remarkable is the facile attack at the carbocationic carbon in $Fe_2(CO)_6(C{\equiv}CR)(PPh_2)$ by an isonitrile, normally an excellent ligand for low-oxidation-state metals. The products of nucleophilic attack at carbon (*see* Scheme VIII) are dipolar with the formal positive charge presumably localized on isonitrile nitrogen. The negative charge on the original β-carbon of the acetylide is delocalized presumably into the $Fe_2(CO)_6(PPh_2)$ framework. X-ray analysis (*see* Figure 13) of the *t*-BuNC derivative showed a C(Bu^tNC)–C (acetylide) bond length of 1.37(2) Å, which is indicative of a strong, partially multiple bond. Significantly the carbon-coordinated isonitrile undergoes addition reactions with secondary amines much in the same manner as metal–isonitrile complexes do to generate the carbon-coordinated carbenes. It is clear from this work that sideways-bound acetylides have a distinct chemistry based on the extreme electrophilicity of the carbocationic carbon atoms. In our opinion such behavior is directly relevant to the expected chemistry of carbidic carbon atoms in cluster carbides and may be pertinent to the C–C coupling mechanism involving CO.

The μ_3-η^2-acetylides as in $M_3(CO)_9(C{\equiv}CR)(PPh_2)$ might be expected to exhibit an even greater propensity for nucleophilic attack. Studies of these molecules have only just begun, but preliminary work shows that $Ru_3(CO)_9(C{\equiv}CPr^i)(PPh_2)$ readily adds ethylamine affording the novel enamine complex $Ru_3(CO)_8\{C(NHEt){=}CHPr^i\}PPh_2$ whose X-ray structure (*see* Figure 14) shows the enamine double bond coordinated in η^2-fashion to ruthenium(3) with the amino group attached to ruthenium(1). The ligand thus retains the five-electron donor capacity of the acetylide in the parent but one molecule of CO is lost to form a closed cluster. As illustrated in Scheme IX, the stabilization of the enamine in the Ru_3 cluster contrasts with the trapping of the iminium ion valence isomer in the binuclear species $Fe_2(CO)_6\{C(\overset{\oplus}{N}HR)\text{-}\overset{\ominus}{C}Ph\}(PPh_2)$. Further studies on the M_3 cluster are in progress.

A Cautionary Note on Phosphido-Bridge Reactivity. Throughout this discussion of cluster and acetylide reactivity, the phosphido

Scheme VIII.

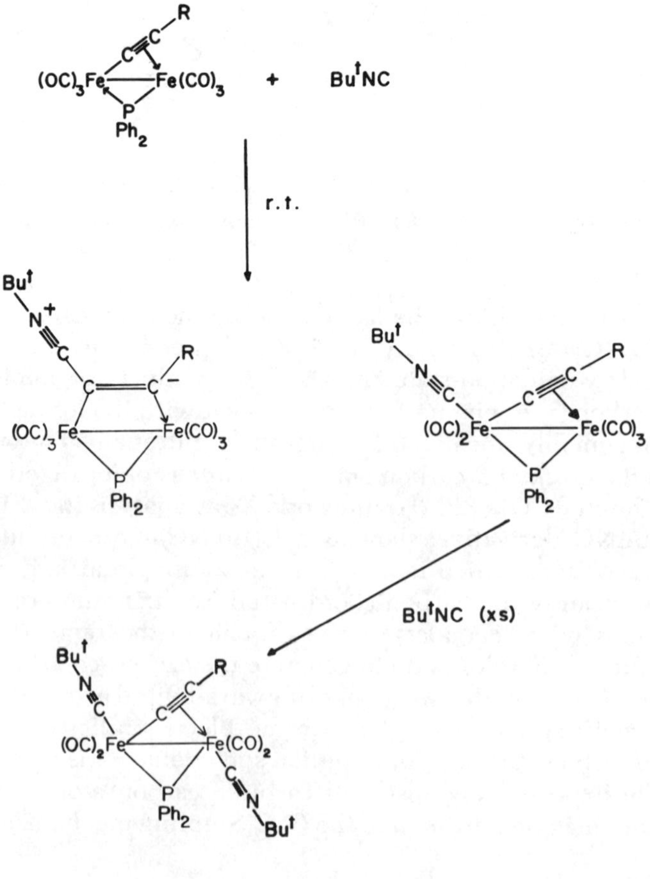

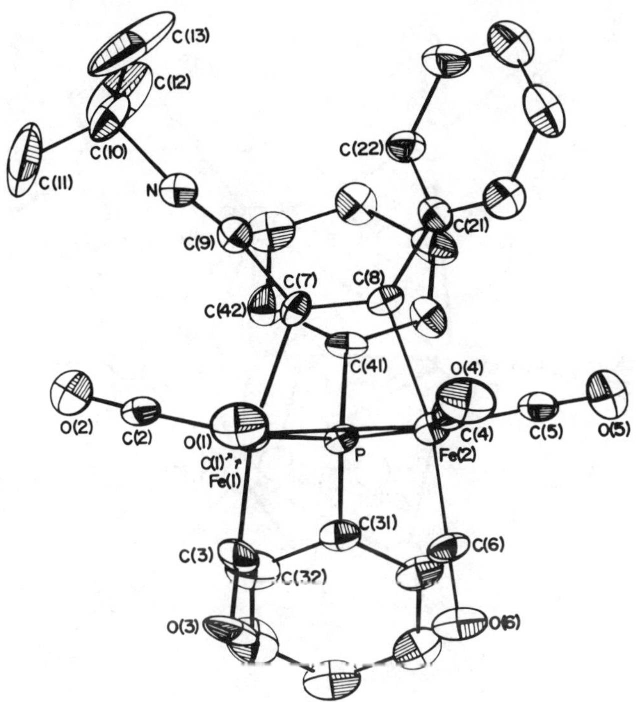

Figure 13. The X-ray crystal structure of $Fe_2(CO)_6\{C(CNBu^t)CPh\}$-$(PPh_2)$ illustrating attachment of the isonitrile carbon atom to the α-carbon atom of the acetylide

bridge has remained intact although the nature of the bridge has varied. This stability of PPh_2 bridges is an important factor in their application as supporting ligands. Nevertheless, for certain reagents, there is evidence for reaction at the metal–P bond of a phosphido bridge. For example Scheme X illustrates that the reaction of $Fe_2(CO)_6(C{\equiv}CBu^t)$-$(PPh_2)$ with disubstituted acetylenes results in Fe–PPh_2 bond cleavage and the generation of phosphine-containing products via acetylide–acetyline coupling. All three product types in Scheme X have been characterized fully by single-crystal X-ray diffraction (*44*). The PPh_2 bridge clearly is relatively labile toward acetylenes.

Conclusions: Structural and Chemical Implications

1. Phosphinoalkynes are a useful source of phosphido and acetylido groups for cluster synthesis.
2. The phosphido bridge is an exceedingly versatile bridging ligand in clusters. Metal—metal bond making and

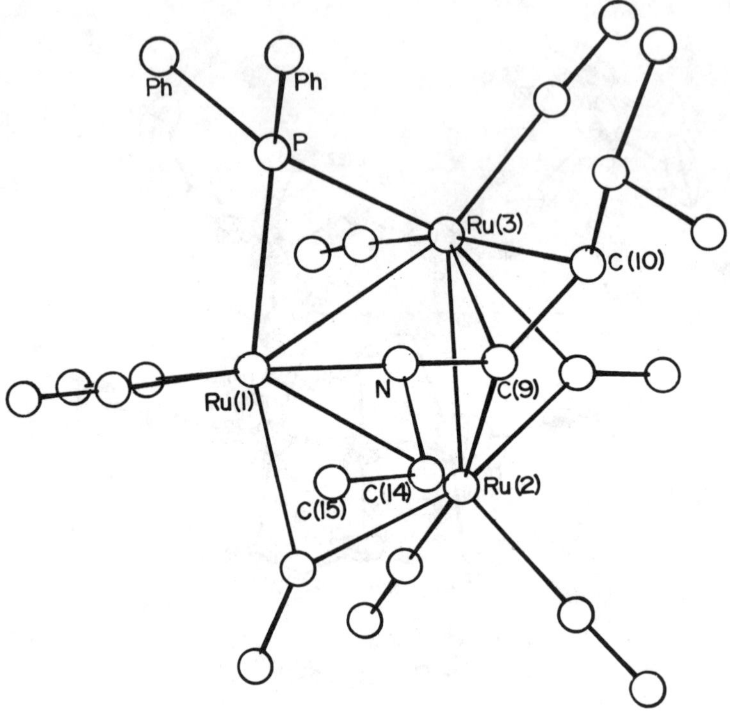

Figure 14. A perspective view of the molecular structure of $Ru_3(CO)_8\{C(NHEt) = CHPr^i\}PPh_2$

Scheme IX. Comparison of reactions of μ_2-$C{\equiv}CR$ *and* μ_3-$C{\equiv}CR$ *with amines*

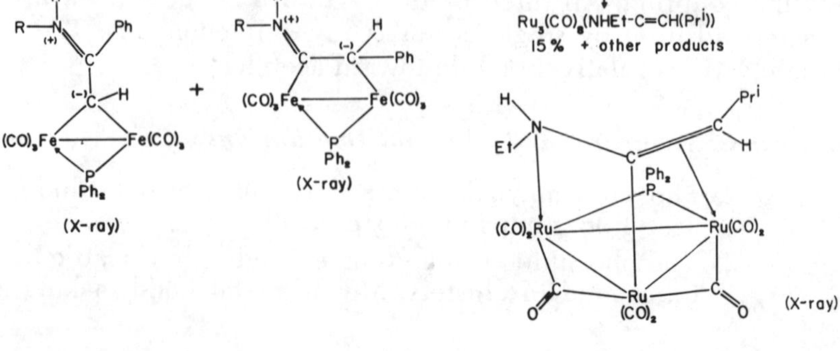

Ph
C
C
$(CO)_3Fe$ — $Fe(CO)_3$
P
Ph_2

RNH_2

H
R—N$^{(+)}$ Ph
C
$^{(-)}$C—H
$(CO)_3Fe$ — $Fe(CO)_3$
P
Ph_2
(X-ray)

+

H
R—N$^{(+)}$ H
C — $^{(-)}$C — Ph
$(CO)_3Fe$ — $Fe(CO)_3$
P
Ph_2
(X-ray)

Pr^i
C≡C
$Ru(CO)_3$
$(CO)_3Ru$
$Ru(CO)_3$
P
Ph_2

$EtNH_2$, heptane
RT, 24 hrs.

$Ru_3(CO)_8(NHEt$-$C{=}CH(Pr^i))$
15 % + other products

H
N
Et
C
C
Pr^i
H
Ph_2
P
$(CO)_2Ru$
$Ru(CO)_2$
C
O
Ru
$(CO)_2$
C
O
(X-ray)

Scheme X. Phosphido-bridge reactivity: insertion into the bridge.

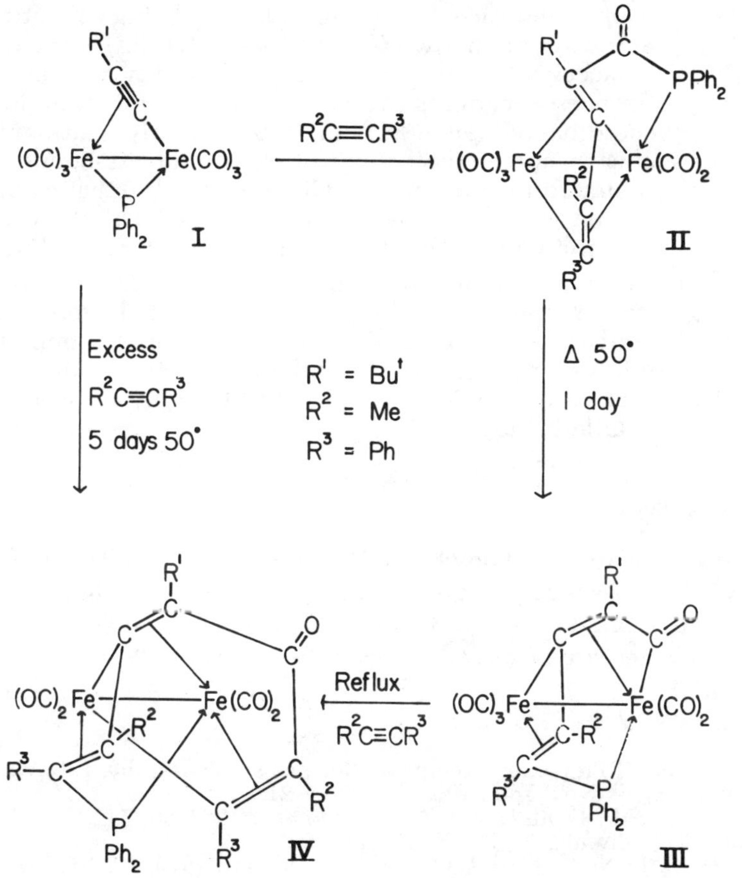

breaking can be accommodated easily because of the wide variation in metal–P–metal angles tolerated by the PPh_2 group.

3. It is clear that P-31 chemical shifts for diphenylphosphido bridges sensitively reflect changes in metal–P–metal angle. Although the number of compounds examined to data is relatively limited, for specific compound types δ (P-31) increases (moves increasingly downfield) as metal–P–metal angle increases for PPh_2 bridging a bonding metal–metal interaction. The δ (P-31) shifts of PPh_2 groups bridging nonbonding metal—metal systems appear at a much higher field and appear to move upfield as the metal–P–metal angle widens. These trends have diagnostic structural significance.

4. Acetylides are the most versatile of the hydrocarbyl groups with terminal, symmetrical μ_2, unsymmetrical μ_2-η^2, and μ_3-η^2-bonding modes being identified. Sideways-bound μ_2-η^2- and μ_3-η^2-acetylides, being readily accessible, offer an exciting opportunity to probe differences in patterns of reactivity resulting from multisite coordination. Such information is highly pertinent to cluster-mediated reactions of other multisite-bound, unsaturated ligands such as CO. It is notable that carbidic carbon atoms in cluster carbides may exhibit extreme sensitivity to nucleophilic attack as do σ-π-acetylides.
5. Observations of triply and doubly bridging hydroxo groups in otherwise low-valent carbonyl clusters is particularly exciting in view of the potentially important role of such ligands in catalytic reductions. Molecules such as $Ru_4(CO)_{10}(C = CHPr^i)(OH)(PPh_2)$ are worthy of detailed study.

Acknowledgments

It is a pleasure to acknowledge the considerable contributions of my co-workers, S. A. McLaughlin, G. N. Mott, W. F. Smith, and N. J. Taylor to this work. I also would like to thank the Natural Sciences and Engineering Research Council for its generous support.

Literature Cited

1. Hayter, R. G. "Preparative Inorganic Reactions"; W. L. Jolly, Ed.; Wiley–Interscience: New York, 1965; Vol. 2, p. 211.
2. Coleman, J. M.; Dahl, L. F. *J. Am. Chem. Soc.* **1967,** *89,* 542.
3. Keller, E.; Vahrenkamp, H. *Chem. Ber.* **1977,** *110,* 430.
4. Haines, R. J.; Nolte, C. R. *J. Organometal. Chem.* **1972,** *36,* 163.
5. Rottinger, E.; Muller, R.; Vahrenkamp, H. *Angew. Chem. Int. Ed.* **1977,** *16,* 332.
6. Ginsburg, R. E.; Rothrock, R. K.; Finke, R. G.; Collman, J. P.; Dahl, L. F. *J. Am. Chem. Soc.* **1979,** *101,* 6550.
7. Matthew, M.; Palenik, G. J.; Carty, A. J.; Paik, H. N. *J. Chem. Soc. Chem. Commun.* **1974,** 25.
8. Carty, A. J.; Ferguson, G.; Paik, H. N.; Restivo, R. *J. Organometal. Chem.* **1974,** *74,* C14.
9. Petersen, J. L.; Stewart, R. P., Jr. *Inorg. Chem.* **1980,** *19,* 186.
10. Collman, J. P.; Rothrock, R. K.; Finke, R. G.; Rose–Munch, F. *J. Am. Chem. Soc.* **1977,** *99,* 7381.
11. Werner, H.; Hofmann, W. *Angew. Chem. Int. Ed.* **1979,** *18,* 158.
12. Carty, A. J.; Mott, G. N.; Taylor, N. J. *J. Am. Chem. Soc.* **1979,** *101,* 3131.
13. Carty, A. J.; Smith, W. F.; Taylor, N. J. *J. Organometal. Chem.* **1978,** *146,* C1.
14. Carty, A. J.; McLaughlin, S. A.; Taylor, N. J., unpublished data.
15. Carty, A. J.; Taylor, N. J.; Smith, W. F. *J. Chem. Soc. Chem. Commun.* **1979,** 750.
16. Carty, A. J.; Mott, G. N.; Taylor, N. J.; Yule, J. E. *J. Am. Chem. Soc.* **1978,** *100,* 3051.

17. Carty, A. J.; Mott, G. N.; Taylor, N. J.; Ferguson, G.; Khan, M. A.; Roberts, P. J. *J. Organometal. Chem.* **1978,** *149,* 345.
18. Wong, Y. S.; Paik, H. N.; Chieh, P. C.; Carty, A. J. *J. Chem. Soc. Chem. Commun.* **1975,** 309.
19. Carty, A. J.; Taylor, N. J.; Smith, W. F.; Lappert, M. F.; Pye, P. L. *J. Chem. Soc. Chem. Commun.* **1978,** 1017.
20. Sappa, E.; Gambino, O.; Milone, L.; Cetini, G. *J. Organometal. Chem.* **1972,** *39,* 169.
21. Catti, M.; Gervasio, G.; Mason, S. A. *J. Chem. Soc. Dalton Trans.* **1977,** 2260.
22. Smith, W. F.; Yule, J. E.; Taylor, N. J.; Paik, H. N.; Carty, A. J. *Inorg. Chem.* **1977,** *16,* 1593.
23. Morosin, B.; Howatson, J. *J. Organometal. Chem.* **1971,** *29,* 7.
24. ten Hoedt, R. W. M.; Noltes, J. G.; Van Koten, G.; Spek, A. L. *J. Chem. Soc. Dalton Trans.* **1978,** 1800.
25. Bruce, M. I.; Swincer, A. G.; Wallis, R. C. *J. Organometal. Chem.* **1979,** *171,* C5.
26. Davison, A.; Selegue, J. P. *J. Am. Chem. Soc.* **1978,** *100,* 7763.
27. Dykstra, C. E.; Schaefer, H. F. *J. Am. Chem. Soc.* **1978,** *100,* 1378.
28. Bellerby, J. M.; Mays, M. J. *J. Organometal. Chem.* **1976,** *117,* C21.
29. King, R. B.; Saran, M. S. *J. Am. Chem. Soc.* **1973,** *95,* 1817.
30. Kolobova, N. E.; Antonova, A. B.; Khitrova, O. M.; Antipin, M.; Struchkov, Y. T. *J. Organometal. Chem.* **1977,** *137,* 69.
31. Brenner, A.; Hucul, D. A. *J. Am. Chem. Soc.* **1980,** *102,* 2484.
32. Fischer, E. O.; Kiener, V. *J. Organometal. Chem.* **1970,** *23,* 215.
33. Blickensderfer, J. R.; Knobler, C. B.; Kaesz, H. D. *J. Am. Chem. Soc.* **1975,** *97,* 2686.
34. Azam, K. A.; Deeming, A. J. *J. Chem. Soc. Chem. Commun.* **1977,** 472.
35. Keller, E.; Trenkle, A.; Vahrenkamp, H. *Chem. Ber.* **1977,** *110,* 441.
36. Szostak, R.; Strouse, C. E.; Kaesz, H. D. *J. Organometal. Chem.* **1980,** *191,* 243.
37. Dixon, K. R.; Rattray, A. R. *Inorg. Chem.* **1978,** *17,* 1099.
38. Eaborn, C.; Odell, K. J.; Pidcock, A. *J. Organometal. Chem.* **1979,** *170,* 105.
39. Nie Wahner, J., Meek, D. W. In "Catalytic Aspects of Metal Phosphine Complexes," *Adv. Chem. Ser.* 1981, *196,* 257.
40. Mott, G. N., Ph.D. Thesis, Univ. of Waterloo, 1980.
41. Taylor, N. J.; Mott, G. N.; Carty, A. J. *Inorg. Chem.* **1980,** *19,* 560.
42. Gorenstein, D. G. *J. Am. Chem. Soc.* **1975,** *97,* 898.
43. Bradley, J. S.; Ansell, G. B.; Hill, E. W. *J. Am. Chem. Soc.* **1979,** *101,* 7419.
44. Smith, W. F.; Taylor, N. J.; Carty, A. J., unpublished data.

RECEIVED August 6, 1980.

11

Chemistry of Tertiary Phosphine Complexes of Rhodium, Iridium, and Platinum

D. P. ARNOLD, M. A. BENNETT, G. T. CRISP, and J. C. JEFFERY
Research School of Chemistry, The Australian National University, P.O. Box 4, Canberra, A.C.T., Australia 2600

The products of oxidative addition of acyl chlorides and alkyl halides to various tertiary phosphine complexes of rhodium(I) and iridium(I) are discussed. Features of interest include (1) an equilibrium between a five-coordinate acetylrhodium(III) cation and its six-coordinate methyl(carbonyl) isomer which is established at an intermediate rate on the NMR time scale at room temperature, and (2) a solvent-dependent secondary- to normal-alkyl-group isomerization in octahedral alkyliridium(III) complexes. The chemistry of monomeric, tertiary phosphine-stabilized hydroxoplatinum(II) complexes is reviewed, with emphasis on their conversion into hydrido–alkyl or –aryl complexes. Evidence for an electronic cis-*PtP bond-weakening influence is presented.*

The stoichiometric reaction of acyl halides with $RhCl(PPh_3)_3$ has been studied in detail by a number of groups in relation to the potentially useful process of decarbonylation of acyl halides catalyzed by various planar d^8 metal complexes (*1, 2, 3, 4, 5*). The reaction of acetyl chloride with $RhCl(PPh_3)_3$ involves initial oxidative addition to give a five-coordinate acetylrhodium(III) complex, subsequent methyl migration to give a six-coordinate methylrhodium(III) complex, and final reductive elimination of methyl chloride (*see* Scheme I) (*5*). A similar sequence of reactions occurs in the corresponding iridium system, although the six-coordinate alkyliridium(III) complexes do not undergo reductive elimination as readily as the corresponding rhodium(III) complexes (*6, 7, 8*). The

0065-2393/82/0196-0195$05.00/0

Scheme I. Reaction of $RhCl(PPh_3)_3$ with acetyl chloride

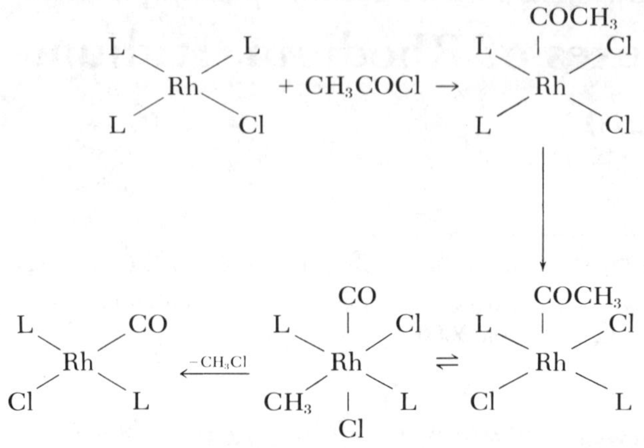

$$L_3RhCl + CH_3COCl \rightarrow Rh(COCH_3)(Cl)_2L_2$$

$$\downarrow$$

$$RhCl(CO)L_2 \xleftarrow{-CH_3Cl} Rh(CO)(CH_3)(Cl)_2L_2 \rightleftharpoons Rh(COCH_3)(Cl)_2L_2$$

α-branched acyl chlorides react either with $IrCl(PPh_3)_3$ or $IrCl(PPh_3)_2$ (prepared in situ from the cyclooctene complex $[IrCl(C_8H_{14})_2]_2$ + $2Ph_3P$) to give *n*-alkyliridium(III) complexes. Thus $(CH_3)_2CHCOCl$, $CH_3CH_2CH(CH_3)COCl$, $(CH_3CH_2)_2CHCOCl$, and $C_6H_5CH(CH_3)COCl$ give, respectively, the *n*-propyl, *n*-butyl, *n*-pentyl, and 2-phenethyl complexes **Ia–Id** (*8*). We have assumed that an undetected octahedral *sec*-alkyliridium(III) complex is formed initially, following Scheme I, and that this isomerized by a β-hydride shift to the *n*-alkyl. In the oxidative addition of α-branched acyl chlorides to the dimeric cyclooctene–iridium(I) complex $[IrCl(CO)(C_8H_{14})_2]_2$, intermediate sec-alkyls can be detected. For example, 2-methylpropanoyl chloride and butanoyl chloride give, respectively, the dimeric isopropyl- and *n*-propyl–iridium(III) complexes **IIa** and **IIb** but, on heating, a 3:2 equilibrium mixture of **IIb** and **IIa** is obtained (*8*).

The secondary-to-normal alkyl-group isomerization undoubtedly proceeds via an olefin–hydride intermediate similar to that invoked for the thermal decomposition via β-elimination of transition-metal alkyls. What factors determine the direction of addition of the iridium(III) hydride to the coordinated olefin and, in particular, why does the presence of triphenylphosphine in the coordination sphere favor anti-Markownikov addition to give exclusively the *n*-alkyl? The answer to this question may be relevant to hydroformylation since it is well known that phosphine-modified cobalt and rhodium catalysts give high ratios of straight- to branched-chain alcohols (*9*). It has been suggested that steric repulsion between the secondary alkyl and the phenyl rings of the mutually trans-triphenylphosphine ligands in complexes of Type **I** may favor the *n*-alkyl. An additional possibility is that the hydride ligand becomes more hydridic in character in tertiary phosphine complexes ($IR^{\delta+} - H^{\delta-}$), which would favor anti-Markownikov addition to the olefin. A second problem is that, unless iridium(III) achieves an unusual coordination number of seven, one of the ligands in **I** must dissociate reversibly to provide a site for the olefin; although the bulky triphenylphosphine seems a likely candidate, no information is available. In the case of the dimeric complexes **II**, reversible cleavage of the chlorine bridges may fulfill this purpose.

Oxidative Addition of Acyl Chlorides to $MClL_3$ (M = Ir, L = $PMePh_2$, PMe_2Ph; M = Rh, L = PMe_2Ph)

We were interested in preparing secondary alkyl complexes similar to **I** containing tertiary phosphines which are less sterically demanding and more basic than triphenylphosphine. Unfortunately, the approach which has proved successful in the case of L = PPh_3 proved to have only limited application. The reaction of $IrCl(PMePh_2)_3$ with 2-methylpropanoyl chloride in hot benzene gives the *n*-propyliridium(III) complex Ie(L = $PMePh_2$; R = $Ch_2CH_2CH_3$), but as in the case of triphenylphosphine the presumed intermediate isopropyl complex could not be detected or isolated. Adding of straight-chain acyl chlorides RCOCl (R = CH_3, C_2H_5, *n*-C_3H_7) to $IrCl(PMePh_2)_3$ differs somewhat from that shown in Scheme I for corresponding additions to $RhCl(PPh_3)_3$ (*10*). The first product that can be isolated is an octahedral alkyliridium(III) complex **III** containing mutually cis-phosphine ligands. In the case of R = C_2H_5 or *n*-C_3H_7, this is in equilibrium in solution with a five-coordinate acyliridium(III) complex **IV** also containing cis-phosphine ligands, which is presumably the initial product of oxidative addition (*see* Scheme II). In the case of R = CH_3, only the six-coordinate methyl complex can be detected, both in solution and in the solid state. Isomerization to the thermodynamically stable product containing mutually trans-phosphine ligands occurs slowly in dichloromethane, but is accelerated markedly by adding methanol or salts such as NH_4PF_6 or $LiClO_4$ and is repressed by the addition of LiCl. These observations suggest that initial loss of Cl^- is required for isomerization to proceed (*10*).

Adding acyl chlorides RCOCl to solutions of $MCl(PMe_2Ph)_3$ (M = Rh, Ir) prepared in situ from $[RhCl(C_2H_4)_2]_2$ or $[IrCl(C_8H_{14})_2]_2$ and 6 equiv of dimethylphenylphosphine gives six-coordinate acyl complexes $MCl_2(COR)(PMe_2Ph)_3$ (**V**) which do not readily lose PMe_2Ph to give complexes of Type **I**. Attempts to prepare such compounds using only 4 equiv of dimethylphenylphosphine also have been unsuccessful. Alkyl migration can be induced to occur in **V** by adding NH_4PF_6, but only when R is a straight-chain alkyl group. The ammonium hexafluorophosphate abstracts one chloride ligand, probably that which is trans to the acyl group, and creates a vacant site for the migrating alkyl group (*see* Scheme III). When M = Ir, the product

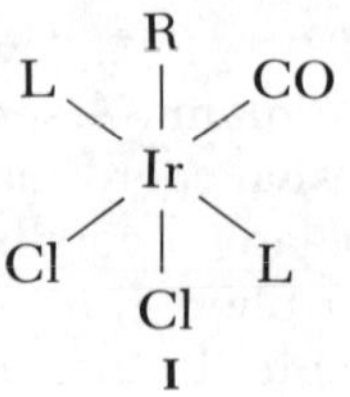

I

$R = CH_2CH_2CH_3$ Ia
$R = CH_2CH_2CH_2CH_3$ Ib
$R = CH_2CH_2CH_2CH_2CH_3$ Ic
$R = CH_2CH_2C_6H_5$ Id
(Ia–Id: $L = PPh_3$)
$R = CH_2CH_2CH_3$ Ie; $L = PMePh_2$

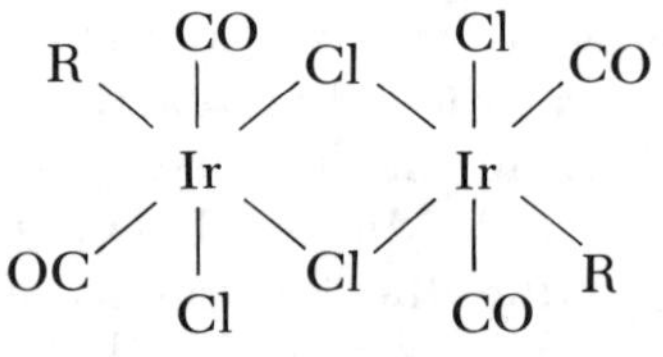

II

$R = CH(CH_3)_2$ IIa
$R = CH_2CH_2CH_3$ IIb

Scheme II. Oxidative addition of acyl chlorides to $IrCl(PMePh_2)_3$

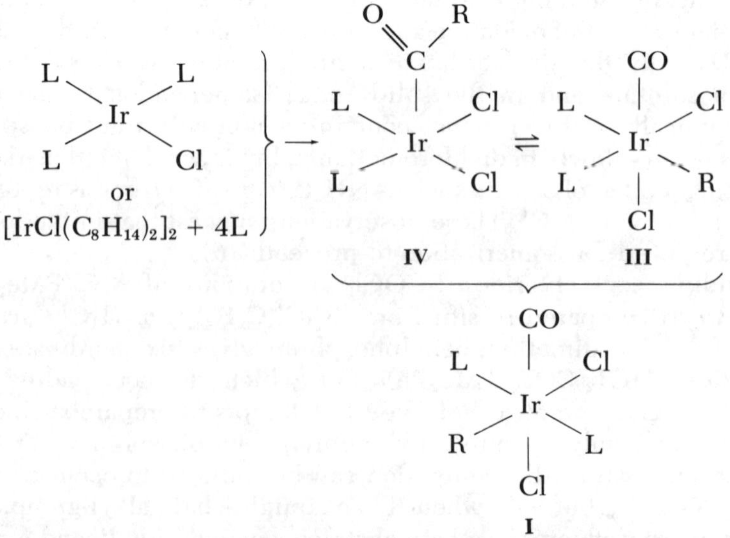

$IrClL_3$ or $[IrCl(C_8H_{14})_2]_2 + 4L$ → IV ⇌ III → I

$L = PMePh_2$; $R = CH_3, C_2H_5, n\text{-}C_3H_7$

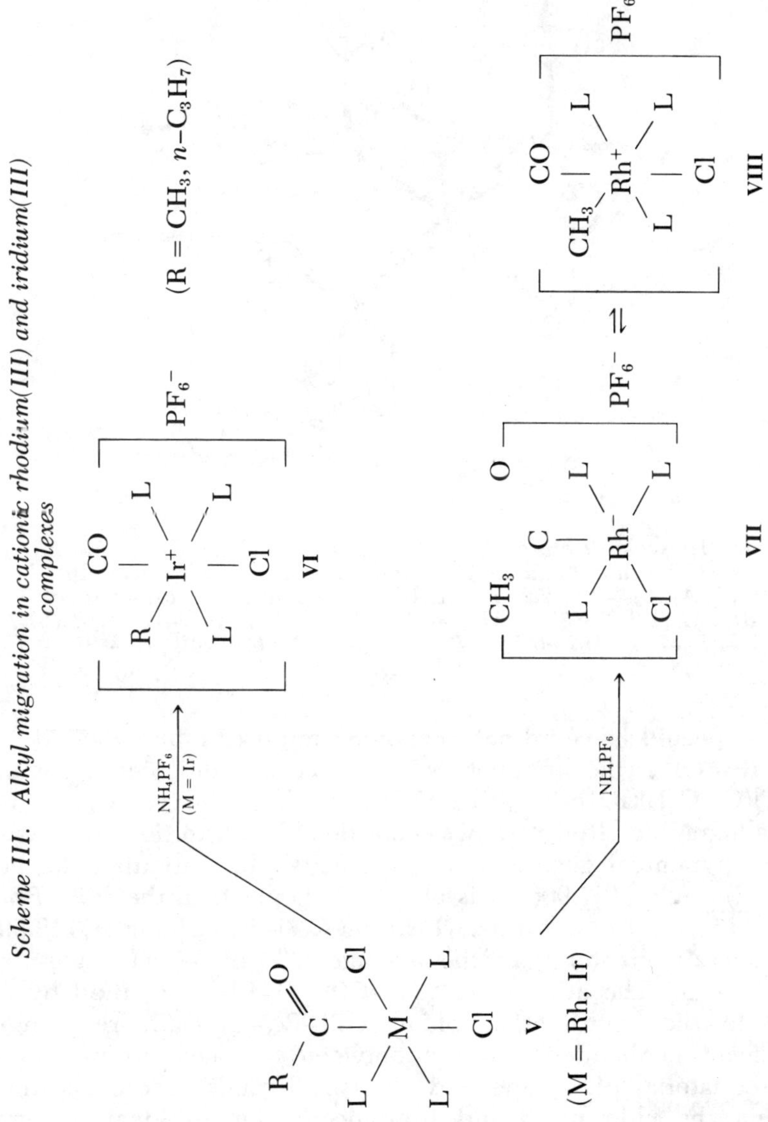

Scheme III. Alkyl migration in cationic rhodium(III) and iridium(III) complexes

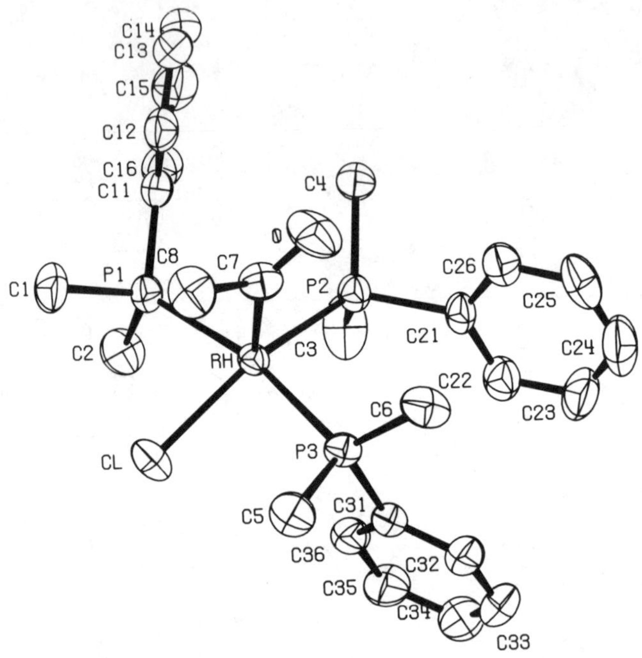

Figure 1. Overall stereochemistry of the $[RhCl(COCH_3)(PMe_2Ph)_3]^+$ cation. Important bond lengths and interbond angles are: Rh—P(1) 2.388(1) Å; Rh—P(2) 2.285(1) Å; Rh—P(3) 2.384(1) Å; Rh—Cl 2.391(1) Å; Rh—C(7) 1.971(5) Å; C(7)—C(8) 1.528(8) Å; C(7)—O 1.184(6) Å. P(1)—Rh—P(3) 167.90(5)°; P(2)—Rh—Cl 156.17(5)°; Cl—Rh—C(7) 112.5(2)°.

is the expected six-coordinate, cationic complex **VI**, but when M = Rh and R = CH_3, the product is a five-coordinate acetyl complex, $[RhCl(COCH_3)(PMe_2Ph)_3]PF_6$ (**VII**), in the solid state. An X-ray crystallographic study shows considerable distortion from ideal square pyramidal geometry (*see* Figure 1). In particular, the P(1)–Rh–P(3) angle [167.90(5)°] is about 12° larger than the P(2)–Rh–Cl angle [156.17(5)°], and the acyl carbon C(7)–Rh–Cl angle [112.5(2)°] is about 20° greater than the average C(7)–Rh–P(n) angle (92.8° for n = 1–3). The acyl plane C(7)–C(8)–O–Rh is inclined by only 10.4° to the coordination plane P(2)–Rh–Cl–C(7), resulting in significant nonbonded contacts between the acyl group and the trans-equatorial phosphine and chlorine ligands. These constraints prevent the chlorine ligand from occupying an ideal equatorial position and cause it to shift towards the vacant coordination site.

The acetyl complex **VII** is in equilibrium in solution with a six-coordinate methylrhodium(III) complex **VIII**, analogous to the iridium complex **VI**. An interesting feature evident from variable-temperature H-1 and P-31 NMR studies is that the methyl group migrates at an intermediate rate on the NMR time scale at room temperature, i.e., ca. 10^2–10^3 s^{-1}, which is 10^7–10^8 times faster than typical rate constants for the forward reaction in the equilibrium $RMn(CO)_5 \rightleftharpoons RCOMn(CO)_4$ (*10, 11*). It seems reasonable that methyl migration involves a concerted movement of the chlorine atom to the vacant apical site of the square pyramid and of the methyl group to the vacated equatorial site. The departures from ideal square pyramidal geometry discussed above suggest that the structure of the acetyl complex is close to that of the transition state for methyl migration, thus accounting for the rapidity of the process. Another important contributing factor may be reduced π-back-bonding to the CO group of **VIII** and to the acetyl group of **VII** relative to that in uncharged complexes, which could destabilize the ground states relative to the transition state for methyl migration.

Isomerization of Secondary Alkyl Iridium(III) Complexes Containing Tertiary Phosphines

A wide range of octahedrally coordinated alkyl–iridium(III) complexes **IX**, including secondary alkyls, can be made in high yield by oxidative addition of the appropriate alkyl iodides, RI, to iridium(I) complexes $IrX(CO)L_2$ (X = Cl, I; L = PMe_3, PMe_2Ph). The complex $IrCl(CO)(PMePh_2)_2$ adds methyl halides but not ethyl iodide, whereas $IrCl(CO)(PEt_3)_2$ adds primarily alkyl iodides but fails to react with secondary alkyl iodides, so that these reactions are clearly sensitive to steric effects. On heating in benzene or dichloromethane, the secondary alkyls isomerize to the corresponding *n*-alkyls, the reaction being accompanied by halide-group scrambling and by decomposition. The same isomerization can be effected at room temperature without complicating side reactions by adding protic solvents such as methanol to a dichloromethane solution of the secondary alkyl. The P-31 chemical shifts of the isomeric secondary and normal alkyl complexes of Type **IX** are sufficiently different and the relaxation times are sufficiently close that the isomerization can be monitored by P-31 NMR spectroscopy. In dichloromethane containing various protic solvents, isomerization of the isopropyl complexes $IrClI\{CH(CH_3)_2\}(CO)L_2$ (L = PMe_3, PMe_2Ph) to the *n*-propyl complexes is first-order, the latter being the sole organo–iridium(III) complex detectable at equilibrium. The order of efficacy of solvents in promoting isomerization is $CF_3CO_2H \gg CH_3OH \gg C_2H_5OH > CH_3CO_2H \sim CH_3CH_2CH_2OH > (CH_3)_2CHOH$ (*see* Table I), the solvents at the top of the list being those that can be expected to most strongly solvate a leaving halide ion. In view of the high trans-influence of σ-bonded alkyls, it seems likely that the departure of the iodide ion in **IX** promotes isomerization by providing a va-

Table I. Rates of Isomerization of $IrClI\{CH(CH_3)_2\}(CO)(PMe_2Ph)_2$ to the Corresponding *n*-Propyl Complex in Various Solvent Mixtures

Solvent A	*Solvent B*[a]	*Rate* ($10^5\ s^{-1}$)[b]
CH_2Cl_2	CF_3CO_2H	[c]
CH_2Cl_2	CH_3OH	78 ± 8
CH_2Cl_2	C_2H_5OH	9 ± 1
CH_2Cl_2	CH_3CO_2H	6 ± 0.5
CH_2Cl_2	n-C_3H_7OH	6 ± 0.5
CH_2Cl_2	$(CH_3)_2CHOH$	3 ± 0.5
THF	CH_3OH	< 0.3
THF	H_2O	7 ± 1

[a] Mole fraction of B = 0.395.
[b] Measured by P-31 NMR spectroscopy at 32°C.
[c] Too fast to measure.

cant site for β-elimination. In agreement, treatment of the isopropyl complexes $IrClI\{CH(CH_3)_2\}(CO)L_2$ with $AgPF_6$ in acetonitrile gives immediately the *n*-propyl cation $[IrCl(CH_2CH_2CH_3)(CO)(CH_3CN)L_2]PF_6$. Qualitatively, adding iodide ion slows down isomerization of $IrClI\{CH(CH_3)_2\}(CO)L_2$ in dichloromethane–methanol, although the kinetics are no longer simple. Isomerization is considerably slower in tetrahydrofuran(THF)–methanol than in dichloromethane–methanol (*see* Table II), perhaps because THF can occupy the site vacated by the iodide ion. Adding water accelerates the isomerization in THF, again because of preferential solvation of the leaving iodide ion by water. The proposed mechanism of isomerization is shown in Scheme IV. All efforts to detect the presumed olefin–hydride intermediate have been unsuccessful.

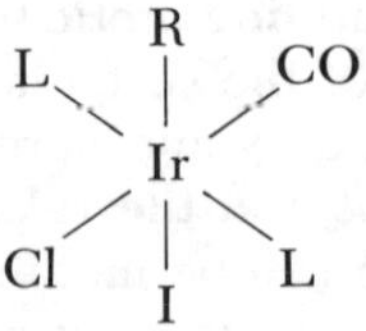

R
L CO
Ir
Cl L
I

IX

L = PMe_3, PMe_2Ph

R = C_2H_5, n-C_3H_7, n-C_8H_{17}, $CH(CH_3)_2$, $CH_2CH(CH_3)_2$, $CH_2CH_2CH(CH_3)_2$, $CH(C_2H_5)_2$, $CH(CH_3)(n$-$C_6H_{13})$, $CH_2C(CH_3)_3$, c-C_6H_{11}

Two observations indicate the importance of steric hindrance in stabilizing a five-coordinate cationic intermediate: (1) isomerization is more rapid for L = PMe_2Ph than for L = PMe_3 (*see* Table II); (2) rates

Table II. Rates of Isomerization of $IrClI\{CH(CH_3)_2\}(CO)L_2$ (L = PMe_2Ph, PMe_3) to the Corresponding *n*-Propyl Complexes in Dichloromethane–Methanol[a]

Mole Fraction of Methanol	Rate (10^5 s^{-1}) L = PMe_2Ph	Rate (10^5 s^{-1}) L = PMe_3
0.787	[b]	32 ± 3
0.395	78 ± 8	9 ± 1
0.316	13 ± 1	5 ± 0.5

[a] Measured by P-31 NMR spectroscopy at 32°C.
[b] Too fast to measure.

of isomerization increase in the order $CH(CH_3)_2 < CH(CH_3)(C_2H_5) < CH(C_2H_5)_2$, but further extension of the chain beyond five carbon atoms has little effect (*see* Table III).

The *n*-alkyliridium(III) complexes of Type **IX** also undergo reversible β-elimination, but much more slowly than do the secondary alkyls. For example, the complex $IrClI(CD_2CH_3)(CO)(PMe_2Ph)_2$ scrambles deuterium atoms between α- and β-carbon atoms in benzene–methanol over 5–6 d at 70°C.

Scheme IV. Isomerization of $IrClI\{CH(CH_3)_2\}(CO)L_2$ to $IrClI(CH_2CH_2CH_3)(CO)L_2$ ($L = PMe_3$, PMe_2Ph)

L = $PMePh_2$; R = CH_3, C_2H_5, *n*–C_3H_7

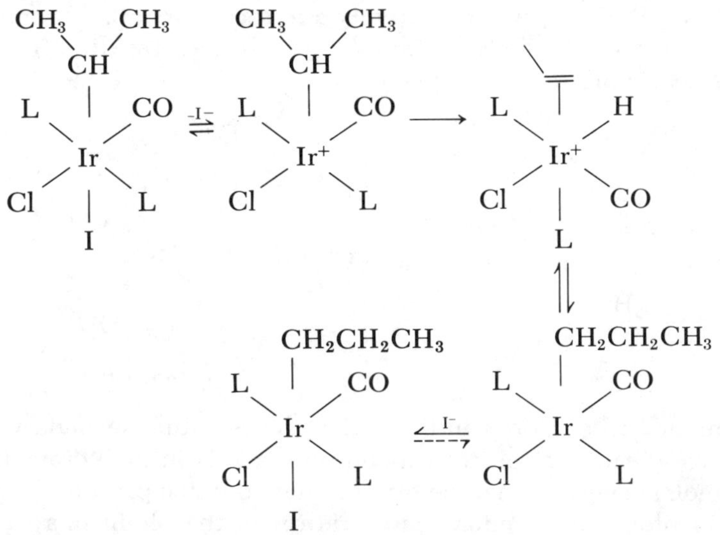

$$IrClI\{CH(CH_3)_2\}(CO)L_2 \underset{I^-}{\overset{-I^-}{\rightleftharpoons}} [IrCl\{CH(CH_3)_2\}(CO)L_2]^+ \longrightarrow [IrClH(CH_2{=}CHCH_3)(CO)L_2]^+ \rightleftharpoons [IrCl(CH_2CH_2CH_3)(CO)L_2]^+ \overset{I^-}{\rightleftharpoons} IrClI(CH_2CH_2CH_3)(CO)L_2$$

Table III. Rates of Isomerization of Various Secondary Alkyliridium(III) Complexes $IrClIR_{sec}(CO)(PMe_3)_2$ to the Corresponding *n*-Alkyls

R_{sec}	Rate (10^5 s^{-1})[a]
$CH(CH_3)_2$	9 ± 1
$CH(CH_3)(C_2H_5)$	15 ± 1
$CH(C_2H_5)_2$	25 ± 2
$CH(CH_3)(n\text{-}C_6H_{13})$	25 ± 2

[a] Measured by P-31 NMR spectroscopy at 32°C in CH_2Cl_2–CH_3OH (mole fraction of CH_3OH = 0.395).

It is noteworthy that the presence of electron-withdrawing substituents such as CN or $CO_2C_2H_5$ on the α- or β-carbon atoms completely inhibits isomerization. Thus, the alkyls obtained separately by adding $CH_3CHBrCO_2C_2H_5$ and $BrCH_2CH_2CO_2C_2H_5$ to $IrCl(CO)L_2$ do not interconvert, even in the presence of CF_3CO_2H or $AgPF_6$ in CH_3CN. This may be a consequence of the low affinity of iridium(III) for electrophilic olefins such as ethyl acrylate or acrylonitrile.

We conclude that in octahedral alkyliridium(III) complexes the presence of tertiary phosphines favors exclusively the *n*-alkyl over the corresponding secondary alkyl, irrespective of the size or basicity of the phosphine. This preference is probably largely electronic in origin, but steric factors cannot be ruled out. A key step that generates a vacant coordination site for both alkyl-group migration and isomerization in octahedral tertiary phosphine complexes of rhodium(III) and iridium(III) is dissociation of halide ion.

Alkyl– or aryl–iridium(I) complexes $IrR(CO)L_2$ (R = CH_3, C_6H_5; L = PMe_3, PMe_2Ph) will oxidatively add secondary alkyl bromides or iodides to give octahedrally coordinated bis(alkyl)iridium(III) complexes, as exemplified in Equation 1.

$$trans\text{-}Ir(C_6H_5)(CO)L_2 + (CH_3)_2CHI \rightarrow Ir[CH(CH_3)_2](C_6H_5)(I)(CO)L_2 \qquad (1)$$

(L = PMe_3)

Preliminary observations indicate that the resulting secondary alkyls isomerize cleanly to the corresponding *n*-alkyls in dichloromethane–methanol, suggesting that the hydride ligand in the presumed cationic hydrido–olefin intermediate must return to the olefin more rapidly than it undergoes reductive elimination with the σ-alkyl or σ-phenyl group. This may well indicate that the hydride and σ-bonded carbon ligand are mutually trans in the intermediate, although it should be noted that planar iridium(I) σ-aryls and σ-alkyls $IrR(CO)L_2$ oxidatively add hydrogen halides to give cis-hydrido–aryls or–alkyls of iridium(III) that are of sufficient thermal stability to be isolated or detected spectroscopically (*12, 13*).

Mononuclear Hydroxo, Methoxo, and Hydrido Complexes of Platinum(II)

Thermally unstable, planar cis-hydrido-alkyls of platinum(II), *cis*-$PtH(R)(PPh_3)_2$, have been prepared recently by the action of the appropriate Grignard reagent with *trans*-$PtHCl(PPh_3)_2$; they decompose by intramolecular reductive elimination of alkane (*14*). Routes to the corresponding transisomers have become available as a consequence of research on mononuclear hydroxo and methoxo complexes of platinum(II). Several years ago we (*15, 16*) observed the irreversible oxidative addition of water and methanol to the cyclohexyne platinum(0) complex $Pt(C_6H_8)(dppe)$(dppe = 1,2-bis(diphenylphosphino) ethane) to give, respectively, mononuclear hydroxo and methoxo complexes of platinum(II) (*see* Equation 2):

$$Pt(C_6H_8)(Ph_2PCH_2CH_2PPh_2) + ROH \rightarrow Pt(C_6H_9)(OR)(Ph_2PCH_2CH_2PPh_2) \quad (2)$$

(R = H, CH_3)

Unexpectedly, the methoxo complex shows no tendency to decompose to a hydride by β-elimination, although it does react immediately with water to give the hydroxo complex. In contrast, cyclohexyne bis(triphenylphosphine)platinum(0) reacts with methanol to give a hydrido complex, presumably via an undetected intermediate methoxide (*see* Equation 3):

$$Pt(C_6H_8)(PPh_3)_2 + CH_3OH \rightarrow \left[Pt(C_6H_9)(OCH_3)(PPh_3)_2\right] \xrightarrow{-CH_2O} Pt(C_6H_9)(H)(PPh_3)_2 \quad (3)$$

A wide range of alkyl- and aryl-hydroxo-bis(tertiary phosphine) complexes of platinum(II) has been prepared now (*17, 18, 19, 20, 21*), generally by the route shown in Equation 4 or some variant of it

$$PtClRL_2 \xrightarrow[\text{solvent}]{AgY} [PtR(solvent)L_2]Y \xrightarrow{OH^-} Pt(OH)RL_2 \quad (4)$$

where $Y = BF_4$ or PF_6; solvent = acetone or methanol; L = various tertiary phosphines.

Hydrido–hydroxo complexes of platinum(II) are believed to be formed by reversible oxidative addition of water to bis(ligand) platinum(0) complexes containing bulky tertiary phosphines (*22*). Although the equilibrium favors the platinum(0) complex, *trans*-$[PtH(OH)(P(iso\text{-}Pr)_3)_2]$ has been isolated and identified spectroscopically (*see* Equation 5):

$$Pt(P(iso\text{-}Pr)_3)_2 + H_2O \rightleftharpoons PtH(OH)(P(iso\text{-}Pr)_3)_2 \quad (5)$$

The hydroxo complexes behave as strong bases in aqueous organic solvents and react with weak acids, including carbon acids such as nitromethane, methyl ketones, acetylacetone, acetonitrile, phenylacetylene and cyclopentadiene; this provides a useful synthesis under mild conditions of platinum(II)–carbon σ-bonded complexes, as exemplified in Equation 6 (*18*):

$$Pt(OH)(CH_3)(dppe) + HX \rightarrow PtX(CH_3)(dppe) + H_2O \quad (6)$$

$$HX = CH_3NO_2, CH_3COR, (CH_3CO)_2CH_2, CH_3CN, C_6H_5C_2H, C_5H_6.$$

In complexes of the type trans-$Pt(OH)RL_2$, the hydroxo group becomes less reactive as R becomes more electron-withdrawing (*17*). The reversibility of the reaction in Equation 6 enables hydroxoplatinum(II) complexes and zerovalent platinum complexes such as $Pt(PEt_3)_3$ to catalyze H–D exchange of organic carbonyl compounds with D_2O (*22*). In addition, these complexes catalyze the addition of water to simple alkane nitriles to give carboxamides. The key species in this catalysis are believed to be *N*-bonded carboxamido complexes, e.g., $Pt(NHCOCH_3)(CH_3)(PPh_3)_2$, which can be formed by nucleophilic attack of the labile hydroxo ligand on a coordinated nitrile (*16, 21, 22*) (*see* Equation 7):

$$Pt\text{—}OH \xrightarrow{RCN} [Pt\text{—}NCR]OH \longrightarrow Pt\text{—}NHCOR \quad (7)$$

(auxiliary ligands omitted)

In the present context, the most interesting reaction of the trans-$Pt(OH)RL_2$ complexes is that with methanol at room temperature, which gives, among other products, trans-hydrido-alkyls or -aryls of platinum(II), $PtH(R)L_2$, presumbaly the β-elimination of formaldehyde from intermediate methoxo-complexes, e.g., Equation (8):

$$trans\text{-}Pt(OH)(CH_3)(P(iso\text{-}Pr)_3)_2 \xrightarrow[\text{3 h, room temperature}]{CH_3OH} trans\text{-}PtH(CH_3)(P(iso\text{-}Pr)_3)_2 \text{ white crystals (55\% yield)} \quad (8)$$

In some cases, these intermediates can be isolated, albeit in an impure state. For example, treating the methanol cation *trans*-$[Pt(CH_3)(CH_3OH)(PPh_3)_2]BF_4$ with methanolic sodium methoxide precipitates the methoxide *trans*-$Pt(OCH_3)(CH_3)(PPh_3)_2$, which can be identified by H-1 NMR spectroscopy, but this compound decomposes in methanol to give *trans*-$PtH(CH_3)(PPh_3)_2$ and other unidentified products. The previously reported methoxide $Pt(OCH_3)(C_6H_5)(PEt_3)_2$ (*23*) decomposes similarly. When the methoxo ligand is trans to a strongly electron-withdrawing σ-bonded carbon ligand, decomposition by β-elimination is retarded, e.g., $Pt(OCH_3)(C_6F_5)(PPh_3)_2$ does not decompose to a hydride on heating in benzene (*17*).

Preparing trans-hydrido-alkyls or -aryls of platinum(II) does not require prior isolation of a hydroxo complex. The main routes we have used are shown in Equations 9–11, and selected spectroscopic data, including some from the literature, are in Table IV.

$$trans\text{-}PtCl(C_6H_5)(Pt\text{-}Bu_2Me)_2 \xrightarrow[\text{reflux 2 h}]{NaOCH_3/CH_3OH} trans\text{-}PtH(C_6H_5)(Pt\text{-}Bu_2Me)_2 \text{ colorless needles from } CH_2Cl_2/CH_3OH \text{ (74\%)} \quad (9)$$

$$trans\text{-}Pt(NHCOCH_3)(CH_3)(PPh_3)_2 \xrightarrow[\text{reflux 1 h}]{NaOCH_3/CH_3OH} trans\text{-}PtH(CH_3)(PPh_3)_2 \text{ colorless solid (75\%)} \quad (10)$$

$$trans\text{-}PtCl(C_6H_5)(PEt_3)_2 \xrightarrow[CH_3OH]{AgBF_4} trans\text{-}[Pt(C_6H_5)(CH_3OH)(PEt_3)_2]^+ \xrightarrow{NaBH_4,\ -40^\circ \text{ to } -20^\circ C} trans\text{-}PtH(C_6H_5)(PEt_3)_2 \text{ colorless needles from isopentane at } -78^\circ C \text{ (55\%).} \quad (11)$$

As expected, the transisomers are thermally more stable and less reactive than the corresponding cis isomers. Thus, *trans*-$PtH(C_6H_5)(PEt_3)_2$

Table IV. Selected Spectroscopic Data for *trans*-PtH(R)L_2 Complexes

	ν_{PtH}(cm^{-1}) Nujol (CH_2Cl_2)	δH [J_{PtH}]
PtH(Ph)$(PPh_3)_2$[a]	1890(1966)	−5.71(600)
PtH(Ph)$(PEt_3)_2$[a]	1940	−6.67(648)
PtH(Me)$(PPh_3)_2$[a]	1936(1968)	−3.77(656)
PtH($CH_2CH_2CH_2CN$)$(PPh_3)_2$[b]	1950	−4.51(636)
PtH(C_6H_9)$(PPh_3)_2$[c]	1920	−4.64(608)
PtH(CH_2CN)$(PPh_3)_2$[b]	2027	−7.32(746)
PtH(CF_3)$(PPh_3)_2$[d]	2073	−8.23(544)
PtHCl$(PEt_3)_2$[e,f]	2220	−16.8(1275)
PtH(ONO_2)$(PEt_3)_2$[e,f]	2242	−23.6(1322)

[a] This work.
[b] Ref. 28.
[c] Ref. 16.
[d] Ref. 29.
[e] Ref. 30.
[f] Ref. 31.

is stable at room temperature but decomposes in *n*-decane at 70°C giving benzene, $Pt(PEt_3)_3$, platinum metal, and an unidentified platinum-containing complex; the decomposition is retarded by triethylphosphine. The compound is inert towards ethylene, acrylonitrile, or dimethyl maleate, but is decomposed rapidly by carbon monoxide, giving benzene and unidentified cluster complexes. In marked contrast with the cis isomers, the *trans*-[PtH(R)$(PPh_3)_2$] compounds react only sluggishly on heating with diphenylacetylene to give Pt($C_6H_5C_2C_6H_5$)$(PPh_3)_2$. Dimethyl acetylene-dicarboxylate inserts into the Pt—H bond of *trans*-PtH(C_6H_4Br-*p*)$(PEt_3)_2$ to give the expected σ-vinyl (*see* Equation 12):

$$trans\text{-}PtH(C_6H_4Br\text{-}p)(PEt_3)_2 + CH_3O_2CC{\equiv}CCO_2CH_3 \rightarrow$$

$$\text{Br–}C_6H_4\text{–}Pt(PEt_3)_2\text{–}C(CO_2CH_3){=}C(H)(CO_2CH_3) \qquad (12)$$

Hydroxo complexes containing cis-bidentate tertiary diphosphines, such as Pt(OH)R(dppe) (R = CH_3, C_6H_9) or Pt(OH)(CH_3)(dppp) (dppp = 1,3-bis(diphenylphosphino)propane) are inert towards methanol at room temperature. It seems likely that the equilibrium shown in Equation 13 strongly favors the hydroxo complex, but that rapid decomposition of the methoxide in the trans series shifts the equilibrium in its favor. On heating in methanol, the complex $Pt(OH)(CH_3)(dppe)$ gives a dimeric complex of apparent formula $[Pt(dppe)]_2$, which is probably cyclometalated and which may form via an intermediate methoxide or hydride. Treating $Pt(OH)(CH_3)(\widehat{P\ P})$ ($\widehat{P\ P}$ = dppe, dppp) with formic acid at −70°C gives initially the formato complexes $Pt(OCHO)(CH_3)(\widehat{P\ P})$, which have been identified spectroscopically. They eliminate CO_2, even at −70°C, to give thermally unstable cis-hydrido-methyls (*see* Scheme V), which in turn decompose to give as yet unidentified cluster compounds. The H-1 and P-31 NMR parameters for the hydrido-methyls $PtH(CH_3)(\widehat{P\ P})$ are similar to those reported for *cis*-$PtH(CH_3)(PPh_3)_2$ (*14*).

$$Pt(OH)RL_2 + CH_3OH \rightleftharpoons Pt(OCH_3)RL_2 + H_2O \qquad (13)$$

It is worth noting that the hydroxo complexes $Pt(OH)(CH_3)(\widehat{P\ P})$ react with CO to give C-bonded carboxylic acids or hydroxycarbonyl complexes $Pt(CO_2H)(CH_3)(\widehat{P\ P})$ that are isomeric with the formates discussed above. In contrast with the formates and other hydroxycarbonyls such as $PtCl(CO_2H)(PEt_3)_2$ (*24*) and $IrCl_2(CO_2H)(CO)(PMc_2Ph)_2$ (*25*), the complexes $Pt(CO_2H)(CH_3)(\widehat{P\ P})$ do not eliminate CO_2 on heating to give $PtH(CH_3)(\widehat{P\ P})$. This represents another example of the ability of the $RPt(\widehat{P\ P})$ moiety to stabilize groups that normally

Scheme V. Preparation of cis-hydrido–methyls *of platinum(II) containing dppe*

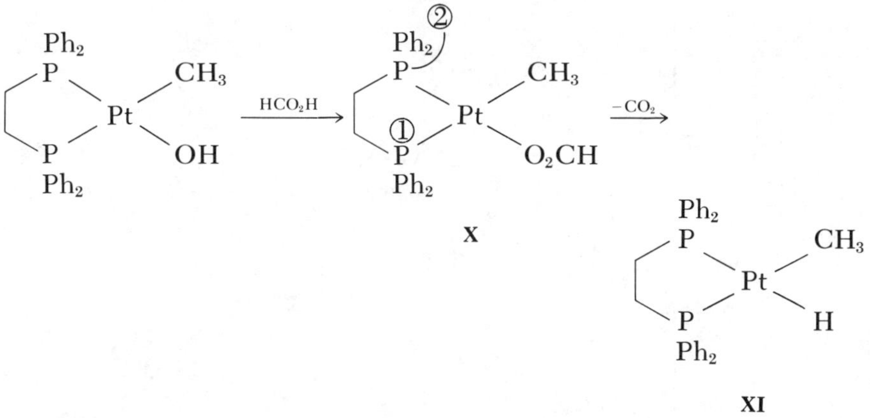

$$(\mathrm{Ph_2P}\frown\mathrm{PPh_2})\mathrm{Pt(CH_3)(OH)} \xrightarrow{HCO_2H} (\mathrm{Ph_2P^{2}}\frown\mathrm{P^{1}Ph_2})\mathrm{Pt(CH_3)(O_2CH)}\ \mathbf{X} \xrightarrow{-CO_2} (\mathrm{Ph_2P}\frown\mathrm{PPh_2})\mathrm{Pt(CH_3)(H)}\ \mathbf{XI}$$

X: P-31 NMR (CH_2Cl_2, −74°C): $\delta(P^1)$ 47.7, J_{PtP} 1841 Hz
$\delta(P^2)$ 34.9, J_{PtP} 4141 Hz

XI: H-1 NMR (CD_2Cl_2, −70°C): $\delta(PtH)$ −0.71, dd, J_{PH} 195, 15 Hz
J_{PtH} 1214 Hz

P-31 NMR (CH_2Cl_2, −75°C): δ_P 48.9, 47.5, J_{PtP} 1792, 1836 Hz

undergo facile β-elimination reactions (cf., $Pt(OCH_3)(C_6H_9)(dppe)$, as discussed earlier), though it is not clear why β-elimination of the formate should take place so easily in this system.

Spectroscopic Data and the Cis Effect

In general, the data summarized in Table IV show an expected trend: the Pt—H stretching frequencies (ν_{PtH}), chemical shifts (δ_{PtH}) and coupling constants ($^1J_{PtH}$) for the trans-hydrido-alkyls and -aryls are less than those for the well-known compounds trans-$PtHXL_2$ (X = Cl, ONO_2). Some anomalies are evident, e.g., for *trans*-$PtHX(PPh_3)_2$ (X = CH_3, CF_3), $\nu_{PtH}(CF_3) > \nu_{PtH}(CH_3)$ and $\delta_{PtH}(CF_3) > \delta_{PtH}(CH_3)$, whereas $^1J_{PtH}(CF_3) < {}^1J_{PtH}(CH_3)$. In a series of meta- and para-substituted aryl hydrides, *trans*-$PtH(C_6H_4X$-*m* or -*p*$)(PEt_3)_2$, we find good linear correlations between Taft $\bar{\sigma}$-parameters and both ν_{PtH} and δ_{PtH}, the correlation for both parameters being somewhat better for the meta series than for the para series. In contrast, correlation between

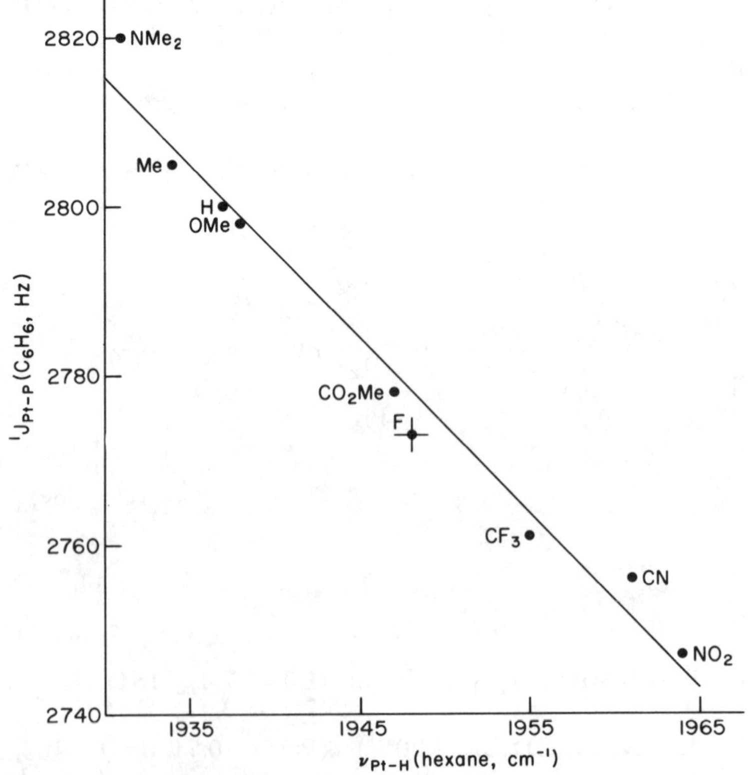

Figure 2. Plot of $^1J_{PtP}$ vs. ν_{PtH} *for* trans-$PtH(C_6H_4X$-m$)(PEt_3)_2$

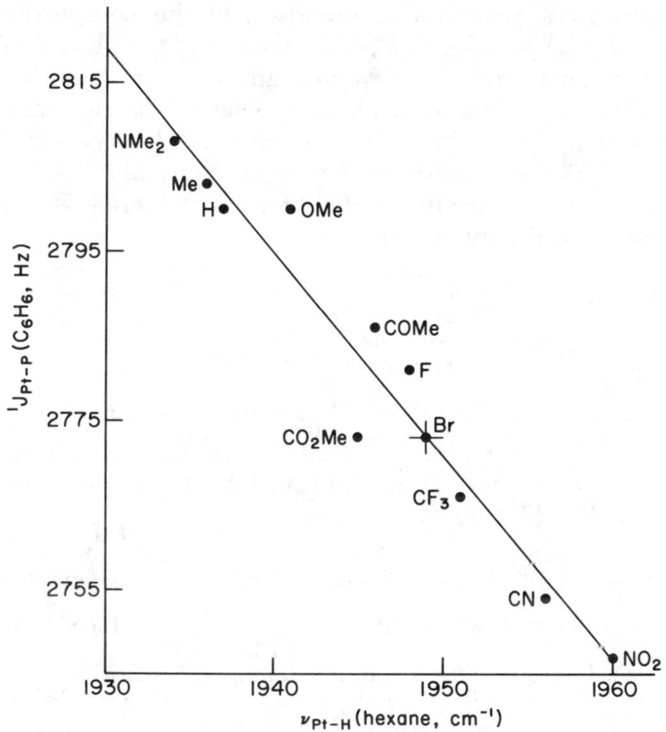

Figure 3. Plot of $^1J_{PtP}$ vs. ν_{PtH} *for* trans-$PtH(C_6H_4X\text{-p})(PEt_3)_2$

Taft $\bar{\sigma}$-parameters and $^1J_{PtH}$ is poor and the slopes of the plots are in opposite directions for the meta and para series. In the trans-hydrido-alkyls and -aryls of platinum(II), therefore, attempts to relate $^1J_{PtH}$ values with the s character of orbitals used by platinum(II) and the covalency of the platinum–ligand bond trans to hydride are unlikely to succeed. This is in marked contrast with the extensive information about the trans influence series derived from $^1J_{PtH}$, $^2J_{PtCH_3}$, $^1J_{PtC}$, and $^1J_{PtP}$ in other complexes (*26*). It may be that, when two strongly σ-donating ligands such as hydride and aryl are mutually trans, variations in $^1J_{PtH}$ arise both from changes in s character and in s-electron density at platinum.

In view of the poor correlation of $^1J_{PtH}$ with ν_{PtH} or δ_{PtH} in the series *trans*-$PtH(C_6H_4X$-*m* or -*p*$)(PEt_3)_2$, we were surprised to find an excellent linear correlation between $^1J_{PtP}$ and ν_{PtH} for both meta and para series, the smaller J_{PtP} values being associated with the more electron-withdrawing meta or para substituents (*see* Figs 2 and 3). We believe that this trend probably reflects small changes in bond lengths (and, presumably, bond strengths), such that the more electron-withdrawing aryl groups give rise to shorter Pt—H bonds and longer Pt—P bonds. In a structural comparison of the complexes *trans*-$PtClX(PMePh_2)_2$ ($X = CH_3, C_2F_5$) (*27*), we found the pentafluoroethyl to have significantly longer Pt—P bonds and we tentatively have attributed this observation to a steric effect associated with the significantly shorter Pt—C bond length in the pentafluoroethyl. In view of the correlation shown in Figures 2 and 3, however, it seems more likely that we are dealing with an electronic cis effect for which there is at present no adequate theoretical explanation.

Literature Cited

1. Tsuji, J.; Ohno, K. *Synthesis* **1969**, *1*, 157.
2. Baird, M. C.; Mague, J. T.; Osborn, J. A.; Wilkinson, G. *J. Chem. Soc. A* **1967**, 1347.
3. Stille, J. K.; Regan, M. T.; Fries, R. W.; Huang, F.; McCarley, T. In "Resins for Aerospace," *Adv. Chem. Ser.* **1974**, *132*, 181.
4. Dunham, N. A.; Baird, M. C. *J. Chem. Soc., Dalton Trans.* **1975**, 774.
5. Egglestone, D. L.; Baird, M. C.; Lock, C. J. L.; Turner, G. *J. Chem. Soc., Dalton Trans.* **1977**, 1576.
6. Kubota, M.; Blake, D. M.; Smith, S. A. *Inorg. Chem.* **1971**, *10*, 1430.
7. Blake, D. M.; Wilkelman, A.; Chung, Y-L. *Inorg. Chem.* **1975**, *14*, 1326.
8. Bennett, M. A.; Charles, R.; Mitchell, T. R. B. *J. Am. Chem. Soc.* **1978**, *100*, 2737.
9. Evans, D.; Osborn, J. A.; Wilkinson, G. *J. Chem. Soc.* **1968**, 3133.
10. Wojcicki, A. *Adv. Organomet. Chem.* **1973**, *11*, 87.
11. Calderazzo, F. *Angew. Chem. Int. Ed. Engl.* **1977**, *16*, 299.
12. Dahlenburg, L.; Nast, R. *J. Organomet. Chem.* **1976**, *110*, 395.
13. Bennett, M. A.; Crisp, G. T., unpublished work.
14. Abis, L.; Sen, A.; Halpern, J. *J. Am. Chem. Soc.* **1978**, *100*, 2915.
15. Bennett, M. A.; Robertson, G. B.; Whimp, P. O.; Yoshida, T. *J. Am. Chem. Soc.* **1973**, *95*, 3028.
16. Bennett, M. A.; Yoshida, T. *J. Am. Chem. Soc.* **1978**, *100*, 1750.
17. Yoshida, T.; Okano, T.; Otsuka, S. *J. Chem. Soc., Dalton Trans.* **1976**, 993.
18. Appleton, T. G.; Bennett, M. A. *Inorg. Chem.* **1978**, *17*, 738.
19. Fakley, M. E.; Pidcock, A. *J. Chem. Soc., Dalton Trans.* **1977**, 1444.
20. Michelin, R. A.; Napoli, M.; Ros, R. *J. Organomet. Chem.* **1979**, *175*, 239.
21. Arnold, D. P.; Bennett, M. A. *J. Organomet. Chem.*, in press.
22. Yoshida, T.; Matsuda, T.; Okano, T.; Kitani, T.; Otsuka, S. *J. Am. Chem. Soc.* **1979**, *101*, 2027.
23. Coulson, D. R. *J. Am. Chem. Soc.* **1976**, *98*, 3111.
24. Catellani, M.; Halpern, J. *Inorg. Chem.* **1980**, *19*, 566.
25. Deeming, A. J.; Shaw, B. L. *J. Chem. Soc. A* **1969**, 443.
26. Appleton, T. G.; Clark, H. C.; Manzer, L. E. *Coord. Chem. Rev.* **1973**, *10*, 335.
27. Bennett, M. A.; Chee, H-K.; Robertson, G. B. *Inorg. Chem.* **1979**, *18*, 1061.
28. Ros, R.; Bataillard, R.; Rorlet, R. J. *J. Organometal. Chem.* **1976**, *118*, C53.
29. Michelin, R. A.; Belluco, U.; Ros, R. *Inorg. Chim. Acta* **1976**, *24*, L33.
30. Chatt, J.; Shaw, B. L. *J. Chem. Soc.* **1962**, 5075.
31. Powell, J.; Shaw, B. L. *J. Chem. Soc.* **1965**, 3879.

RECEIVED July 20, 1980.

12

Palladium–Triarylphosphine Complexes as Catalysts for Vinylic Halide Reactions

R. F. HECK
Department of Chemistry, University of Delaware, Newark, DE 19711

Palladium acetate–triarylphosphine complexes catalyze the addition of vinylic groups from vinylic halides to olefinic compounds in the presence of amines. Conjugated dienes are major products from α,β-unsaturated acids, esters, or nitriles while unactivated olefinic compounds react best in the presence of secondary amines where allylic amines are major products. The reactions are usually regio- and stereospecific. The synthetic utility of the reaction is illustrated with a wide variety of examples.

Palladium–triarylphosphine complexes are catalysts for the reaction of aryl, heterocyclic, benzyl, and vinylic halides with olefins and amines (*1*). The usual catalyst is a combination of palladium(II) acetate with two equivalents of a triarylphosphine. The complex that is formed is reduced in situ in the reaction mixtures to an active Pd(0) phosphine complex (*see* Equations 1, 2, 3, and 4 which depict catalyst formation). This species then oxidatively adds the organic halide (*see* Equation 2) and the adduct in turn adds to the olefinic reactant (*see* Equation 3). The organic group adds mainly or often exclusively to the less-substituted olefinic carbon. This addition product is generally unstable because sp^3 carbon–hydrogen bonds usually will be in positions beta to the palladium group, and β-hydride elimination of the palladium hydride group will occur. The elimination produces olefin-π-complexes initially. After dissociation or displacement of the hydridopalladium halide group from the π complex, the hydride decomposes into the Pd(0) phosphine complex and hydrogen halide. The hydrogen halide is neutralized by the amine present. The Pd(0) complex then reacts with the organic halide again and the cycle is repeated.

0065–2393/82/0196–0213$05.00/0

$$Pd(OAc)_2L_2 \xrightarrow[L,\ -HOAc]{Olefin} PdL_n + \text{olefinic acetates} \quad n = 2, 3, \text{ or } 4 \qquad (1)$$

$$RX + PdL_n \rightarrow RPdL_2X + (n - 2)L \qquad (2)$$

$$RPdL_2X + \underset{}{\overset{H}{\;}}\!\!>C{=}C< \;\rightarrow\; \left[R{-}\overset{H}{\underset{|}{\overset{|}{C}}}{-}\underset{|}{\overset{|}{C}}{-}PdL_2X \right] \xrightarrow{-L} \qquad (3)$$

$$\left[\begin{array}{c} L \\ | \\ H{-}Pd{-}X \\ \uparrow \\ R\!>C{=}C< \end{array} \right] \xrightarrow{L} \; R\!>C{=}C< \; + [HPdL_2X]$$

$$HPdL_2X + R'_3N \xrightarrow{nL} PdL_n + R'_3NH^+X^- \qquad (4)$$

X = Br, I
L = PAr_3 or if X = I, solvent, olefin, etc.
R = aryl, heterocyclic, benzyl, or vinylic group

We are presently concentrating on using vinylic halides in this reaction. The vinylic halide reactions are often more complicated than the basic reaction shown in Equations 1 through 4 (2). The first complication occurs in the oxidative addition step. If the vinylic halide exists in isomeric forms, the oxidative addition occurs with complete retention of stereochemistry. Adding the vinylic palladium complex to the olefin occurs to place the organic group on the least-substituted olefinic carbon or on the more distant carbon if one is substituted with a strongly electron-withdrawing group (*see* Equations 5, 6, and 7). The β-hydride elimination takes place largely or exclusively to form the conjugated diene π-complex. Dissociation of the π-complex or displacement of the diene from the π-complex now may occur, especially if carbonyl, nitrile, or aryl groups are conjugated with the diene system (*see* Equation 6). If loss of the hydridopalladium group occurs, the double bond originating from the vinylic halide will retain the stereochemistry of that halide and the second double bond will be formed with stereochemistry resulting from a syn addition of the vinylpalladium group followed by a syn elimination of the hydridopalladium group. If there is a choice of hydrogens for the syn elimination, the thermodynamically more stable double bond isomers are preferred. However the dissociation of the hydride group or the displacement of the diene from the π-complexes is often less favorable than the

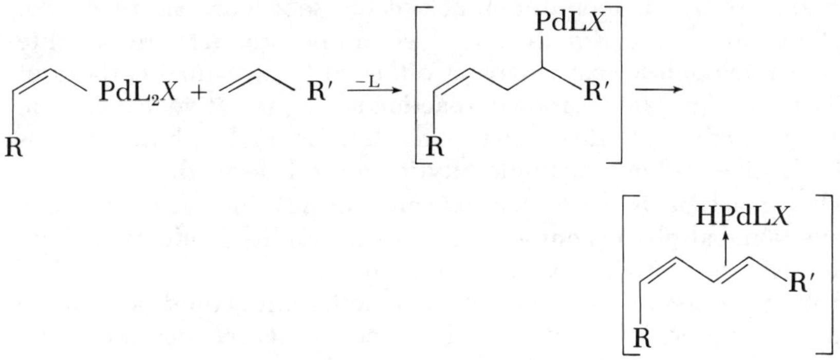

(5)

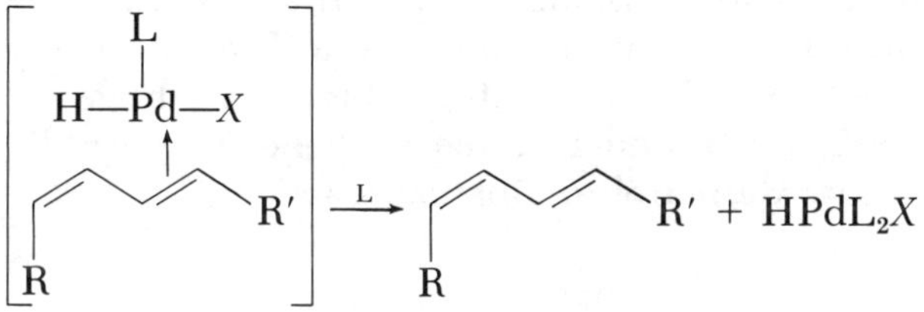

(6)

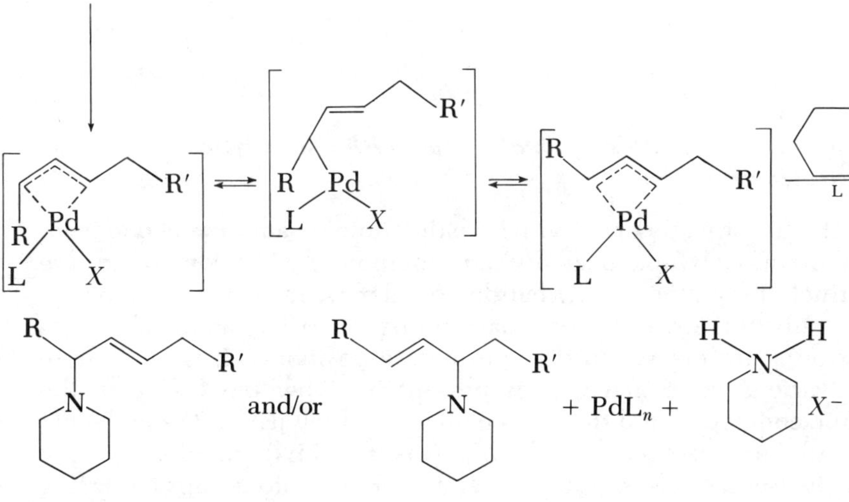

(7)

readaddition of the hydride to the diene group to form π-allylic palladium halide complexes. Only when the dienes are conjugated with ester, nitrile, carboxylate, or aryl groups does the loss of the hydridopalladium group from the π-complex occur easily. In the other cases, the π-allylic palladium halide intermediates are usually resistant to further reaction with our normal base triethylamine. However we can cause a catalytic reaction to occur if a nucleophilic secondary amine is used instead of triethylamine. This variation produces tertiary allylic amines by nucleophilic attack of the secondary amine on the π-allylic complex (*see* Equation 7). Two amine products are possible since the nucleophile may attack at either of the terminal carbons of the allylic system. The position of reaction is very sensitive to the steric environment of the allylic carbons and often the nucleophilic attack is quite selective and only a single allylic amine is formed.

This reaction is very general and useful for forming many polyfunctional aliphatic compounds since the amine products are converted easily into other types of compounds.

Before discussing examples of the reactions, it should be pointed out that most types of vinylic halides used in this reaction are easily available. The UV-catalyzed addition of HBr to terminal acetylenes forms the 1-bromo-1-alkenes. The cis isomer is formed almost exclusively if the addition is carried out at dry ice temperature. The 2-bromo or 2-iodo-1-alkenes are obtained from reacting aqueous hydrogen halides with alkynes. The 2-substituted-1-bromo-1-alkenes are available by the bromination–base dehydrobromination reactions.

+ Br_2 → Br Br $\xrightarrow{KOH}$ Br

$E/Z \sim 10$

Chain Extension and Functionalization with the Vinylic Substitution Reaction

The major value of the vinylic halide–olefin–amine reaction is for synthesizing polyfunctional aliphatic compounds. The compounds are constructed from the vinylic halide, the olefinic reactant with the necessary substituents, and a secondary amine (or tertiary amine if activating groups are present in the appropriate positions). It is possible to have various functional groups present in either the halide or the olefinic compound and there is a choice of which part of the molecule is the olefinic reactant and which part is the vinylic halide.

It is useful to classify functionalized olefins into groups according to the number of carbon atoms they add to the vinylic double bond of the halide in the reaction. Useful reactants are listed in Table I.

Some success has been achieved using all of the above olefinic compounds in reactions with vinylic halides and amines to add the number of carbons indicated, selectively to form one isomeric product or one which is separated easily from a mixture that may have been formed. Numerous other similar reagents may be imagined, but either they have not been tried or successful reactions were not found.

Table II shows a similar classification of functionalized vinylic halides.

Table I. Functionalized Olefins Useful in the Vinylic Substitution Reaction

2 Carbons	*3 Carbons*	*4 Carbons*
$CH_2{=}CH_2$	$CH_2{=}CHCO_2R(H)$	
$CH_2{=}CHSR$ (3)	$CH_2{=}CHCN$	
$CH_2{=}CHN$ (phthalimide: O, O)	$CH_2{=}CHCONH_2$	N (phthalimide: O, O)
	$CH_2{=}CH CH(OR)_2$	
	$CH_2{=}CHCH_2OH$	
	$CH_2{=}CHCH_2NR_2$	

5 Carbons	*6 Carbons*
	N
CO_2H	
N	

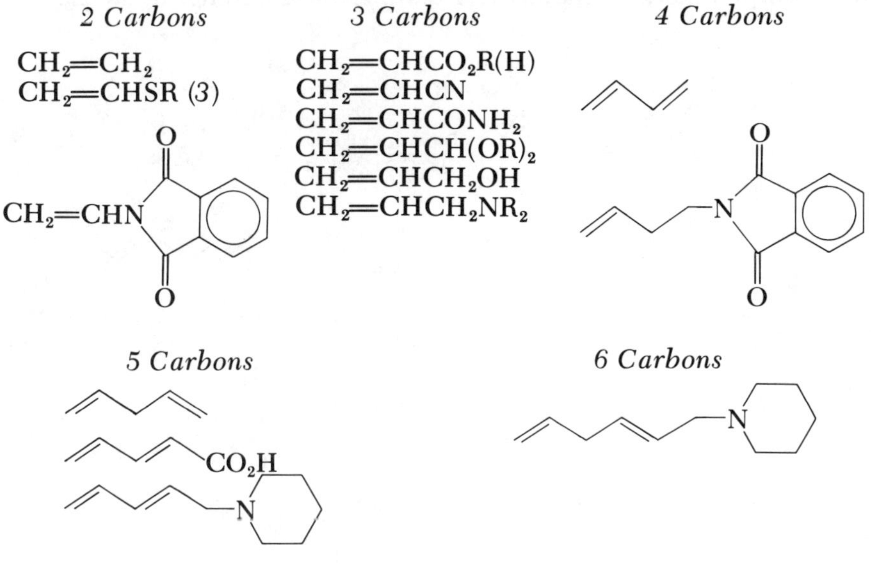

The choice as to which fragment of the molecule to be synthesized should be the vinylic halide and which should be the olefin will depend on several factors. In the cases where elimination to form conjugated dienes is the favored reaction, either possible combination of vinylic halide and olefin may produce the same diene; however, different intermediates are involved and in some instances different products may be formed. The situation is more complex when allylic amines are produced since these products always will be different from the two different combinations of reactants. For example, Z-

Table II. Functionalized Vinylic Bromides Useful in the Vinylic Substitution Reaction

2 Carbons	*3 Carbons*	*4 Carbons*	*5 Carbons*
Br	CH_3O_2C, Br	HO, Br	Br
	$(CH_3O)_2CH$, Br		
	$ROCH_2CH{=}CHBr$		

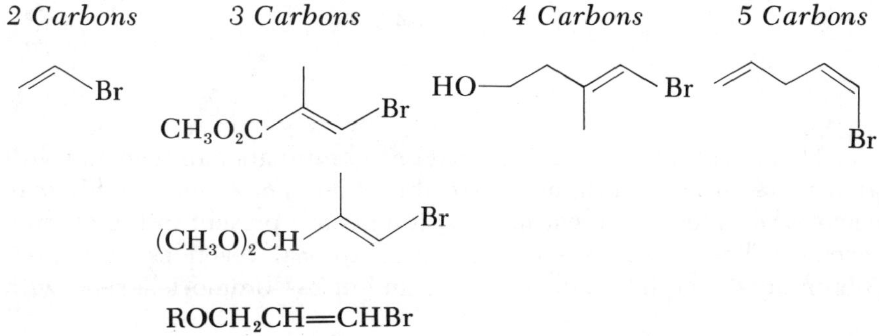

1-bromo-1-hexene and ethylene with morpholine produce mainly Z-2-morpholino-3-octene (84%) while vinyl bromide and 1-hexene with morpholine give mainly *E*-1-morpholino-2-octene (84%) (2).

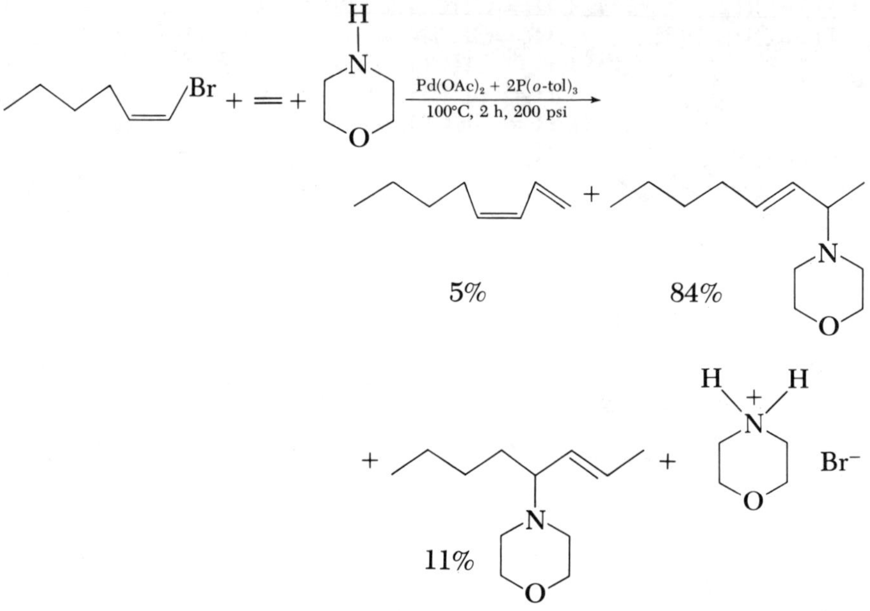

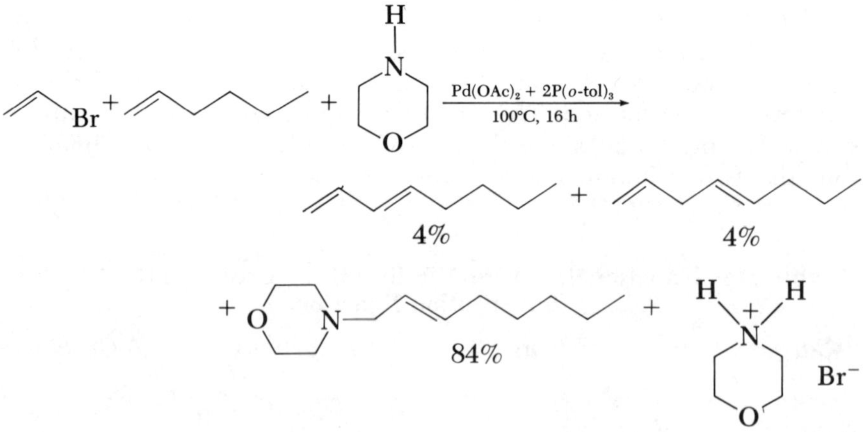

Br + = + morpholine (H–N) $\xrightarrow[\text{100°C, 2 h, 200 psi}]{\text{Pd(OAc)}_2 + 2\text{P}(o\text{-tol})_3}$

5% + 84% (N-morpholino)

+ 11% (N-morpholino) + morpholinium (H_2N^+) Br^-

Br + 1-hexene + morpholine (H–N) $\xrightarrow[\text{100°C, 16 h}]{\text{Pd(OAc)}_2 + 2\text{P}(o\text{-tol})_3}$

4% + 4%

+ morpholino-N– 84% + morpholinium (H_2N^+) Br^-

There is also the possibility that one combination of reactants will give more isomeric products than the other. For example, this may occur when electron-donating substituents are present in the olefinic reactant. This effect is seen in the reciprocal reactions of methyl β-bromomethacrylate with 1-hexene and of Z-1-bromo-1-hexene with

Br + CO_2CH_3 + Et_3N $\xrightarrow[100°C,\ 20\ h]{Pd(OAc)_2\ +\ 2P(o\text{-}tol)_3}$

CO_2CH_3 (~60%) + CO_2CH_3 + CO_2CH_3 + [72%]

CO_2CH_3 + CO_2CH_3 + CO_2CH_3 [16%]

Br + CO_2CH_3 + Et_3N $\xrightarrow[100°C,\ 15\ days]{Pd(OAc)_2\ +\ 2P(o\text{-}tol)_3}$

CO_2CH_3 (49%) + CO_2CH_3 (14%) + CO_2CH_3 (5%)

methyl methacrylate. In the first combination, three more isomers are formed than in the second combination because the addition to 1-hexene occurs about 80% on the terminal carbon and 20% on the second carbon, while Z-1-bromo-1-hexene adds exclusively to the terminal carbon of methyl methacrylate (*4*). Other factors to consider in choosing reactants are relative rates of reaction (note the large difference in the above examples) and the availability of reactants.

Control of the selectivity of the addition often is best achieved by choosing the proper reactants. The major influence is steric. The organic portion of the organopalladium intermediate normally adds to the less-hindered, double-bond carbon of the olefin. The secondary electronic effect is to cause the organic group to add to the more electron-deficient double-bond carbon. The structure of the vinylic halide also may influence the selectivity of the addition. Both vinyl bromide and 2-bromopropene (and other internal vinylic halides) add selectively to the terminal double-bond carbon of 1-hexene. However β-alkyl-substituted vinylic halides give mixtures of adducts with the terminal adducts generally predominating (*2*) (*see* Table III).

Table III. Vinylic Substitution Productions from 1-Hexene and Various Vinylic Bromides and Morpholine

Bromide	*Terminal/Internal Addition of Vinylic Groups to 1-Hexene*	*% Yield of Terminal Adducts*	
		Amines	*Dienes*
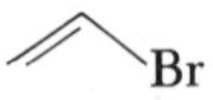 Br	> 20	84	4
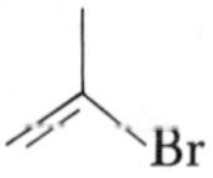 Br	> 20	64	16
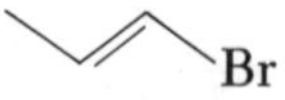 Br	1.8	34	18
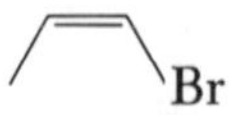 Br	1.8	42	15
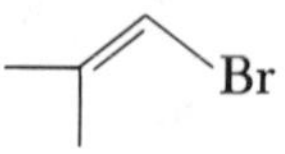 Br	1.7	31	13
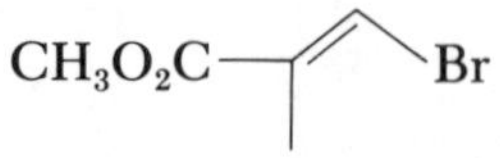 CH_3O_2C— Br (Et_3N)	4.6	—	72

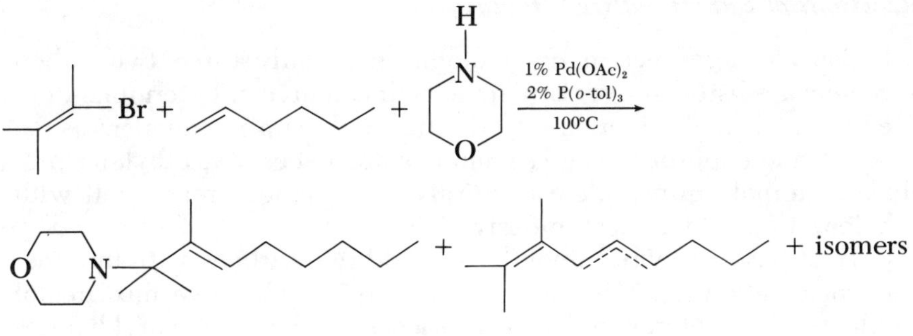

1% $Pd(OAc)_2$
2% $P(o\text{-tol})_3$
100°C

+ isomers

It is interesting and of preparative value that although 1-bromo-2-methyl-1-propene and similar halides add to 1-hexene at both double-bond carbons, only one of the two π-allylic intermediates reacts with the amine. The result is that a mixture of six isomeric dienes is formed, but only one allylic amine is produced. Therefore the reaction is useful since the dienes and the amine are separated easily (2).

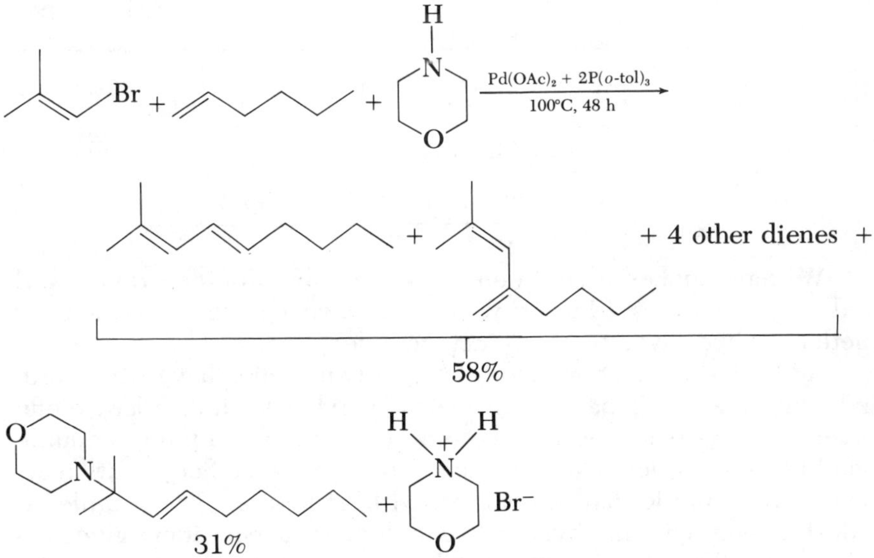

$Pd(OAc)_2 + 2P(o\text{-tol})_3$
100°C, 48 h

+ 4 other dienes +

58%

31%

Br^-

The reason for the selective formation of the highly hindered *N*-tertiary alkyl amine isomer appears to be that the tertiary π-allylic carbon is more susceptible to nucleophilic attack by the amine than the secondary carbon, presumably because of the weaker palladium–tertiary carbon bond. Similar results have been observed in several related reactions.

Reactions of Functionalized Olefins

Considering functionalized olefins first, only three two-carbon reactants are listed above. Ethylene itself can give only terminal double bonds when elimination is facile and triethylamine functions as the base. If nucleophilic secondary amines are necessary, ethylene produces internal amines almost entirely. Aryl halides react well with ethylene to produce styrene derivatives.

Trost has used vinyl thioethers to add two carbons with terminal functionalization (*3*). Vinyl ethers and vinyl acetate give mixtures of products while isopropenyl acetate reacts selectively with aryl halides and at least one vinylic halide.

CH_3OCO Br + OAc + Et_3N $\xrightarrow[125°C,\ 27\ h]{Pd(OAc)_2\ +\ 6P(o\text{-tol})_3}$

CH_3O_2C OAc + CH_3O_2C OAc +

42%

CH_3O_2C OAc + $Et_3NH^+Br^-$

We have looked at additions to *N*-vinylphthalimide briefly. Aryl and vinylic halides add easily, but we have not yet found a good method to hydrolyze these adducts to aldehydes.

Additions to functionalized three-carbon olefins have been studied extensively. We have used methyl acrylate as a standard olefin since it always reacts only at the terminal carbon and the α,β-double bond in the product is always trans. The stereospecificity of its reactions with vinylic halides varies with structure. The simple 1-halo-1-alkenes with methyl acrylate under normal conditions give mixtures of *E,Z*- and *E,E*-dienoates. The reaction is more selective with the bromides than with the iodides and the stereoselectivity increases with increasing triphenylphosphine concentration. This occurs because the excess phosphine displaces the hydridopalladium halide group from the diene π-complex before readdition to form the π-allylic species occurs (*see* Equation 6). The disubstituted vinylic bromides react stereospecifically with methyl acrylate (*4*).

Acrylic acid may be used in place of its esters in these reactions with little or no difference in yield provided an extra equivalent of the

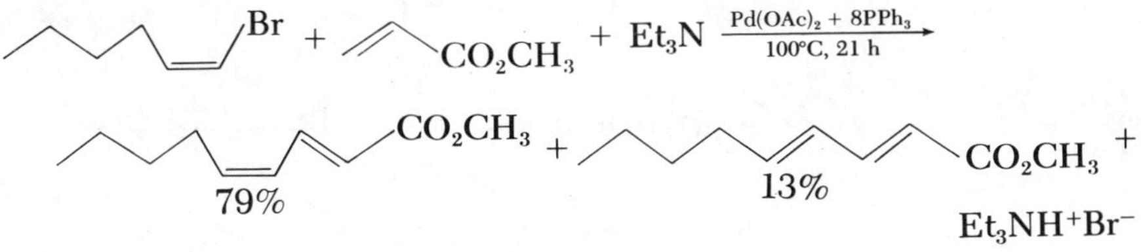

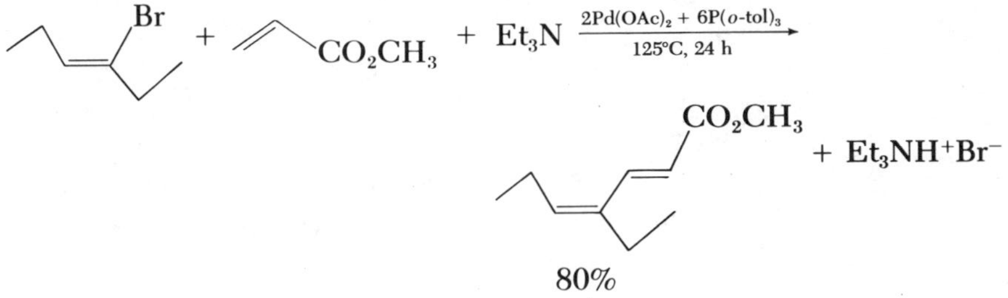

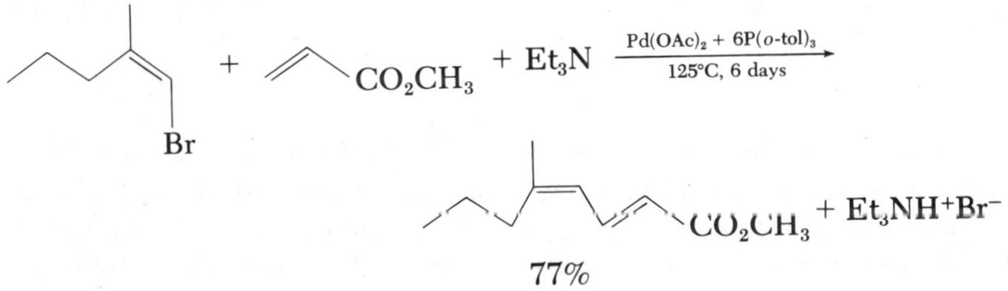

Br + CO_2CH_3 + Et_3N $\xrightarrow[100°C,\ 21\ h]{Pd(OAc)_2 + 8PPh_3}$

CO_2CH_3 (79%) + CO_2CH_3 (13%) + $Et_3NH^+Br^-$

Br + CO_2CH_3 + Et_3N $\xrightarrow[125°C,\ 24\ h]{2Pd(OAc)_2 + 6P(o\text{-}tol)_3}$

CO_2CH_3 (80%) + $Et_3NH^+Br^-$

Br + CO_2CH_3 + Et_3N $\xrightarrow[125°C,\ 6\ days]{Pd(OAc)_2 + 6P(o\text{-}tol)_3}$

CO_2CH_3 (77%) + $Et_3NH^+Br^-$

amine is added to neutralize the carboxylic acid. Methacrylate esters and acid react similarly but more slowly than the acrylates and an additional isomer is formed generally. In these reactions the α-methylene isomers often are formed. Crotonate esters (or acid) also react slowly giving the three-substituted esters.

Acrylonitrile reacts easily with aryl and vinylic halides, but product mixtures are generally more complex than the ones from the related reactions with acrylate esters. A significant amount of nitrile isomers with cis-α,β-double bonds are formed and these isomers are not seen in the acrylate ester reactions. Acrylamide appears similar to the acrylate esters, however we have not studied its reactions in detail.

A very useful three-carbon olefin is acrolein dimethyl acetal (5). Acrolein itself cannot be used because it polymerizes and/or reacts with amines under the normal reaction conditions. With piperidine or morpholine as the base, acrolein acetals react in good yield with a wide variety of vinylic bromides to give dienal acetals and/or aminoenal acetals. These product mixtures, after being treated with excess aqueous oxalic acid and being steam distilled, yield *E,E*-conjugated dienals, usually in good yields. Methacrolein acetals and 3-buten-2-one ethylene ketal also react well, but the crotonaldehyde acetals do not.

[Reaction scheme: Z-1-bromo-1-hexene + $CH_2{=}CHCH(OCH_3)_2$ + piperidine; $Pd(OAc)_2 + 2P(o\text{-tol})_3$, 100°C, 24 h; $(CO_2H)_2$, H_2O, 100°C → 2,4-decadienal (CHO), 76%]

[Reaction scheme: 1-bromo-2-methyl-1-propene (Br) + $CH_2{=}CHCH(OCH_3)_2$ + piperidine; $Pd(OAc)_2 + 2P(o\text{-tol})_3$, 100°C, 18 h; $(CO_2H)_2$, H_2O, 100°C → 5-methyl-2,4-hexadienal (CHO), 76%]

Allylic alcohols are another group of valuable three-carbon olefins (6). These reactants generally are useful for preparing 4-enals and 4-enones. Eliminating the hydrido substituent on the carbon bearing the hydroxyl group in the initial organopalladium complex–allylic alcohol adduct is usually more favorable than eliminating the allylic hydrido group. Mixtures of products are obtained usually, but often the carbonyl products are separated easily so that the reactions are preparatively useful. For example, Z-1-bromo-1-propene, methallyl alcohol, and morpholine react to form a mixture of products containing the morpholine enamine of Z-2-methyl-4-hexenal. It is not necessary to isolate the enamine, but the entire reaction mixture can be added to excess 5% aqueous oxalic acid and steam distilled. In this way a 38% yield of Z-2-methyl-4-hexenal is obtained based on the vinylic halide. The product is entirely the Z-isomer.

[Reaction scheme: Z-1-bromo-1-propene (Br) + methallyl alcohol (OH) + morpholine; $Pd(OAc)_2 + 2P(o\text{-tol})_3$, 100°C, 40 h; $(CO_2H)_2$, H_2O, Δ → Z-2-methyl-4-hexenal (CHO), 38%]

Other examples follow:

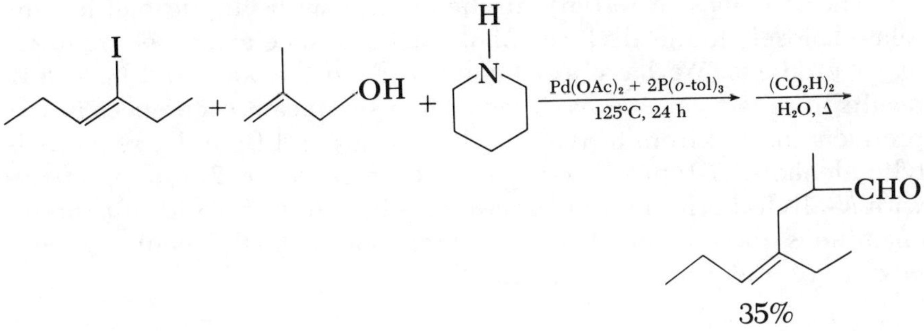
I
OH
H
N
Pd(OAc)₂ + 2P(o-tol)₃
125°C, 24 h
(CO₂H)₂
H₂O, Δ
CHO
35%

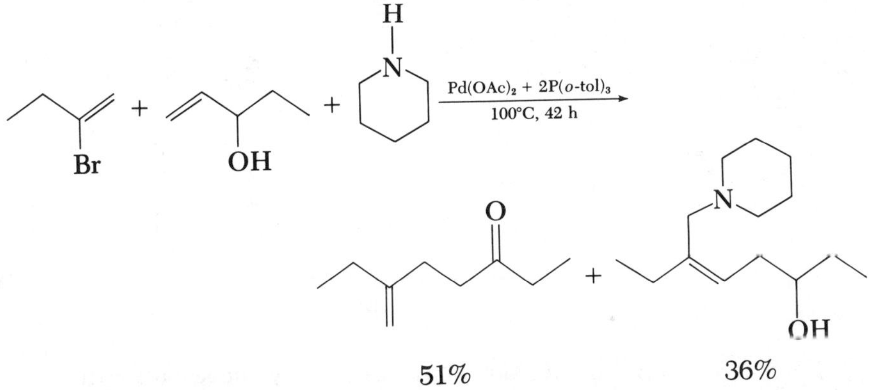
Br
OH
H
N
Pd(OAc)₂ + 2P(o-tol)₃
100°C, 42 h
O
N
OH
51%
36%

We have encountered isomer problems with the allylic alcohol reactions, as would be expected, with some reactants. For example, crotyl alcohol and 1-bromo-1-propene with piperidine give a mixture of two isomeric amino alcohols and two enamines (*6*).

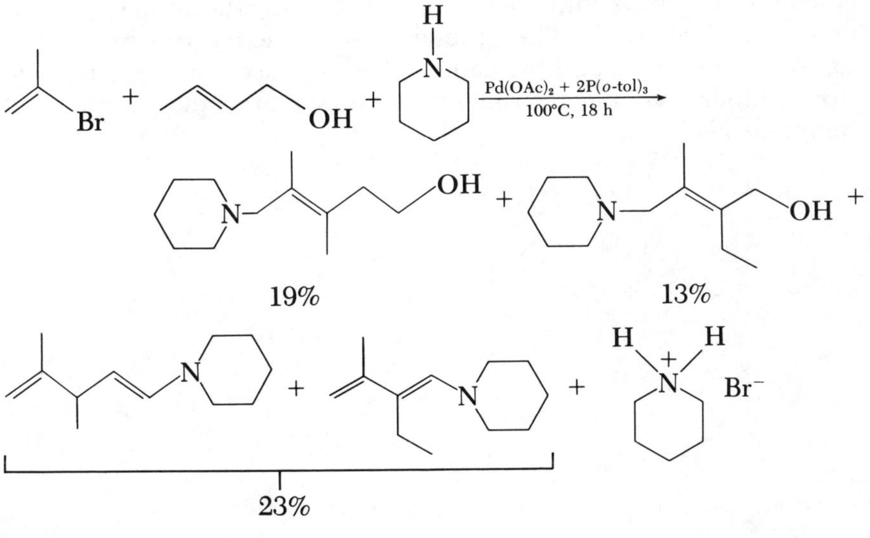
Br
OH
H
N
Pd(OAc)₂ + 2P(o-tol)₃
100°C, 18 h
N
OH
19%
N
OH
13%
N
N
H H
N
Br⁻
23%

The reactions of tertiary allylic amines with vinylic halides are related closely to the allylic alcohol reactions since enamines are often major products. We have just begun work in this area and have few results to report yet. We have seen some significant differences in the products formed from tertiary allylic amines and from the related allylic alcohols. A typical example is the reaction of 2-bromopropene with *N*-allylpiperidine and piperdine where a 42% yield of a single enamine is obtained (*6*). The related reaction with allyl alcohol gives a mixture of regioisomeric enamines.

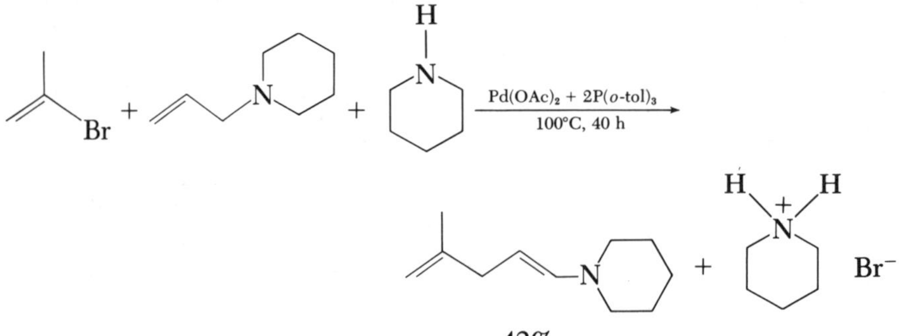

$Pd(OAc)_2$ + $2P(o\text{-tol})_3$

100°C, 40 h

Br^-

42%

Four-carbon-chain extensions have been very successful with conjugated dienes as the functionalized olefins. We have used a few other compounds also, but they are of limited value, such as *N*-3-butenylphthalimide. The last compound is only useful with aromatic or certain vinyl halides where mixtures of allylic amines would not be formed. A typical diene example is the reaction of vinyl bromide with butadiene and piperidine which gives *E*-*N*-(2,5-hexadienyl)-piperidine in 70% yield (*7*). The product of this reaction can be reacted again and used to extend the carbon chains by six atoms (*see* below). The reactions of conjugated dienes can be used to produce conjugated trienes also (*4*).

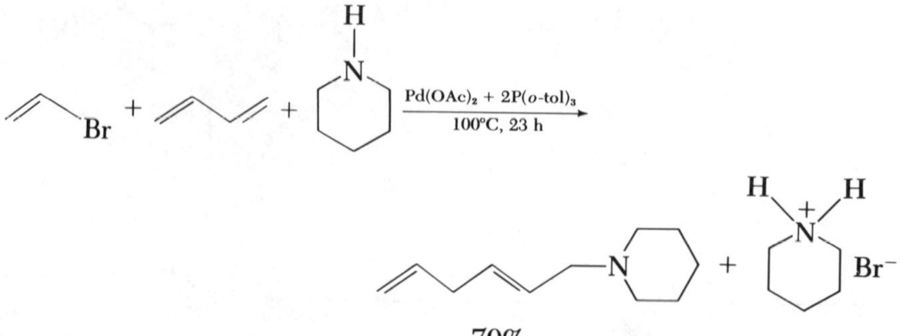

$Pd(OAc)_2$ + $2P(o\text{-tol})_3$

100°C, 23 h

Br^-

70%

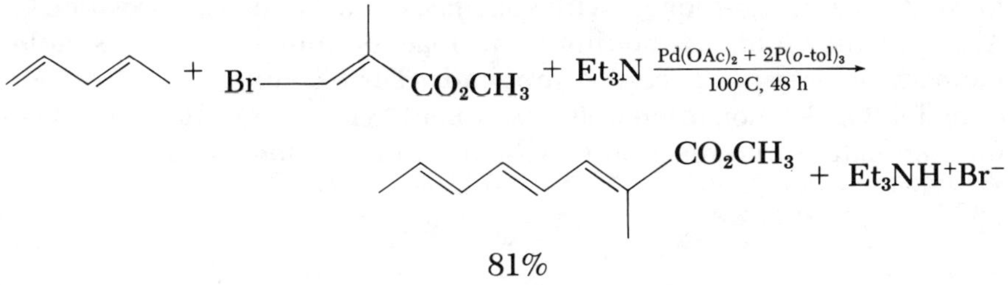

+ Br CO_2CH_3 + Et_3N $\xrightarrow[100°C,\ 48\ h]{Pd(OAc)_2\ +\ 2P(o\text{-}tol)_3}$

CO_2CH_3 + $Et_3NH^+Br^-$

81%

Five-carbon-atom-chain extensions have been achieved with three kinds of unsaturated compounds: 1,4-dienes, 2,4-dienoic acids or esters, and *N*-2,4-dienylamines. The 1,4-dienes are generally useful with aromatic halides, but give mixtures of isomeric amines with most vinylic halides because of elimination and readdition of palladium hydride in both possible directions to produce two different π-allylic intermediates. Exceptions occur in cases where symmetrical compounds are being formed and where both directions of elimination lead to the same product (*8*). For example:

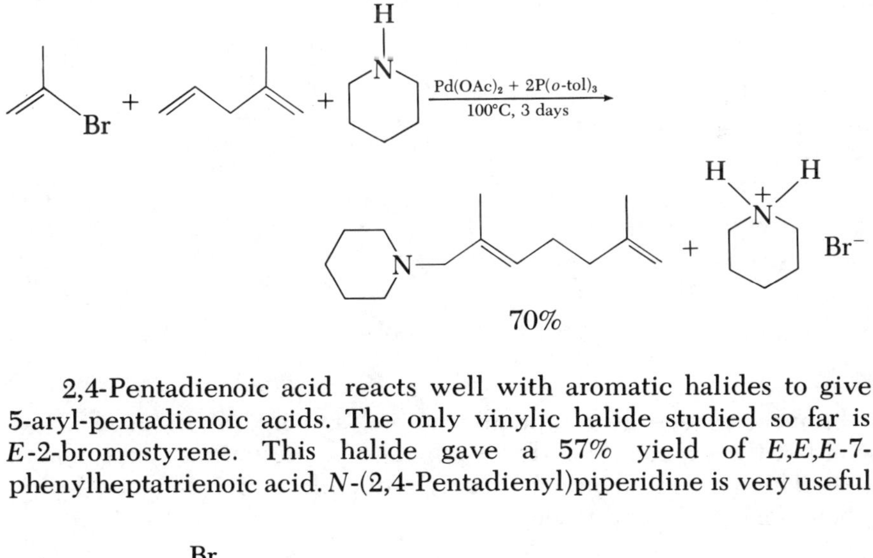

Br + + H–N $\xrightarrow[100°C,\ 3\ days]{Pd(OAc)_2\ +\ 2P(o\text{-}tol)_3}$

N + H_2N^+ Br^-

70%

2,4-Pentadienoic acid reacts well with aromatic halides to give 5-aryl-pentadienoic acids. The only vinylic halide studied so far is *E*-2-bromostyrene. This halide gave a 57% yield of *E,E,E*-7-phenylheptatrienoic acid. *N*-(2,4-Pentadienyl)piperidine is very useful

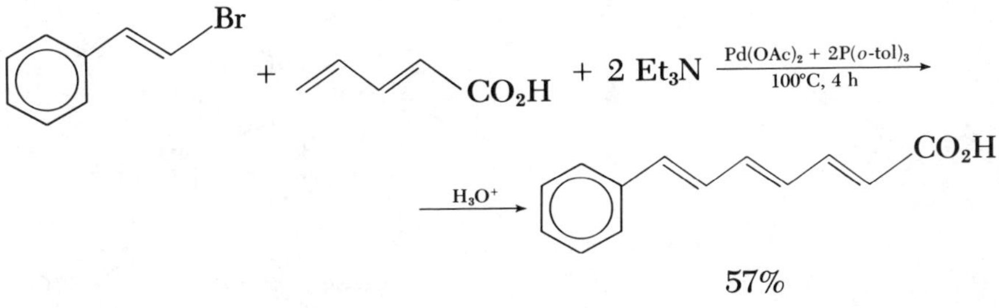

Br + CO_2H + 2 Et_3N $\xrightarrow[100°C,\ 4\ h]{Pd(OAc)_2\ +\ 2P(o\text{-}tol)_3}$

$\xrightarrow{H_3O^+}$ CO_2H

57%

for synthesizing 2,6-dienals with specific stereochemistry at Position 6. Although the yields are not high, this reaction provides a very simple method for preparing these compounds. For example, violet leaf aldehyde, 2-*E*,6-*Z*-nonadienal, is easily made via this reaction (*10*). The intermediate in the reaction is believed to be a dienamine.

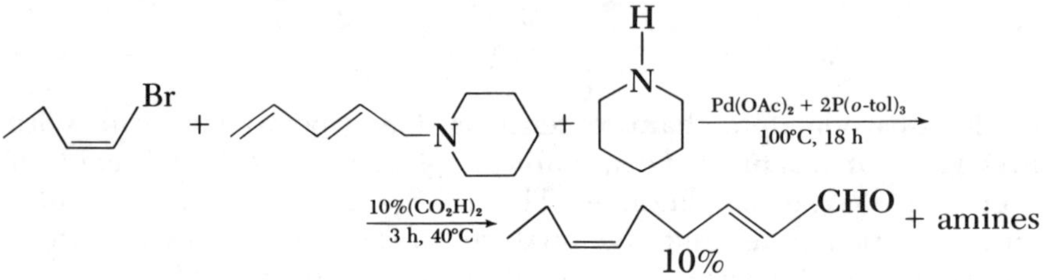

Br + N + H–N $\xrightarrow[100°C,\ 18\ h]{Pd(OAc)_2 + 2P(o\text{-tol})_3}$

$\xrightarrow[3\ h,\ 40°C]{10\%(CO_2H)_2}$ CHO + amines

10%

Only one, six-carbon unit for chain lengthening has been studied. This is *E*-(*N*-2,5-hexadienyl)piperidine, prepared as described earlier in this chapter. This compound reacts with bromobenzene to give a 55% yield of the 6-phenyl derivative (*8*). The reaction with Z-

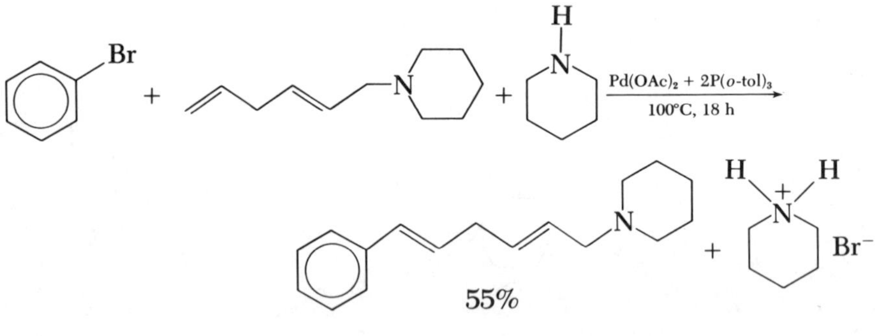

Br + N + H–N $\xrightarrow[100°C,\ 18\ h]{Pd(OAc)_2 + 2P(o\text{-tol})_3}$

N + H H N+ Br⁻

55%

1-bromo-1-hexene also appeared normal and hydrolysis gave a low yield of the expected 2,7-dodecadienal in a preliminary experiment. (*10*). This product should be easily reducible to Z-7-dodecen-1-ol, a pheromone of several insects.

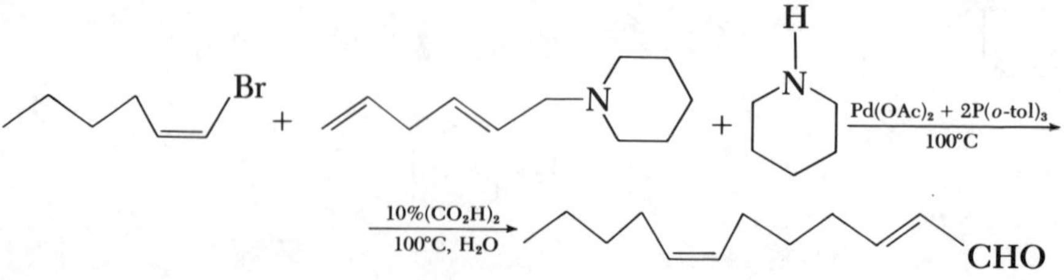

Br + N + H–N $\xrightarrow[100°C]{Pd(OAc)_2 + 2P(o\text{-tol})_3}$

$\xrightarrow[100°C,\ H_2O]{10\%(CO_2H)_2}$ CHO

Reactions of Functionalized Vinylic Bromides

Much less has been done with the functionalized halides than with the functionalized olefins. The only two-carbon compound in this group is vinyl bromide (or iodide). This is a very useful reactant with secondary amines because terminal allylic amines are usually major or exclusive products. We have used this halide for synthesizing the pheromone of the Red Bollworm Moth, 9,11-docecadien-1-ol as follows (*9*). Of the four three-carbon functionalized vinylic halides listed,

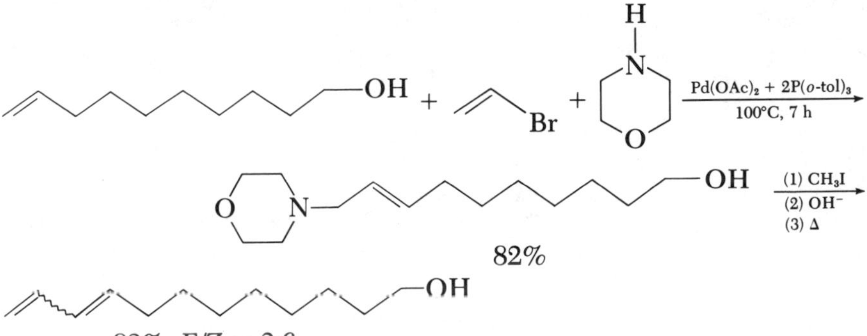

OH + Br + H–N(morpholine)O $\xrightarrow[100°C,\ 7\ h]{Pd(OAc)_2\ +\ 2P(o\text{-}tol)_3}$

O(morpholine)N– ... –OH $\xrightarrow[(2)\ OH^-,\ (3)\ \Delta]{(1)\ CH_3I}$

82%

OH

83%, *E*/*Z* = 2.6

only the first three have been studied so far and these have been studied only briefly. An example of the use of *E*-methyl 3-bromo-2-methylpropenoate has been given earlier in this chapter. The 3-bromoacrylate esters undergo elimination under our reaction conditions and cannot be used; however the related bromoaldehyde acetals can be used. The bromoacrolein acetal is unstable and not very useful, but bromomethacrolein acetal reacts normally (*4*). The pyranyl ether of 3-bromoallyl alcohol has been used to produce a 50% yield of the pyranyl ether of methyl 6-hydroxysorbate (*9*).

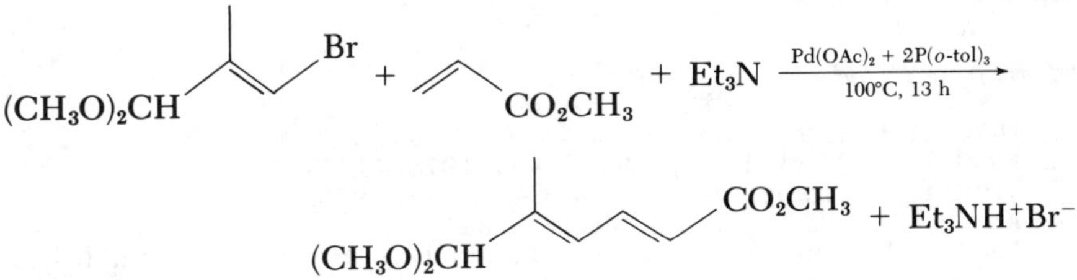

$(CH_3O)_2CH$ ... Br + CO_2CH_3 + Et_3N $\xrightarrow[100°C,\ 13\ h]{Pd(OAc)_2\ +\ 2P(o\text{-}tol)_3}$

$(CH_3O)_2CH$ ... CO_2CH_3 + $Et_3NH^+Br^-$

90%

THPO ... Br + CO_2CH_3 + Et_3N $\xrightarrow[100°C,\ 20\ h]{Pd(OAc)_2\ +\ 2PPh_3}$

THPO ... CO_2CH_3 + $Et_3NH^+Br^-$

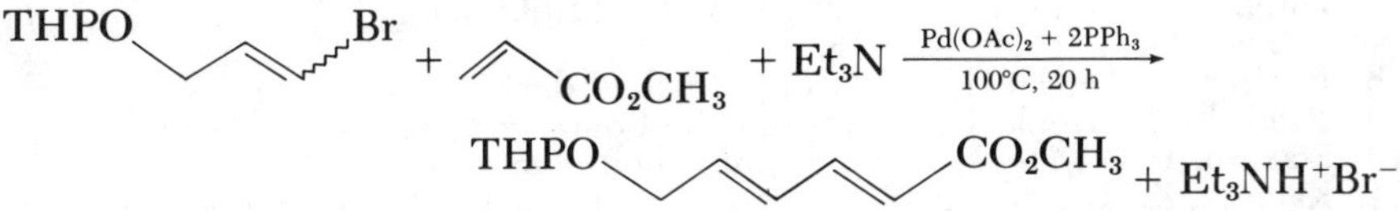

50%

We have investigated 4-bromo-3-methyl-3-buten-1-ol as a four-carbon-chain extender that incorporates an isoprene unit. First experiments suggest this reagent will be useful for preparing sesquiterpenes from myrcene.

$Pd(OAc)_2 + 2P(o\text{-tol})_3$, 100°C, 48 h

+ tetraenols + Br⁻

41%

The five-carbon unit, 1-bromo-1,4-pentadiene, is easily available from the palladium-catalyzed reaction of allyl bromide with acetylene (*11*), but we have not yet studied its reactions.

This very brief survey of the palladium–triarylphosphine-catalyzed reaction of vinylic halides with olefins and amines is intended to show the wide applicability of the reaction to synthesizing numerous types of polyfunctional aliphatic (and aromatic) compounds. We have much more to do in several areas, but it is already clear that the reaction will be very valuable for synthesizing numerous types of organic compounds.

Literature Cited

1. Heck, R. F. *Acct. Chem. Res.* **1979,** *12,* 146.
2. Patel, B. A.; Heck, R. F. *J. Org. Chem.* **1978,** *43,* 3898.
3. Trost, B. M.; Tanigawa, Y. *J. Am. Chem. Soc.* **1979,** *101,* 4743.
4. Kim, J-I.; Heck, R. F., unpublished data.
5. Patel, B. A.; Kim, J-I.; Bender, D. D.; Kao, L-C.; Heck, R. F., unpublished data.
6. Kao, L-C.; Heck, R. F., unpublished data.
7. Patel, B. A.; Kao, L-C.; Cortese, N. C.; Minkiewicz, J. V.; Heck, R. F. *J. Org. Chem.* **1979,** *44,* 918.
8. Bender, D. D.; Heck, R. F., unpublished data.
9. Patel, B. A.; Heck, R. F., unpublished data.
10. Stakem, F. G.; Heck, R. F., unpublished data.
11. Kaneda, K.; Uchiyama, T.; Fujiwara, Y.; Imanaka, T.; Teranishi, S. *J. Org. Chem.* **1979,** *44,* 55.

RECEIVED July 10, 1980.

13

Binuclear Organoplatinum Complexes as Models for Catalytic Intermediates

M. P. BROWN, J. R. FISHER, and S. J. FRANKLIN
Donnan Laboratories, University of Liverpool, U.K.
R. J. PUDDEPHATT and M. A. THOMSON
Department of Chemistry, University of Western Ontario, London, Canada, N6A 5B7

Attempts have been made to mimic proposed steps in catalysis at a platinum metal surface using well-characterized binuclear platinum complexes. A series of such complexes, stabilized by bridging bis(diphenylphosphino)methane ligands, has been prepared and structurally characterized. Included are diplatinum(I) complexes with Pt–Pt bonds, complexes with bridging hydride, carbonyl or methylene groups, and binuclear methylplatinum complexes. Reactions of these complexes have been studied and new binuclear oxidative addition and reductive elimination reactions, and a new catalyst for the water gas shift reaction have been discovered.

Mononuclear platinum complexes often have been used as models for catalytic intermediates since systematic studies of synthesis, reactivity, and mechanism are often convenient and because metallic platinum is a very important catalyst. However, using binuclear or polynuclear platinum complexes as models for proposed intermediates in heterogeneous catalysis has not been studied, probably because planned routes to such complexes have not been available. This chapter describes our first studies in this area.

Desirable Functional Groups in Binuclear Complexes

In an oversimplification, we can identify the following two ways in which the nature of heterogeneous catalytic reactions may differ from homogeneous catalytic reactions with mononuclear catalysts.

0065–2393/82/0196–0231$05.00/0

Table I. Bonding Modes of Some Important Functional Groups in Terminal or Bridging Situations

Ligand	*Terminal Ligand*	*Bridging Ligand*
Hydride	Pt–H	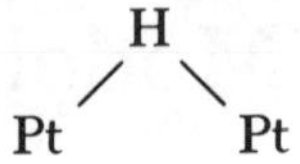 Pt(μ-H)Pt: H bridging two Pt
Carbon Monoxide	Pt–C≡O	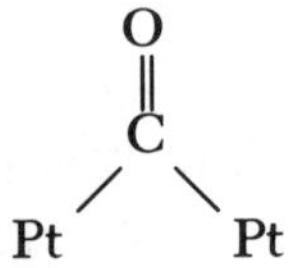 Pt(μ-C=O)Pt: C(=O) bridging two Pt
Carbene	$Pt{=}CH_2$	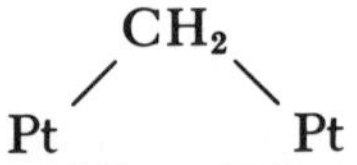 Pt(μ-CH_2)Pt: CH_2 bridging two Pt
Alkene	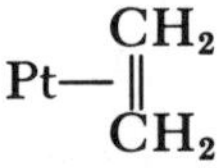 Pt—(CH_2=CH_2), side-on	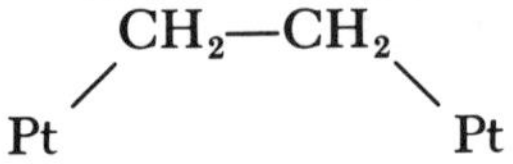 Pt—CH_2—CH_2—Pt
Alkyne	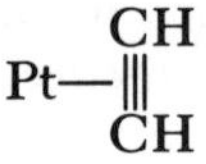 Pt—(CH≡CH), side-on	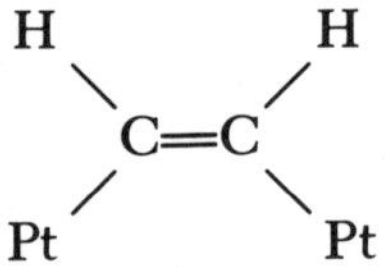 H(Pt)C=C(Pt)H
Alkyl	$Pt{-}CH_3$	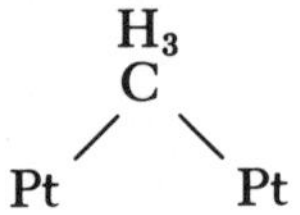 Pt(μ-CH_3)Pt: H_3C bridging two Pt

1. Ligands may bind to more than one metal center by bridging and the reactivity of the coordinated ligand will therefore be different. We set out to prepare complexes with the bridging functional groups shown in Table I and to compare their reactivity with those having terminal ligands.
2. Cooperative effects in which, for example, a reagent is activated by coordination to one metal center and then induced to react with a ligand on an adjacent metal center or in a bridging position can occur. A search for such model reactions has been initiated.

Some Diplatinum(I) Complexes

We have used the ligand bis(diphenylphosphino)methane (dppm) to stabilize binuclear platinum complexes. When chelating, this ligand forms a strained four-membered ring and it therefore tends to bridge between metal atoms. Our first platinum(I) complex was prepared by the following reaction (Equation 1), PP = dppm (*1, 2*).

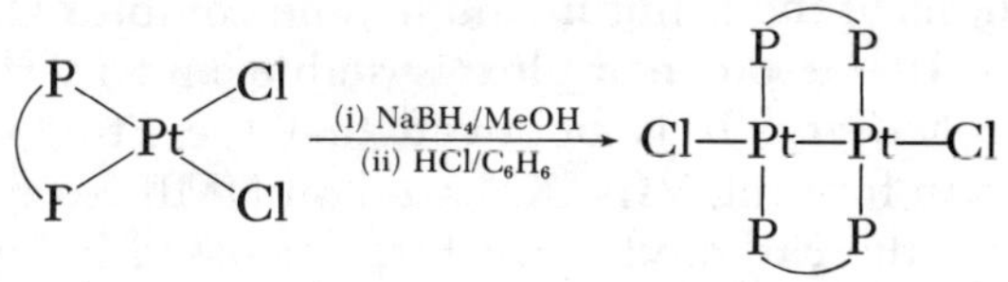

P Cl / Pt / P Cl —(i) $NaBH_4$/MeOH; (ii) HCl/C_6H_6→ Cl—Pt—Pt—Cl (bridged by two P P) (1)

I II

Complex II was originally prepared by Glockling and Pollock by a different route but was characterized incorrectly as $[Pt_2(\mu\text{-}Cl)_2(dppm)_2]$. We developed criteria to distinguish between chelating and bridging dppm ligands in such complexes based on the H-1 and P-31 NMR spectra. The criterion is particularly simple based on the intensities of satellites associated with the CH_2P_2 protons in the H-1 NMR spectrum due to coupling to ^{195}Pt ($I = \frac{1}{2}$, natural abundance 33.8%). For a terminal ligand the normal intensities are 1:4:1 but for a bridging ligand intensities are 1:8:18:8:1, and the inner satellites have almost half the intensity of the central peak (*2*). The Raman spectrum gives a strong band at 150 cm^{-1} due to the Pt–Pt bond. The presence of this bond was demonstrated by an X-ray structure determination (*3, 4*).

The Pt–Cl bonds in Complex **II** are labile and several derivatives were prepared by metathesis (e.g., *see* Equation 2) (*2, 5*).

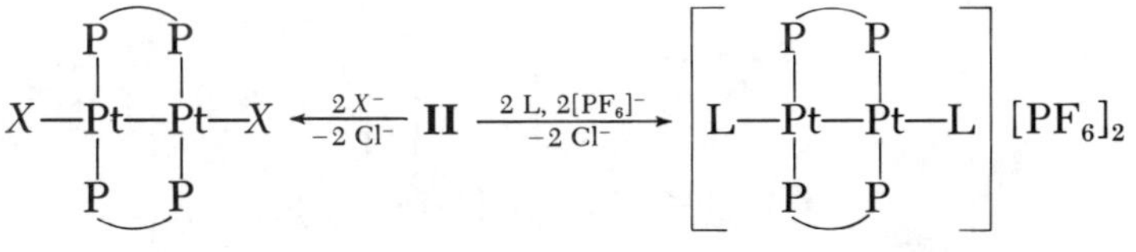

X—Pt—Pt—X (bridged by P P) ←2 X^- / −2 Cl^-— II —2 L, 2$[PF_6]^-$ / −2 Cl^-→ [L—Pt—Pt—L (bridged by P P)] $[PF_6]_2$ (2)

III, *X* = Br
IV, *X* = I

V, L = NH_3
VI, L = PMe_2Ph

Many small unsaturated reagents react by adding to the Pt–Pt bond of Complex II rather than displacing chloride (*see* Equation 3) (*6*).

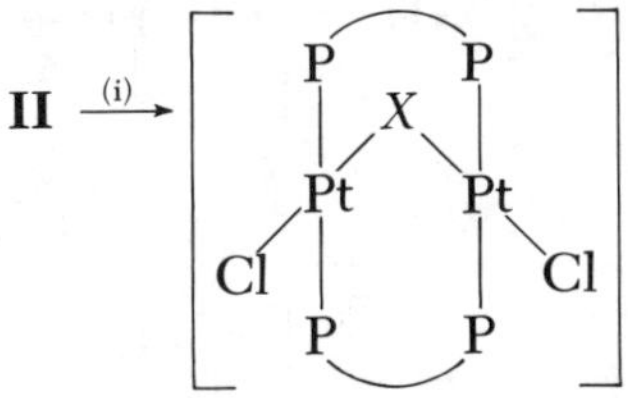

II —(i)→ [Cl–Pt(μ-X)Pt–Cl, bridged by P P / P P] (3)

VII, *X* = SO_2, (i) = SO_2
VIII, *X* = S, (i) = S_8
IX, *X* = CH_2, (i) = CH_2N_2

The synthesis of the bridging methylene complex (**IX**) is noteworthy (*see* Table I) but the complex is stable and inert, showing no carbene-like reactivity. It is believed that the Pt–Pt bond is completely broken in forming **VII–IX**, based on NMR parameters (*8*) and by analogy with the carbonyl derivative discussed below, which was characterized by X-ray structure determination.

Carbonyl Derivatives

Carbon monoxide can react with Complex **II** either by adding to the Pt–Pt bond or by displacing chloride (*see* Scheme 1).

In nonpolar solvents such as CH_2Cl_2 the bridging carbonyl (**X**) is formed preferentially. It has a remarkably low carbonyl stretching frequency of 1638 cm^{-1} and an X-ray structure determination of the corresponding bis(diphenylarsino)methane complex shows that there is no Pt–Pt bond (*3, 9, 10*). The complex therefore is considered as a dimetallated formaldehyde derivative.

In polar solvents such as methanol the ionic derivative (**XI**) is preferred in the equilibrium mixture (*9*). An X-ray structural investigation of the PF_6 salt (**XII**) has confirmed the presence of a Pt–Pt bond and the Pt–Pt bond appears to be slightly stronger than in the parent (**II**) (*11*). There is also evidence for the dicarbonyl dicationic complex

Scheme 1.

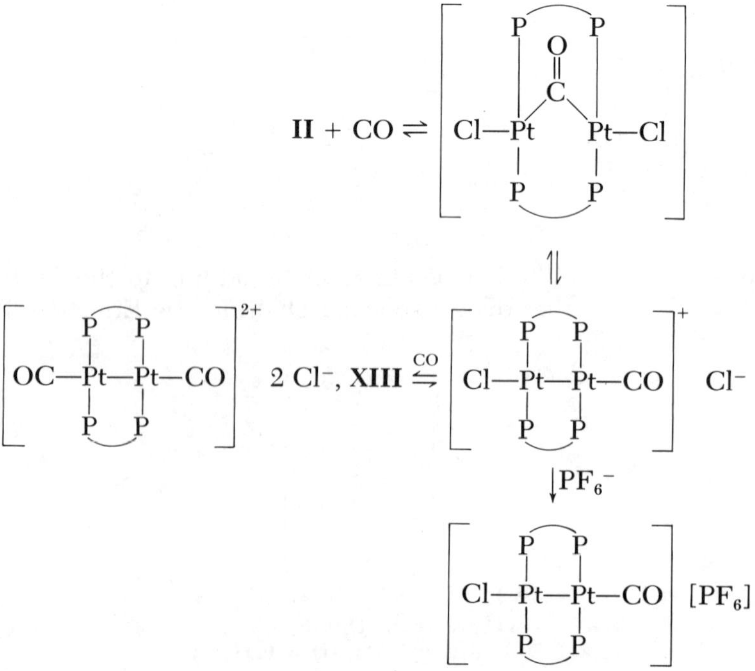

II + CO ⇌ [Cl—Pt(μ-C=O)Pt—Cl, each Pt bridged by two P⌒P ligands] **X**

⥮

[OC—Pt—Pt—CO, bridged by two P⌒P]$^{2+}$ 2 Cl^-, **XIII** $\overset{CO}{\leftrightharpoons}$ [Cl—Pt—Pt—CO, bridged by two P⌒P]$^{+}$ Cl^- **XI**

↓ PF_6^-

[Cl—Pt—Pt—CO, bridged by two P⌒P] [PF_6] **XII**

[XIII] in this system, but this can be prepared in pure form more readily as the hexafluorophosphate salt by displacement of ammonia from Complex **V** by carbon monoxide (*5*).

The easy isomerization of **X** $\rightleftharpoons$ **XI** is particularly interesting since bridging and terminally bonded carbon monoxide adsorbed on platinum surfaces may be expected to undergo an analogous reaction even more easily.

Hydride Derivatives

Reduction of monomeric Complex **I** with sodium borohydride gives the binuclear cationic hydride $[Pt_2H_2(\mu\text{-}H)(\mu\text{-dppm})_2][PF_6]$ (Equation 4, PP = dppm) (*12, 13*).

I $\xrightarrow{NaBH_4}$ $[Pt_2H_2(\mu\text{-}H)(\mu\text{-}PP)_2]^+$ Cl^- **XIV**

$\downarrow PF_6^-$ (4)

$[Pt_2H_2(\mu\text{-}H)(\mu\text{-}PP)_2]$ $[PF_6]$ **XV**

At room temperature Complex **XV** is not fluxional on the NMR time scale, and separate signals are seen for the bridging and terminal hydride ligands (*see* Figure 1). However, at higher temperatures the signals broaden and finally coalesce due to the exchange between bridging and terminal hydrides. Other binuclear hydrides have been isolated recently but these have structures in which the bridging and terminal hydrides are mutually cis and exchange is then much faster than in Complex **XV** and related complexes (*4*). These exchange processes could be related to the process in which hydrogen atoms migrate on a platinum surface.

Hydride ligands in Complex **XV** can be replaced consecutively by chloride ligands by reacting with HCl or with chlorinated solvents (*see* Equation 5).

It is interesting that Complex **XVI** has a bridging chloride and terminal hydride ligands, whereas Complex **XVII** has bridging hydride and terminal chloride ligands. Complex **XVII** can be reversibly deprotonated to give the diplatinum(I) complex (**II**) (*12, 13*).

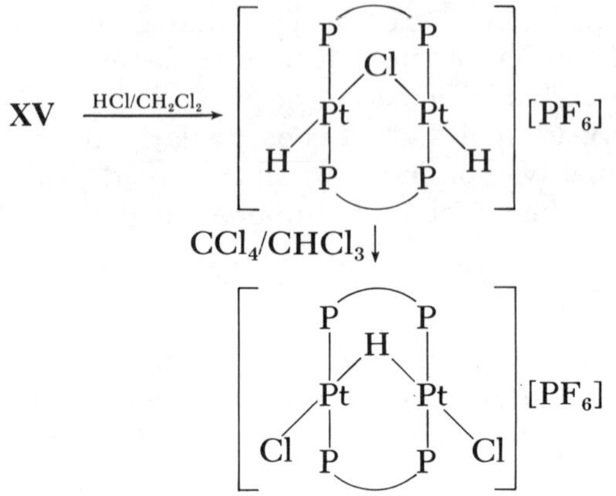

XV $\xrightarrow{HCl/CH_2Cl_2}$ [Pt₂(μ-Cl)H₂(μ-P—P)₂] $[PF_6]$ **XVI**

$CCl_4/CHCl_3$ ↓ (5)

[Pt₂(μ-H)Cl₂(μ-P—P)₂] $[PF_6]$ **XVII**

Complex **XV** undergoes an interesting binuclear reductive elimination reaction losing H_2 upon reacting with ligands such as tertiary phosphines (*see* Equation 6) (*15*), but pyridine and simple alkenes such as ethylene will not induce this reaction.

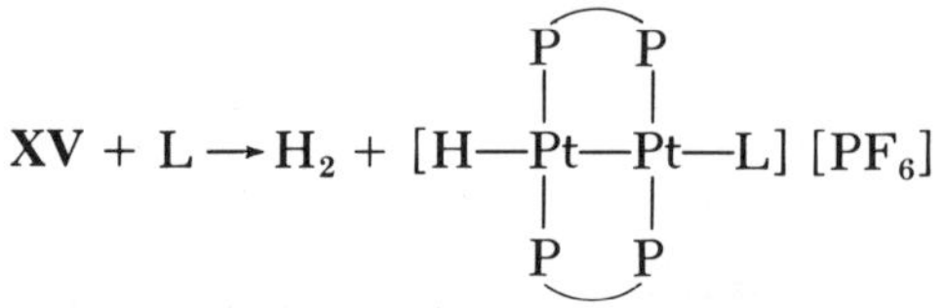

XV + L → H_2 + [H—Pt—Pt—L] $[PF_6]$ (6)

XVIII, L = PMe_2Ph **XX,** L = η'-dppm
XIX, L = PPh_3 **XXI,** L = 4-MeC_6H_4NC

When *L* = PPh_3, Reaction 6 follows overall second-order kinetics, first order in both Complex **XV** and PPh_3. Thus an associative mechanism is indicated but no intermediates could be detected. Adding PPh_3 to Complex **XV** should occur yielding a complex that would not be electron deficient and therefore would not contain a bridging hydride ligand. It is thus not possible to determine if the reductive elimination of H_2 occurs at a single platinum center or in a true binuclear process. The term binuclear reductive elimination is used to describe the overall reaction of Equation 6 and is not intended to infer a mechanism of reaction, and this applies also to Reactions 7, 8, and 9 discussed below.

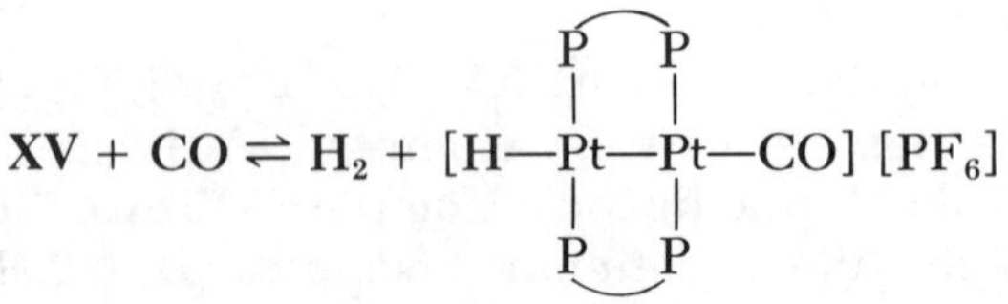

XV + CO ⇌ H_2 + [H—Pt—Pt—CO] $[PF_6]$ (7)

XXII

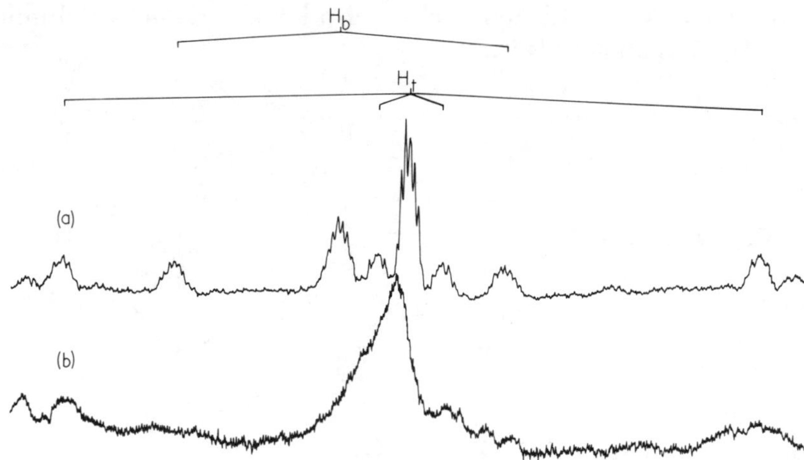

Figure 1. H-1 NMR spectra of Complex **XV** *in the PtH region: (a) spectrum at −15°C; (b) spectrum at 65°C; H_b = bridging hydride with 8:18:8 intensity pattern; $\delta(H_b)$ − 5.76 ppm, $^1J(PtH)$ 545 Hz; H_t = terminal hydride with 1:1:4:1:1 intensity pattern; $\delta(H_t)$ − 6.85 ppm, $^1J(PtH)$ 1150 Hz, $^2J(PtH)$ 108 Hz.*

The reversible reaction with carbon monoxide (Equation 7) (*16, 17*) is particularly interesting.

Since several cationic carbonyls react with water to give carbon dioxide and a metal hydride (*18*), it is predicted that Complex **XV** or **XXII** should be catalysts for the water gas shift reaction. Preliminary results indicate that this is correct and catalysis occurs slowly at 100°C in aqueous methanol, with the slow step in the catalytic cycle being the reaction of Complex **XXII** with water to regenerate Complex **XV** with the evolution of CO_2.

The slow step is deduced from the reluctance of Complex **XXII** to react with water or OH^- in model reactions to regenerate **XV** under mild conditions. This is probably because the single positive charge in Complex **XXII** is spread over two platinum centers and the carbonyl group is therefore not sufficiently activated towards nucleophilic attack. In contrast the dication $[Pt_2(CO)_2(\mu\text{-dppm})_2]^{2+}$, **XIII**, reacts readily with water to give **XXII** and carbon dioxide.

Binuclear reductive eliminations followed by oxidative addition may also be involved in reactions such as Equation 8.

$$\mathrm{XV} + \mathrm{HX} \longrightarrow \left[\begin{array}{ccccc} & \mathrm{P} & \frown & \mathrm{P} & \\ & | & X & | & \\ & \mathrm{Pt} & & \mathrm{Pt} & \\ \mathrm{H} & | & & | & \mathrm{H} \\ & \mathrm{P} & \smile & \mathrm{P} & \end{array}\right][\mathrm{PF_6}] + \mathrm{H_2} \qquad (8)$$

XXIII, X = SMe **XXIV,** X = PPh_2

Reactions with activated alkynes also involve reductive elimination of H_2 (Equation 9) (*17*).

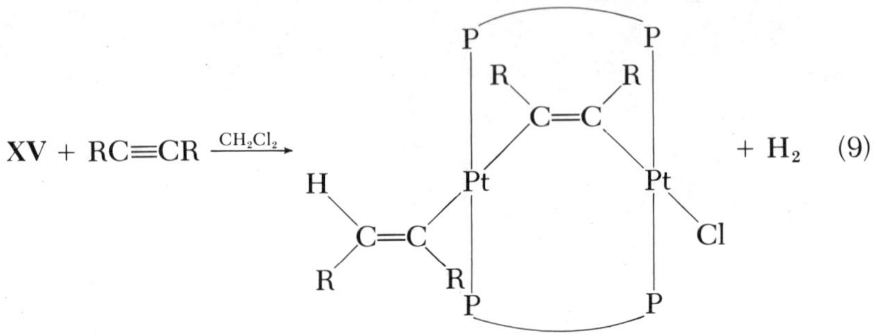

$$XV + RC{\equiv}CR \xrightarrow{CH_2Cl_2} [\text{structure}] + H_2 \quad (9)$$

P P
R R
C=C
H Pt Pt
C=C Cl
R R
P P

XXV, R = CF_3
XXVI, R = CO_2Me

In this case a reaction sequence involving reductive elimination of H_2 from Complex **XV** followed by insertion of alkyne into the remaining Pt–H bond, addition of alkyne to the Pt–Pt bond of the proposed diplatinum(I) intermediate, and then abstraction of chloride from the solvent is probable. Work is in progress to isolate and characterize proposed intermediates in this sequence, as well as to study the reactions in nonchlorinated solvents.

Methylplatinum Complexes

Both mononuclear and binuclear forms of [$PtMe_2$(dppm)] and [PtClMe(dppm)] can be prepared (*see* Scheme 2). The cis-dimer (**XXVII**) is the less stable form and is converted to monomeric [$PtMe_2$(dppm)] (**XXVIII**) when heated in the presence of dimethylsulfide. The monomethyl derivatives Complexes **XXIX** and **XXX** can be interconverted in solution but Complex **XXX** is more stable in solution (especially in polar solvents) and is isolated easily as either the chloride or the hexafluorophosphate salt. A binuclear methyl(hydrido) derivative (**XXXI**) can be prepared by reducing Complex **XXX** with sodium borohydride.

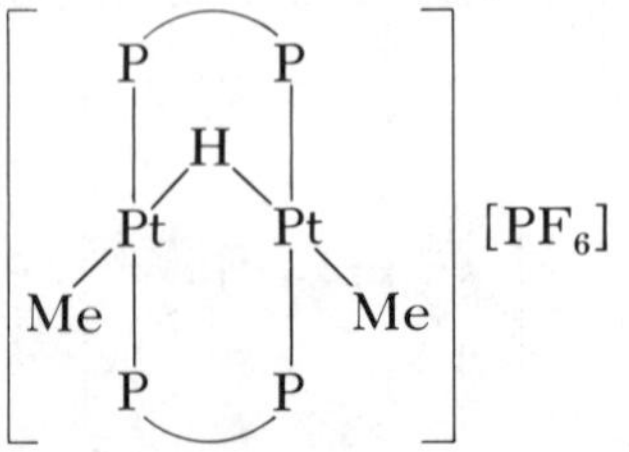

P P
H
Pt Pt [PF_6]
Me Me
P P

XXXI

Scheme 2.

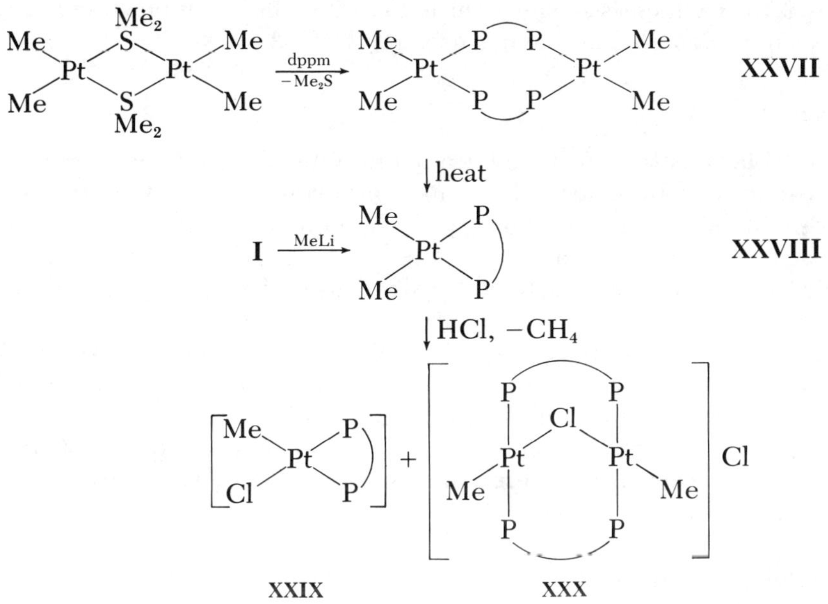

$[PtMe_2(\mu\text{-}SMe_2)]_2 \xrightarrow[-Me_2S]{dppm} [Me_2Pt(\mu\text{-}dppm)_2PtMe_2]$ **XXVII**

↓ heat

I $\xrightarrow{MeLi}$ $[PtMe_2(dppm)]$ **XXVIII**

↓ HCl, $-CH_4$

$[PtMeCl(dppm)]$ + $[Pt_2Me_2(\mu\text{-}Cl)(\mu\text{-}dppm)_2]Cl$

XXIX XXX

This complex has remarkable thermal stability, being indefinitely stable at room temperature, compared with mononuclear methyl(hydrido)platinum complexes such as *cis*-$[PtHMe(PPh_3)_2]$ which decomposes at ca. −20°C (*19*). This may be because the methyl and hydrido groups are held mutually trans in Complex **XXXI**.

Several reactions, such as that of Complex **XXVII** with H^+, have led to formation of an orange complex ion $[Pt_2Me_3(\mu\text{-dppm})_2]^+$. This complex does not contain a bridging methyl group, but probably contains a coordinate Pt→ Pt bond (*20, 21*). The structure is Complex **XXXII**, Figure 2. It is assumed that the electron-rich dimethylplatinum center donates a pair of electrons to the vacant orbital

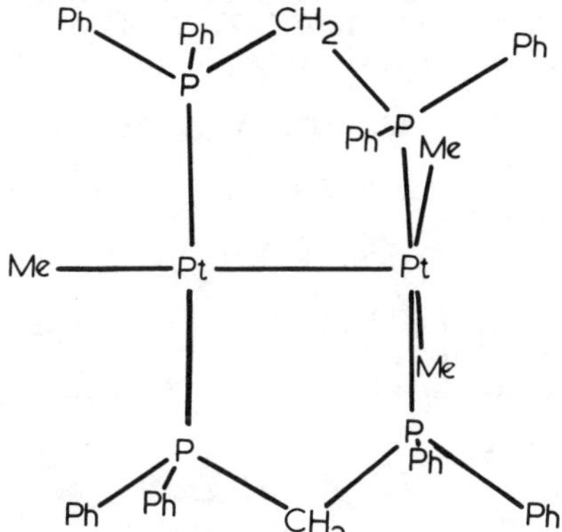

*Figure 2. The structure of the $[Pt_2Me_3(\mu\text{-dppm})_2]^+$ cation, (**XXXII**), as the hexafluorophosphate salt*

of the monomethylplatinum center. The Pt–Pt bond distance (r(Pt–Pt)) is 2.79 Å which is longer than is found in the platinum(I) complexes such as Complex **II**, having r(Pt–Pt) 2.651 Å (*21*).

Conclusions

Many of the functional groups in Table I have been identified in reactions of binuclear platinum complexes stabilized by the ligand dppm, and several reactions (including one catalytic reaction) involving both platinum centers have been discovered. Thus the modeling of intermediates and reactions involved in catalysis by platinum metal has been successful. However, generally we have not found enhanced chemical reactivity of the binuclear complexes compared with mononuclear analogs and, in some cases, reactivity is considerably lower. Thus useful catalysis by binuclear platinum complexes appears unlikely. The most encouraging aspect is that many new structural types have been discovered and there is every chance that a greater number await discovery.

Acknowledgments

Thanks are due to NSERC for financial support and Johnson–Matthey Ltd. for the generous loan of platinum. Special thanks go to Lj. Manojlović–Muir and K. W. Muir and their coworkers who have completed the X-ray structure determinations, K. R. Seddon whose help with spectroscopic studies was invaluable and the able research students whose names are given in the references.

Literature Cited

1. Brown, M. P.; Puddephatt, R. J.; Rashidi, M. *Inorg. Chim. Acta* **1976,** *19*, L33.
2. Brown, M. P.; Puddephatt, R. J.; Rashidi, M.; Seddon, K. R. *J. Chem. Soc. Dalton*, **1977,** 951.
3. Brown, M. P.; Puddephatt, R. J.; Rashidi, M.; Manojlović-Muir, Lj.; Muir, K. W.; Solomun, T.; Seddon, K. R. *Inorg. Chim. Acta* **1977,** *23*, L33.
4. Manojlović-Muir, Lj.; Muir, K. W.; Solomun, T. *Acta Crystallogr.* **1979,** *B35*, 1237.
5. Brown, M. P.; Franklin, S. J.; Puddephatt, R. J.; Thomson, M. A.; Seddon, K. R. *J. Organometal. Chem.* **1979,** *178*, 281.
6. Brown, M. P.; Fisher, J. R.; Franklin, S. J.; Puddephatt, R. J.; Seddon, K. R. *J. Chem. Soc. Chem. Comm.* **1978,** 749.
7. Brown, M. P.; Fisher, J. R.; Puddephatt, R. J.; Seddon, K. R. *Inorg. Chem.* **1979,** *18*, 2808.
8. Brown, M. P.; Fisher, J. R.; Franklin, S. J.; Puddephatt, R. J.; Seddon, K. R. *J. Organometal. Chem.* **1978,** *161*, C46.
9. Brown, M. P.; Puddephatt, R. J.; Rashidi, M.; Seddon, K. R. *J. Chem. Soc. Dalton* **1978,** 1540.
10. Brown, M. P.; Keith, A. N.; Manojlović-Muir, Lj.; Muir, K. W.; Puddephatt, R. J.; Seddon, K. R. *Inorg. Chim. Acta* **1979,** *34*, L223.

11. Manojlović-Muir, Lj.; Muir, K. W.; Solomun, T. *J. Organometal. Chem.* **1979**, *179*, 479.
12. Brown, M. P.; Puddephatt, R. J.; Rashidi, M.; Seddon, K. R. *Inorg. Chim. Acta* **1977**, *23*, L27.
13. Brown, M. P.; Puddephatt, R. J.; Rashidi, M.; Seddon, K. R. *J. Chem. Soc. Dalton* **1978**, 516.
14. Tulip, T. H.; Yamagata, T.; Yoshida, T.; Wilson, R. D.; Ibers, J. A.; Otsuka, S. *Inorg. Chem.* **1979**, *18*, 2239.
15. Brown, M. P.; Fisher, J. R.; Manojlović-Muir, Lj.; Muir, K. W.; Puddephatt, R. J.; Thomson, M. A.; Seddon, K. R. *J. Chem. Soc. Chem. Comm.* **1979**, 931.
16. Brown, M. P.; Fisher, J. R.; Puddephatt, R. J.; Thomson, M. A. *Inorg. Chim. Acta* **1980**, *44*, L271.
17. Puddephatt, R. J.; Thomson, M. A. *Inorg. Chim. Acta* **1980**, *45*, L281.
18. Clark, H. C.; Jacobs, W. J. *Inorg. Chem.* **1970**, *9*, 1229.
19. Abis, L.; Sen, A.; Halpern, J. *J. Am. Chem. Soc.* **1978**, *100*, 2915.
20. Brown, M. P.; Cooper, S. J.; Puddephatt, R. J.; Thomson, M. A.; Seddon, K. R. *J. Chem. Soc. Chem. Comm.* **1979**, 1117.
21. Frew, A. A.; Manojlović-Muir, Lj.; Muir, K. W. *J. Chem. Soc. Chem. Commun.* **1980**, 624.

RECEIVED July 14, 1980.

14

Novel Reactions of Dinuclear Metal Complexes

Reactions and Structures of Palladium Complexes Containing Bis(diphenylphosphino)methane as a Bridging Ligand

ALAN L. BALCH
Department of Chemistry, University of California, Davis, CA 95616

The reaction chemistry interconverting various binuclear palladium complexes involving bis(diphenylphosphino)methane (dpm) is summarized. Small molecules (CO, CNR, SO_2, electronegatively substituted acetylenes, and diazonium ions) readily insert into the Pd–Pd bond of the Pd(I) dimers, $Pd_2(dpm)_2X_2$ (X = halide or pseudohalide) to form molecular A-frames. Adding geminal dihalides to $Pd_2(dpm)_3$ occurs by two-center, three-fragment oxidative addition, e.g., the reaction of diiodomethane gives $Pd_2(dpm)_2(\mu\text{-}CH_2)I_2$. Dimers, $Pd_2(dpm)_2X_2Y_2$, are formed by adding halogens or methyl halides to $Pd_2(dpm)_3$.

The current high level of interest in binuclear metal complexes arises from the expectation that the metal centers in these complexes will exhibit reactivity patterns that differ from the well-established modes of reactivity of mononuclear metal complexes. The diphosphine, bis(diphenylphosphino)methane (dpm), has proved to be a versatile ligand for linking two metals while allowing for considerable flexibility in the distance between the two metal ions involved (*1*). This chapter presents an overview of the reaction chemistry and structural parameters of some palladium complexes of dpm that display the unique properties found in some binuclear complexes. Palladium complexes of dpm are known for three different oxidation states. Palladium(O) is present in $Pd_2(dpm)_3$ (*2*). Although the structure of this molecule is unknown, it exhibits a single P-31 NMR reso-

0065–2393/82/0196–0243$05.00/0

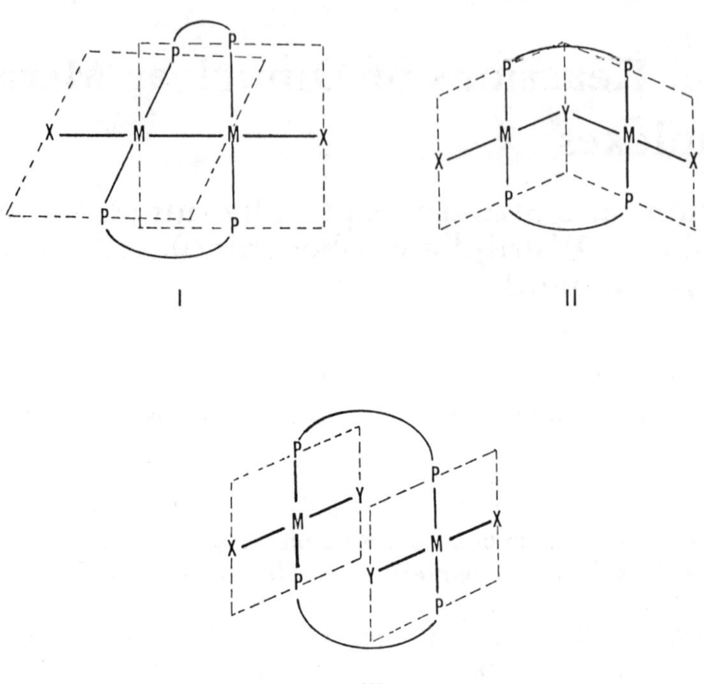

Figure 1. Idealized structural types for planar metal centers bridged by bis(diphenylphosphino)methane (4); I, side-by-side dimer; II, molecular A-frame, III, face-to-face dimer.

nance at 14.5 ppm in chloroform-d_1 relative to 85% H_3PO_4 (the same solvent and reference are used throughout unless noted) and the dpm ligands are assumed to occupy bridging sites. Palladium(I) dimers of the type $Pd_2(dpm)_2X_2$ (X = halide or pseudohalide) adopt the side-by-side, metal–metal bonded geometry (3) which is one of the three characteristic structures (*see* Figure 1) that are found for two planar metal centers bridged by *trans*-dpm ligands (4). The Pd–Pd distance in $Pd_2(dpm)_2Br_2$ is 2.699 Å and it is consistent with the presence of a single bond between the two metals (3). Palladium(II) complexes include monomeric complexes with chelating dpm, $Pd(dpm)Cl_2$ (5), and $Pd(dpm)(SCN)_2$ (6), and binuclear complexes that adopt either the face-to-face or A-frame structures shown in Figure 1. Some of these will be described later in this chapter.

Insertion into the Pd–Pd Bond in $Pd_2(dpm)_2X_2$

A variety of small molecules insert into the metal–metal bond of $Pd_2(dpm)_2X_2$ via Reaction 1. The diarsine analog $Pd_2(dam)_2Cl_2$, formed from bis(diphenylarsino)methane (dam), reacts similarly. The prod-

```
Ph   CH2   Ph                       Ph   CH2   Ph
  \ /  \  /                           \ /  \  /
Ph—P    P—Ph                        Ph—P    P—Ph
   |    |                           X  |    |  X
   |    |                            \ |    | /
X—Pd——Pd—X   + A  ⇌                    Pd    Ph            (1)
   |    |                              | \  / |
   |    |                              |  A   |
Ph—P    P—Ph                        Ph—P    P—Ph
  / \  / \                            / \  / \
Ph   CH2   Ph                       Ph   CH2   Ph
```

ucts have the structure of a molecular A-frame, a typical example of which is shown in Figure 2. Substrates that insert into the Pd–Pd bond include carbon monoxide (*4, 7, 8*), isocyanides (*4, 7*), sulfur dioxide (*9, 10*), atomic sulfur (from cyclooctasulfur or episulfides) (*10*), electronegatively substituted acetylenes (*11*), and diazonium ions (*12*). Reaction 1 causes the palladium ions to move about 0.5 Å further apart in forming the A-frame. We interpret this as a rupture of the Pd–Pd bond. The variability in the Pd–Pd separation can be seen by inspect-

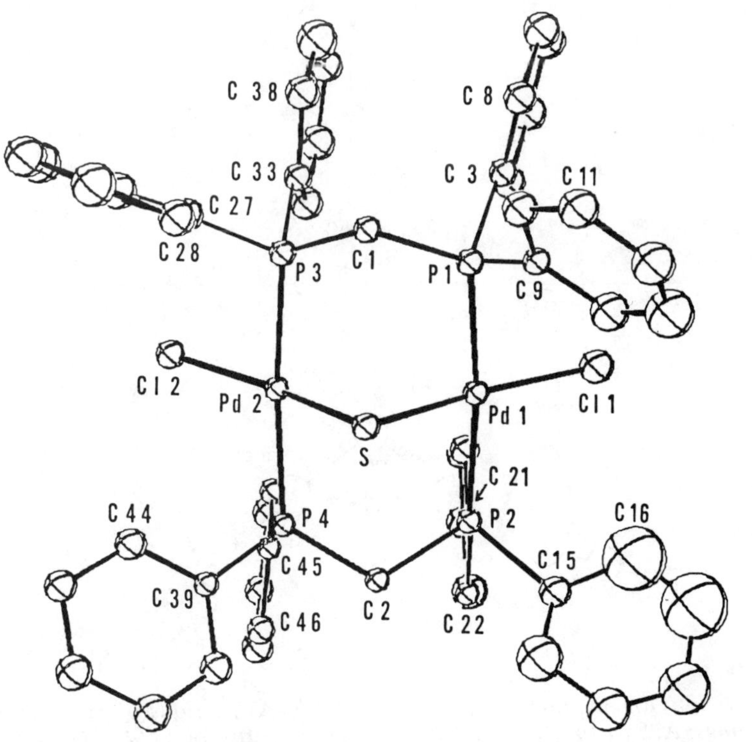

Inorganic Chemistry

Figure 2. The molecular structure of $Pd_2(dpm)_2(\mu\text{-}S)Cl_2$ (10)

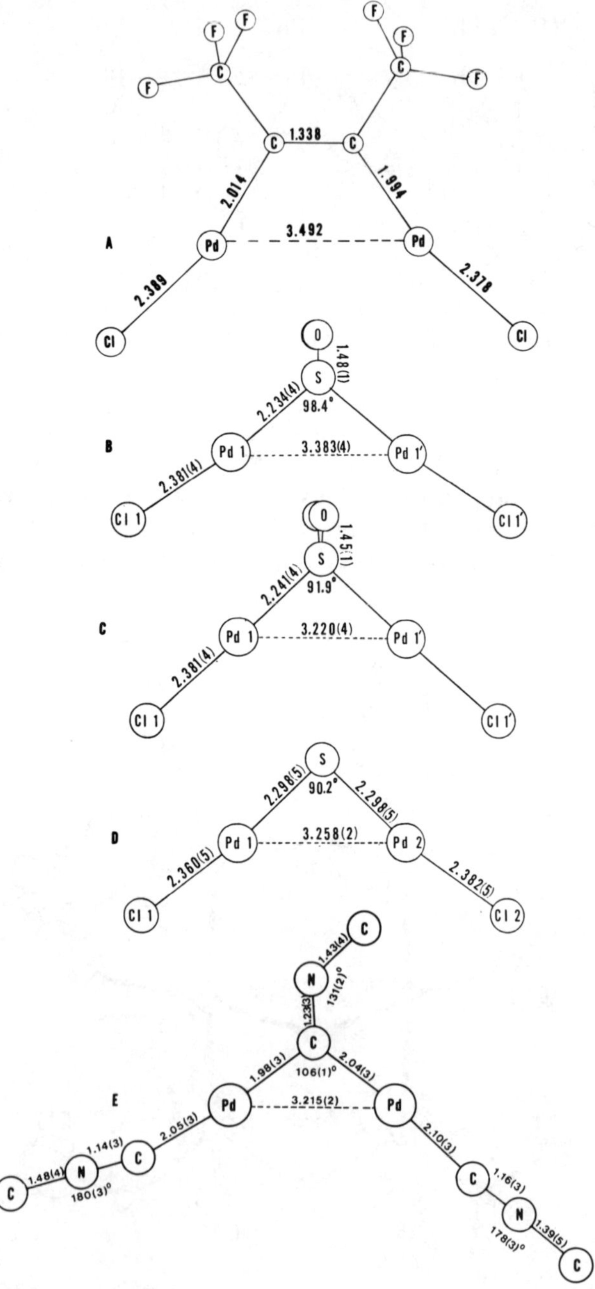

Figure 3. Dimensions of the planar $X_2Pd_2(\mu\text{-}A)$ portions of molecular A-frames: A, $Pd_2(dpm)_2(\mu\text{-}C_2\{CF_3\}_2)Cl_2$; B, $Pd_2(dpm)_2(\mu\text{-}SO_2)Cl_2$, molecule A; C, $Pd_2(dpm)_2(\mu\text{-}SO_2)Cl_2$, molecule B; D, $Pd_2(dpm)_2(\mu\text{-}S)Cl_2$; E, $[Pd_2(dpm)_2(\mu\text{-}CNCH_3)(CNCH_3)_2]^+$.

ing Figure 3, which shows the dimensions of planar $X_2Pd_2(\mu\text{-}A)$ portions of a number of insertion products that have been characterized by X-ray crystallography. The sulfur dioxide adduct, $Pd_2(dpm)_2(\mu\text{-}SO_2)Cl_2$, is particularly interesting. It crystallizes with two independent molecules in the lattice. These two species have generally similar dimensions except for the Pd–S–Pd bond angles and the Pd · · · Pd separations. We attribute the variation in these parameters to packing forces that can deform this portion of the molecule and take the plasticity of the molecule as an additional indication of the weakness of any Pd · · · Pd interaction. Unlike the other A-frames which have a single atom bridge between the two metals, the acetylene adduct, $Pd_2(dpm)_2(\mu\text{-}C_2\{CF_3\}_2)Cl_2$, has a two-atom bridge between the metals. In order to accommodate this longer bridge, the palladium atoms have moved further apart than any other pair of dpm-bridged metal ions. The structural parameters of the bridging ligand itself indicate that the acetylene has been reduced to a cis-dimetalated olefin.

Assigning formal oxidation states in these A-frame species is a matter of some controversy and may best be left to personal preferences. X-ray photoelectron spectra of a group of complexes of the type $Pd_2(dpm)_2X_2$ and $Pd_2(dpm)_2(\mu\text{-}A)X_2$ reveal no major changes in binding energies of the metal or ligand atoms (*13*). It is most important to emphasize the observation that both the reactants and products of Reaction 1 behave as 16-electron metal complexes exhibiting planar geometry and facile ligand substitution.

Some aspects of the reactivity of the A-frames formed by Reaction 1 have been explored. Carbon monoxide and sulfur dioxide are readily lost from the respective adducts upon mild heating or exposure to vacuum. The insertions of isocyanides or sulfur have not been reversed. However the oxidation of $Pd_2(dpm)_2(\mu\text{-}S)Cl_2$ to $Pd_2(dpm)_2(\mu\text{-}SO_2)Cl_2$ can be effected by using *m*-chloroperbenzoic acid as oxidant. Acetylene addition is photoreversible; photolysis of $Pd_2(dpm)_2(\mu\text{-}C_2\{CO_2CH_3\}_2)Cl_2$ forms dimethylacetylene dicarboxylate and $Pd_2(dpm)_2Cl_2$ (*14*). $Pd_2(dpm)_2X_2$ is a catalyst for converting dimethylacetylene dicarboxylate into hexamethyl mellitate, and $Pd_2(dpm)_2(\mu\text{-}C_2\{CO_2CH_3\}_2)X_2$, which forms during the reaction, is presumed to be an intermediate.

$Pd_2(dpm)_2Cl_2$ shows unique specificity of reactivity toward small molecules. It does not react with dinitrogen or nitriles and is unaffected by molecular oxygen even in refluxing benzene solution. Tin(II) chloride, which inserts into other metal–metal bonds, adds to $Pd_2(dpm)_2Cl_2$ according to Reaction 2. The structure of the product, shown in Figure 4, indicates that the presence of the $SnCl_3^-$ ligand has little effect on the Pd–Pd bond (*15*).

The ability of $Pd_2(dpm)_2Cl_2$ to insert small molecules into its metal–metal bond is not characteristic of most other metal–metal

$$\mathrm{Cl{-}Pd(\mu\text{-}Ph_2PCH_2PPh_2)_2Pd{-}Cl} + \mathrm{SnCl_2} \longrightarrow \mathrm{Cl{-}Pd(\mu\text{-}Ph_2PCH_2PPh_2)_2Pd{-}SnCl_3} \qquad (2)$$

bonded compounds. The platinum analog, $Pt_2(dpm)_2Cl_2$, undergoes similar reactions (*7, 16, 17, 18*). Because both of these molecules are coordinatively unsaturated, substrates have ready access to the metal–metal bond. Nevertheless other species possessing coordination unsaturation adjacent to a metal bond do not experience similar insertions. For example **I** and **II**, which is obtained from **I** by adding $Ph_2P(CH_2)_2PPh(CH_2)_2PPh_2$, are unreactive toward insertion of carbon monoxide, isocyanides, or sulfur dioxide. The P-31 NMR spectrum of **II**, shown in Figure 5, establishes the bridging nature of the triphosphine ligand (*19*).

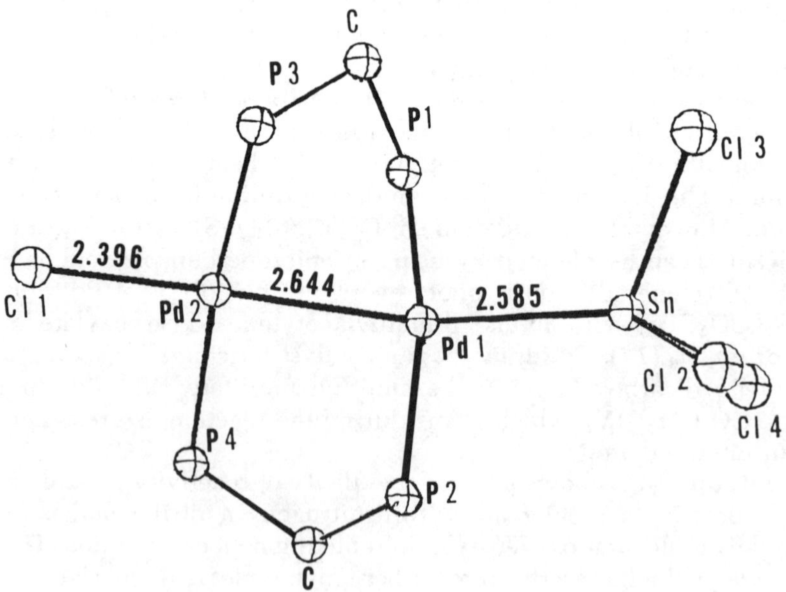

Figure 4. The molecular structure of $Pd_2(dpm)_2(SnCl_3)Cl$. The phenyl rings have been removed in order to clarify the figure.

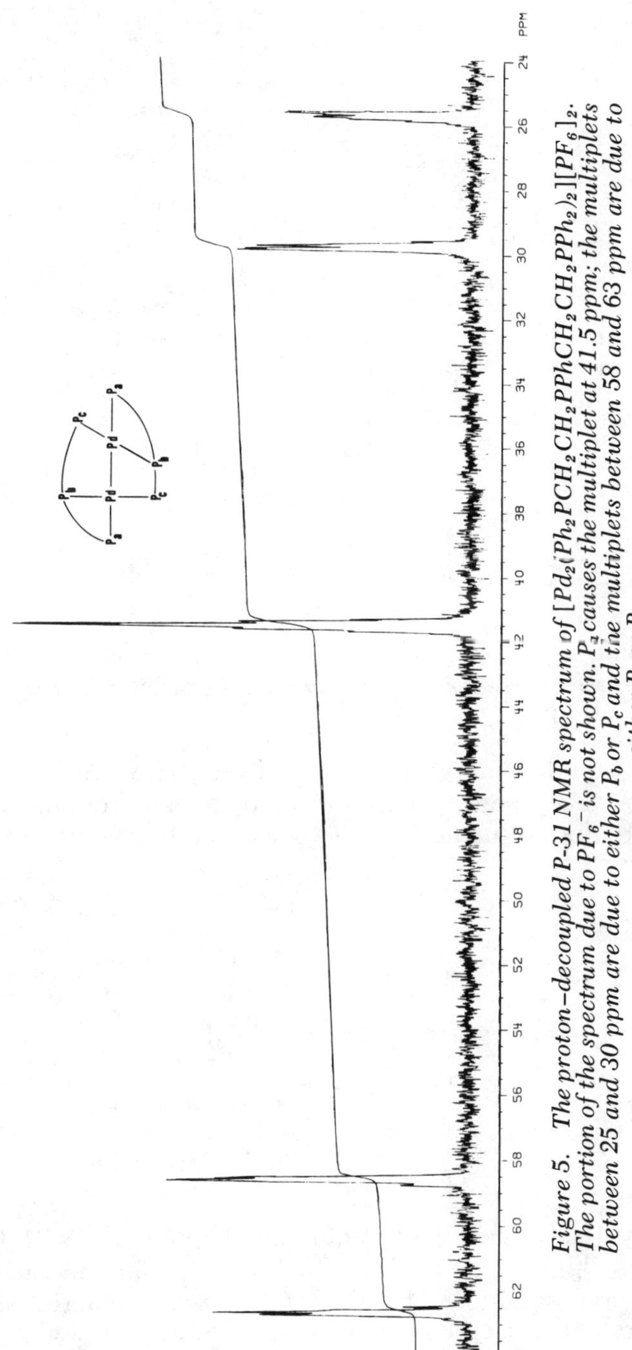

Figure 5. The proton-decoupled P-31 NMR spectrum of $[Pd_2(Ph_2PCH_2CH_2PPhCH_2CH_2PPh_2)_2][PF_6]_2$*. The portion of the spectrum due to* PF_6^- *is not shown.* P_a *causes the multiplet at 41.5 ppm; the multiplets between 25 and 30 ppm are due to either* P_b *or* P_c *and the multiplets between 58 and 63 ppm are due to either* P_c *or* P_b*.*

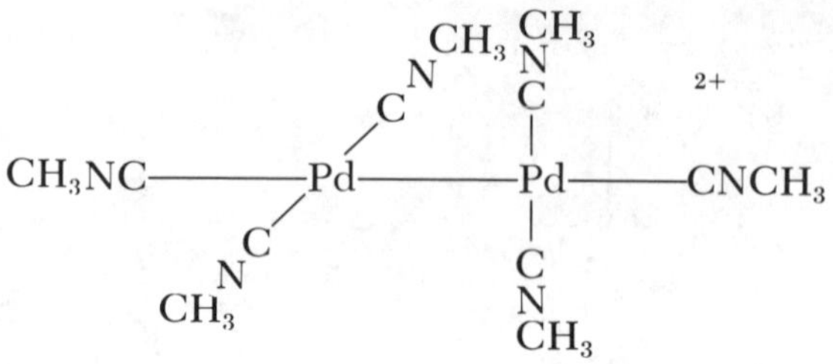

I

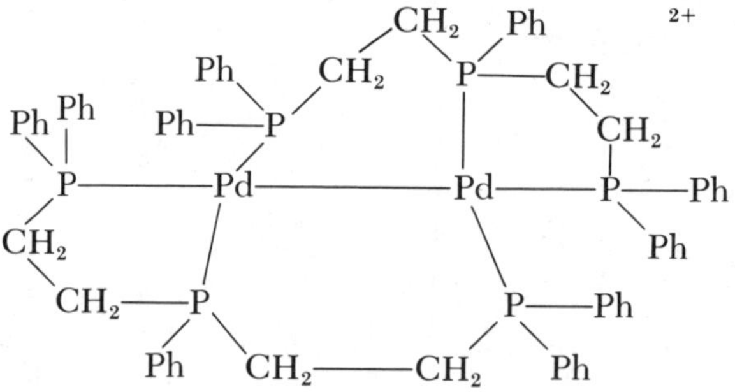

II

Two-Center, Three-Fragment Oxidative Addition Reactions of $Pd_2(dpm)_3$ (20)

$Pd_2(dpm)_3$ undergoes a series of unusual oxidative additions via Reaction 3. In these reactions three fragments from the substrate are added to the two metal ions and one of the fragments serves to bridge

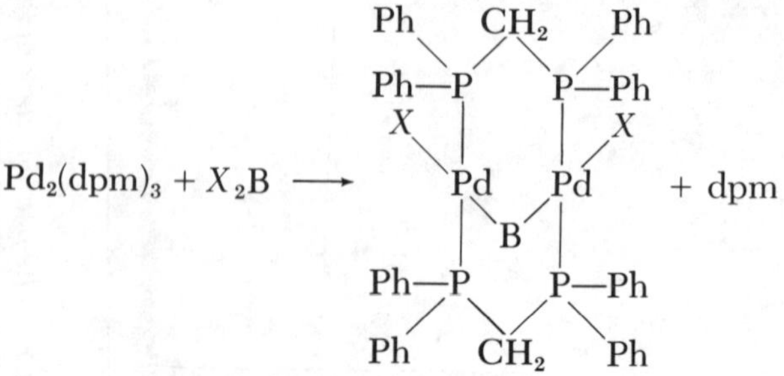

$$Pd_2(dpm)_3 + X_2B \longrightarrow Pd_2(dpm)_2(\mu\text{-}B)X_2 + dpm \qquad (3)$$

the two metals. Phenyl isocyanide dichloride reacts with $Pd_2(dpm)_3$ to form $Pd_2(dpm)_2(\mu\text{-CNPh})Cl_2$, which may be obtained also by adding phenyl isocyanide to $Pd_2(dpm)_2Cl_2$. Oxalyl chloride reacts with $Pd_2(dpm)_3$ to form carbon monoxide and $Pd_2(dpm)_2(\mu\text{-CO})Cl_2$, which is identical to the product formed by adding carbon monoxide to $Pd_2(dpm)_2Cl_2$.

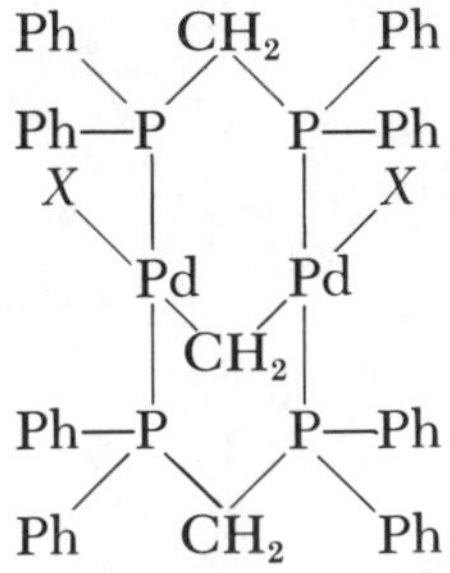

Ph CH$_2$ Ph
Ph—P P—Ph
X X
Pd Pd
CH$_2$
Ph—P P—Ph
Ph CH$_2$ Ph

III

More interestingly diiodomethane reacts with $Pd_2(dpm)_3$ to form the methylene-bridged complex **III** (X = **I**). Dibromomethane reacts similarly to form **III** (X = Br), but fortunately, under these conditions, dichloromethane does not react with $Pd_2(dpm)_3$, so it may be used as the solvent for these reactions. Complex **III** is expected to possess an A-frame structure similar to that of $Pd_2(dpm)_2(\mu\text{-}SO_2)Cl_2$ and $Pd_2(dpm)_2(\mu\text{-}S)Cl_2$. Unlike most other μ-methylene complexes such as $(\eta^5\text{-}C_5H_5)_2Rh_2(CO)_2(\mu\text{-}CH_2)$ (*21*) and $(CO)_8Fe_2(\mu\text{-}CH_2)$ (*22*), **III** is not expected to possess a metal–metal bond. Structure **III** (X = **I**), which has been isolated as yellow crystals, shows a singlet in its P-31 NMR spectrum at 18.5 ppm and exhibits the H-1 NMR spectrum shown in Figure 6. In this spectrum the methylene bridge between the two palladium ions occurs as the upfield quintet with coupling of the four equivalent phosphorus nucleii. The two methylene protons of the dpm ligands are inequivalent; consequently they occur as an AB pattern further split by coupling to two pairs of virtually coupled phosphorus nuclei. These features of the dpm methylene protons are diagnostic for the existence of a molecular A-frame (*10*). An apparently similar platinum complex, $Pt_2(dpm)_2(\mu\text{-}CH_2)Cl_2$, has been obtained by reacting diazomethane with $Pt_2(dpm)_2Cl_2$ (*17*).

These methylene-bridged complexes **III** are extremely robust, air-stable substances. We have sought without success to effect insertions into the Pd–C bonds. The complexes are unreactive toward carbon monoxide (at 5 atm at 30°C) or sulfur dioxide. Reaction with methyl isocyanide or pyridine results in displacement of the terminal halide ions and produces cations that have been isolated as hexafluorophosphate salts: $[Pd_2(dpm)_2(\mu\text{-}CH_2)(CNCH_3)_2][PF_6]_2$($\nu$(CN) = 2217 cm^{-1}) and $[Pd_2(dpm)_2(\mu\text{-}CH_2)(py)_2][PF_6]_2$. Treatment of **III** with fluoroboric or trifluoroacetic acid slowly results in the protonation of the methylene group which is converted into a terminal methyl group. The resulting brown complex, which has been isolated as its tetrafluoroborate salt has been shown by H-1 and P-31 NMR spectroscopy and X-ray crystallography to have Structure **IV**.

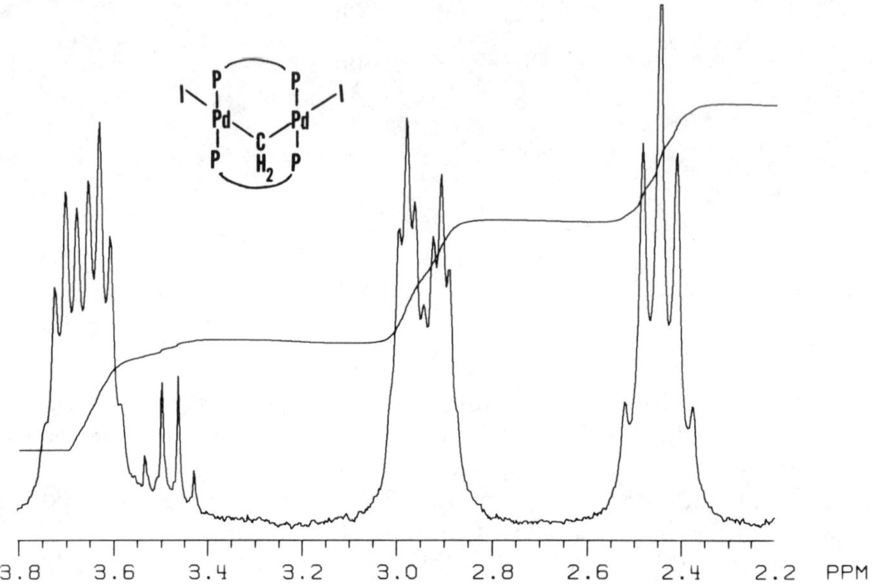

Figure 6. The H-1 NMR spectrum of $Pd_2(dpm)_2(\mu\text{-}CH_2)I_2$. The quintet at 245 ppm is due to the Pd–CH_2–Pd group and the multiplet at 2.8–3.1 and 3.6–3.8 ppm are due to the methylene protons of the dpm ligands. The quartet at 3.84 ppm is due to ethyl ether which is an impurity.

```
                                +
  Ph      CH2      Ph
Ph—P             P—Ph
         I
     Pd·····Pd
  I                CH3
Ph—P             P—Ph
  Ph      CH2      Ph
```

IV

Addition of Halogens and Methyl Halides to $Pd_2(dpm)_3$ (20)

Adding halogens to $Pd_2(dpm)_3$ occurs stepwise. The case of iodine addition is representative and may be summarized in Reactions 4, 5, and 6. Additionally, $Pd(dpm)I_2$ reacts with $Pd_2(dpm)_3$ via Equation 7. While $Pd_2(dpm)_3$, $Pd_2(dpm)_2I_2$, and $Pd(dpm)I_2$ are stable compounds

$$Pd_2(dpm)_3 + I_2 \rightarrow Pd_2(dpm)_2I_2 + dpm \qquad (4)$$

$$Pd_2(dpm)_2I_2 + I_2 \rightarrow Pd_2(dpm)_2I_4 \qquad (5)$$

$$Pd_2(dpm)_2I_4 \rightarrow 2Pd(dpm)I_2 \quad (6)$$

$$Pd_2(dpm)_3 + 2Pd(dpm)I_2 \rightarrow 2Pd_2(dpm)_2I_2 + dpm \quad (7)$$

that may be isolated in crystalline form, $Pd_2(dpm)_2I_4$ is an unstable intermediate that has been observed and identified by low-temperature H-1 and P-31 NMR spectroscopic studies. Our ability to detect this intermediate occurs because the rate of Reaction 5 exceeds that of Reaction 6. Table I shows the spectroscopic characteristics of these species. Observing J_{P-H} at ~5 Hz and the coupling of the methylene protons to two pairs of trans and therefore virtually coupled phosphorus nuclei indicates that this intermediate has Structure V, a face-to-face dimer.

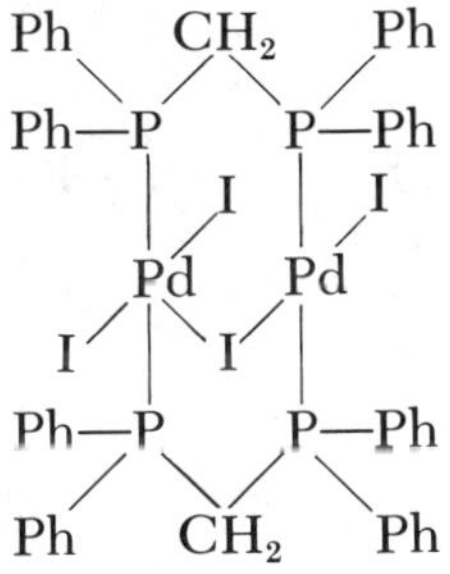

Ph CH₂ Ph
Ph—P P—Ph
I I
Pd Pd
I I
Ph—P P—Ph
Ph CH₂ Ph

V

Table I. NMR Parameters for Palladium Compounds

Compound	^1H-(CH_2)[a]		^{31}P[b] *δ ppm*
	δ ppm	$J_{(P-H)}$ *(Hz)*	
$Pd_2(dpm)_3$	3.0	broad singlet	14.5[c]
$Pd_2(dpm)_2I_2$	4.27	4.05 (q)[d]	−10.3
$Pd_2(dpm)_2I_4$	5.2	4.0 (q)	− 0.7
$Pd(dpm)I_2$	4.55	10.3 (t)	−62.7

[a] In CD_2Cl_2 solution.
[b] In $CHCl_3$ solution, proton decoupled.
[c] In C_6H_6 solution.
[d] t = triplet, q = quintet.

Adding methyl iodide or methyl bromide to $Pd_2(dpm)_3$ produces the $Pd_2(dpm)_2(CH_3)_2X_2$ as the only product isolated so far. The H-1 NMR parameters for $Pd_2(dpm)_2(CH_3)_2I_2$, δ 4.1 (quintet) with J_{P-H} 4.1 Hz, and the observation of a single P-31 resonance at 17.0 ppm, indicates that this complex has a dimeric structure rather than a monomeric structure with a chelating dpm ligand. In contrast to V which readily rearranges to monomeric Pd(dpm)I_2, **VI** shows no tendency to produce the monomer Pd(dpm)I(CH_3), a species that we have not as yet been able to detect. In dichloromethane solution $Pd_2(dpm)_2(CH_3)_2X_2$ behaves as a 1:1 electrolyte. Treatment of $Pd_2(dpm)_2(CH_3)_2X_2$ with ammonium hexafluorophosphate produces the crystalline salts $[Pd_2(dpm)_2(\mu\text{-}X)(CH_3)_2][PF_6]$. The H-1 NMR spectra of these hexafluorophosphate salts indicate that they possess the A-frame Structure **VI**. In order to explain the H-1 NMR spectra of $Pd_2(dpm)_2$-

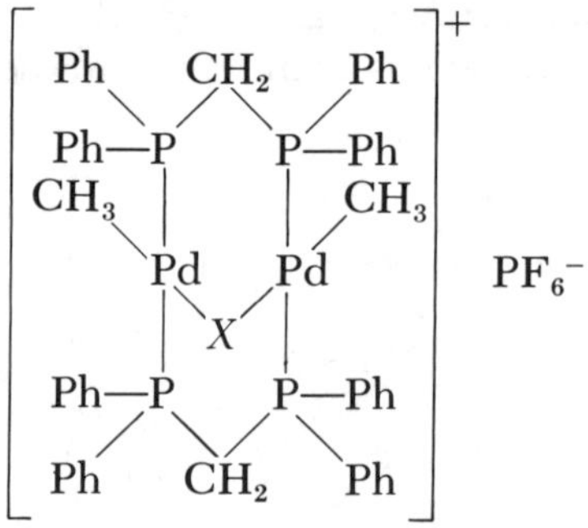

```
[ Ph     CH2     Ph            ]+
     \  /   \   /
  Ph—P         P—Ph
  CH3 |        | CH3
      \|       |/
       Pd     Pd                  PF6⁻
       | \   / |
       |   X   |
  Ph—P         P—Ph
     /  \   /   \
[ Ph     CH2     Ph            ]
```

VI

$(CH_3)_2X_2$ and their temperature dependence, we propose that the A-frame $Pd_2(dpm)_2(\mu\text{-}X)(CH_3)_2^+$ undergoes the rapid halogen exchange illustrated in Reaction 8. The occurrence of this reaction is

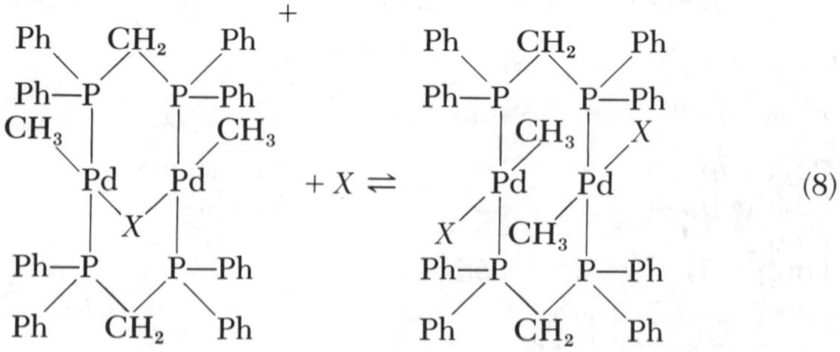

```
                         +
Ph    CH2    Ph                     Ph    CH2    Ph
   \ /   \  /                          \ /   \  /
Ph—P       P—Ph                     Ph—P       P—Ph
CH3 |      | CH3                        |CH3 |  X
    \|     |/                           |    | /
     Pd   Pd          + X ⇌             Pd   Pd          (8)
     | \ / |                           /|    |/
     |  X  |                          X |CH3 |
Ph—P       P—Ph                     Ph—P       P—Ph
   / \   /  \                          / \   /  \
Ph    CH2    Ph                     Ph    CH2    Ph
     VI
```

supported by the observation that trace amounts of halide ion render the methylene protons of **VI** equivalent on the NMR time scale. Precedent for such a reaction exists in the related chemistry of rhodium dinuclear compounds where we have observed Reaction 9 (*1*).

Further studies on the reaction chemistry of these binuclear complexes are in progress.

```
                                                              +
 Ph    CH2    Ph              Ph    CH2    Ph
   \  /   \  /                  \  /   \  /
 Ph—P      P—Ph               Ph—P      P—Ph
    |  Cl  |  CO              OC  |      |  CO
    | /    | /                  \ |      | /
    Rh     Rh         ⇌           Rh     Rh        + Cl⁻        (9)
   /|    / |                      | \   /|
 OC |  Cl  |                      |  Cl  |
 Ph—P      P—Ph               Ph—P      P—Ph
   /  \   /  \                  /  \   /  \
 Ph    CH2    Ph              Ph    CH2    Ph
```

Acknowledgments

I am grateful for the expert experimental assistance of Linda Benner, Chung–Li Lee, Katie Lindsay, and Marilyn Olmstead and for the financial support of the National Science Foundation and the UCD NMR Facility.

Literature Cited

1. Olmstead, M. M.; Lindsay, C. H.; Benner, L. S.; Balch, A. L. *J. Organometal. Chem.* **1979,** *179,* 289.
2. Stern, E. W.; Maples, P. K. *J. Catal.* **1972,** *27,* 134.
3. Colton, R.; McCormick, M. J.; Pannan, C. D. *J. Chem. Soc. Chem. Commun.* **1977,** 823.
4. Benner, L. S.; Balch, A. L. *J. Am. Chem. Soc.* **1978,** *100,* 6099.
5. Steffen, W. L.; Palenik, G. J. *Inorg. Chem.* **1976,** *15,* 2432.
6. Palenik, G. J.; Mathew, M.; Steffen, W. L.; Beran, G. *J. Am. Chem. Soc.* **1975,** *97,* 1059.
7. Olmstead, M. M.; Hope, H.; Benner, L. S.; Balch, A. L. *J. Am. Chem. Soc.* **1977,** *99,* 5502.
8. Holloway, R. G.; Penfold, B. R.; Colton, R.; McCormick, M. J. *J. Chem. Soc. Chem. Commun.* **1976,** 485.
9. Benner, L. S.; Olmstead, M. M.; Hope, H.; Balch, A. L. *J. Organometal. Chem.* **1978,** *153,* C31.
10. Balch, A. L.; Benner, L. S.; Olmstead, M. M. *Inorg. Chem.* **1979,** *18,* 2996.
11. Balch, A. L.; Lee, C-L.; Lindsay, C. H.; Olmstead, M. M. *J. Organometal. Chem.* **1979,** *177,* C22.
12. Rattray, A. D., Sutton, D. *Inorg. Chim. Acta* **1978,** *27,* L85.
13. Brant, P.; Benner, L. S.; Balch, A. L. *Inorg. Chem.* **1979,** *18,* 3422.
14. Lee, C.-L.; Balch, A. L., unpublished data.
15. Olmstead, M. M.; Benner, L. S.; Hope, H.; Balch, A. L. *Inorg. Chim. Acta* **1979,** *32,* 193.
16. Brown, M. P.; Puddephatt, R. J.; Rashidi, M.; Seddon, K. R. *J. Chem. Soc. Dalton* **1978,** 1540.
17. Brown, M. P.; Fisher, J. R.; Puddephatt, R. J.; Seddon, K. R. *Inorg. Chem.* **1979,** *18,* 2808.
18. Brown, M. P.; Keith, A. N.; Manojlovic-Muir, Lj.; Muir, K. W.; Puddephatt, R. J.; Seddon, K. R. *Inorg. Chim. Acta* **1979,** *34,* L223.
19. Lindsay, C. H.; Benner, L. S.; Balch, A. L. *Inorg. Chem.* **1980,** *19,* 3503.
20. Lindsay, C. H.; Balch, A. L., unpublished data.
21. Herrmann, W. A.; Krüger, C.; Goddard, R.; Bernal, I. *J. Organometal. Chem.* **1977,** *140,* 73.
22. Sumner, C. E., Jr.; Riley, P. E.; Davis, R. E.; Pettit, R. *J. Am. Chem. Soc.* **1980,** *102,* 1752.

RECEIVED August 10, 1980.

15

Hydrogenation Catalysis of Olefins by a Rhodium Hydride Complex of $PhP(CH_2CH_2CH_2PPh_2)_2$

In Situ Generation of RhH(ttp)

JAMES NIEWAHNER[1] and DEVON W. MEEK
Department of Chemistry, The Ohio State University, Columbus, OH 43210

The complex RhCl(ttp), where ttp = $PhP(CH_2CH_2CH_2$-$PPh_2)_2$, in the presence of either triethylaluminum or diethylaluminum chloride, is an effective homogeneous catalyst for hydrogenation of 1-olefins and 1-octyne. The rates of hydrogenation of substituted olefins are considerably slower than for terminal olefins. H-1 and P-31 NMR spectra were used to identify several different chemical species [including RhH(ttp)] in these catalytically active solutions. The observed rate of hydrogenation of 1-octene to n-octane *at 20 ± 0.3°C and under a constant H_2 pressure of 750 torr is $6.4 \times 10^4 M^{-1}\ min^{-1}$, i.e. 25 times more rapid than the Wilkinson catalyst, $RhCl(PPh_3)_3$, under comparable conditions. The rate expression is first order in the rhodium complex, first order in H_2, and zero order in the olefin. A mechanism involving RhR(ttp), RhH(ttp) associated with an ethylaluminum species, and H_2 is proposed to account for the observations. The analogous system containing Et_2AlCl and the rhodium complex of $PhP(CH_2CH_2CH_2$-$PCy_2)_2$, where Cy = the cyclohexyl group, also catalyzes 1-octene at a rate comparable with that of RhCl(ttp) + Et_2AlCl; however, in this case, the rate depends on the concentration of the olefin.*

[1] Current address: Department of Physical Sciences, Northern Kentucky University, Highland Heights, KY 41076.

0065–2393/82/0196–0257$05.00/0

A great deal of research has centered around catalysis by transition-metal complexes (*1*) since Wilkinson and co-workers reported the rapid homogeneous hydrogenation of olefins catalyzed by $RhCl(PPh_3)_3$ (*2*). It now is generally accepted that the formation of a coordinatively unsaturated metal hydride is a necessary condition for hydrogenation by transition-metal complexes (*3, 4*). Dubois and Meek (*5*) have shown that the complex RhCl(ttp), where ttp is $PhP(CH_2CH_2CH_2PPh_2)_2$, catalyzed the hydrogenation of 1-octene when $NaBH_4$ and ethanol were added. The complex RhCl(ttp) differs significantly from Wilkinson's catalyst in that the tridentate ligand remains bonded (*6*), whereas some triphenylphosphine dissociates when Wilkinson's catalyst is used (*7*). The catalytic activity of RhCl(ttp) was presumed (*5*) to be due to RhH(ttp), which was generated in situ. In addition to $NaBH_4$, compounds such as $AlEt_3$ and $LiN(CH_3)_2$ also have been used to generate metal hydrides from halide complexes via the β-elimination mechanism (*8, 9*). Also, the effect of Lewis acids on the homogeneous hydrogenation of olefins catalyzed by transition-metal complexes has been examined (*8, 10, 11*).

In this chapter, we describe the nature of the species generated in solution when $AlEt_3$ or $AlEt_2Cl$ is added to RhCl(ttp) in toluene. We also describe the kinetics of the catalytic hydrogenation of various organic substrates by RhCl(ttp) and $AlEt_2Cl$ and propose a mechanism that is consistent with the observations.

Experimental

General Procedure. Solutions containing the aluminum alkyls and RhCl(ttp) are air-sensitive, and they were handled under a nitrogen atmosphere using either Schlenk glassware or a glovebag. Toluene and 1-octene were distilled from sodium benzophenoneketyl into a dry flask under an atmosphere of nitrogen. Other olefins were stirred at least 24 h with $LiAlH_4$ and were vacuum distilled into a dry flask. Cyclohexanone and 1-octyne were stored over Linde 4A molecular sieves for 1 week, transferred via a syringe to a flask and freeze-pump-thawed three times using dry ice and 1-propanol. Triethylaluminum, $AlEt_3$, was syringed from stock bottles under an atmosphere of nitrogen in a glovebag. Diethylaluminum chloride, $AlEt_2Cl$, used in the kinetic study was transferred in a dry box from stock bottles into vials that contained a septum port and valve (available from Pierce Chemical Co.). The reagent then was syringed into the reaction flask under an atmosphere of nitrogen in a glovebag.

Materials. The complex RhCl(ttp) was prepared as described previously (*6*). Cyclohexanone was obtained from Mallinkrodt; cyclohexene and 1-pentene were obtained from MCB. All other olefins and 1-octyne were purchased from ChemSampCo. The aluminum alkyls were obtained from Aldrich as 25% w/w solutions in toluene. Neat liquids used in the H-1 NMR work were obtained from these solutions by removing the toluene under vacuum. Lithium dimethylamide, $LiN(CH_3)_2$, was obtained from Alpha/Ventron as a 10% w/w slurry in hexane. Hydrogen was passed through an Englehard Deoxo purifier and activated alumina before use.

Spectra. P-31 and H-1 NMR spectra were obtained with a Bruker HX90 Fourier transform spectrometer using spinning 10-mm tubes, a deuterium lock from the deuterated solvent, and external 85% H_3PO_4 as the reference. IR spectra were obtained using a Perkin–Elmer Model 337 grating spectrometer that covered the range from 4000 cm^{-1} to 400 cm^{-1}.

Kinetics. Most hydrogenation experiments were carried out at constant hydrogen pressure using an automatic gas-measuring instrument designed by Robert Fagan of the Department of Chemistry, at The Ohio State University. Some experiments were carried out using simple manometric methods wherein the pressure of hydrogen decreased as the reaction proceeded. Solutions were thermostated at 20 ± 0.3°C. The system volume for reactions carried out at constant pressure was about 125 mL—about 0.5m*M* in rhodium, 640 m*M* in olefin, and 25 m*M* in $AlEt_2Cl$. The total pressure was kept at about 775 torr; $p(H_2)$ = 750 torr, p(solvents) = 25 torr. For reactions carried out at constant volume, the volume of the system was about 400 mL.

No hydrogen consumption was observed for solutions containing all of the reagents except RhCl(ttp), thereby excluding any significant catalysis by $AlEt_2Cl$ itself under these conditions. Solutions were prepared under a nitrogen atmosphere in a glovebag, closed to the atmosphere, and then transferred to the hydrogenation apparatus and vacuum line. The solutions then were freeze (−196°C)-pump-thawed twice before the hydrogenation experiment.

Results and Discussion

P-31 and H-1 NMR Spectra. SOLUTIONS OF RhCl(ttp) AND $AlEt_3$. The P-31{H-1} NMR spectrum of a toluene solution of RhCl(ttp) and $AlEt_3$ is shown in Figure 1. This pattern is typical of complexes in which the Rh(ttp) portion is planar and the three phosphorus atoms of the ttp ligand are in a meridional arrangement. Spin–spin soupling between the terminal phosphorus, the central phosphorus, and the rhodium atom, all of which have a spin of ½, accounts for the observed spectrum (*12*). Table I gives the P-31{H-1} NMR data for this and other solutions containing RhCl(ttp) and the aluminum alkyls. The H-1 NMR spectrum for this solution, given in Figure 2, definitely shows a rhodium hydride resonance at −5.2 ppm. Also, there are two sets of ethyl resonances due to the ethyl groups on aluminum; however, the positions of these resonances are not the same as those of free $AlEt_3$ or $AlEt_2Cl$. Table II summarizes the H-1 NMR data of these ethyl resonances, which are thought to arise from either $AlEt_3$ or $AlEt_2Cl$ coordinated to the RhH(ttp) species through the hydride as well as the rhodium atom. The hydride is presumed to come from a β-hydride elimination of ethylene from RhEt(ttp), which is formed by an initial alkylation reaction. Since the upfield set of ethyl resonances is not present after adding ethylene, it is assigned to an ethylaluminum species that is coordinated to the hydrido ligand. This also accounts for the resonance position of the hydride (−5.2 ppm) as compared with that observed for $RhH(PPh_3)_3$ (−7.9 ppm). (Strauss and Shriver in Ref. 7 report (Rh–H) = −7.9 ppm. We observe the same value.) The downfield set of ethyl resonances is assigned to an ethylaluminum com-

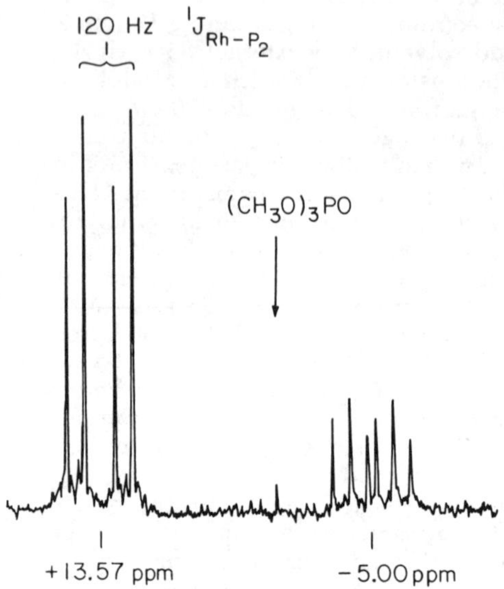

Figure 1. P-31, broad-band, proton-decoupled spectrum of RhCl(ttp) in the presence of a twentyfold molar ratio of Et_3Al in toluene. The $(CH_3O)_3PO$ signal is the secondary standard, which is superimposed on the spectrum.

pound coordinated to the rhodium. This ethylaluminum compound must be in equilibrium with $AlEt_3$ since the position of the ethyl resonances moves closer to that of free $AlEt_3$ on adding more $AlEt_3$. It has been shown that the internal chemical shift of the ethyl resonances, i.e. the differences in chemical shift between the methyl and methylene protons, decreases upon coordination to Lewis bases (*13, 14*). Aliphatic ethers and tetrahydrofuran (THF) (*13, 14*) gave an internal chemical shift of −1.25 ppm as compared with −0.78 ppm for $AlEt_3$ itself at 0.5*M*. We calculate internal chemical shifts of −1.09 ppm and −1.24 ppm for the two sets of ethyl resonances observed for solutions of RhCl(ttp) and $AlEt_3$ in toluene. These are assigned to the ethylaluminum species which is associated with the rhodium atom and the hydrido ligand, respectively. A value of −0.84 ppm is obtained for $AlEt_3$ at 5m*M* in toluene. Thus it appears that a toluene solution of RhCl(ttp) and $AlEt_3$ can be characterized by Equation 1. The exact nature of the ethylaluminum species is not known because of the

$$\begin{matrix} \text{EtAl}\langle \\ | \\ \text{(ttp)Rh–H–}\overset{\diagdown\diagup}{\text{AlEt}} \rightleftharpoons \overset{\diagdown\diagup}{\text{EtAl}} + \text{(ttp)Rh–H–}\overset{\diagdown\diagup}{\text{AlEt}} \end{matrix} \qquad (1)$$

Table I. P-31{H-1} NMR Data[a]

	δP_1 ppm	δP_2 ppm	$^2J_{P_1-P_2}$ Hz	$^1J_{Rh-P_1}$ Hz	$^1J_{Rh-P_2}$ Hz
Rh + 2 $AlEt_3$	−4.6	14.2	43.8	108.5	121.3
Rh + 2 $AlEt_3$ + C_2H_4	−12.8	11.5	49.9	141.7	114.7
Rh + 2 $AlEt_3$ + l-Octene	−12.4	12.0	49.3	141.9	114.7
	−4.6	14.4	43.0	108.1	121.3
Rh + 2 $AlEt_3$ + l-Pentene	−12.6	11.7	49.3	141.9	114.7
Rh + 2 $AlEt_2Cl$	−12.8	12.2	50.0	139.7	115.8
	20.5	8.4	53.7	185.3	126.5 minor
	—	16.1	42.7	—	117.7 minor
Rh + 2 $AlEt_2Cl$ + l-Octene	−12.8	12.2	50.0	139.7	115.8
Rh + 2 $AlEt_2Cl$ + 1-Pentene	−12.1	12.3	50.0	140.5	116.2

[a] RhCl(ttp) + 2 $AlEt_{3-n}Cl_n$ + olefin → RhR(ttp).

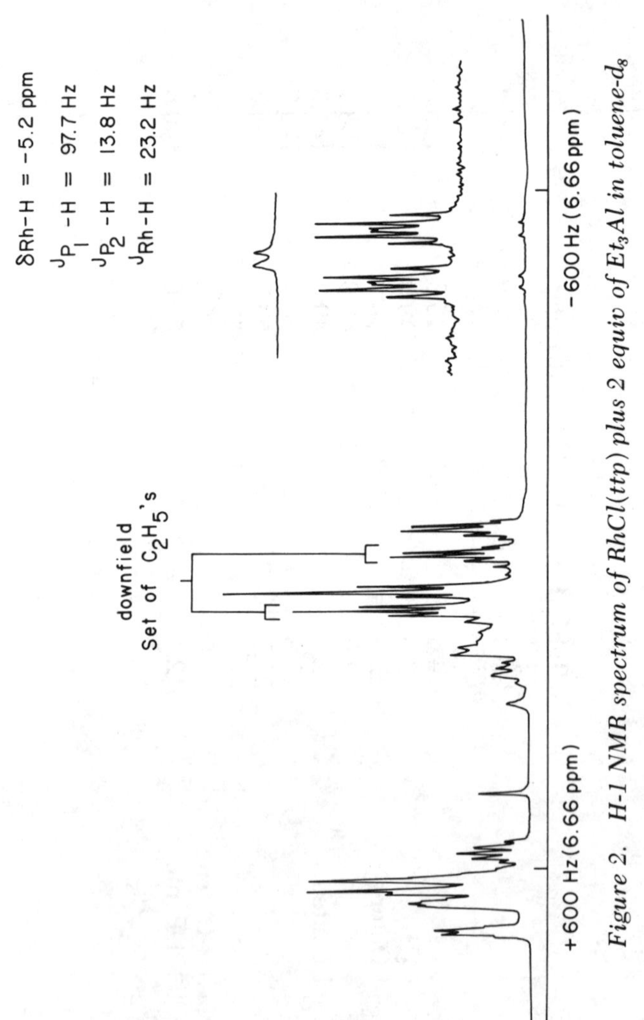

Figure 2. H-1 NMR spectrum of RhCl(ttp) plus 2 equiv of Et_3Al in toluene-d_8

Table II. H–1 NMR Parameters of Selected Peaks in Spectra of Solutions Containing RhCl(ttp) and $AlEt_3$ or $AlEt_2Cl$

Reactants	*Set A* *Triplet*	*Set A* *Quartet*	*Set B* *Triplet*	*Set B* *Quartet*	*Rh–H*
RhCl(ttp) + 2 $AlEt_3$	1.742	0.701	1.384	0.191	−5.147
	1.655	0.614	1.297	0.104	−5.418
	1.569	0.527	1.211	0.017	
		0.440		−0.069	
RhCl(ttp) + 3 $AlEt_3$ Set A	1.580	0.614	1.319	0.147	−5.321
increased in intensity	1.493	0.516	1.221	0.060	−5.581
relative to Set B	1.406	0.429	1.135	−0.025	
		0.343		—	
RhCl(ttp) + 4 $AlEt_3$ Set A	1.428	0.538	Not	0.126	−5.353
increased in intensity	1.341	0.451	distinct	0.050	−5.613
relative to Set B	1.254	0.353		−0.036	
		0.267		—	
RhCl(ttp) + 2 $AlEt_3$ + C_2H_4.	1.818	0.777	Not observed		Not observed
Set B and Rh–H	1.737	0.690			
disappeared.	1.645	0.603			
		0.516			
$AlEt_3$ Only	Triplet	1.200	0.397	Quartet	
		1.113	0.321		
		1.025	0.223		
			0.136		
RhCl(ttp) + 5 $AlEt_3$ in THF	Triplet	1.363	0.197	Quartet	Not observed
		1.268	0.102		
		1.187	0.021		
			—		
RhCl(ttp) + 2 $AlEt_2Cl$	Triplet	1.418	0.570	Quartet	Not observed
		1.330	0.482		
		1.242	0.394		
			0.299		

equilibrium between coordinated and free ethylaluminum species and the likelihood that $AlEt_2Cl$ is one of the products in the reaction between RhCl(ttp) and $AlEt_3$. Hence, the ethylaluminum species is indicated by $\overset{\backslash\,/}{\text{EtAl}}$.

Adding ethylene to a solution of RhCl(ttp) and $AlEt_3$ results in a new set of P-31{H-1} NMR parameters as shown in Table I. Furthermore, the H-1 NMR spectrum no longer exhibits the Rh–H resonance at −5.2 ppm, and the set of ethyl resonances assigned to Rh–H–$\overset{\backslash\,/}{\text{AlEt}}$ is not observed. Thus, this new rhodium species is thought to be $\overset{\backslash\,/}{\text{EtAl}}$–RhEt(ttp).

Adding either 1-pentene or 1-octene to a toluene solution of RhCl(ttp) and $AlEt_3$ results in a mixture of two compounds that are shown by the P-31{H-1} NMR spectra. One compound is assigned as $\overset{\backslash\,/}{\text{EtAl}}$–RhH(ttp)–$\overset{\backslash\,/}{\text{AlEt}}$, whereas the other is $\overset{\backslash\,/}{\text{EtAl}}$–RhR(ttp). As the alkyl group R is varied, we did not observe significant differences in the P-31 NMR parameters; hence, the pentene solutions may contain both $Rh(C_2H_5)(ttp)$ and $Rh(C_5H_{11})(ttp)$. Analogously, the octene solutions may contain both $Rh(C_2H_5)(ttp)$ and $Rh(C_8H_{17})(ttp)$. In all of these cases, association between the ethylaluminum species and the rhodium complex is proposed. The chemical equations representing these reactions are as follows:

$$\mathrm{RhCl(ttp)} + \mathrm{xs\ AlEt_3} \rightleftharpoons \mathrm{(ttp)}\underset{}{\overset{\overset{\text{EtAl}<}{|}}{\mathrm{Rh}}}\text{–H–}\overset{\backslash\,/}{\mathrm{AlEt}} + \mathrm{C_2H_4} \qquad (2)$$

$$\mathrm{(ttp)}\overset{\overset{\text{EtAl}<}{|}}{\mathrm{Rh}}\text{–H–}\overset{\backslash\,/}{\mathrm{AlEt}} + \mathrm{olefin} \rightleftharpoons \mathrm{(ttp)}\overset{\overset{\text{EtAl}<}{|}}{\mathrm{Rh}}\text{–R} + \overset{\backslash\,/}{\mathrm{AlEt}} \qquad (3)$$

SOLUTIONS OF RhCl(ttp) AND $AlEt_2Cl$. The P-31{H-1} NMR spectrum of a solution of RhCl(ttp) and $AlEt_2Cl$ shows three species, cf. Table I. The major component has the same spectral parameters as the compound thought to be $\overset{\backslash\,/}{\text{EtAl}}$–RhEt(ttp). Since the H-1 NMR spectrum of such a solution does not show a hydride resonance upfield from tetramethylsilane (TMS), probably neither of the other components is a rhodium hydride species. The reason for not observing the rhodium hydride resonance in the presence of $AlEt_2Cl$ is not understood at this time. However we find that $RhH(PPh_3)_3$ in the presence of $AlEt_3$ shows no hydride resonance upfield from TMS. Adding 1-pentene or 1-octene to solutions containing RhCl(ttp) and $AlEt_2Cl$ results in a clean NMR spectrum that contains only one set of parameters, which is consistent with rhodium–alkyl compounds of the type $\diagdown\diagup$ EtAl–RhR(ttp), vida supra. The ethylaluminum resonances in the H-1 NMR spectrum result in an internal chemical shift of −0.89 ppm suggesting only slight coordination between aluminum and rhodium. This is consistent with greater steric hindrance between the aluminum species and rhodium when ethyl is attached to rhodium than when hydride is attached.

Based on the P-31{H-1} and the H-1 NMR spectra and the fact that solutions of RhCl(ttp) and either $AlEt_3$ or $AlEt_2Cl$ catalyze the hydrogenation of olefins, the following catalytic cycle is thought to be involved. The association of the ethylaluminum species with the rhodium complex is omitted for clarity.

$$\mathrm{RhCl(ttp)} + \mathrm{AlEt_3} \rightarrow \mathrm{RhH(ttp)} + \mathrm{AlEt_2Cl} + \mathrm{C_2H_4} \quad (4)$$

$$\mathrm{RhCl(ttp)} + \mathrm{AlEt_2Cl} \rightarrow \mathrm{RhEt(ttp)} + \mathrm{AlEtCl_2} \quad (5)$$

$$\mathrm{RhEt(ttp)} \rightleftharpoons \mathrm{RhH(ttp)} + \mathrm{C_2H_4} \quad (6)$$

alkene → RhH(ttp) → alkane

RhR(ttp) ← H_2

Carbon-13 NMR Spectrum. A solution of RhCl(ttp) and $AlEt_2Cl$ to which C_2H_4 has been added gives a C-13 NMR spectrum with a low-intensity, broad resonance at 76.61 ppm downfield from TMS. This position corresponds to a coordinated ethylene (*15*). The broad resonance suggests that the ethylene is undergoing exchange. This conclusion, in conjunction with the H-1 NMR data, suggests an equilibrium involving ethylene, the rhodium species, and $AlEt_2Cl$, i.e.

$$\overset{\diagdown\,\diagup}{\mathrm{EtAl}}\text{–}\mathrm{RhEt(ttp)} + \mathrm{C_2H_4} \rightleftharpoons \begin{matrix}\mathrm{CH_2}\\ \Vert\\ \mathrm{CH_2}\end{matrix}\text{—}\mathrm{RhEt(ttp)} + \mathrm{EtAl}\!\begin{smallmatrix}\diagup\\ \diagdown\end{smallmatrix} \quad (7)$$

Kinetics and Mechanism. The kinetic study was carried out using $AlEt_2Cl$ since we had difficulty in obtaining reproducible results with $AlEt_3$. This problem is thought to arise from the greater air sensitivity of $AlEt_3$ as compared with $AlEt_2Cl$. Figure 3 shows a typical plot of hydrogen consumption vs. time for the hydrogenation of 1-octene under constant-pressure conditions. The fact that the rate does not tail off toward the end of the reaction, but proceeds steadily at a constant rate until the olefin is hydrogenated completely—then stops immediately—is indicative of zero-order olefin dependence. This behavior was observed for both 1-octene and 1-octyne, but not for 2-olefins.

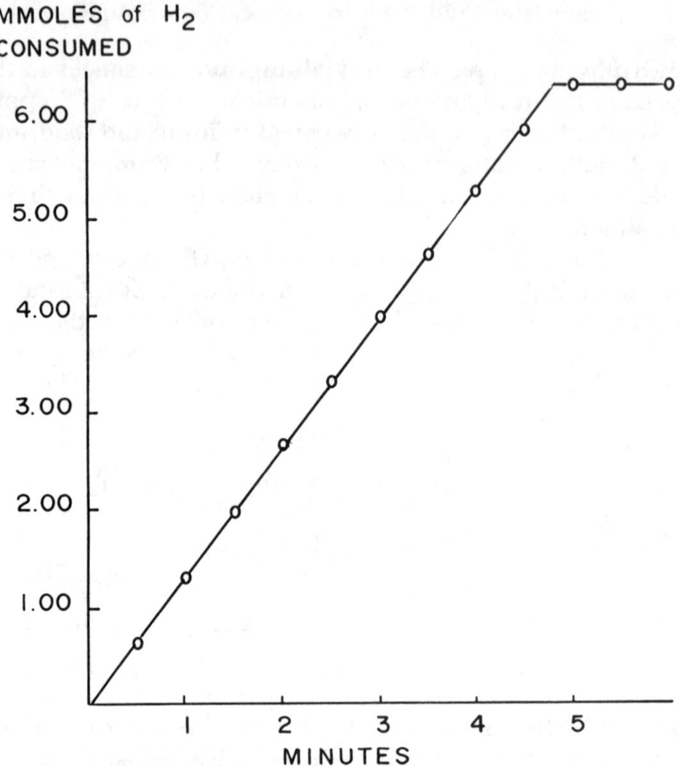

Figure 3. A plot of hydrogen consumption vs. time for the hydrogenation of 1-octene at constant pressure

Effect of Hydrogen Pressure. Reactions carried out at different pressures of hydrogen resulted in essentially the same second-order rate constant, $k = 6.2 \times 10^4 M^{-1}$ min^{-1} (first order in rhodium complex and first order in hydrogen), as shown in Table III. The solubility of hydrogen in toluene was calculated from data available in the International Critical Tables. All three reactions resulted in linear

Table III. Effect of Pressure of Hydrogen[a]

$[RhCl(ttp)] \times 10^4$	$P(H_2)_i$ (torr)	$P(H_2)_f$ (torr)	$k, M^{-1} min^{-1} \times 10^{-4}$
4.9	481	240	5.9
5.4	651	414	7.3
5.4	767	537	5.5
			avg = 6.2

[a] $[AlEt_2Cl] = 2.5 \times 10^{-2}$, [1-Octene] = 0.637$M$, 20°C.

plots of the type shown in Figure 4, which show that the hydrogen dependence is first order. The rate constants calculated from Ln P (H_2) vs. time plots are the same as those obtained from plots of olefin concentration vs. time. The latter method is less complicated.

EFFECT OF OLEFIN CONCENTRATION. Table IV lists data for reactions carried out at different concentrations of 1-octene. These data are consistent with the zero-order behavior of olefin which is implied in Figure 4.

EFFECT OF RHODIUM CONCENTRATION. Increasing the concentration of the rhodium complex, RhCl(ttp), increases the rate of the reaction as shown in Table V. These data (*see* Figure 5) give a linear plot that passes through the origin. Thus, the reaction is first order in total rhodium concentration.

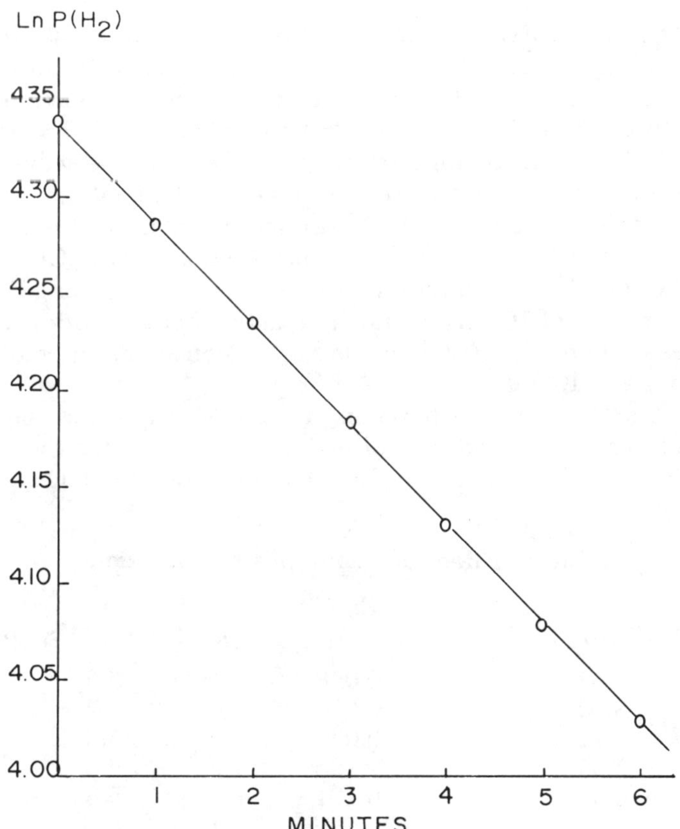

Figure 4. A plot of Ln P(H_2) vs. time for hydrogenation of 1-octene at constant volume

Table IV. Effect of Olefin Concentration[a]

[l-Octene]	*k, M^{-1} min^{-1} $\times 10^{-4}$*
0.127	7.3
0.319	5.3
0.319	7.8
0.637	6.3
0.637	5.6
0.956	7.1
0.956	7.0
1.27	7.0
1.27	5.5
	avg = 6.5

[a] [RhCl(ttp)] = 5.0×10^{-4}, [$AlEt_2Cl$] = 2.5×10^{-4}; $P(H_2)$ = 750 torr, 20°C.

EFFECT OF $AlEt_2Cl$ CONCENTRATION. The air sensitivity of the aluminum alkyls has caused difficulty in quantitatively determining the effect of the concentration of the $AlEt_2Cl$ on the reaction rate. Using new bottles of $AlEt_2Cl$, concentrations of $AlEt_2Cl$ greater than $3.7 \times 10^{-3}M$, and an Al : Rh ratio > 3 produced no observable change in the rate with increasing concentrations of $AlEt_2Cl$. Older supplies of $AlEt_2Cl$, in the presence of RhCl(ttp) and 1-olefins, did not catalyze hydrogenation of 1-octene until the concentration of $AlEt_2Cl$ was about $2 \times 10^{-3}M$ and the Al : Rh ratio was about 10 : 1. Thus, using RhCl(ttp) at concentrations of $10^{-3}M$, the rate of reaction became independent of the concentration of $AlEt_2Cl$, as long as its concentration was greater than about $2 \times 10^{-2}M$.

MECHANISM. A consideration of the NMR data and the kinetic data leads us to propose the following mechanism for the catalytic hydrogenation of olefins by RhCl(ttp) and excess $AlEt_2Cl$.

Table V. Effect of Rhodium Concentration[a]

[RhCl(ttp)] $\times 10^4$	*Rate (M min^{-1})*	*k, M^{-1} min^{-1} $\times 10^{-4}$*
1.43	0.028	5.6
3.00	0.074	7.2
4.71	0.090	5.6
5.85	0.126	6.3
6.56	0.131	5.8
		avg = 6.1

[a] [1-Octene] = 0.637, [$AlEt_2Cl$] = 2.5×10^{-2}; $P(H_2)$ = 750 torr, 20°C.

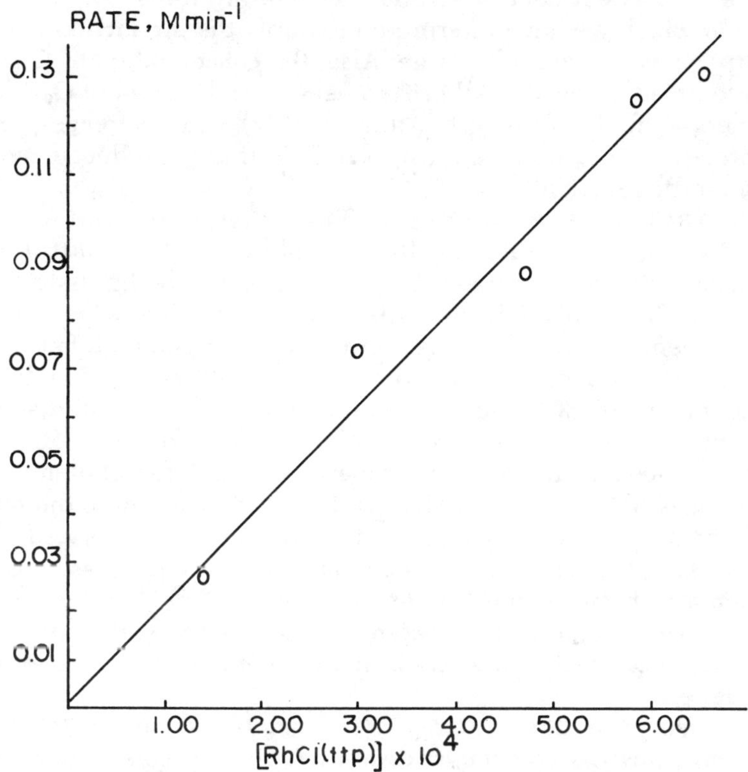

Figure 5. A plot of the total rhodium concentration vs. the rate of the reaction

$$\mathrm{RhCl(ttp) + AlEt_2Cl \overset{K_1}{\rightleftharpoons} RhEt(ttp) + AlEtCl_2}$$
$$\mathrm{RhEt(ttp) \overset{K_2}{\rightleftharpoons} RhH(ttp) + C_2H_4}$$
$$\mathrm{RhH(ttp) + alkene \overset{K_3}{\rightleftharpoons} RhR(ttp)}$$
$$\mathrm{RhR(ttp) + H_2 \overset{k}{\rightarrow} RhH(ttp) + alkane}$$

The resultant rate law is:

$$\frac{-d\,[\mathrm{H_2}]}{dt} = \left[\frac{K_3[\text{olefin}]}{\dfrac{[\mathrm{AlEtCl_2}][\mathrm{C_2H_4}]}{K_1K_2[\mathrm{AlEt_2Cl}]} + \dfrac{[\mathrm{C_2H_4}]}{K_2} + 1 + K_3[\text{olefin}]} \right] k\,[\mathrm{Rh}]_T[\mathrm{H_2}].$$

The observed rate law, $-d[\mathrm{H_2}]/dt = k_{\mathrm{obsd}}[\mathrm{Rh}]_T[\mathrm{H_2}]$, requires that the above term in brackets be either one or a constant. This is consistent with the P-31 NMR data of solutions containing RhCl(ttp), $AlEt_2Cl$, and olefin which we have interpreted to imply the production of only RhEt(ttp); thus K_3 would be large. Also, the concentration of C_2H_4 is known to be either very small or zero because only 1 mol of C_2H_4 can be formed per mole of RhCl(ttp) (thus, 0.5mM), and the freeze-pump-thaw process would remove most, if not all, that is produced by the stoichiometric reaction.

SUBSTRATES HYDROGENATED. Table VI lists the various substrates hydrogenated catalytically by solutions of RhCl(ttp) and $AlEt_2Cl$ and the relative rates based on 1-octene being assigned a value of 100. The rates indicate, as is expected, that internal and more hindered olefins react slower than terminal, unhindered olefins. After exposure to hydrogen, the solutions were analyzed by gas–liquid chromatography (GLC), and they contained only the saturated alkane in all but one case. In the case of 1-octene, 96% of the olefin was reduced to *n*-octane and 4% was isomerized to an internal olefin. The *n*-octane was obtained quantitatively from 1-octyne, even though it should pass through 1-octene and possibly be isomerized like 1-octene. Also, the rate of hydrogenation of 1-octyne decreased as the time increased between preparation of the solution and hydrogenation. These two facts suggest that 1-octyne forms an intermediate with the rhodium complex that is different from the intermediate that is formed by 1-octene.

LIFETIME OF THE CATALYST. The rate of reactions using solutions of previously used catalyst decreased with successive additions of 1-octene. These observations, along with those reported in the previous section on the effect of $AlEt_2Cl$ concentration, indicate that the life of the catalyst is related more to the air sensitivity of the $AlEt_2Cl$ than to the instability of the rhodium complex.

Table VI. Relative Rates of Hydrogenation of Various Substrates

Substrate	*Relative Rate*
1-Octene	100
2,3-Dimethyl-1-butene	5.2
2,3-Dimethyl-2-butene	no reaction
cis-2-Octene	1.2
trans-2-Octene	0.84
trans-2-Pentene	1.1
Cyclohexene	no reaction
1,3-Octadiene (cis + trans)	no reaction
1-Octyne	> 28
Cyclohexanone	no reaction
1-Octene (Wilkinson's catalyst)	3.8

USE OF A MORE HINDERED CATALYST. Using the more sterically hindered catalyst RhCl(Cyttp), where Cyttp represents the tetra-substituted cyclohexyl ligand $Cy_2P(CH_2)_3P(Ph)(CH_2)_3PCy_2$, resulted in a calculated second-order rate constant of $6.5 \times 10^4 M^{-1}\ min^{-1}$. However, the reaction is definitely dependent on the concentration of the olefin, in contrast to the RhCl(ttp) system. The third-order rate constant for the equation $r = k[H_2][Rh]_T[\text{octene}]$ is about $1.2 \times 10^4 M^{-2}\ min^{-1}$. The olefin dependence is understandable in terms of the mechanism proposed for catalysis by RhCl(ttp). In the case of the more hindered catalyst RhCl(Cyttp), the equilibrium constant K_3 would not be expected to be as large; thus, the hydrogenation mechanism with this catalyst may be more complex and may involve the olefin.

SOLID OBTAINED FROM TREATING RhCl(ttp) WITH EITHER $LiN(CH_3)_2$ OR $AlEt_3$. In either toluene or THF, RhCl(ttp) gives a yellow–orange solid on adding $LiN(CH_3)_2$. The solid resulting from toluene is darker orange than that produced in THF; however, the IR spectra of these materials are identical. Furthermore, based on the IR spectra, the same product can be obtained by treating an excess of $AlEt_3$ with RhCl(ttp) in toluene, followed by adding ether to precipitate a solid. The spectra show a peak at 1790 cm^{-1} assigned to ν(Rh–H). Elemental analyses of several different reaction products obtained from $LiN(CH_3)_2$ in THF gave inconsistent and uninterpretable results. Thus, we are unable to propose a satisfactory formula that is consistent with the IR data that show the Rh–H moiety but no O–H stretching frequency.

Acknowledgments

We are grateful to the Northern Kentucky University for a sabbatical leave which was granted to J. Niewahner during 1978–1979, to P. Kreter, T. Mazanec, and R. Waid for collecting the P-31 NMR spectra, and to Johnson Matthey and Ingelhard Industries for loans of small amounts of rhodium trichloride.

Literature Cited

1. White, C. "Organometallic Chemistry"; The Chemical Society, London, 1976, 5.
2. Osborn, J. A.; Jardine, F. H.; Young, J. F.; Wilkinson, G. *J. Chem. Soc.* 1966, *A*, 1711.
3. James, B. R. "Homogeneous Hydrogenation"; Wiley: New York, 1973.
4. Muetterties, E. L., Ed. "Transition Metal Hydrides"; Marcel Dekker: New York, 1971.
5. Dubois, D. L.; Meek, D. W. *Inorg. Chim. Acta.* 1976, *19*, L29–30.

6. Nappier, T. E., Jr.; Meek, D. W.; Kirchner, R. M.; Ibers, J. A. *J. Amer. Chem. Soc.* 1973, *95*, 4194.
7. Tolman, C. A.; Meakin, P. Z.; Lindner, D. L.; Jesson, J. P. *J. Amer. Chem. Soc.* 1974, *96*, 2763.
8. Strauss, S. H.; Shriver, D. F. *Inorg. Chem.* 1978, *17*, 3069.
9. Diamond, S. E.; Mares, F. *J. Organomet. Chem.* 1977, *142*, C55–57.
10. Porter, R. A.; Shriver, D. F. *J. Organomet. Chem.* 1975, *90*, 41.
11. Sloan, M. F.; Matlack, A. S.; Breslow, D. S. *J. Am. Chem. Soc.* 1963, *85*, 4014.
12. Tiethof, J. A.; Peterson, J. L.; Meek, D. W. *Inorg. Chem.* 1976, *15*, 1365.
13. Hatada, K.; Juki, H. *Tetrahedron Lett.* 1968, 213–217.
14. Yamamoto, O. *Bull. Chem. Soc. Japan* 1963, *36*, 1463.
15. Levy, G. C.; Nelson, G. L. "Carbon-13 Nuclear Magnetic Resonance for the Organic Chemist"; Wiley–Interscience: New York, 1972, p. 143.

RECEIVED July 28, 1980.

16

The Preparation, Characterization, and Properties of Highly Soluble Transition-Metal Complexes of Long-Chain Tertiary Phosphines

S. FRANKS and F. R. HARTLEY

Department of Chemistry and Metallurgy, The Royal Military College of Science, Shrivenham, England

J. R. CHIPPERFIELD

Department of Chemistry, The University of Hull, Hull, England

Two series of phosphines, I (PR_3R = n-$C_{10}H_{21}$ − n-C_{19}-H_{39}) and II ($P(C_6H_4R')_3$ R' – C_2H_5 − n-C_9H_{19}), which form very soluble complexes, have been prepared, characterized, and used to prepare cis-$[PtL_2Cl_2]$, trans-$[PdL_2Cl_2]$, trans-$[PtL_2'HCl]$, $[PtL_4]$, *and* trans-$[RhL_2Cl(CO)]$ *where L = I and II and L′ = I only. The problems of isolating and purifying the complexes are discussed. The influence of the two series of phosphines on (1) selective hydrogenation catalysts of the type $[PtL_2Cl_2]$ or $[PdL_2Cl_2]$ in association with $SnCl_2$ for polyunsaturated olefins and (2) the oxidative addition of methyl iodide to* trans-$[RhL_2Cl(CO)]$ *is reported. The mechanism of the latter reaction in the presence of triarylphosphines is more complex than suggested previously.*

Phosphine complexes have been used widely as homogeneous catalysts, and in many of their reactions the solvent plays a very important part. Accordingly we decided to prepare phosphine complexes with rather different solubility properties to those previously available. In particular two series of tertiary phosphine complexes were prepared that would confer extreme solubility in hydrocarbon solvents such as alkanes. Part of the rationalization for doing this was the view that if alkanes were ever going to be activated using homo-

0065–2393/82/0196–0273$05.00/0

geneous catalysts, the activation would have to be effected in the presence of solvents that were chemically more inert than the alkanes themselves. No such solvents are, of course, available and if homogeneous catalysts are to be used then those catalysts must dissolve freely in the alkanes themselves.

Two series of tertiary phosphine ligands that are very soluble in hydrocarbon solvents were prepared. The first series, $P(n\text{-}C_mH_{2m+1})_3$, where $m = 10$ to 19, were prepared by treating the Grignard reagent of the n-alkyl bromide with phosphorus trichloride in tetrahydrofuran (THF) (*1*). They were purified by recrystallization from a mixture of chloroform and ethanol. The lower members of the series were waxy materials, C_{10}–C_{12} being soft waxes and C_{13}–C_{15} being hard waxes, while the higher members of the series were crystalline. All were fairly low melting—the melting points increasing steadily from 37°–40°C for $P(n\text{-}C_{10}H_{21})_3$ to 60°–62°C for $P(n\text{-}C_{19}H_{39})_3$. They were extremely soluble in hydrocarbon solvents such as hexane and chlorinated solvents such as dichloromethane, chloroform, carbon tetrachloride, and 1,2-dichloroethane and moderately soluble in THF, benzene, and other aromatic solvents, hot alcohols such as methanol and ethanol, and warm acetone. They were, however, insoluble in cold methanol, ethanol, and acetone. On exposure to air they were oxidized readily to a complex mixture of products of the type $P(OR)_{3-n}R_n$ ($n = 0$, 1, or 2) and $P(O)(OR)_{3-n}R_n$ ($n = 0$, 1, 2, or 3). However treatment with a slight excess of hydrogen peroxide (6% w/v) resulted in a smooth oxidation to the trialkylphosphine oxides which were stable in air and more crystalline than the corresponding tertiary phosphines. These phosphine oxides had very similar solubility characteristics to the parent trialkylphosphines. In contrast to trialkylphosphine oxides below $PO(n\text{-}C_8H_{17})_3$, the long-chain trialkylphosphine oxides were not deliquescent.

The second series of phosphines prepared were a series of tris(p-alkylaryl)phosphines with alkyl groups ranging from $n\text{-}C_3H_7$ to $n\text{-}C_9H_{19}$. They were prepared by treating the Grignard reagent of the corresponding p-alkylbromobenzene with phosphorus trichloride in THF (*2*). However in order to use these routes it was necessary to devise a preparative route for synthesizing p-alkylbromobenzenes. Many of the methods for preparing alkylbromobenzenes yield a mixture of isomers. Since these are liquids with similar boiling points, their separation is difficult. A three-stage synthesis was adopted eventually (*see* Reactions 1, 2, and 3) which gave the pure para isomer in overall yields of 46 to 60%. The compound p-propylbromobenzene also can be prepared in this way but it is prepared more easily by reacting allylbromide with the mono-Grignard reagent of p-dibromobenzene followed by hydrogenating the resulting p-propenylbromobenzene to yield the desired product.

$$\text{Br}\text{—}\langle\bigcirc\rangle\text{—CHO} + \text{RCH}_2\text{MgBr} \rightarrow \text{Br}\text{—}\langle\bigcirc\rangle\text{—}\underset{\text{OH}}{\overset{\text{H}}{\text{C}}}\text{—CH}_2\text{R} \quad (1)$$

$$\text{Br}\text{—}\langle\bigcirc\rangle\text{—}\underset{\text{OH}}{\overset{\text{H}}{\text{C}}}\text{—CH}_2\text{R} \xrightarrow[\text{fused KHSO}_4]{\text{heat for 8 h over}} \text{Br}\text{—}\langle\bigcirc\rangle\text{—CH}{=}\text{CHR} \quad (2)$$

$$\text{Br}\text{—}\langle\bigcirc\rangle\text{—CH}{=}\text{CHR} \xrightarrow[\text{catalyst}]{\text{H}_2\text{, Adams}} \text{Br}\text{—}\langle\bigcirc\rangle\text{—CH}_2\text{CH}_2\text{R} \quad (3)$$

After completing this work our attention was drawn to the synthetic procedure of Manassen and Dror (*3*) based on reacting bromobenzene with an acyl chloride in the presence of aluminum trichloride. This then is followed by a Wolff–Kischner reduction of the acyl group to an alkyl group using alkaline hydrazine. Although these authors do not give overall yields, the acylation step (*4*) gives only 30% of the desired product together with 25% of C_6H_5COR. Thus our procedure gives much higher overall yields with only minor quantities of side products to be separated off in the final vacuum distillation.

The tris(*p*-alkylaryl)phosphine with a propyl group was a crystalline solid, whereas the *n*-butyl to *n*-octyl derivatives were viscous oils. The *n*-nonyl derivative was a viscous, waxy solid. The solids were purified by recrystallization from ethanol and the oils by vacuum distillation. All of the tris(*p*-alkylaryl)phosphines were extremely soluble in hydrocarbon solvents such as hexane and chlorinated solvents such as dichloromethane, chloroform, and carbon tetrachloride, and moderately soluble in THF, diethyl ether, and aromatic solvents such as benzene and toluene. The lower members of the series were soluble in ethanol but this decreased as the alkyl chain length increased. The higher members were insoluble in cold ethanol, methanol, and other polar solvents. P-31 NMR chemical shifts were measured in $CDCl_3$ solution relative to a trimethylphosphate external standard and were found to be as follows: *p*-alkyl = C_2H_5, +10.6; C_3H_7, +10.5; C_4H_9, +10.6; C_5H_{11}, +10.6; C_7H_{15}, +10.8; C_8H_{17}, +10.7, and C_9H_{19} +10.8 ppm. All of the tris(*p*-alkylaryl)phosphines were sensitive to oxygen both as solids or oils and in solution. This contrasts with triphenylphosphine which is fairly resistant to oxidation both as a solid and in solution. Unlike the trialkylphosphines which yield a complex mixture of products on oxidation in the air, the tris(*p*-alkylaryl)phosphines give a single product, the phosphine oxide. The sensitivity to air oxidation increased as the alkyl chain on the phenyl ring increased in length. Oxidation with hydrogen peroxide (6% w/v) gave a convenient preparative route to the tris(*p*-alkylaryl)phosphine oxides.

Five series of transition-metal complexes of the new tertiary phosphines have been prepared: *cis*-[$Pt(PR_3)_2Cl_2$], *trans*-[$Pd(PR_3)_2Cl_2$], *trans*-[$Pt(PR_3')_2HCl$], [$Pt(PR_3)_4$], and *trans*-[$Rh(PR_3)_2Cl(CO)$] where R = *n*-alkyl or *p*-alkylaryl and R′ = *n*-alkyl (5). All of the complexes displayed essentially the same solubility characteristics as the parent phosphines. Accordingly the synthetic routes that were used were chosen because they gave the cleanest reactions with largely volatile side products and a minimum of isomers.

cis-[$Pt(PR_3)_2Cl_2$]

When [$Pt(COD)Cl_2$] was treated with two equivalents of the tertiary phosphine, cyclooctadiene was displaced and *cis*-[$Pt(PR_3)_2Cl_2$] was formed. The displaced cyclooctadiene was removed easily in vacuo. The lower triarylphosphine complexes were recrystallized from a 60:40 mixture of ethanol and chloroform. The higher triarylphosphine complexes could be obtained only as either waxes or viscous oils. These were purified chromatographically on alumina (Brockman activity I, neutral) eluting with a 60:40 mixture of chloroform and methanol. The trialkylphosphine complexes were recrystallized from a mixture of chloroform and ethanol except for the trioctylphosphine complex which was purified chromatographically to yield a yellow wax. The melting points of the trialkylphosphine complexes are interesting (*see* Figure 1) in that the long-chain alkyl groups clearly disrupt the packing in the crystal structure so that the melting point initially decreases sharply until the tri-*n*-octylphosphine is reached after which a slight rise occurs, presumably due to increasing molecular weight. The initial sharp drop in Figure 1 has been observed previously for platinum(II) complexes with *n* in the 2–4 range as well as for palladium(II) complexes with *n* in the 2–5 range (6, 7). On heating the cis isomers decompose but they do not either isomerize or give any metallation. When [$Pt(MeCN)_2Cl_2$] is used as the starting platinum(II) complex, a mixture of *cis*- and *trans*-[$Pt(PR_3)_2Cl_2$] is formed. This mixture is virtually impossible to separate because of the very similar physical characteristics of the two isomers.

trans-[$Pd(PR_3)_2Cl_2$]

When [$Pd(COD)Cl_2$] was treated with two equivalents of the tris(*p*-alkylaryl) phosphines, PAr_3, *trans*-[$Pd(PAr_3)_2Cl_2$] complexes were formed. Apart from their trans geometry they were very similar to their platinum(II) analogues, although less crystalline so that purifica-

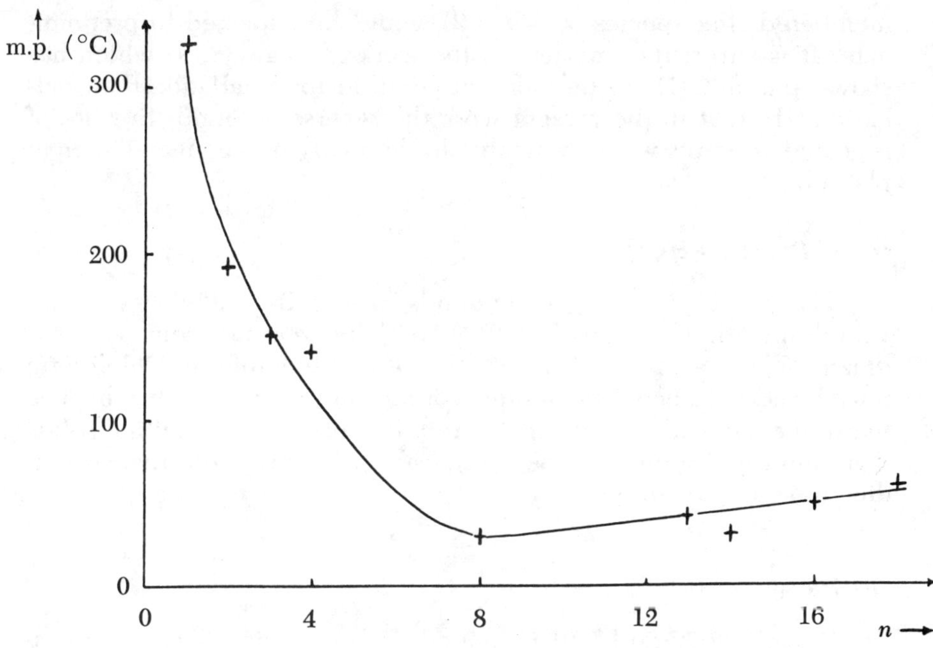

Figure 1. Melting points of cis-*[Pt{P(*n-*$C_mH_{2m+1})_3\}_2Cl_2$] complexes*

tion depended more on column chromatography. With the trialkylphosphines, PR_3', two products were obtained. The first was a yellow material, *trans*-[$Pd(PR_3')_2Cl_2$], which was very similar to the *trans*-[$Pd(PAr_3)_2Cl_2$] complexes. However a second, darker product was also present and in an increasing amount as the chain length increased. H-1, C-13, and P-31 NMR analyses and IR spectroscopy combined to suggest that this darker material is a mixture of chloride-bridged, dimeric Species **I** and **II**. In spite of repeated attempts it was impossible to separate these species by column chromatography. The metallated species, **II**, appear to be a complex mixture of products; the five-

$R_3'P$ Cl Cl
Pd Pd
Cl Cl PR_3'

I

$R_2'P$ Cl $CH—(CH_2)_aCH_3$
$(CH_2)_n$ Pd Pd $(CH_2)_b$
CH Cl PR_2'
$CH_3(CH_2)_m$

II

membered ring species ($n = b = 2$) would be expected to predominate. It is surprising, in view of the work of Shaw (*8, 9*) which has shown platinum(II) to be more susceptible to metallation than palladium(II), that in the present work the reverse is found. In spite of repeated attempts we were unable to thermally induce metallation at platinum(II).

trans-[$Pt(PR_3')_2HCl$]

The *cis*-[$Pt(PR_3')_2Cl_2$] compounds, where R′ = alkyl, were converted smoothly to *trans*-[$Pt(PR_3')_2HCl$] by treating them with hydrazine hydrate in a mixture of ethanol and chloroform. Chloroform must be added since the complexes are insoluble in pure ethanol. The hydride complexes are cream-colored, low-melting, crystalline solids that closely resemble the physical properties of the parent dichloro complexes.

[$Pt(PR)_3$]$_4$

When [$Pt(COD)_2$] (*10*) is treated with 4 equiv of tertiary phosphine in a hexane solution under nitrogen, [$Pt(PR_3)_4$] is formed. These platinum(0) complexes are yellow, air-sensitive, low-melting, crystalline solids with solubility properties that are very similar to the parent phosphines. When dissolved in solution these complexes reversibly dissociate (*see* Equation 4). Thus the P-31 NMR spectra in benzene

$$[Pt(PR_3)_4] \underset{+PR_3,\ k_{-1}}{\overset{-PR_3,\ k_1}{\rightleftharpoons}} [Pt(PR_3)_3] \underset{+PR_3,\ k_{-2}}{\overset{-PR_3,\ k_2}{\rightleftharpoons}} [Pt(PR_3)_2] \quad (4)$$

solution at ambient temperatures show a broad singlet only with no platinum–phosphorus coupling. A spectrophotometric study of the kinetics of phosphine dissociation at 25°C in a 4×10^{-4} mol L^{-1} benzene solution showed that the rate of the first step decreased in the following order: $P(n\text{-}C_{16}H_{33})_3$ (k_1(obs) too large to measure) $\gg P(-C_6H_4-CH_3)_3$ ($k_1(\text{obs}) = 3.5 \times 10^{-4}\ \text{sec}^{-1}$) > PPh_3 ($k_1(\text{obs}) = 1.5 \times 10^{-3}\ \text{sec}^{-1}$). The rates of the second step were more similar, decreasing in the following order: $P(-C_6H_4-CH_3)_3$ ($k_2(\text{obs}) = 6.7 \times 10^{-5}\ \text{sec}^{-1}$) > $P(n\text{-}C_{16}H_{33})_3$ ($k_2(\text{obs}) = 3.7 \times 10^{-5}\ \text{sec}^{-1}$) ≳ PPh_3 ($k_2(\text{obs}) = 3.5 \times 10^{-5}\ \text{sec}^{-1}$). The significantly faster rate of dissociation of the tetrakis(trialkylphosphine)platinum(0) complex than the triarylphosphine complexes is probably due to the unfavorable build-up of electron density at platinum(0) effected by the more strongly electron-donating trialkylphosphine.

trans-[$Rh(PR_3)_2Cl(CO)$]

The *trans*-[$Rh(PR_3)_2Cl(CO)$] complexes were prepared by treating [$RhCl(CO)_2]_2$ under nitrogen in chloroform solution with 2 equiv of tertiary phosphine per rhodium atom. The only side product was carbon monoxide so that purification by recrystallization from ethanol or ethanol–chloroform was relatively simple. The lower-molecular-weight trialkylphosphine and higher-molecular-weight triarylphosphine complexes were orange, viscous oils; the others were cream-colored, low-melting crystalline solids.

Olefin Hydrogenation

Bailar and his co-workers have studied using mixtures of [$M(PR_3)_2Cl_2$] and stannous chloride as catalysts for selectively hydrogenating the polyolefins present in vegetable oils to monoolefins that subsequently may be used as components of margarine (*11*). Selectivity is very important in this hydrogenation because diolefins of the type —CH═CH—CH_2—CH═CH— are very susceptible to aerial oxidation at the activated methylene group resulting in the fat becoming rancid, while the fully saturated products, although stable to storage, are very difficult to digest. From previous work it is likely that the solvent plays a very important role in the selectivity of these hydrogenations. Since the present tertiary phosphines convey unusual and somewhat extreme solubility properties, it seemed desirable to examine them as possible catalysts. Accordingly their effectiveness in promoting the selective hydrogenation of two of the common components of vegetable oils, linoleic esters **III** and linolenic esters **IV**, was examined using the methyl esters.

III **IV**

When methyl linoleate was hydrogenated in a 3:2 mixture of benzene and methanol at 90°C under 40 atm of hydrogen pressure in the presence of 0.65 mmol [$M(PR_3)_2Cl_2$] and 5.7 mmol stannous chloride (*see* Table I), it was apparent that of the platinum(II) complexes, that with P(—⬡—$C_4H_9)_3$ was the most selective giving 87.2% monoolefin while P(—⬡—$C_2H_5)_3$ and P($C_8H_{17})_3$ were close behind. P(—⬡—$C_9H_{19})_3$ was the most selective (87.8% monoolefin) with palladium(II). All of these complexes were much more specific than the triphenylphosphine complex under comparable conditions. It

Table I. The Homogeneous Hydrogenation of Methyl Linoleate (4 g) in the Presence of 0.65 mmol [ML_2CL_2] and 5.7 mmol $SnCl_2$ in a Mixture of Benzene (30 mL) and Methanol (20 mL) at 90°C Under 40 atm of Hydrogen Pressure for 3 h

		Products Formed (%)				
					Conjugated Diolefin	
M	*L*	*Methyl Stearate*	*Mono-olefin*	*Non-conjugated Diolefin*	*cis–trans*	*trans–trans*
None[a]			2.4	97.6	—	—
Platinum	PPh_3	—	41.0	24.7	8.5	25.8
	P(—⬡—Et)$_3$	3.7	83.7	12.6	—	—
	P(—⬡—Bu)$_3$	2.0	87.2	10.8	—	—
	$P(n\text{-}C_4H_9)_3$	—	11.8	52.8	16.8	18.6
	$P(n\text{-}C_8H_{17})_3$	—	78.2	21.8	—	—
	$P(n\text{-}C_{14}H_{29})_3$	—	4.4	64.5	20.2	10.9
Palladium	PPh_3	—	29.0	39.5	11.2	20.3
	P(—⬡—Et)$_3$	—	50.2	25.0	6.9	17.9
	P(—⬡—Bu)$_3$	—	17.2	40.9	13.3	28.6
	P(—⬡—C_9H_{19})$_3$	—	87.8	12.2	—	—
	$P(n\text{-}C_4H_9)_3$	—	35.3	29.3	11.0	24.4
	$P(n\text{-}C_8H_{17})_3$	—	56.2	22.4	5.3	16.1

[a] Control hydrogenation carried out in the absence of catalyst.

is very obvious from Table I that the selectivity is very dependent on the phosphine used. Thus while $P(C_8H_{19})_3$ gives high selectivity on platinum(II), $P(C_4H_9)_3$ and $P(C_{14}H_{29})_3$ give only 11.8 and 4.4% monoolefin, respectively. A similar result occurs with palladium(II), with monoolefin yields varying apparently randomly as the R′ group in $P(—C_6H_4—R')_3$ is varied systematically in the following order: R′ = H (29.0%), C_2H_5 (50.2%), C_4H_9 (17.2%), and C_9H_{19} (87.8%). Since it is hard to see how variations in either steric or electronic factors or even in both could account for this result on their own, we suggest that the interaction of the solvent with the active catalyst has a profound and subtle effect on the reaction.

When the same catalyst precursors and the same conditions were used to hydrogenate the triolefin methyl linolenate we found (*see* Table II) that platinum(II) complexes were more active under the present conditions than palladium(II) complexes, and that the introduction of a para-ethyl group into triphenylphosphine increased the activity of the platinum(II) catalyst markedly and that of the palladium(II) catalyst slightly. Introducing a para-nonyl group into the palladium(II) system markedly depressed both activity and selectivity, again suggesting that a complex set of factors probably including interaction with the solvent are involved in determining catalytic activity. The greater activity of the platinum(II) over the palladium(II) complexes under present conditions well may be a reflection of their greater stability. No signs of decomposition were detected when platinum(II) complexes were used, whereas deposition of metal was apparent when palladium(II) complexes were used.

Since the nature of the solvent plays an important part in these hydrogenations (*11*) and since methanol coordinates to palladium(II) and platinum(II) (*12*) and will therefore compete with the substrate, it was interesting to examine the effectiveness of the present catalysts in the absence of any solvent other than the substrate itself. When comparing the results obtained in the absence of solvent (*see* Table III) with those in Tables I and II, we must emphasize that in the former case the amount of substrate has been increased by a factor of 2.5, the amount of catalyst has been halved, and the amount of hydrogen has remained unaltered. Qualitative observation of the rate of fall of hydrogen pressure in the autoclave suggested that the initial reaction in the absence of solvent was faster than in its presence. Further work is currently in progress to confirm this observation and to modify the autoclave, enabling hydrogenations to be carried out under constant hydrogen pressure rather than merely constant hydrogen volume as is being done presently. This will ensure that hydrogenations can be totally completed.

Table II. The Homogeneous Hydrogenation of Methyl Linolenate (4 g) in the Presence of 0.65 mmol [ML_2Cl_2] and 5.7 mmol $SnCl_2$ in a Mixture of Benzene (30 mL) and Methanol (20 mL) at 90°C Under 40 atm of Hydrogen Pressure for 3 h

		Products Formed (%)						
					Conjugated Diolefin			
M	*L*	*Methyl Stearate*	*Monoolefin*	*Nonconjugated Diolefin*	*cis–trans*	*trans–trans*	*Nonconjugated Triolefin*	*Conjugated Triolefin*
None[a]	—	—	2.0	19.9	—	—	78.1	—
Platinum	PPh_3	—	55.6	38.8	1.9	3.7	—	—
	P(—⬡—Et)$_3$	5.4	75.0	19.6	—	—	—	—
Palladium	PPh_3	—	21.4	61.8	4.8	12.0	—	—
	P(—⬡—Et)$_3$	—	26.6	59.3	5.3	8.8	—	—
	P(—⬡—C_9H_{19})$_3$	—	12.2	44.9	17.9	10.5	14.6	—

[a] Control hydrogenation carried out in the absence of catalyst.

Experimental

Olefin hydrogenations were performed in a 250-mL stainless steel autoclave equipped with a glass liner using a stirring speed of 500 rpm. The autoclave was charged with reactants, assembled, and gas was introduced to the desired initial pressure. All valves remained closed throughout the reaction. A thermostat was provided to enable temperature control. Product analysis was carried out via gas chromatography using a 20% diethylene glycol succinate on Chromasorp P (60–80 mesh) stationary phase in a glass column at 200°C.

The reaction of *trans*-[Rh(PR_3)$_2$Cl(CO)] with methyl iodide was studied by monitoring the IR spectra of solutions of the complex (2×10^{-2} mol L^{-1}) in methyl iodide (redistilled) in a thermostatted solution IR cell provided with sodium chloride windows. A Perkin–Elmer model 577 spectrometer was used. The data was analyzed using the FACSIMILE computer program (*14*).

Oxidative Addition to trans-[*Rh*(*PR*$_3$)$_2$*Cl*(*CO*)]

Since one of the key steps in the activation of alkanes is believed to be oxidative addition, and since this reaction is very important in the mechanism of many homogeneous catalysts, it was interesting to determine the influence of the present series of phosphines on this type of reaction. Therefore we decided to study the oxidative addition of methyl iodide to *trans*-[Rh(PR_3)$_2$Cl(CO)] because this reaction has been studied previously (*13*) with $PR_3 = PPh_3$, P(C_6H_4OMe-*p*)$_3$, P(C_6H_4F-*p*)$_3$, and $AsPh_3$, and also because it is relatively easy to independently monitor the loss of *trans*-[Rh(PR_3)$_2$Cl(CO)], the formation and later decay of [Rh(PR_3)$_2$Cl(CO)(CH_3)I], and the subsequent formation of [Rh(PR_3)$_2$Cl($COCH_3$)I] by recording the C=O stretching region of the IR spectrum as a function of time. The reaction is reported to occur by the following mechanism:

$$\begin{array}{ccc} [\mathrm{Rh}(\mathrm{PR_3})_2\mathrm{Cl}(\mathrm{CO})] + \mathrm{CH_3I} & \underset{k_{-1}}{\overset{k_1}{\rightleftharpoons}} & [\mathrm{Rh}(\mathrm{PR_3})_2\mathrm{Cl}(\mathrm{CO})(\mathrm{CH_3})\mathrm{I}] \\ \nu_{\mathrm{CO}} = 1980\ \mathrm{cm^{-1}} & & \nu_{\mathrm{CO}} = 2060\ \mathrm{cm^{-1}} \\ \mathbf{V} & & \mathbf{VI} \\ & & \downarrow \\ & & [\mathrm{Rh}(\mathrm{PR_3})_2\mathrm{Cl}(\mathrm{COCH_3})\mathrm{I}] \\ & & \nu_{\mathrm{CO}} = 1705\ \mathrm{cm^{-1}} \\ & & \mathbf{VII} \end{array} \quad (5)$$

which in the presence of excess methyl iodide can be examined in terms of a model:

$$\mathbf{V} \underset{k_{-1}}{\overset{k_1}{\rightleftharpoons}} \mathbf{VI} \overset{k_2}{\rightarrow} \mathbf{VII} \quad (6)$$

The FACSIMILE computer program (*14*) was used to analyze data obtained when solutions of *trans*-[Rh(PR_3)$_2$Cl(CO)] (2×10^{-2} mol L^{-1})

Table III. The Homogeneous Hydrogenation of Methyl Linolenate in the Presence of 0.33 mmol [ML_2Cl_2] and 2.9 mmol

M	*L*	*H_2 Pressure (atm)*	*Methyl Stearate*
Methyl Linolenate			
None[a]		40	—
Platinum	$P(n\text{-}C_8H_{17})_3$	40	—
Palladium	PPh_3	40	—
Methyl Linolenate + Methyl Linoleate (Mixture 1)			
None[a]	—	40	—
Platinum	$P(-C_6H_4-C_4H_9)_3$	40	—
Palladium	$P(-C_6H_4-C_9H_{19})_3$	40	—
Palladium	$P(-C_6H_4-C_9H_{19})_3$	20	—
Methyl Linolenate + Methyl Linoleate (Mixture 2)			
None[a]	—	40	—
Platinum	PPh_3	40	—

[a] Control hydrogenation carried out in the absence of catalyst.

in methyl iodide were allowed to react at various temperatures. An analysis of the data obtained with $P(C_6H_4Bu\text{-}p)_3$ according to Equation 6 gave a rather poor fit to the data (*see* Figure 2). In particular the rate of formation of $[Rh\{P(C_6H_4Bu\text{-}p)_3\}_2(COCH_3)ClI]$ at times very close to the start is more rapid than expected from Equation 6. However, when a number of alternative models were tried, we found that the following model in which **VI** is formed but is not an intermediate in the formation of **VII** gave the best fit. The calculated

$$\mathbf{V} \underset{k_{-1}}{\overset{k_1}{\rightleftharpoons}} \mathbf{VI}, \qquad \mathbf{V} \xrightarrow{k_3} \mathbf{VII} \tag{7}$$

absorbances obtained when the data is analyzed according to Equation 7 and the observed absorbances are shown in Figure 3. For comparison the sum of the squares of errors (observed absorbance − calculated absorbance) in Figure 2 is 210 whereas in Figure 3 it is 65.

A mechanism that is consistent with the scheme in Equation 7 is given in Scheme I. If k_2 were small relative to k_4, k_5, k_6, and k_7 then the reaction mechanistically would follow Equation 7. Halide ions

(10 g) and Mixtures of Methyl Linolenate and Methyl Linoleate (10 g) $SnCl_2$ in the Absence of Solvent at 90°C for 3 h

Products Formed (%)

Mono-olefin	*Non-conjugated Diolefin*	*Conjugated Diolefin*		*Non-conjugated Triolefin*	*Con-jugated Triolefin*
		cis–trans	*trans–trans*		
—	—	—	—	100	—
15.8	45.4	16.5	22.3	—	—
8.8	45.0	16.3	29.9	—	—
2.0	19.9	—	—	78.1	—
36.9	32.4	7.7	23.0	—	—
47.1	33.4	5.2	14.3	—	—
3.5	20.5	11.5	18.5	30.0	16.0
1.2	12.0	—	—	86.8	—
11.6	29.4	16.9	42.1	—	—

catalyze the oxidative addition of alkyl halides to rhodium(I) (*15*) so that k_5 undoubtedly will be quite large. However, is there any evidence for the presence of any halide ions to effect the k_4 step? There is, although the mechanism by which they are formed is uncertain because both we and others (*13*) observe that there is an increase in the conductance during the course of the reaction. In addition both we and others (*13*) observe an initial induction period during which these halide ions may be formed. Both k_6 and k_7 are believed to be even larger than k_5 because previous work with anionic rhodium(I) complexes has shown that the oxidative addition of methyl iodide rather than the combination of methyl and carbonyl ligands is rate determining (*16*). Thus of the rate constants k_4, k_5, k_6, and k_7, k_5 is the smallest and it is larger than k_1. Although the mechanism in Scheme I (and in Equation 7) gives a significantly better fit for the observed data in the case of tris(p-butylphenyl)phosphine than that in Equation 5 (and in Equation 6), the distinction in the case of triphenylphosphine is less clear-cut (*see* Table IV). However we favor Scheme I because it readily accounts for both the observed induction period and the increase in conductance during the reaction.

When *trans*-[$Rh(PR_3)_2Cl(CO)$] complexes in which R was n-C_4H_9 and n-$C_{18}H_{37}$ were dissolved in methyl iodide, rapid oxidative addition

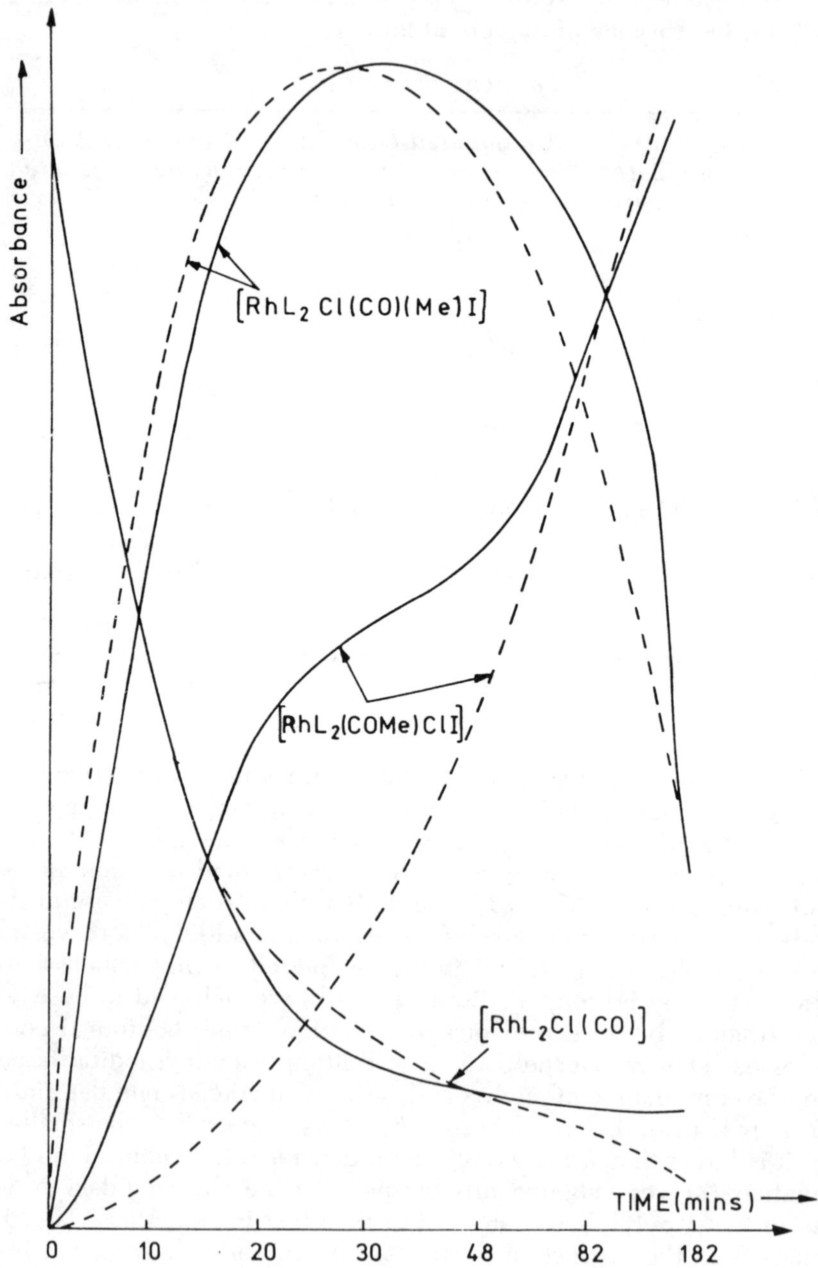

Figure 2. Observed (——) and calculated (---) absorbances for [Rh{P(— ⬡ —Bu)$_3$}$_2$Cl(CO)] + MeI based on the model in Equation 6

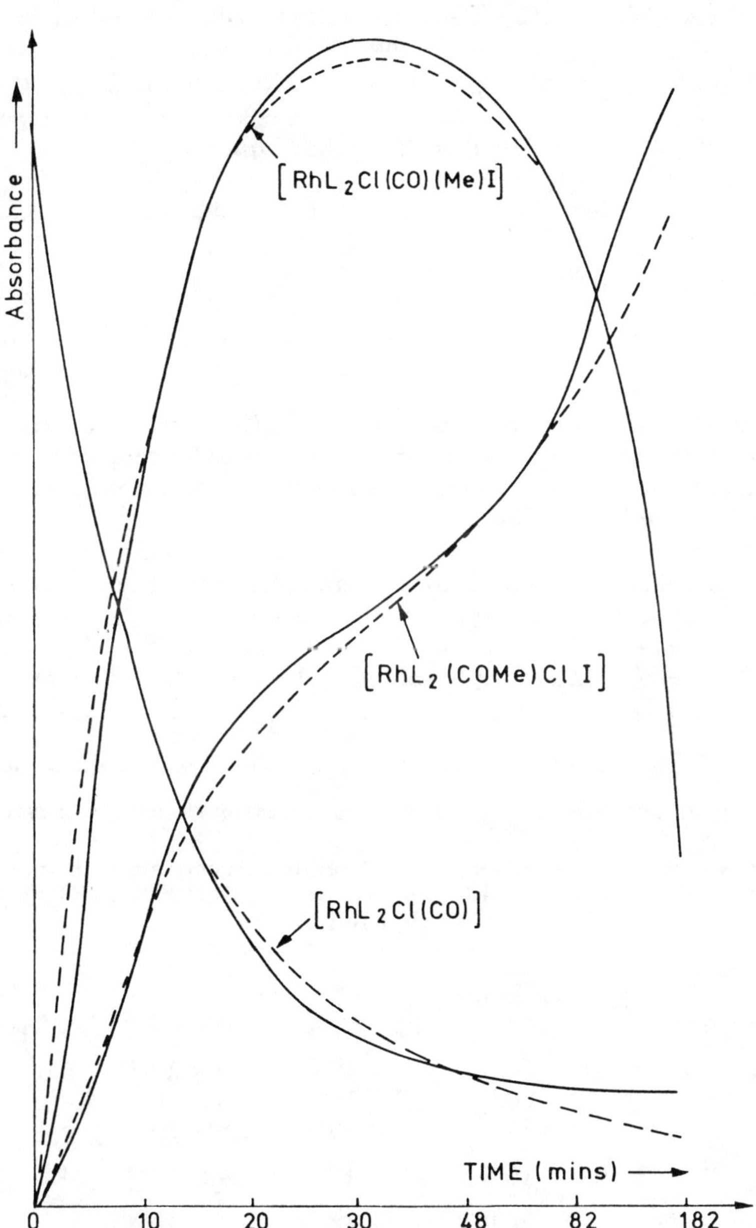

Figure 3. Observed (——) and calculated (---) absorbances for [Rh{P(—⬡—Bu)$_3$}$_2$Cl(CO)] + MeI based on the model in Equation 7

Table IV. Analysis of the Absorbance Data for the Reaction of *trans*-[$Rh(PR_3)_2Cl(CO)$] with Methyl Iodide According to Equations 6 and 7

Phosphine	Temperature (°C)	Sum of Squares of Errors: Equation 6	Equation 7
P(—⬡—Bu)$_3$	16.5	640	72
	28.5	690	107
	36.3	270	65
PPh_3	20.5	1510	780
	26.5	1610	740
	33.3	1480	620

of methyl iodide to yield *trans*-[$Rh(PR_3)_2Cl(CO)(CH_3)I$] occurred, but subsequent combination of methyl and carbonyl ligands to form an acyl complex did not take place. These reactions accordingly followed simple first-order rate laws.

The observed rate constants are given in Table V. In order to compare rate constants, values at 18°C were determined from plots of ln k vs. T^{-1} (*see* Table V). It is apparent that the rate of oxidative addition of methyl iodide to *trans*-[$Rh(PR_3)_2Cl(CO)$] as measured by the values of k_1 at 18°C increases as the phosphine is altered in the following order: PPh_3 < P(—⬡—Bu)$_3$ < P(n-$C_{18}H_{37}$)$_3$ < P(n-C_4H_9)$_3$ < P(n-C_8H_{17})$_3$. The order PPh_3 < P(—⬡—Bu)$_3$ < P(n-alkyl)$_3$ follows the expected order of electron-donating ability. The electron-donor abilities of all

Table V. Rate Constants k_1 and k_3 Defined According to Equation 7 for the Reaction of 2×10^{-2} mol L^{-1} *trans*-[$Rh(PR_3)Cl(CO)$] with Methyl Iodide

PR_3	Temperature (°C)	$k_1(sec^{-1})$	$k_3(sec^{-1})$	$k_1^{18°C}(sec^{-1})$[a]
PPh_3	20.5	7.2×10^{-5}	9.9×10^{-5}	5.0×10^{-5}
	26.5	1.7×10^{-4}	1.3×10^{-4}	
	33.3	5.2×10^{-4}	2.8×10^{-4}	
P(—⬡—Bu)$_3$	16.5	4.0×10^{-4}	1.4×10^{-4}	4.2×10^{-4}
	28.5	9.0×10^{-4}	4.1×10^{-4}	
	36.3	2.4×10^{-3}	6.1×10^{-4}	
P(n-$C_{18}H_{37}$)$_3$	10.5	5.6×10^{-3}	—	7.8×10^{-3}
	19.0	7.9×10^{-3}	—	
P(n-C_8H_{17})$_3$	18.0	1.3×10^{-2}	—	1.3×10^{-2}
P(n-C_4H_9)$_3$	18.3	1.2×10^{-2}	—	1.2×10^{-2}

[a] The k_1 values at 18°C obtained from plots of ln k_1 vs. T^{-1}.

three trialkylphosphines are expected to be very similar. The larger steric bulk of $P(n\text{-}C_{18}H_{37})_3$ may be responsible for its significantly lower ability at promoting oxidative addition. However the relative order of the trialkylphosphines suggests that subtle solvation effects are very important. In the previous paragraph we suggested that the k_3 values in Table V largely reflect the k_5 step of Scheme I which is itself the oxidative addition of methyl iodide to an anionic rhodium(I) complex. Again the rate constants (k_3) in Table V show that P(—⬡—Bu)$_3$ is more effective than triphenylphosphine at promoting this oxidative addition.

Scheme I

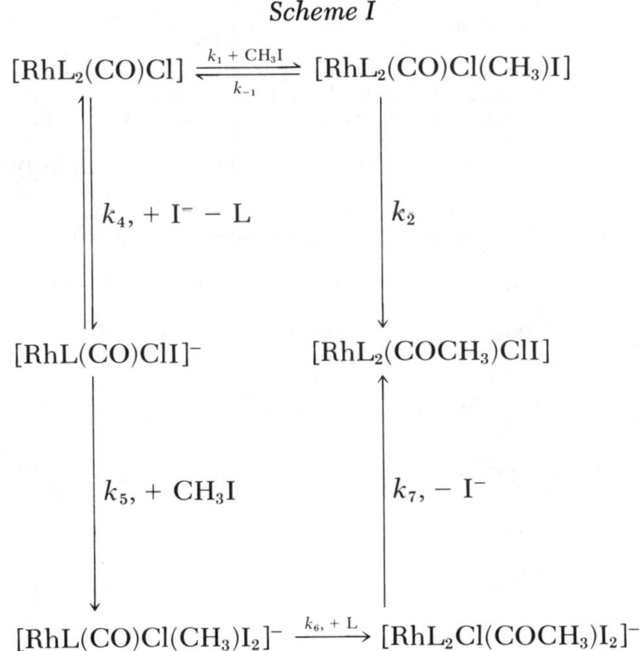

Conclusions

Tertiary phosphine groups with long alkyl chains bound directly to phosphorus or substituted at the para position of triphenylphosphine give rise to a range of interesting and potentially useful complexes. In particular these may be used to prepare polyolefin hydrogenation catalysts based on platinum(II) and palladium(II) complexes that are both more active and more selective towards reduction to monoolefins than previous catalysts based on these systems. The platinum(II) complexes are better than the palladium(II) complexes. Additionally the new phosphines are more effective than triphenylphosphine in promoting the oxidative addition of methyl iodide to *trans*-$[Rh(PR_3)_2Cl(CO)]$.

Acknowledgments

We thank S. G. Murray and W. P. Griffith for useful discussions. We also gratefully acknowledge financial assistance by the European Office of the U.S. Army under the auspices of Grant DAERO-79-G-0033 and the loan of precious metal salts by Johnson Matthey and Co. Ltd.

Literature Cited

1. Franks, S.; Hartley, F. R.; McCaffrey, D. J. A. *J. Am. Chem. Soc., Perkin Trans. 1* **1979**, 3029.
2. Franks, S.; Hartley, F. R. *J. Chem. Soc., Perkin Trans. 1* **1980**, 2233.
3. Manassen, J.; Dror, Y. *J. Mol. Catal.* **1977/78**, *3*, 227.
4. Sulzbacher, M.; Bergmann, E. *J. Org. Chem.* **1948**, *13*, 303.
5. Franks, S.; Hartley, F. R. *Inorg. Chim. Acta* **1981**, *47*, 235.
6. Jensen, K. A. *Z. Anorg. Allg. Chem.* **1936**, *229*, 225.
7. Mann, F. G.; Purdie, D. *J. Chem. Soc.* **1935**, 1549.
8. Cheney, A. J.; Mann, B. E.; Shaw, B. L.; Slade, R. M. *J. Chem. Soc. (A)* **1971**, 3833.
9. Empsall, H. D.; Heys, P. N.; Shaw, B. L. *J. Chem. Soc., Dalton Trans.* **1978**, 257.
10. Spencer, J. L. *Inorg. Synth.* **1979**, *19*, 213.
11. Bailar, J. C. In "Inorganic Compounds with Unusual Properties—II," *Adv. Chem. Ser.* **1979**, *173*, 1.
12. Davies, J. A.; Hartley, F. R.; Murray, S. G. *J. Chem. Soc., Dalton Trans.* **1980**, 2246.
13. Douek, I. C.; Wilkinson, G. *J. Chem. Soc. (A)* **1969**, 2604.
14. Chance, E. M.; Curtis, A. R.; Jones, I. P.; Kirby, C. R. "Report 8775," Atomic Energy Research Establishment, U.K., 1977.
15. Forster, D. *J. Am. Chem. Soc.* **1975**, *97*, 951.
16. Ibid., **1976**, *18*, 846.

RECEIVED August 10, 1980.

Homogeneous Catalytic Oxidation with Phosphine-Substituted Complexes of Rhodium Carbonyl Clusters

$Rh_6(CO)_{16}$- and $Re_2(CO)_{10}$-Catalyzed Autoxidation of Ketones and Alcohols

D. MAX ROUNDHILL, MARK K. DICKSON, NAGARAJ S. DIXIT, and B. P. SUDHA–DIXIT

Department of Chemistry, Washington State University, Pullman, WA 99164

$Rh_6(CO)_{16}$ catalyzes the oxidation of PPh_3, $PMePh_3$, and $AsPh_3$ with molecular oxygen to $OPPh_3$, $OPMePh_2$, and $OAsPh_3$. $Rh_6(CO)_{16}$ is reformed when CO is added. The compounds $Rh_4(CO)_8(PPh_3)_4$ and $Rh_2(CO)_6(PPh_3)_2$ are likely intermediates. $Rh_6(CO)_{16}$ and $Re_2(CO)_{10}$ catalyze the autoxidation of ketones and cyclic alcohols to dicarboxylic acids. The metal carbonyls catalyze the decomposition of hydroperoxides, hydrogen peroxide, and peracids. Catalyzed peracid decomposition is shown by the decreased yield of ε-caprolactone when cyclohexanone is oxidized to adipic acid. Increased acid yield with $Rh_6(CO)_{16}$ under conditions of increased CO and decreased O_2 pressure suggest that lower nuclearity rhodium carbonyls are active intermediates.

Using transition metal carbonyl cluster compounds for a variety of homogeneously catalyzed reactions is a field that is experiencing increased interest and activity (*1*, *2*, *3*). The large clusters are particularly interesting because of their potential to involve multiple metal sites, or their latent function to dissociate to coordinately unsaturated reactive species. A compound that has a demonstrated function as a homogeneous catalyst is hexarhodium hexadecacarbonyl, $Rh_6(CO)_{16}$. The hydrogenation of alkenes (*4*) and aldehydes (*5*), the

0065–2393/82/0196–0291$05.00/0

hydroformylation of alkenes (*6, 7*), the conversion of synthesis gas into glycol (*8, 9*), the water gas shift reaction (*10, 11*), and the co-oxidation of CO and ketones with molecular oxygen (*12*) are among the reported catalytic reactions using rhodium carbonyl clusters. In all of these reactions the nature of the catalytically active species is a major point of interest. In the aldehyde hydrogenation it is considered likely that the active compound is the monomer $RhH(CO)_3$, but in the glycol formation from synthesis gas there is involvement of the higher aggregates $[Rh_{12}(CO)_{30}]^{2-}$ and $[Rh_{12}(CO)_{34}]^{2-}$.

Oxidation of Triphenylphosphine

Our interest in the catalytic chemistry of $Rh_6(CO)_{16}$ has been in the use of the compound as a homogeneous catalyst for oxidizing phosphines and organic compounds with molecular oxygen. The compound $Rh_6(CO)_{16}$ is an effective catalyst for converting triphenylphosphine, methyldiphenylphosphine, and triphenylarsine to their corresponding oxides. The reaction is carried out at room temperature in dry benzene solvent under a 1-atm pressure of CO and O_2. The compound $Rh_6(CO)_{16}$ itself has negligible solubility in benzene, but complexation occurs in the presence of triphenylphosphine and the mixture becomes homogeneous. Thus a stirred mixture of $Rh_6(CO)_{16}$ (23 mg, 0.02 mmol) and triphenylphosphine (310 mg, 1.2 mmol) in benzene (30 mL) results in the complete conversion to triphenylphosphine oxide after 2 h. The reaction mixture is maintained in contact with CO. During the catalytic cycle this CO is oxidized to CO_2. This only occurs in conjunction with the oxidation of the phosphine, and after the triphenylphosphine is oxidized completely CO_2 formation apparently ceases and $Rh_6(CO)_{16}$ is reformed. This regeneration of $Rh_6(CO)_{16}$ completes the catalytic cycle and avoids the ultimate formation of rhodium metal or rhodium oxide after the triphenylphosphine has been converted completely to triphenylphosphine oxide.

It is apparent, however, that $Rh_6(CO)_{16}$ is not the species involved in activating oxygen. Early in the catalytic cycle the solution is red, and the IR spectrum in the carbonyl region corresponds to that found for $Rh_4(CO)_8(PPh_3)_4$ (ν_{CO} 1975sh, 1955s, 1755w cm^{-1} (C_6H_6/-$CHCl_3$)) (*13*). (Other possible red-colored compounds that have been considered are $Rh_4(CO)_{11}(PPh_3)$ (*13*), $Rh_4(CO)_{10}(PPh_3)_2$ (*13, 14*), $Rh_6(CO)_{10}(PPh_3)_6$ (*13*).) The former two can be discounted from IR spectral evidence, but not the latter compound. We feel that the compound $Rh_4(CO)_8(PPh_3)_4$ has been characterized more adequately than $Rh_6(CO)_{10}(PPh_3)_4$, and hence we assign this former structure to our data.) During the latter stages of the reaction the color changes to yellow. From this solution a compound has been isolated that corresponds to $Rh_2(CO)_6(PPh_3)_2$ (ν_{CO} 1960s, 1955s, 1910w cm^{-1} (Nujol mull))

(*14*) by IR spectroscopy. We conclude that $Rh_6(CO)_{16}$ undergoes substitution with triphenylphosphine leading to lower nuclearity rhodium(O) carbonyl triphenylphosphine complexes. It is likely that one (or more) of these compounds reacts with oxygen to form an oxygen complex that initiates the oxidation of triphenylphosphine. The reaction is inhibited by α-naphthol and 2,6-bis(*t*-butyl)-*p*-cresol, and the mechanism is probably similar to that found for the catalyst $Pt(PPh_3)_3$ (*15*).

The transformations that are proposed to occur during the oxidation of triphenylphosphine are outlined in Scheme I. The catalyzed oxidation of triphenylphosphine is not a general one for binary metal carbonyls. Attempted oxidation under identical conditions with $Cr(CO)_6$, $Mo(CO)_6$, $W(CO)_6$, $Fe(CO)_5$, $Mn(CO)_{10}$, $Re_2(CO)_{10}$, and $Ru_3(CO)_{12}$ gives negligible amounts of triphenylphosphine oxide.

Scheme I. Intermediates in $Rh_6(CO)_{16}$-catalyzed oxidation of PPh_3

$OPPh_3$ $Rh_6(CO)_{16}$
PPh_3
CO
$O_2 + PPh_3$
$Rh_2(CO)_6(PPh_3)_2$ $Rh_4(CO)_8(PPh_3)_4$
CO
$OPPh_3$ $O_2 + PPh_3$

Oxidation of Organic Compounds

In earlier publications we described using the complex $Rh_6(CO)_{16}$ as a homogeneous catalyst for the oxidation, with molecular oxygen, of carbon monoxide to carbon dioxide, and of ketones to carboxylic acids (*12*). In that work we concluded that the oxidation of ketones was a free-radical process, but the role of the transition metal carbonyl complex was not identified. Two roles appear to have been identified for metal complexes in oxidations with molecular oxygen. The first is the coordination and activation of molecular oxygen (*16*). This role has been claimed in converting phosphines to phosphine oxides (*15*), alkenes to ketones (*17, 18, 19, 20*), alkenes to epoxides (*21*), and isocyanides to isocyanates (*22*). A further role identified for transition metal compounds in catalyzed oxidations is the decomposition of preformed peroxides. Examples of using simple metal salts are in the oxidation of alkenes, aromatics, alkanes (*23*), and cyclohexanone (*24*). Similarly, complexes of heavy-metal ions in low oxidation states have been used in oxidizing aldehydes to carboxylic acids (*25, 26*), and of cyclohexene to 2-cyclohexene-1-one (*27, 28*).

Our earlier work with $Rh_6(CO)_{16}$ was the first example of using a transition metal carbonyl compound for catalyzed oxidations. We now find that $Rh_6(CO)_{16}$ is not unique in functioning as a homogeneous catalyst for oxidizing ketones. The dimer $Re_2(CO)_{10}$ is equally effective. The reaction mixture is homogeneous throughout the catalytic reaction. The carbonyl is not decomposed and can be recovered quantitatively at the end of the oxidation. This latter situation even prevails when the oxidations are carried out in the absence of carbon monoxide. Species characterization is done by isolation methods or by IR spectroscopy in the carbonyl region.

In a brief survey of other simple binary carbonyls we find that the compounds $M(CO)_6(M = Cr, Mo, W)$ and $Ru_3(CO)_{12}$ have only minimal catalytic activity for autooxidizing alcohols or ketones. The compounds $Fe(CO)_5$ and $Fe_3(CO)_{12}$ are decomposed completely when we try to use them as catalysts. When the compound $Mn_2(CO)_{10}$ is used, there is a considerable enhancement in acid formation. During this reaction there is extensive decomposition to manganese dioxide, and we believe that this compound is the one primarily involved in the catalytic oxidation.

Oxidation of Cyclohexanone

Our data on the catalyzed oxidation of cyclohexanone are shown in Tables I, II, and III. The solvent effects on adipic acid yield require comment. Although solvent effects often are considered to be small in free-radical chemistry, we find that our data is consistent with the published solvent effects on the decomposition rate of benzoyl

Table I. Solvent Effects in the Oxidation of Cyclohexanone to Adipic Acid (Reaction Conditions: Cyclohexanone, 5 mL; Oxygen, 500 lb in.$^{-2}$; Temperature, 96°C Yield Based on Cyclohexanone)

Solvent	*Time (h)*	*Adipic Acid (mmol)*	*Yield (%)*	*Catalyst (mg)*
CH_2Cl_2 (10 mL)	20	15.3	30	$Re_2(CO)_{10}$ (25)
$(CH_3)_2CO$ (10 mL)	25	11.1	22	$Re_2(CO)_{10}$ (25)
C_6H_{12} (5 mL)	24	18.4	36	$Re_2(CO)_{10}$ (25)
t-BuOH	24	2.1	4	$Re_2(CO)_{10}$ (25)
CH_3CN	24	2.2	4	$Re_2(CO)_{10}$ (25)
C_6H_6 (10 mL)	24	2.9	5	$Re_2(CO)_{10}$ (25)
CH_2Cl_2 (10 mL)	24	0.34	0.7	None
$(CH_3)_2CO$	24	0.48	1.0	None
None	18	10.6	21	$Re_2(CO)_{10}$ (17)[a]

[a] This reaction is with CO (600 lb in.$^{-2}$) and O_2 (300 lb in.$^{-2}$).

Table II. Temperature and Oxygen Pressure Effects in Oxidation of Cyclohexanone to Adipic Acid (Reaction Conditions: Cyclohexanone, 5 mL; $Re_2(CO)_{10}$, 50 mg; Solvent, 10 mL)

Solvent	*Time (h)*	*Temperature (°C)*	*Oxygen Pressure (lb in.$^{-2}$)*	*Adipic Acid (mmol)*
$(CH_3)_2CO$	20	115°	300	10.7
$(CH_3)_2CO$	18	115°	400	15.4
$(CH_3)_2CO$	15.5	115°	500	20.8
$(CH_3)_2CO$	5	130°	500	17.9
C_6H_{12}	22	100°	500	12.8
C_6H_{12}	5	122°	500	18.7

peroxide (*29*). Our high yield of adipic acid correlates with the faster rates of peroxide decomposition. Thus our data in Table I showing smaller yields of adipic acid in the solvent sequence cyclohexane > dichloromethane > benzene correlate with the respective decreasing rates for benzoyl peroxide in these solvents. From Table I the highest yield of adipic acid is formed in the absence of solvent. This observation may be simply due to chain termination reactions occurring in the solvent media. Of the volatile side-products of cyclohexanone autoxidation, one of the few components to show significant quantitative change in the presence of $Rh_6(CO)_{16}$ is 2-(1-cyclohexenyl)-cyclohexanone. The yield of this condensation product increases by a factor of four in the catalyzed reaction. This product has been reported previously in the autoxidation of cyclohexanone, and arises from a self-condensation reaction of the ketone (*1*).

The reaction pathway and product distribution observed in the $Re_2(CO)_{10}$- and $Rh_6(CO)_{16}$-catalyzed autoxidation of cyclohexanol and cyclohexanone are shown in Scheme II. An important intermediate is the peracid. In this sequence the peracid is the final intermediate;

$$2\ \text{(cyclohexanone)} \rightarrow \text{(2-cyclohexylidenecyclohexanone)} + H_2O \qquad (1)$$

adipic acid results from the final reaction between this compound and the aldehyde. We have assayed lactone formation both as a means of identifying peracid as an intermediate, and of probing any changes in peracid concentration caused by the presence of $Rh_6(CO)_{16}$. In the case of cyclohexanone we have used the yield of ϵ-caprolactone as a measure of this intermediate peracid. In both the absence and presence of

Table III. Acid Yields from Cyclic Ketones

Ketone (mL)	*Solvent (mL)*	*Acid (mmol)*
Cyclopentanone (5)	Acetone (20)	10.1
Cyclohexanone (5)	None	21.2
Cycloheptanone (10)	None	9.3
Acetone (10)	None	0.8

$Rh_6(CO)_{16}$, the product mixture contains ϵ-caprolactone. Analysis of the oxidation products for the uncatalyzed and $Rh_6(CO)_{16}$-catalyzed oxidation of cyclohexanone shows the respective rations of ϵ-caprolactone to be 0.43 and 1.00 when measured against an added aliquot of dodecane. Thus there is a decrease in ϵ-caprolactone formation in the presence of $Rh_6(CO)_{16}$. This observation is consistent with the premise that the metal carbonyl accelerates the decomposition of peroxides. In the presence of $Rh_6(CO)_{16}$, the increased decomposition rate of peracid leading to a lower steady-state concentration of this species, and hence to its reduced transformation to ϵ-caprolactone is expected. These data support a reaction pathway proceeding via hydroperoxide and its sub-

Scheme II. Reaction pathway in the $Re_2(CO)_{10}$- and $Rh_6(CO)_{16}$-catalyzed autoxidation of cyclohexanol and cyclohexanone

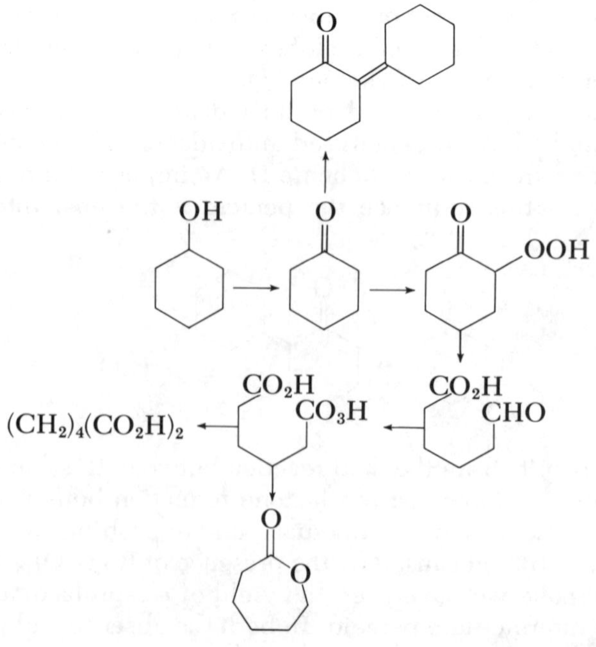

O
OH O O
OOH
CO_2H CO_2H
CO_3H CHO
$(CH_2)_4(CO_2H)_2$
O
O

(Reaction Conditions: Temperature, 98°C)

Time (h)	*Catalyst (mg)*	*O_2 (lb in.$^{-2}$)*	*CO (lb in.$^{-2}$)*
17	$Re_2(CO)_{10}$ (50)	450	0
18	$Re_2(CO)_{10}$ (17)	500	0
22	$Re_2(CO)_{10}$ (22)	500	0
29	$Rh_6(CO)_{16}$ (44)	170	330

sequent conversion to the peracid. Adipic acid results from final reaction between peracid and aldehyde. The latter steps in the chemistry thus resemble those found in Baeyer–Villiger oxidations.

Oxidation of Alcohols

The compounds $Rh_6(CO)_{16}$ and $Re_2(CO)_{10}$ are also effective homogeneous catalysts for autoxidating cyclic alcohols to dicarboxylic acids. Solvent effect data for cyclohexanol are shown in Table IV. Again low yields are found in benzene solvent, and considerably higher conversions in cyclohexane. The yields of carboxylic acids obtained from both cyclic and acyclic alcohols are shown in Table V. It is apparent that the acid yields are small for acyclic alcohols. There is no difference in catalytic activity whether the compound $Rh_6(CO)_{16}$ or $Re_2(CO)_{10}$ is used and low yields are obtained from both primary and secondary alcohols.

Decomposition of Hydrogen Peroxide

From our high-pressure data it is apparent that the carbonyls are involved mechanistically in the decomposition of peroxide intermedi-

Table IV. Solvent Effects in the Oxidation of Cyclohexanol to Adipic Acid (Reaction Conditions: Cyclohexanol, 5 mL; Oxygen, 500 lb. in.$^{-2}$; Temperature, 98°C)

Solvent (5mL)	*Time (h)*	*Adipic Acid (mmol)*	*Catalyst (40 mg)*
None	22	7.8	$Re_2(CO)_{10}$
None	22	1.1	None
$(CH_3)_2CO$	22	10.0	$Re_2(CO)_{10}$
$(CH_3)_2CO$	22	1.8	None
C_6H_{12}	22	14.2	$Re_2(CO)_{10}$
C_6H_{12}	22	4.4	None
C_6H_6	22	4.1	$Re_2(CO)_{10}$
C_6H_6	22	1.0	None

Table V. Acid Yields from Cyclic and Acyclic Alcohols (Reaction Conditions: Alcohol, 10 mL; Time, 22 h; Temperature, 98°C)

Alcohol	*Acid (mequiv)*	*Catalyst (mg)*	*Oxygen (lb in.$^{-2}$)*	*CO (lb in.$^{-2}$)*
Cyclohexanol	15.6	$Re_2(CO)_{10}$ (40)	500	0
Ethanol	0.2	$Re_2(CO)_{10}$ (14)	500	0
Ethanol	0.15	$Rh_6(CO)_{16}$ (15)	300	500
Diethylcarbinol	0.7	$Re_2(CO)_{10}$ (17)	500	0
Diethylcarbinol	0.4	$Rh_6(CO)_{16}$ (11)	300	600
Isopropanol	0.08	$Rh_6(CO)_{16}$ (10)	330	320[a]
1-Phenylethanol	0.05	$Rh_6(CO)_{16}$ (10)	325	315[b]

[a] Time = 12 h, Temperature = 86°.
[b] Time = 24 h, Temperature = 86°.

ates. Therefore we have investigated catalytic effects of these metal carbonyls on hydrogen peroxide decomposition. With $Rh_6(CO)_{16}$ reaction occurs to form a product not yet characterized, but with $Re_2(CO)_{10}$ the compound is recovered unchanged. From data in Table VI it is

Table VI. Catalyzed Decomposition of H_2O_2. (The Data are Expressed as Percent Decomposition for a Solution of 30% H_2O_2 (2 mL) in Solvent (15 mL) and a Reaction Time of 2 h at 90°C)

	$Re_2(CO)_{10}$ (25 mg)	*None*
Benzene	64	19
Acetone	9	2
Cyclohexane	57	26

Table VII. Acid Produced Using $Rh_6(CO)_{16}$ with Changing Pressures of CO and O_2

Pressure of CO (atm)	*Pressure of O_2 (atm)*	*Pressure of CO / Total Pressure*	*Acid (mmol)*	*Time (h)*
340	330	0.51	1.10	10
440	220	0.67	1.37	10
520	140	0.79	1.71	10
540	100	0.84	1.94	10
555	70	0.89	2.04	10
280	380	0.42	1.76	12
500	160	0.76	2.65	12
520	110	0.83	2.84	12
550	80	0.87	2.96	12

apparent that these compounds are catalysts for the decomposition of hydrogen peroxide, and as such gives further credence that the mode of action in the catalytic cycle is in hydroperoxide decomposition.

Pressure Effects on $Rh_6(CO)_{16}$-Catalyzed Cyclohexanone Oxidation

In an attempt to obtain some information concerning the catalytically active species in the $Rh_6(CO)_{16}$-catalyzed cyclohexanone autoxidation, we have carried out a series of experiments with changing CO and O_2 pressures under isobaric conditions. The data in Table VII have been obtained at 85 ± 1° under the isobaric conditions of 650 ± 20 lb in.$^{-2}$. This condition is created by changing the partial pressure of CO in a mixture of CO and O_2 between 0.42 and 0.89. The data in Table VII are arranged to show the effect on acid yield of changing the CO and O_2 pressure in a solution containing $Rh_6(CO)_{16}$ and cyclohexanone. Data for both 10- and 12-h reaction times are given, and in Figure 1 we plot the 12-h data for acid yield in millimoles against both O_2 pressure and the CO partial pressure. These data show that as the oxygen pressure decreases, or concurrently as the CO pressure increases, there is an increase in the quantity of carboxylic acid formed under isobaric conditions. This result it unexpected since lowering the partial pressure of O_2 creates a situation where its concentration becomes increasingly deficit for its function as a reactant for converting both CO and CO_2 and cyclohexanone to adipic acid. Therefore it is apparent that there is an increase in catalytic activity as the CO pressure is raised.

The most reasonable explanation for these data is that under the increasing CO pressure there is a corresponding increase in the concentration of a lower nuclearity rhodium carbonyl compound which is

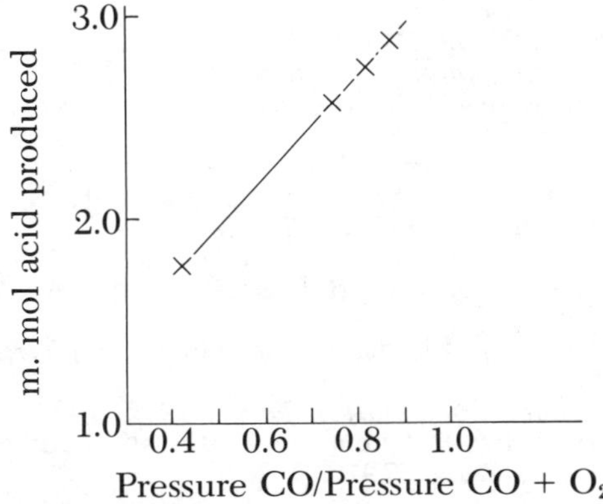

Figure 1. Plot of acid yield (in millimoles) against the pressure of CO/total pressure: $Rh_6(CO)_{16}$; 86° ± 1°C; t = 12 h.

the catalytically active species in the reaction mixture. This concept agrees with previous ideas that smaller aggregate carbonyl clusters are formed under high CO pressure (*30, 31*). In particular for rhodium(0), Whyman has shown that at low temperatures and under an extremely high pressure of CO there is a conversion to the dimeric compound $Rh_2(CO)_8$ (*32*). Under our experimental conditions it is not possible to directly observe such intermediates, and no attempt has been made yet to do so. However we previously have used the lower cluster compound $Rh_4(CO)_{12}$ as a catalyst for oxidizing cyclohexanone. The results show that the compound is effective as a catalyst, but is converted to $Rh_6(CO)_{16}$ and is not recoverable as $Rh_4(CO)_{12}$.

The chemistry and function of $Rh_6(CO)_{16}$ and $Re_2(CO)_{10}$ as oxidation catalysts for organic compounds is under continuing investigation. In particular, we are studying the role of $Rh_6(CO)_{16}$ as a labile multisubstrate oxidation catalyst for oxidizing CO and triphenylphosphine (*33, 34*).

Acknowledgments

We thank the Division of Basic Chemical Sciences of the Department of Energy for support of this work. Thanks are due to K. Alexander and C. Craven for their capable experimental assistance.

Literature Cited

1. Muetterties, E. L.; Rhodin, T. N.; Band, E.; Brucker, C. F.; Pretzer, W. R. *Chem. Rev.* **1979,** *79,* 91.
2. Muetterties, E. L. *Angew. Chem. Int. Ed. Engl.* **1978,** *17,* 545.
3. Muetterties, E. L. *Science* **1977**, 196, 839.
4. Imyanitov, N. S.; Rudkovskii, D. M. *Kinet. Katal.* **1967**, 1240; *Chem. Abstr.* **1968,** *68,* 86867b.
5. Heil, B.; Marko, L. *Acta Chim. Acad. Sci. Hung.* **1968,** *55,* 107; *Chem. Abstr.* **1968,** *68,* 77430b.
6. Heil, B.; Marko, L. *Chem. Ber.* **1968,** *101,* 2209.
7. Pruett, R. L. *Adv. Organometal. Chem.* **1979,** *17,* 1.
8. Walker, W. E.; Brown, E. S.; Pruett, R. L. U.S. Patent 3 878 292; *Chem. Abstr.* **1975,** *83,* 45426y.
9. Masters, C. *Adv. Organometal. Chem.* **1979,** *17,* 61.
10. Ford, P. C.; Rinker, R. G.; Ungermann, C.; Laine, R. M.; Landis, V.; Moya, S. A. *J. Am. Chem. Soc.* **1978,** *100,* 4595.
11. Laine, R. M. *J. Am. Chem. Soc.* **1978,** *100,* 6451.
12. Mercer, G. D.; Beaulieu, W. B.; Roundhill, D. M. *J. Am. Chem. Soc.* **1977,** *99,* 6551.
13. Booth, B. L.; Else, M. J.; Fields, R.; Haszeldine, R. *J. Organometal. Chem.* **1971,** *27,* 119.
14. Whyman, R. *J. Chem. Soc., Dalton Trans.* **1972,** 1375.
15. Sen, A.; Halpern, J. *J. Am. Chem. Soc.* **1977,** *99,* 8337.
16. Vaska, L. *Accts. Chem. Res.* **1976,** *9,* 175.
17. Read, G.; Walker, P. J. C. *J. Chem. Soc., Dalton Trans.* **1977,** 883.
18. Read, G. *J. Mol. Catal.* **1978,** *4,* 83.
19. Tang, R.; Mares, R.; Smith, D. E. *J. Chem. Soc., Chem. Commun.* **1979,** 274.

20. Mimoun, H.; Machiran, M. M. P.; de Roch, I. S. *J. Am. Chem. Soc.* **1978,** *100,* 5437.
21. Chen, M. J. Y.; Kochi, J. K. *J. Chem. Soc., Chem. Commun.* **1977**, 204.
22. Otsuka, S.; Nakamura, A.; Tatsuno, Y. *J. Am. Chem. Soc.* **1969,** *91,* 6994.
23. Sheldon, R. A.; Kochi, J. K. *Adv. Catal.* **1976,** *25,* 272.
24. Kamiya, Y.; Kotake, M. *Bull. Chem. Soc. Jpn.* **1973,** *46,* 2780.
25. Hojo, J.; Yuasa, S.; Yamazoe, N.; Mochida, I.; Seiyama, T. *J. Catal.* **1975,** *36,* 93.
26. Sakamoto, H.; Funabiki, T.; Tarama, K. *J. Catal.* **1977,** *48,* 427.
27. Fusi, A.; Ugo, R.; Fox, F.; Pasini, A.; Cenini, S. *J. Organometal. Chem.* **1971,** *26,* 417.
28. Cenini, S.; Fusi, A.; Porta, F. *Gazz. Chim. Ital.* **1978,** *108,* 109.
29. Nozaki, K.; Bartlett, P. D. *J. Am. Chem. Soc.* **1946,** *68,* 1686.
30. Chini, P. *Inorg. Chim. Acta Rev.* **1968,** *2,* 31.
31. Johnston, R. D. *Adv. Inorg. Chem. Radiochem.* **1970,** *43,* 470.
32. Whyman, R. *J. Chem. Soc., Chem. Commun.* **1970,** 1194.
33. Roundhill, D. M.; Dickson, M. K.; Dixit, N. S.; Sudha-Dixit, B. P. *J. Am. Chem. Soc.* **1980,** *102,* 5538.
34. Dickson, M. K.; Sudha, B. P.; Roundhill, D. M. *J. Organometal. Chem.* **1980,** *190,* C43.

RECEIVED July 10, 1980.

Metal Complexes of Iminophosphine and Iminoarsine Chelating Agents

Structure, Reactivity, and Stereochemistry

JOHN E. HOOTS, THOMAS B. RAUCHFUSS[1], and STEVEN P. SCHMIDT
School of Chemical Sciences, University of Illinois, Urbana IL 61801

JOHN C. JEFFERY and PAUL A. TUCKER
Research School of Chemistry, The Australian National University, Canberra, The A.C.T., Australia

The coordination chemistry of iminophosphine and iminoarsine ligands is surveyed with emphasis on our recent results with tetradentate diiminodiarsines and diiminodiphosphines derived from 1,2-diaminoalkanes. The ligand en = P_2 prepared from ethylenediamine and o-*diphenylphosphinobenzaldehyde forms tetrahedral complexes with copper(I) (X-ray structure) and silver(I), both of which are configurationally labile. The same ligand forms diamagnetic, four- and five-coordinate nickel(II) complexes of the formula [Ni(en = P_2)](BF_4)$_2$ and [NiBr(en = P_2)]Br. Two tridentate en = P_2 complexes have been prepared, Mo(en = P_2)(CO)$_3$ which contains an uncoordinated phosphine moiety, and [Cu-(en = P_2)(*t-*BuNC)]ClO_4 which contains an uncoordinated imine. Chiral diiminodiarsines prepared from (*R*)-1,2-diaminopropane and (*R,R*)-1,2-diaminocyclohexane are uniquely stereospecific in their binding of tetrahedral copper(I) affording complexes in exclusively the $\Delta(\lambda)$ configuration.*

Chelating phosphine ligands continue to serve coordination chemists in their efforts to manipulate the electronic and reactivity patterns of metal complexes. These studies have received an addi-

[1] To whom correspondence should be addressed.

0065-2393/82/0196-0303$05.00/0

tional impetus from the promised and sometimes proven desirable catalytic properties of these compounds. Synthetic routes to most organophosphines rely almost exclusively on a P-C bond coupling as the essential and usually final synthetic step in their synthesis. Our interest in the phosphinobenzaldehydes stems from the prospect of developing alternative, synthetically more flexible strategies to chelating agents bearing tertiary phosphine donors.

Imine formation, probably the most generally useful route to multidentate ligands, has been applied to the preparation of only a limited number of arsenic- (*1*) and phosphorus- (*2, 3*) containing chelating agents via Equation 1. We have since described the alternative and

$$R_2E\rightsquigarrow NH_2 + R'\text{–}CHO \rightarrow R_2E\rightsquigarrow N{=}CHR' \quad (1)$$

more generally useful route to this class of chelating agents via Equation 2, E = P, As (*4*). Inasmuch as linear tetradentate ligands derived from ethylenediamine, e.g. salen and acacen, have proved to be excep-

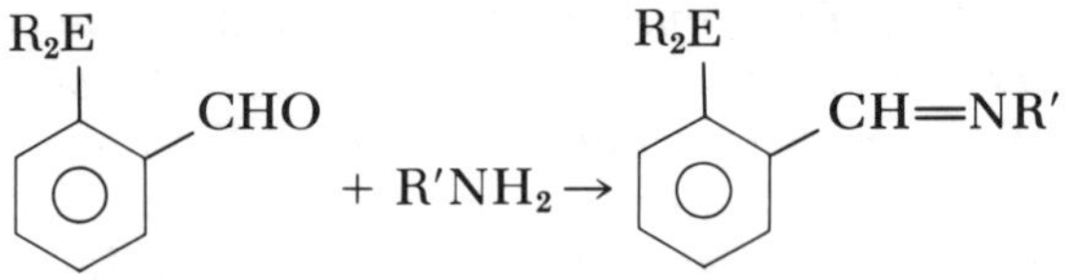

$$\text{o-}(R_2E)C_6H_4CHO + R'NH_2 \rightarrow \text{o-}(R_2E)C_6H_4CH{=}NR' \quad (2)$$

tionally versatile in coordination chemistry and catalysis, we have focused much of our effort on the corresponding diiminodiphosphines and diiminodiarsines derived from phosphino- and arsinobenzaldehydes and a variety of diamines.

Discussion

The Carbonylphosphines. The arsino- and phosphinobenzaldehydes are the essential starting materials for the present work and deserve comment. These ligands, many of which were prepared first

Scheme I.

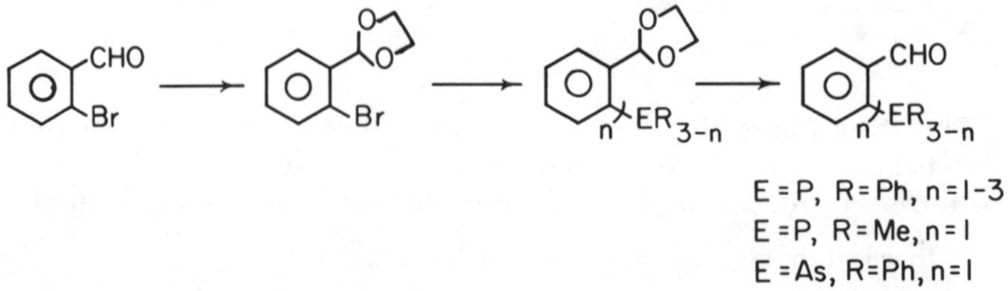

o-BrC₆H₄CHO → o-BrC₆H₄(1,3-dioxolan-2-yl) → [(dioxolanyl)C₆H₄]$_n$ER$_{3-n}$ → [o-OHCC₆H₄]$_n$ER$_{3-n}$

E = P, R = Ph, n = 1–3
E = P, R = Me, n = 1
E = As, R = Ph, n = 1

Scheme II.

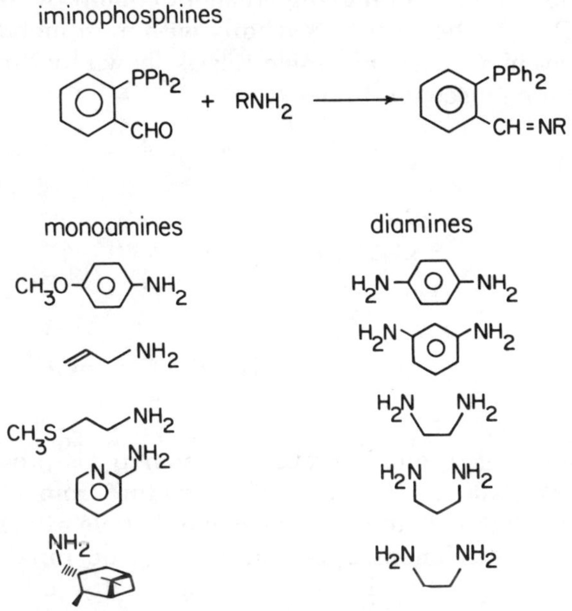

iminophosphines

PPh2 CHO + RNH2 ⟶ PPh2 CH=NR

monoamines

diamines

by Schiemenz and Kaack, can be synthesized in ca. 50% overall yield from the commercially available *o*-bromobenzaldehyde in three steps (*see* Scheme I) (*5*). The products are isolatable as air-stable, bright yellow crystalline materials ($Ph_2AsC_6H_4CHO$ is pale yellow) that are characterized readily by IR ($\nu_{co} \simeq 1690\ cm^{-1}$) and H-1 NMR ($J$(P,C*H*O) $\simeq$ 7 Hz).

Monoiminophosphines. As illustrated in Scheme II, PCHO condenses with a wide variety of simple and functionalized monoamines. Studies of the Mo(O) derivatives of several of these ligands demonstrate that they efficiently displace CO from $Mo(CO)_6$ affording (chel)$Mo(CO)_4$ or *fac*-(chel)$Mo(CO)_3$ derivatives (*4*). Similar, but bimetallic, complexes can be obtained from the ligands prepared by condensations with nonchelating diamines such as *m*- or *p*-diaminobenzene.

In more recent studies, we have demonstrated that this condensation reaction allows one to readily introduce chiral functionalities into the phosphine system. For instance, (−)-3-pinane methylamine readily condenses with PCHO and AsCHO to afford the chiral imines; these imines also may be reduced to the corresponding secondary amines using $NaBH_4$. The efficacy of these ligands in asymmetric catalysis is being explored currently, and in view of the wide variety of naturally occurring chiral amines, our methodology is extremely promising.

Diiminodiphosphines. The aldehyde PCHO cleanly condenses with ethylenediamine in refluxing ethanol to afford the off-white diimine, en = P_2. This ligand reacts readily with most metal ions affording derivatives in a variety of geometries as shown by X-ray crystallographic and spectroscopic studies (6).

$$2\ o\text{-}(PPh_2)C_6H_4CHO + H_2NCH_2CH_2NH_2 \rightarrow o\text{-}(PPh_2)C_6H_4CH{=}NCH_2CH_2N{=}CHC_6H_4\text{-}o\text{-}(PPh_2) \quad (3)$$

An example of (pseudo)tetrahedral geometry is provided by the Cu(I) complex, [Cu(en = P_2)]ClO_4. The structure from an X-ray study is depicted in Figure 1 from a perspective that clearly shows the expected skewed ethylenediamine backbone (vide infra). The corresponding Ag(I) derivative has the same geometry judging from the similarity of the H-1 NMR and IR spectra of the Cu(I) and Ag(I) complexes.

Square planar and five-coordinate derivatives of en = P_2 are found for the diamagnetic yellow [Ni(en = P_2)]$^{+2}$ and brown [Ni(en = P_2)Br]$^+$. A most interesting feature of these compounds is their elec-

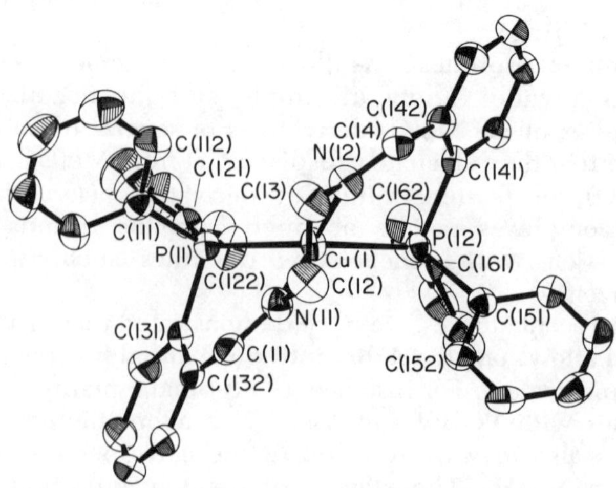

Inorganic Chemistry

Figure 1. Structure of [Cu(en = P_2)]$^+$ with thermal ellipsoids drawn at the 50% probability level

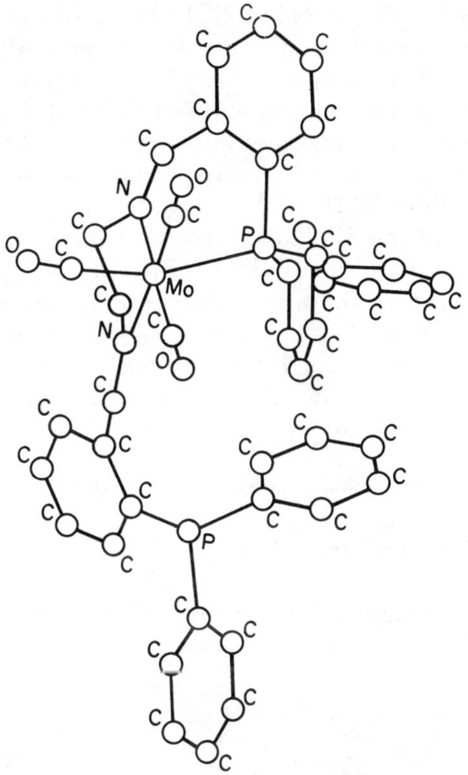

Inorganic Chemistry

Figure 2. Structure of Mo(en = P_2)($CO)_3$

trochemistry, where it is seen by cyclic voltametry that both the nickel(I) and nickel(0) derivatives of en = P_2 are accessible via reversible couples in acetonitrile.

In an attempt to prepare complexes where the en = P_2 ligand is coordinated to an octahedral metal ion, we examined its chemistry with molybdenum(0). The red molybdenum carbonyl derivative Mo(en = P_2)$(CO)_3$ is formed in high yield from $Mo(CO)_6$ (*see* Equation 4). We have shown now that this complex contains one uncoordinated phosphine moiety in addition to the facially bound P–N–N segment of the chelating agent. The P-31{H-1} NMR clearly shows two single resonances at 34.86 and −16.15 ppm vs. H_3PO_4 attributed to the bound and uncoordinated phosphines, respectively. The structure (*see* Figure 2) is in interesting contrast to another tridentate en = P_2 complex, [Cu(en = P_2)(*t*-BuNC)]ClO_4 prepared from the reaction of [Cu(en = P_2)]ClO_4 with a slight excess of *t*-BuNC. IR spectroscopic data indicated that the imines were inequivalent in this compound. The X-ray structure (*see* Figure 3) revealed that the isonitrile has displaced one imine with the formation of the tridentate en = P_2 chelate which contains a *nine-membered chelate ring.* This complex is labile and the *t*-BuNC exchange can be brought within the NMR time frame only at −60°C. Cessation of the exchange between the *N* sites, which would result in a nonequivalence of the phosphorus donors, is not observed even at −90°C by P-31 NMR.

The geometry of the chelates in the two tridentate en = P_2 complexes can be rationalized neatly by considering the spacial restraints of the octahedron and tetrahedron. In the former, the coordination sites are accomodated comfortably by adjacent six- and five-membered chelate rings whereas the greater ligand–metal–ligand bite angle of the tetrahedron stabilizes the more flexible six- and nine-membered chelate ring sequence (*see* Figure 4).

$$Mo(CO)_6 + en = P_2 \rightarrow Mo(en = P_2)(CO)_3 + 3CO \qquad (4)$$

Chiral Diiminodiarsines. The facility of the imine formation allows for straightforward introduction of chiral substituents into multidentate ligands containing soft donors. This synthetic route is prom-

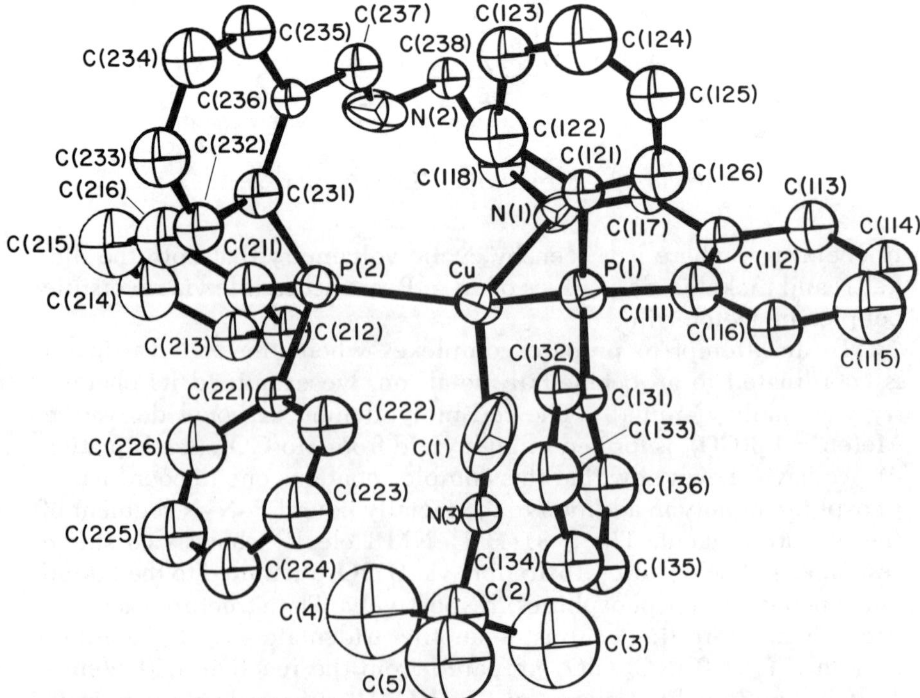

Figure 3. Structure of [Cu(en = P_2)(t-BuNC)]$^+$ with the thermal ellipsoids drawn at the 50% probability level

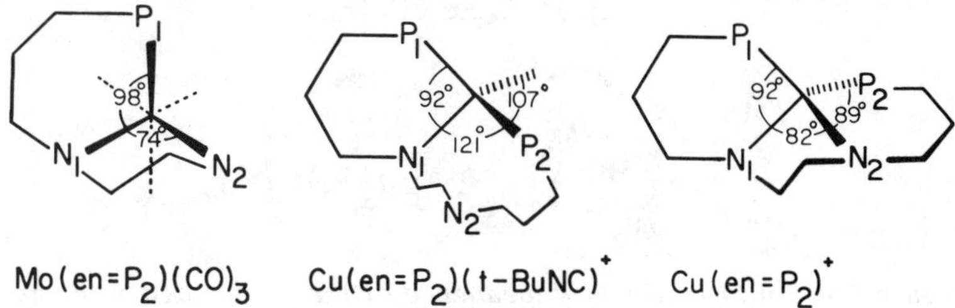

Figure 4. Selected angular parameters for en = P_2 structures

ising for preparing stereospecific chelating agents for low-valency metal ions. The H-1 NMR spectra of [Cu(en = P_2)]ClO_4 (35° to −53°C) and [Cu(en = As_2)]ClO_4 (35°C) indicate that these complexes are undergoing rapid inversion about the metal ion as indicated in Equation 5. Inasmuch as the disposition of the E · · · N (E = As, P) chelate

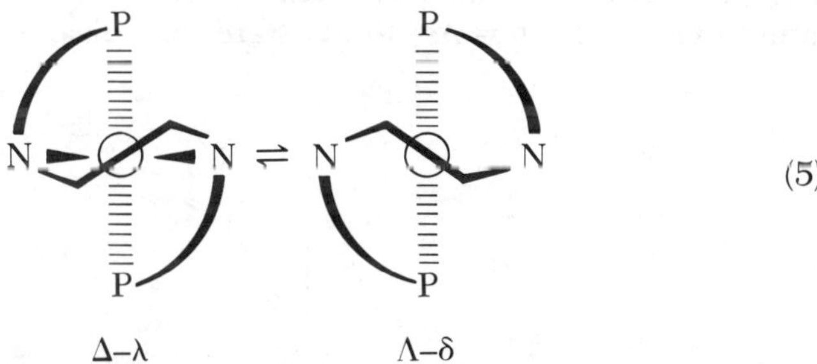

(5)

rings is dictated by the conformation of the ethylene backbone, it should be possible to fix the overall chirality of the complex by locking the conformation of that backbone. We have approached this problem by examining the conformation behavior of two analogues of en = As_2.

Both the X-ray structure of [Cu(en = P_2)]ClO_4 and the Fieser molecular models indicate that placement of a methyl group in an axial site on the ethylene backbone of Cu(en = E_2)$^+$ would result in significantly unfavorable nonbonded interactions with the aromatic substituents, and preliminary experiments confirm this prediction (*see* Figure 5). (*R*)-1,2-Diaminopropane efficiently condenses with AsCHO to afford the diimine R-pn = As_2 which was converted directly to its crystalline copper(I) derivative, [Cu(R-pn = As_2)]ClO_4. For references, the ligand en = As_2 and its copper(I) complex have both been isolated and characterized. The H-1 NMR spectrum of [Cu(R-pn = As_2)]ClO_4 displays the complexities expected for its low symmetry (*see* Figure 6), where the most important feature is the pair of resonances for the nonequivalent imino protons. If this spectrum were the weighted average of rapidly interconverting diastereomers (*see* Equation 6), the

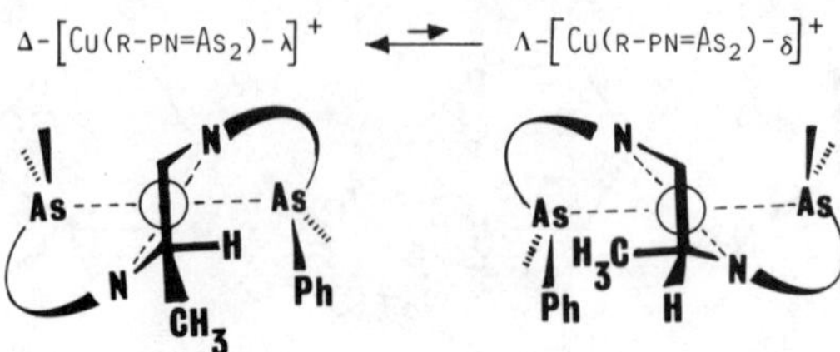

Figure 5. Conformational dependence of the nonbonded interactions in tetrahedral R-pn = As_2 complexes

chemical shifts would be temperature dependent in keeping with the temperature dependence of the corresponding equilibrium constant.

$$\Delta\text{-}[Cu(R\text{-}pn = As_2)\text{-}\lambda]^+ \rightleftarrows \Lambda\text{-}[Cu(R\text{-}pn = As_2)\text{-}\delta]^+ \quad (6)$$

However we find that the H-1 NMR spectrum of [Cu(R-pn = As_2)]ClO_4 (CD_2Cl_2 solution) is invariant from 35° to −49°C, thus indicating that the R-pn = As_2 binds stereospecifically to the metal ion.

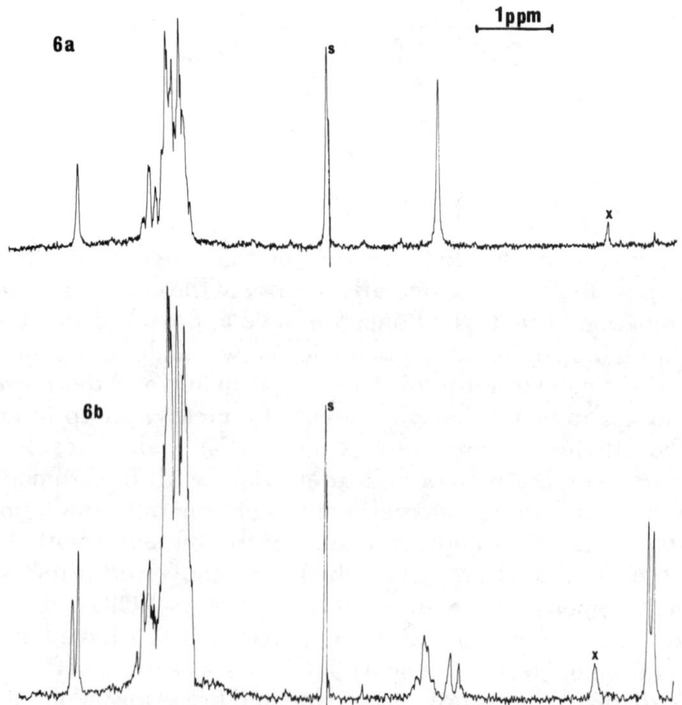

Figure 6. 90-MHz H-1 NMR spectra of [Cu(en = As_2)]ClO_4 (a) and [Cu(R-pn = As_2)]ClO_4 (b) in CD_2Cl_2 solution

It is known that (*R,R*)-1,2-diaminocyclohexane (*R,R*)-chxn is restricted to the λ-conformation when coordinated as a bidentate ligand, and it is for this reason that (*R,R*)-chxn can be used as a reference ligand in studying the stereochemistry of coordination compounds. (For a general discussion of the conformational properties of diaminoalkanes, *see* C. J. Hawkins' "Absolute Configuration of Metal Complexes," Wiley–Interscience, New York, 1971.) We find that (*R,R*)-chxn efficiently condenses with AsCHO to afford the diimine (*R,R*)-chxn = As_2 which we have isolated and characterized. The corresponding copper(I) complex, [Cu(*R,R*)-chxn = As_2)]ClO_4, has been isolated recently as a crystalline compound whose H-1 NMR indicates equivalent imine sites consistent with its dissymmetric C_2 symmetry. Comparative circular dichroism (CD) studies of [Cu(R-pn = As_2)]ClO_4 and [Cu(*R,R*)-chxn = As_2)]ClO_4 confirm the proposed stereospecificities of these ligands.

Acknowledgments

Acknowledgment is made to the donors of the Petroleum Research Fund, administered by the American Chemical Society, for partial support of this work. We also thank the Australian National University Computer Services Centre for the use of their facilities. We also thank G. B. Robertson for the use of his X-ray equipment and Climax Molybdenum for the gift of the molybdenum hexacarbonyl.

Literature Cited

1. Chiswell, B.; Lee, K. W. *Inorg. Chim. Acta.* **1973**, *7*, 49, 509, 517.
2. Riker-Nappier, J.; Meek, D. W. *J. C. S. Chem. Comm.* **1974**, 442.
3. deO. Cabral, J.; Cabral, M. F.; Drew, M. G. B.; Nelson, S. M.; Rogers, A. *Inorg. Chim. Acta* **1977**, *25*, L77.
4. Rauchfuss, T. B. *J. Organomet. Chem.* **1978**, *162*, C19.
5. Schiemenz, G. P.; Kaack, H. *Liebigs Ann. Chem.* **1973**, 1480.
6. Jeffery, J. C.; Rauchfuss, T. B.; Tucker, P. A. *Inorg. Chem.* **1980**, *19*, 3306.

RECEIVED July 10, 1980.

19

Recent Advances in Polyphosphine Synthesis

R. B. KING
Department of Chemistry, University of Georgia, Athens, GA 30602

The base-catalyzed addition of phosphorus–hydrogen compounds to vinylphosphines is a useful method for synthesizing polyphosphines containing PCH_2CH_2P structural units. Applications of this method to the syntheses of polyphosphines containing phosphorus–hydrogen bonds, polyphosphines containing terminal dialkylamino or alkoxy groups, and chiral polyphosphines containing terminal neomenthyl groups are reviewed.

During the past 25 years trivalent phosphorus derivatives have become important ligands in coordination chemistry and molecular catalysis. Such ligands are useful for stabilizing both high- and low-transition metal oxidation states, metal hydrides, metal dinitrogen complexes, and unusual coordination numbers. Furthermore, trivalent phosphorus ligands are used in homogeneous catalysts for hydrogenation and hydroformylation reactions. Using chiral phosphorus ligands in such catalyst systems can lead to effective and useful asymmetric catalysts.

In connection with the application of trivalent phosphorus ligands in coordination chemistry and molecular catalysis, the synthesis of such ligands having diverse steric and electronic properties is very important. In 1968 we became interested in developing new methods for synthesizing chelating poly(tertiary phosphines). Shortly thereafter we discovered the base-catalyzed addition of phosphorus–hydrogen bonds to vinylphosphorus compounds according to the following general scheme (*1*):

$$\rangle P\text{—}H + CH_2 = CHP\langle \longrightarrow \rangle PCH_2CH_2P\langle \qquad (1)$$

0065–2393/82/0196–0313$05.00/0

The mechanism of this reaction appears to be closely related to a Michael addition (*2, 3*). Suitable catalysts are potassium *t*-butoxide and phenyllithium. Shortly after our original discovery, we exploited this synthetic method for preparing diverse chelating poly(tertiary phosphines) containing terminal phenyl groups including the tritertiary phosphine **I** ($R = R' = C_6H_5$), the tripod tetratertiary phosphine **II** ($R = C_6H_5$), the linear tetratertiary phosphine **III** ($R = C_6H_5$), and the hexatertiary phosphine **IV** ($R = C_6H_5$) (*1*). A brief review article (*4*) summarizes the general aspects of our work in this area up to the middle of 1971.

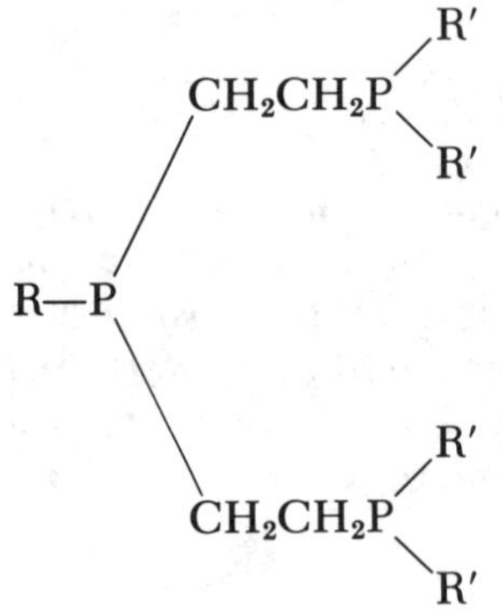

$$R{-}P\begin{cases} CH_2CH_2P(R')_2 \\ CH_2CH_2P(R')_2 \end{cases}$$

I

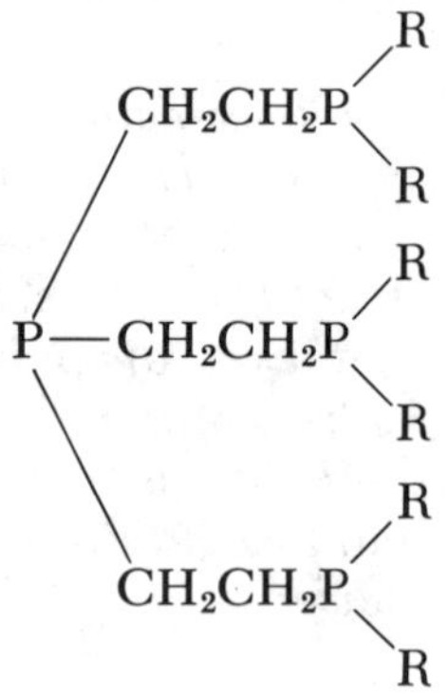

$$P\begin{cases} CH_2CH_2PR_2 \\ CH_2CH_2PR_2 \\ CH_2CH_2PR_2 \end{cases}$$

II

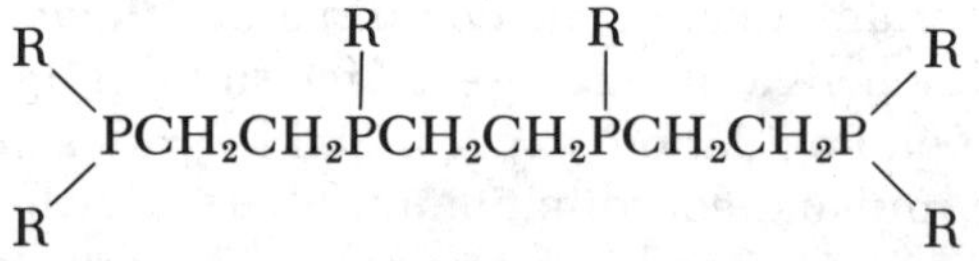

$$R_2PCH_2CH_2P(R)CH_2CH_2P(R)CH_2CH_2PR_2$$

III

$$(R_2PCH_2CH_2)_2PCH_2CH_2P(CH_2CH_2PR_2)_2$$

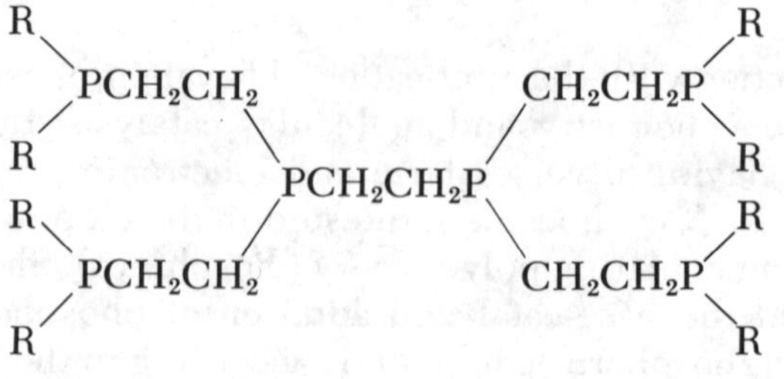

IV

Other research groups developed related reactions for synthesizing polyphosphines. Thus in 1970 Grim, Molenda, and Keiter (*5*) reported the addition of $(C_6H_5)_2PLi$ to $(C_6H_5)_2PCH = CH_2$ to give

$(C_6H_5)_2PCH_2CH_2P(C_6H_5)_2$ and subsequently have (6) exploited this method to prepare mixed alkyl–aryl ditertiary phosphines of the type $(C_6H_5)_2PCH_2CH_2P(C_6H_5)R$. In 1971 Issleib and Weichmann (7) reported the free-radical addition of primary phosphines to vinylphosphines $CH_2 = CHPR'_2$ to give the secondary–tertiary diphosphines $RPHCH_2CH_2PR'_2$. In 1975 Meek (8) and co-workers (8, 9) reported more extensive applications of the free-radical addition of phosphorus–hydrogen bonds across carbon–carbon double bonds in vinylphosphines.

This chapter summarizes the highlights of our research in polyphosphine synthesis at the University of Georgia since we last reviewed our work in 1972 (*4*). The major advances in our work in this area since then have involved the extension of our methods to the synthesis of polyphosphines containing terminal groups of interest because of their chemical reactivity (e.g., hydrogen and dialkylamino), their electronic properties (e.g., methoxy in chelating polyphosphines with relatively strong π-acceptor properties), or their stereochemical properties (e.g., neomenthyl in chiral chelating polyphosphines for asymmetric catalysis).

Polyphosphines Containing Phosphorus–Hydrogen Bonds

Two general methods are available for preparing polyphosphines containing PCH_2CH_2P structural units and phosphorus–hydrogen bonds:

(1) adding phosphorus–hydrogen compounds to vinylphosphonates followed by $LiAlH_4$ reduction, e.g.,

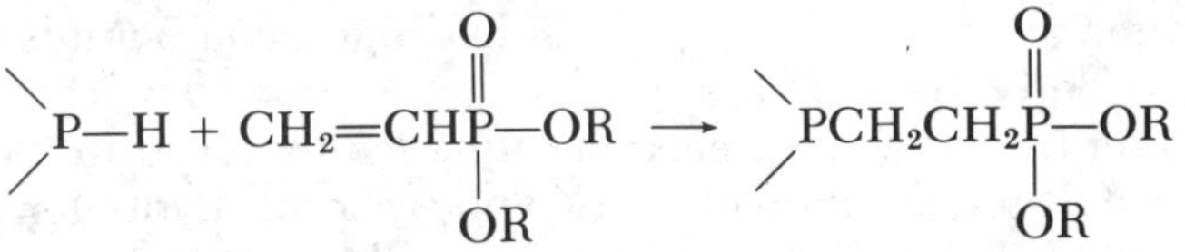

$$\rangle P{-}H + CH_2{=}CH\overset{\overset{\displaystyle O}{\|}}{\underset{\underset{\displaystyle OR}{|}}{P}}{-}OR \longrightarrow \rangle PCH_2CH_2\overset{\overset{\displaystyle O}{\|}}{\underset{\underset{\displaystyle OR}{|}}{P}}{-}OR \qquad (2a)$$

$$\rangle PCH_2CH_2\overset{\overset{\displaystyle O}{\|}}{\underset{\underset{\displaystyle OR}{|}}{P}}{-}OR \xrightarrow{LiAlH_4} \rangle PCH_2CH_2P\langle^{H}_{H} \qquad (2b)$$

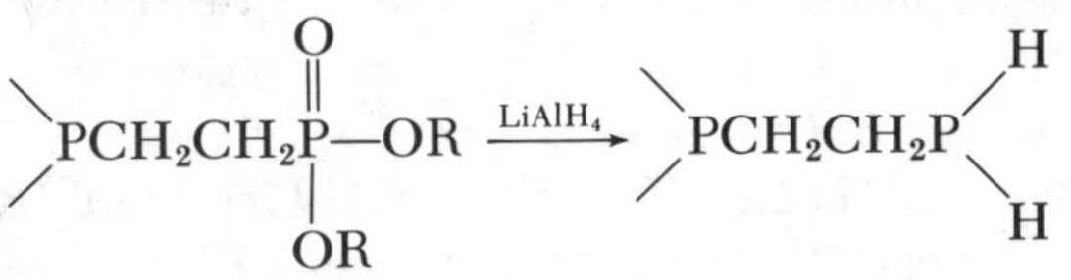

(2) a 1:1 addition of a primary phosphine to a vinylphosphorus compound, e.g.,

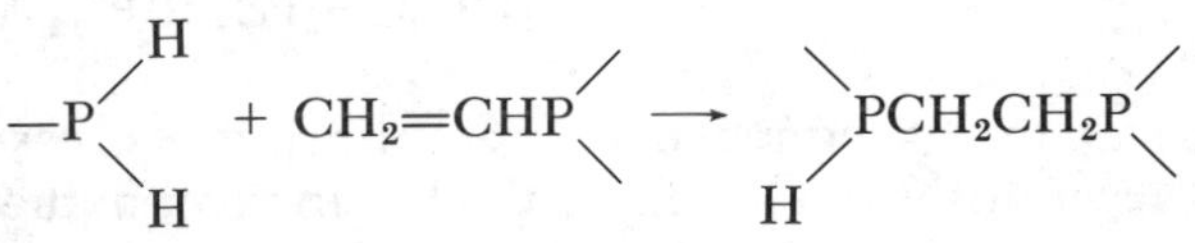

$$-P\langle^{H}_{H} + CH_2{=}CHP\langle \longrightarrow {}_{H}^{\rangle}PCH_2CH_2P\langle \qquad (3)$$

The first of these methods is useful for preparing tertiary–primary diphosphines such as $(C_6H_5)_2PCH_2CH_2PH_2$ (*10, 11*) and $(CH_3)_2PCH_2CH_2PH_2$ (*12*). Using a free-radical catalyst, the second method can be used to prepare tertiary–secondary diphosphines of the type $R_2PCH_2CH_2PHR$ (*7*). However, under base-catalyzed conditions it is difficult to stop the addition of primary phosphine to the vinylphosphine at the stage of the tertiary–secondary phosphine 1:1 adduct; formation of the tritertiary phosphine $RP[CH_2CH_2PR_2]_2$ with recovery of unreacted primary phosphine is observed instead (*12*).

The synthetic principles in Equations 2a, 2b, and 3 can be combined to synthesize secondary–primary phosphines by the following sequence of reactions (R = C_6H_5, n-C_6H_{13}, and $(CH_3)_3CCH_2$) (*12*):

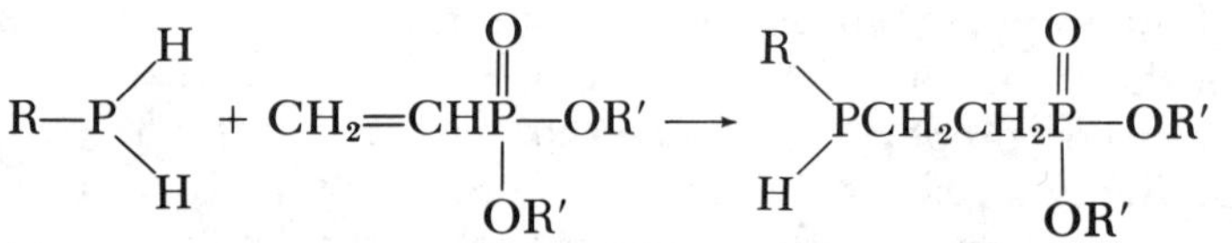

$$\text{RPH}_2 + CH_2{=}CH\overset{O}{\overset{\|}{P}}(OR')\text{—OR}' \longrightarrow \text{R(H)PCH}_2\text{CH}_2\overset{O}{\overset{\|}{P}}(OR')\text{—OR}' \qquad (4a)$$

$$\text{R(H)PCH}_2\text{CH}_2\overset{O}{\overset{\|}{P}}(OR')\text{—OR}' \xrightarrow{\text{LiAlH}_4} \text{R(H)PCH}_2\text{CH}_2\text{PH}_2 \qquad (4b)$$

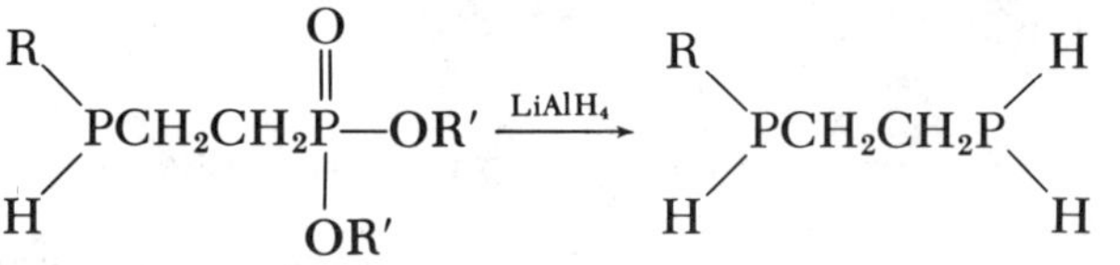

The 1:1 addition of RPH_2 to a vinylphosphonate (Equation 4a) proceeds effectively under base-catalyzed conditions in contrast to the 1:1 addition of RPH_2 to a vinylphosphine (Equation 3). This point can be related to the basicity of the anions involved as intermediates in the Michael-type addition of phosphorus–hydrogen compounds to vinylphosphorus derivatives (*12*).

Consider the following reaction sequence involved in the successive base-catalyzed addition of the two phosphorus–hydrogen bonds in a primary phosphine to a vinylphosphorus derivative:

$$RPH_2 + \text{base} \rightarrow RPH^- + \text{base } H^+ \qquad (5a)$$
$$RPH^- + CH_2{=}CHP(Z)R'R'' \rightarrow RP(H)CH_2CH^-P(Z)R'R'' \qquad (5b)$$
$$RP(H)CH_2CH^-P(Z)R'R'' \rightleftarrows RP^-CH_2CH_2P(Z)R'R'' \qquad (5c)$$
$$RP^-CH_2CH_2P(Z)R'R'' + CH_2{=}CHP(Z)R'R'' \rightarrow RP[CH_2CH_2P(Z)R'R''][CH_2CH^-P(Z)R'R''] \qquad (5d)$$
$$RP[CH_2CH_2P(Z)R'R''][CH_2CH^-P(Z)R'R''] + \text{base } H^+ \rightarrow RP[CH_2CH_2P(Z)R'R'']_2 + \text{base} \qquad (5e)$$

The critical step in this sequence is the equilibrium expressed in Equation 5c. If this equilibrium lies far to the left in favor of the carbanion, then it will be relatively easy to stop the base-catalyzed addition at the 1:1 adduct stage. However, if this equilibrium lies far to the right in favor of the phosphide anion, then Reaction 5d will proceed so easily that it will be relatively difficult to stop the base-catalyzed addition at the 1:1 adduct stage. Thus the ability to stop the base-catalyzed addition of a phosphine RPH_2 to a vinylphosphorus compound $CH_2 = CHP(Z)R'R''$ at the 1:1 adduct stage will depend upon the ability of the $P(Z)R'R''$ group to stabilize an adjacent carbanion. If the $P(Z)R'R''$ group contains tetracoordinate phosphorus (i.e., if Z = O or S), the adjacent carbanion is stabilized to the extent that the base-catalyzed addition of RPH_2 to $CH_2 = CHP(Z)R'R''$ can be stopped at the 1:1 adduct stage, namely $RP(H)CH_2CH_2P(Z)R'R''$. However, this need not be the case if the $P(Z)R'R''$ group contains trivalent phosphorus (i.e., Z is a lone pair). In practice we found that the base-catalyzed additions of $C_6H_5PH_2$ to $CH_2 = CHP(O)(OCHMe_2)_2$ (*12*), $CH_2 = CHP(S)R'_2$ ($R' = CH_3$ and C_6H_5) (*12*), and $CH_2 = CHP[N(CH_3)_2]_2$ (*13*) can be controlled to give good yields of the corresponding 1:1 adducts $C_6H_5P(H)CH_2CH_2P(O)(OCHMe_2)_2$, $C_6H_5P(H)CH_2CH_2P(S)R'_2$, and $C_6H_5P(H)CH_2CH_2P[N(CH_3)_2]_2$, respectively, whereas the base-catalyzed addition of $C_6H_5PH_2$ to $CH_2 = CHP(C_6H_5)_2$ gives only a low yield (~13%) of the corresponding 1:1 adduct $C_6H_5P(H)CH_2CH_2P(C_6H_5)_2$ (*12*).

The relative stabilities of alternative anionic reaction intermediates also account for the observation that the secondary phosphine hydrogen in the secondary–primary diphosphine $C_6H_5P(H)CH_2CH_2PH_2$ reacts preferentially with $CH_2 = CHP(O)(OCHMe_2)_2$ under base-catalyzed conditions to give $C_6H_5P(CH_2CH_2PH_2)_2$ after $LiAlH_4$ reduction (*12*). The phosphide anion $C_6H_5P^-CH_2CH_2PH_2$ is stabilized over the isomeric phosphide anion $C_6H_5PHCH_2CH_2PH^-$ by the phenyl group directly bonded to the negatively charged phosphorus atom. The tertiary–diprimary triphosphine $C_6H_5P(CH_2CH_2PH_2)_2$ obtained in this reaction is prepared more conveniently (*11*) by the 1:2 addition of $C_6H_5PH_2$ to $CH_2 = CHP(O)(OC_2H_5)_2$ followed by $LiAlH_4$ reduction.

A 1:1 addition of the PH_2 group of the tertiary–primary diphosphine $(CH_3)_2PCH_2CH_2PH_2$ to a vinyl phosphonate is the key step in the synthesis of a tertiary–secondary–primary triphosphine by the following sequence of reactions (*12*):

$$(CH_3)_2PCH_2CH_2PH_2 + CH_2{=}CHP(O)(OCHMe_2)_2 \rightarrow (CH_3)_2PCH_2CH_2P(H)CH_2CH_2P(O)(OCHMe_2)_2 \quad (6a)$$

$$(CH_3)_2PCH_2CH_2P(H)CH_2CH_2P(O)(OCHMe_2)_2 \xrightarrow{LiAlH_4} (CH_3)_2PCH_2CH_2P(H)CH_2CH_2PH_2 \quad (6b)$$

The product is the first example of a polyphosphine containing tertiary, secondary, and primary phosphorus atoms in the same molecule.

Polyphosphines Containing Terminal Dialkylamino or Alkoxy Groups

Phosphorus–nitrogen and phosphorus–oxygen bonds are generally more reactive than phosphorus–carbon bonds. Therefore it was interesting to see whether our base-catalyzed addition reactions could be used to synthesize polyphosphines containing these structural features. Such polyphosphines would not only be interesting as ligands in coordination chemistry but they also might be useful intermediates for synthesizing other types of organophosphorus compounds.

Synthesizing polyphosphines containing terminal dialkylamino groups can be performed under basic conditions since the phosphorus–nitrogen bonds in such compounds, although sensitive to acid, are resistant to base. The key vinyl compound for such syntheses is $CH_2 = CHP[N(CH_3)_2]_2$ which can be obtained in 60% yield (*13*) by reacting the readily available $[(CH_3)_2N]_2PCl$ with vinylmagnesium bromide followed by hydrolyzing with aqueous tetrasodium ethylenediamine tetraacetate. On the other hand, dialkylaminophosphines with phosphorus–hydrogen bonds such as $[(CH_3)_2N]_2PH$ do not seem to be available in the uncomplexed state. It is therefore more feasible to introduce the terminal dialkylamino groups into the vinylphosphorus component rather than the phosphorus–hydrogen component of the polyphosphine synthesis.

Adding $R_2PH(R = CH_3$ and $C_6H_5)$, $RPH_2(R = CH_3$ and $C_6H_5)$, and PH_3 to $CH_2 = CHP[N(CH_3)_2]_2$ proceeds easily using a potassium hydride catalyst to give the diphosphines $R_2PCH_2CH_2P[N(CH_3)_2]_2$, triphosphines $RP[CH_2CH_2P[N(CH_3)_2]_2$ (I: $R' = N(CH_3)_2$), and tripod tetraphosphine $P[CH_2CH_2P[N(CH_3)_2]_3$ (II: $R = N(CH_3)_2$), respectively (*13*). In addition, the tertiary–secondary phosphine $C_6H_5P(H)$-$CH_2CH_2P[N(CH_3)_2]_2$ is obtained in a reasonable yield (53%) by carrying out the reaction between $C_6H_5PH_2$ and $CH_2 = CHP[N(CH_3)_2]_2$ in a 1:1 mole ratio. Also the base-catalyzed addition of $(C_6H_5)_2PH$ to $(CH_2 = CH)_2PN(C_2H_5)_2$ can be controlled to add to one or both of the vinyl groups giving the diphosphine $(C_2H_5)_2NP(CH = CH_2)CH_2CH_2$-$P(C_6H_5)_2$ or the triphosphine $(C_2H_5)_2NP[CH_2CH_2P(C_6H_5)_2]_2$, respectively.

The reaction of a primary phosphine RPH_2 with a divinylphosphorus derivative $(CH_2 = CH)_2PR'$ might be expected to give either a polymer or a cyclic derivative of the general formula $[-P(R)CH_2CH_2P$-$(R')CH_2CH_2-]_n$. However, none of our attempts to carry out the base-catalyzed addition of $C_6H_5PH_2$ to $(CH_2 = CH)_2PC_6H_5$ gave tractable products. On the other hand, more favorable results appear to be obtained if one of the phosphorus atoms bears a dialkylamino substituent. Thus the base-catalyzed additions of the primary phosphines RPH_2($R = C_6H_5$, $C_6H_5CH_2$, and $(CH_3)_3CCH_2$) to $(CH_2 = CH)_2PN(C_2H_5)_2$ lead to the corresponding 1,4-diphosphacyclohexane derivatives $RP(CH_2CH_2)_2PN(C_2H_5)_2$ (*13*). The C-13 and P-31 NMR spectra of these 1,4-diphosphacyclohexane derivatives indicate that they are a mixture of the cis and trans isomers **Va** and **Vb**, respectively.

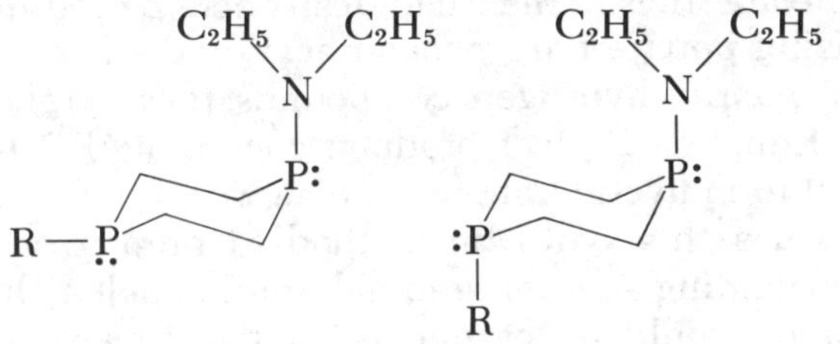

C_2H_5 C_2H_5 N P: R—P **Va**

C_2H_5 C_2H_5 N P: :P R **Vb**

The polyphosphines containing terminal methoxy groups are interesting because they provide potentially chelating ligands with strong π-acceptor properties. The following two approaches can be used to prepare such polyphosphines: (1) base-catalyzed addition of various phosphorus–hydrogen compounds to $CH_2 = CHP(OCH_3)_2$, which can be prepared (*13*) by the methanolysis of $CH_2 = CHP[N(CH_3)_2]_2$ in boiling methanol; (2) methanolysis in boiling toluene of the corresponding polyphosphine containing terminal dimethylamino groups. The latter method appears preferable in most cases and has been used to synthesize the diphosphines $R_2PCH_2CH_2P(OCH_3)_2$ ($R = CH_3$ and C_6H_5), the triphosphines $RP[CH_2CH_2P(OCH_3)_2]_2$ (I: $R' = OCH_3$), and the tripod tetraphosphine $P[CH_2CH_2P(OCH_3)_2]_3$ (II: $R = OCH_3$). In addition, the base-catalyzed addition of $C_6H_5P(H)CH_2CH_2P[N(CH_3)_2]_2$ to $CH_2 = CHP(OCH_3)_2$ gives the triphosphine $C_6H_5P[CH_2CH_2P[N(CH_3)_2]_2][CH_2CH_2P(OCH_3)_2]$, which represents a novel example of a polyphosphine with terminal dimethylamino groups on one arm and terminal methoxy groups on the other arm (*13*). Preliminary studies on the coordination chemistry of the diphosphines and triphosphines containing terminal methoxy groups (*14*) indicate the facile formation of red tetrahedral iron(II) chloride, blue tetrahedral cobalt(II) chloride, and yellow square planar nickel(II) chloride complexes.

Polyphosphines Containing Terminal Neomenthyl Groups

During the past several years several asymmetric chelating di(tertiary phosphines) have been found which give very high optical yields when used as ligands in rhodium(I) asymmetric hydrogenation catalysts. Thus the chiral di(tertiary phosphines) (-)(o-$CH_3OC_6H_4$)-(C_6H_5)PCH_2CH_2P(C_6H_5)(C_6H_4 OCH_3-o) ("dipamp") (*15*), (-)-(2*S*, 3*S*)-$(C_6H_5)_2PCH(CH_3)CH(CH_3)P(C_6H_5)_2$ ("(*S*, *S*)-chiraphos") (*16*), and $(C_6H_5)_2PCH(CH_3)CH_2P(C_6H_5)_2$ ("R-prophos") (*17*) give optical yields approaching 100% in the hydrogenation of prochiral olefins such as α-acetamidoacrylic acid. The high optical yields using catalysts containing these bidentate ligands may relate to the rigidity of the five-membered chelate rings in the catalytically active rhodium(I) species.

Synthesizing poly(tertiary phosphines) by the base-catalyzed addition of phosphorus–hydrogen compounds to vinylphosphorus derivatives (*see* Equation 1) gives products containing PCH_2CH_2P structural units that form five-membered chelate rings. It therefore seemed interesting to use this synthetic method to prepare a poly(tertiary phosphine) containing a chiral terminal group. Such a chiral poly(tertiary phosphine) would be particularly interesting as a ligand in a rhodium(I) asymmetric hydrogenation catalyst.

An inviting chiral terminal group to use for this purpose is the neomenthyl group which is readily accessible from commercial (-)-menthol. Accordingly, we prepared the secondary phosphine (Nmen)-(C_6H_5)PH (**VI**) by the following sequence of reactions (Men = menthyl, Nmen = neomenthyl (*18*)):

$$2\ C_6H_5PH_2 + 2\ Na \xrightarrow{NH_3} 2\ C_6H_5PHNa + H_2 \quad (7a)$$

$$C_6H_5PHNa + MenCl \rightarrow (Nmen)(C_6H_5)PH + NaCl \quad (7b)$$

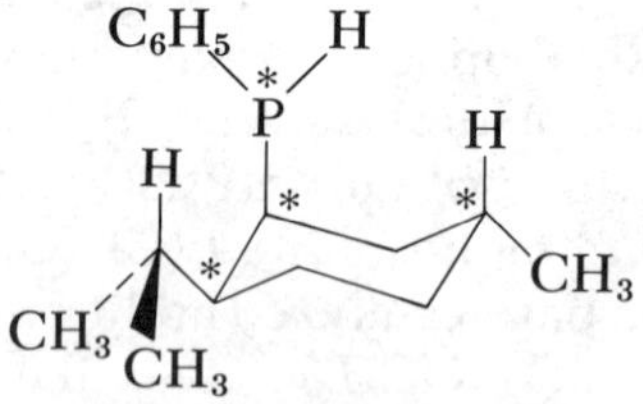

VI

The C-13 NMR spectrum of the product (**VI**) clearly indicates that the menthyl chloride had undergone inversion to give a neomenthyl group in its reaction with C_6H_5PHNa exactly as has been established by previous workers (*19*) in the corresponding reaction of menthyl chloride with $(C_6H_5)_2PNa$. The P-31 NMR spectrum of (Nmen)-(C_6H_5)PH (**VI**) exhibits two resonances of approximately equal relative intensities indicating the presence of the two diastereomers arising from the two configurations of the chiral phosphorus atom in conjunction with the chiral neomenthyl group. No attempt was made to separate these two diastereomers at this point in the synthesis since (Nmen)(C_6H_5)PH is an air-sensitive liquid.

The base-catalyzed additions of (Nmen)(C_6H_5)PH to CH_2 = CHP-(C_6H_5)$_2$,(CH_2 = CH)$_2$PC_6H_5, and CH_2 = CHP(S)(CH_3)$_2$ give the di(tertiary phosphine) (Nmen)(C_6H_5)PCH_2CH_2P(C_6H_5)$_2$ (**VII**), the tri(tertiary phosphine) C_6H_5P[CH_2CH_2P(C_6H_5)(Nmen)]$_2$, and the diphosphine monosulfide (Nmen)(C_6H_5)PCH_2CH_2P(S)(CH_3)$_2$, respectively. Among these three products the di(tertiary phosphine) (**VII**) was investigated in the greatest detail because of its close relationship to di(tertiary phosphines) known to give high optical yields when used as ligands in rhodium(I) asymmetric hydrogenation catalysts.

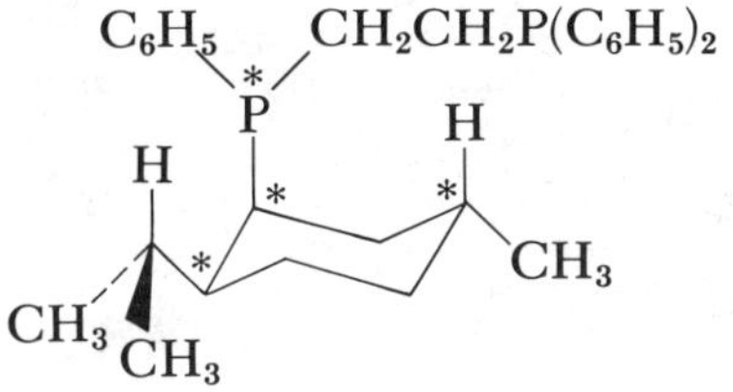

C_6H_5 $CH_2CH_2P(C_6H_5)_2$
P
H H
CH_3
CH_3 CH_3

VII

The P-31 NMR spectrum of (Nmen)(C_6H_5)PCH_2CH_2P(C_6H_5)$_2$ (**VII**) exhibits two resonances for the (Nmen)(C_0H_5)P phosphorus atom indicating the presence of two diastereomers similar to those found in its precursor (Nmen)(C_6H_5)PH (**VI**). Repeated fractional crystallization of (Nmen)(C_6H_5)PCH_2CH_2P(C_6H_5)$_2$ led to separation of the two individual pure diastereomers as crystalline solids with optical rotations $[\alpha]_D$ of +109° and −24°. Thus (Nmen)(C_6H_5)PCH_2CH_2P(C_6H_5)$_2$ (**VII**) may be regarded as a "self-resolving" chiral ditertiary phosphine in which the chiral neomenthyl group provides a basis for resolving the isomers with opposite configurations around the chiral phosphorus atom.

Rhodium(I) complexes of both diastereomers of (Nmen)(C_6H_5)-PCH_2CH_2P(C_6H_5)$_2$ (**VII**) can be generated in situ from [*n*-C_7H_8RhCI]$_2$ and the ditertiary phosphine and have been used as catalysts for the asymmetric homogeneous hydrogenation of the prochiral olefins C_6H_5CH = C(NHCOR)(CO_2R′)(R = CH_3, R′ = H and CH_3; R = C_6H_5,) R′ = H and C_2H_5). Some results are summarized in Table I. Generally the optical yields obtained using these neomenthyl polytertiary phosphines are lower than those previously reported for comparable asymmetric hydrogenations using the chiral di(tertiary phosphines) dipamp (*15*), (*S*, *S*)-chiraphos (*16*), and (*R*)-prophos (*17*). However, the following observations concerning these data can be made:

1. The two diastereomers of (Nmen)(C_6H_5)PCH_2CH_2P(C_6-H_5)$_2$ (**VII**), which have opposite configurations around

Table I. Optical Yields for the Asymmetric Hydrogenation of α-(Acylamido)cinnamic Acid Derivatives Using Rhodium(I) Complexes Containing (Nmen)$(C_6H_5)PCH_2CH_2P(C_6H_5)_2$

Prochiral Olefin	*Hydrogenation Optical Yield*[a], % *(Nmen)* $(C_6H_5)PCH_2CH_2P(C_6H_5)_2$ *Diastereomer* $[\alpha]_D + 109°$	$[\alpha]_D - 24°$
H, CO_2H / C_6H_5, $NHCOCH_3$	46 [*R*]	60 [*S*]
H, CO_2CH_3 / C_6H_5, $NHCOCH_3$	42 [*R*]	58 [*S*]
H, CO_2H / C_6H_5, $NHCOC_6H_5$	49 [*R*]	85 [*S*]
H, $CO_2C_2H_5$ / C_6H_5, $NHCOC_6H_5$	56 [*R*]	31 [*S*]

[a] The absolute configurations of the hydrogenation product (*R* or *S*) are given in brackets.

the chiral phosphorus atom but the same configuration in the neomenthyl group, produce opposite enantiomers for each of the four prochiral α-(acylamido)cinnamic acid derivatives investigated.

2. A given diastereomer of (Nmen)$(C_6H_5)PCH_2CH_2P(C_6H_5)_2$ (**VII**) gives the same absolute configuration of the chiral hydrogenation product for each of the four prochiral α-(acylamido)cinnamic acid derivatives investigated.
3. In the case of the $[\alpha]_D - 24°$ diastereomer but not the $[\alpha]_D + 109°$ diastereomer the product optical yield is very sensitive towards minor changes in the structure of the prochiral olefin.

The first two observations indicate that the chiral phosphorus atom in **VII** dominates over the chiral neomenthyl group in determining the optical yield and absolute configuration of the hydrogenation product.

Acknowledgments

I am indebted to the U.S. Air Force Office of Scientific Research for support of the research on polyphosphine synthesis. In addition the work on the synthesis and catalytic activity of polyphosphines containing neomenthyl terminal groups was part of an United States–Hungarian cooperative science project in collaboration with L. Markó of the Veszprém University of Chemical Engineering and funded jointly by the U.S. National Science Foundation (Grant INT-76-20080) and the Hungarian Institute for Cultural Relations. In connection with these projects I acknowledge the skillful experimental collaboration of Pramesh Kapoor, John Cloyd, William Masler, József Bakos, and Carl Hoff.

Literature Cited

1. King, R. B.; Kapoor, P. N. *J. Am. Chem. Soc.* **1969,** *91,* 5191.
2. Ibid., **1971,** *93,* 4158.
3. Bergmann, E. D.; Ginsburg, D.; Pappo, R. *Org. React.* **1959,** *10,* 179.
4. King, R. B. *Acc. Chem. Res.* **1972,** *5,* 177.
5. Grim, S. O.; Molenda, R. P.; Keiter, R. L. *Chem. Ind. (London)* **1970,** 1378.
6. Grim, S. O.; Del Gaudio, J.; Molenda, R. P.; Tolman, C. A.; Jesson, J. P. *J. Am. Chem. Soc.* **1974,** *96,* 3416.
7. Issleib, K.; Weichmann, H. *Z. Chem.* **1971,** *11,* 188.
8. DuBois, D. L.; Myers, W. H.; Meek, D. W. *J. Chem. Soc., Dalton Trans.* **1975,** 1011.
9. Meek, D. W.; DuBois, D. L.; Tiethof, J. In "Inorganic Compounds With Unusual Properties", *Adv. Chem. Ser.* **1976,** *150,* 335.
10. King, R. B.; Kapoor, P. N. *Angew. Chem. Int. Ed. Engl.* **1971,** *10,* 734.
11. King, R. B.; Cloyd, J. C., Jr.; Kapoor, P. N. *J. Chem. Soc., Perkin Trans. 1,* **1973,** 2226.
12. King, R. B.; Cloyd, J. C., Jr.; *J. Am. Chem. Soc.* **1975,** *97,* 46.
13. King, R. B.; Masler, W. F. *J. Am. Chem. Soc.* **1977,** *99,* 4001.
14. King, R. B.; Bibber, J. W., unpublished data.
15. Knowles, W. S.; Sabacky, M. J.; Vineyard, B. D.; Weinkauff, D. J. *J. Am. Chem. Soc.* **1975,** *97,* 2567.
16. Fryzuk, M. D.; Bosnich, B. *J. Am. Chem. Soc.* **1977,** *99,* 6262.
17. Ibid., **1978,** *100,* 5491.
18. King, R. B.; Bakos, J.; Hoff, C. D.; Markó, L. *J. Org. Chem.* **1979,** *44,* 3095.
19. Aguiar, A. M.; Morrow, C. J.; Morrison, J. D.; Burnett, R. E.; Masler, W. F.; Bhacca, N. S. *J. Org. Chem.* **1976,** *41,* 1545.

RECEIVED November 11, 1980.

Studies of Asymmetric Homogeneous Catalysts

W. S. KNOWLES, W. C. CHRISTOPFEL, K. E. KOENIG, and C. F. HOBBS
Monsanto Company, Corporate Research Laboratories, St. Louis, MO 63166

A number of chiral bisphosphines related to DiPAMP(1) were prepared and evaluated in asymmetric catalysis. Many variants were closely equivalent but none were superior to the parent compound. In addition, some monophosphines containing sulfone substituents were quite effective. These had the particular advantage of being usable in water solution. Several new DIOP derivatives were tried in the hydroformylation of vinyl acetate but only modest enantiomeric excesses were achieved. A 72% enantiomeric excess was achieved on dehydrovaline under relatively forcing conditions using DiCAMP(3). This result was remarkable since these phosphine ligands generally work very poorly, if at all, on tetrasubstituted olefins.

Over the past decade the use of chiral phosphine ligands complexed with rhodium to effect an asymmetric hydrogenation of enamide precursors of α-amino acids has been highly effective (*1*, *2*). Even though the enantiomeric excess (ee) has turned out to be a very sensitive function of ligand structure, a considerable number of highly successful phosphines have been discovered. The field as applied to enamides has matured now to a program of finding ligands with marginal improvements such as water solubility, faster rates, ease of synthesis, and also a study of mechanisms.

When one turns to other prochiral unsaturates not related to enamides, optimizing by varying ligand structure has been only marginally productive, and the 90–95% ee so easily obtained with enamides has remained well out of reach for the most part. In several cases these low results persist in spite of having been the subject of a considerable research effort.

0065–2393/82/0196–0325$05.00/0

Over the past decade we have explored a number of phosphine ligands for incremental improvements of present systems and for studies in new types of reductions and hydroformylations. Even though bisphosphines continue to dominate the field, we have found fairly efficient monophosphines. This chapter deals with these and other systems that we have explored for α-phenylacrylic acid, dehydrovaline, and hydroformylation of vinyl acetate. Even though we have not solved efficiently any of these cases, we've learned a lot about the behavior of a variety of ligands and hope that this knowledge will contribute to these and other unsolved problems in this field.

Experimental

All compounds were characterized by MS and by NMR. In most cases only small amounts were needed for catalytic studies.

In Situ Catalyst. Bisphosphine (.05 mmol) and Rh(COD)AcAc (*17, 18*) (.05 mmol) were weighed into a 6-mL vial. Air was displaced with nitrogen and 5 mL of peroxide-free, 88% isopropyl alcohol (IPA) was added while bubbling through a small stream of nitrogen. The vial was sealed with a septum and supersonically stirred until solution was complete. For a standard run, 0.5 mL (0.005 mmol of complex) was used to hydrogenate 1.0 g (4.88 mmol) of AAC in 25 ml of 88% IPA at 50°C and 3 atm. The ee was measured by diluting to 100 mL and comparing the optical rotation with a blank run in the same manner (*3*).

This in situ catalyst is suitable for reducing acids which provide a proton source to release the AcAc. When reducing neutral substrates, one must either add .005 mmol of HCl or use a preformed solid catalyst of the type $[Rh(COD)(Bisphosphine)]^+ BF_4^-$ (*3*).

(*R,R*)-1,2-Ethanediylbis[*o*-hydroxyphenyl)phenylphosphine] (Compound 12). In a flame-dried, nitrogen-purged 100-mL round bottom flask equipped with a nitrogen inlet, magnetic stirring bar, and syringe septum was placed 3.0 g (16.12 mmol) diphenylphosphine in 50 ml of the THF (freshly distilled from sodium ketyl). A solution of *n*-butyl lithium in hexane (10 mL, 16.1 mmol) was syringed in over several minutes. After stirring at room temperature for 5 min, 2.5 g (5.45 mmol DiPAMP(1) was added in one portion, and the resulting solution stirred at room temperature for 60 h. The solvent was removed under reduced pressure and the resulting material was taken up in H_2O–NaOH–CH_2Cl_2. Sodium hydroxide was added until both layers were homogeneous. The water layer was separated and carefully acidified with HCl, and the phosphine was extracted with CH_2Cl_2. The organic layer was separated and dried with potassium carbonate, and the solvent was removed under reduced pressure to give 1.54 g of white foam. Final purification was accomplished by crystallization from CH_3OH (4 mL)/acetone (2 mL)/deionized water (1 mL) at −3°C. Within 2 h 1.09 g of product were obtained (crystallized with 1.5 mol of acetone per mole of product).

$$\text{mp} = 148\text{–}149.5°\text{C}, [\alpha]_D^{20} = +55.9°\text{C} (0.85, \text{CHCl}_3)$$

The same procedure could be used on DiPAMPO to give (*S,S*)-1,2-ethanediylbis[(*o*-hydroxyphenyl)phenylphosphine oxide.].

$$\text{mp} = 327°\text{C.}, [\alpha]^{\text{D}}_{20} + 20.8°\text{C} \ (1, CF_3COOH)$$

The diacetate (Compound 13) was prepared by acetylating Compound 12, and since it did not crystallize, it was used without purification.

(*R*)-(*o*-Methoxyphenyl)methyl(morpholinosulfonylmethyl)phosphine (Compound 18). To a solution of morpholinomethanesulfonamide (*13*) (14.0 g, 84.9 mmol) in 100 mL dry THF at 15°C was added *n*-butyl lithium (50.9 mL 1.6*M* solution in hexane (81.5 mmol). Then (*S*)*p*-menthyl *o*-methoxyphenylmethylphosphinate (11.0 g, 34.0 mmol) in 50 mL dry THF was added at 15°C and the mass was held at 20°–25°C for 16 h giving a clear yellow solution. The product was purified on a dry silica column by eluting with ethyl acetate.

NMR ($CDCl_3$) S 1.96 (d, 3H, J = 14 Hz) 3.20–3.90 (m, 8H), 3.55 (d, 2H, J = 14.H), 3.92 (S, 3H), and 6.74–8.11 (m, 4H).

The above solid (2.26 g, 6.8 mmol) was dissolved in 150 ml of dry acetonitrile and 6 ml of Si_2Cl_6 slowly were added at 70°C. The mixture then was held for 30 min, cooled, and added to 100 ml of 25% NaOH at 0°–5°C. The layers were separated and the aqueous phase was extracted with CH_2Cl_2. After removal of the solvent, 1.9 g of clear yellow oil was obtained. It was crystallized from ethanol to give a product, mp 76°–77°C, $[\alpha]^{20}_{\text{D}}$ = +123.8°C (0.8, $CHCl_3$). MS and NMR were consistent. Further crystallization did not improve the rotation.

The physical constants for related compounds were made by the same procedure.

(*R*)-(*o*-Methoxyphenyl)methyl(methylsulfonylmethyl)phosphine (Compound 14):

$$\text{mp } 112\text{–}113, [\alpha]^{20}_{\text{D}} = +131.0°\text{C} \ (0.4, \text{EtOH}).$$

(*R*)-(*o*-Methoxyphenyl)methyl(*N,N*-dimethylsulfonylmethyl)phosphine (Compound 16). This compound is not crystalline; it formed a crystalline derivative of the type L_2Rh AcAc, which was used for the catalyst.

Compounds 15 and 17 were made in a similar manner but did not purify readily.

(2*S*,4*S*)-*N*-Methanesulfonyl-4-diphenylphosphino-2-diphenylphosphinomethylpyrolidine (Compound 21). The *t*-BOC group on Achiwa's BPPM (*5, 6*) was removed with formic acid and the free base reacted with CH_3SO_2Cl. During purification on a silica column the bisphosphine was oxidized to the bisphosphine oxide. This product was readily reduced with Si_2Cl_6 in CH_3CN to give the desired bisphosphine (Compound; mp 124°–126°C 21), $[\alpha]^{20}_{\text{D}}$ = −46.7°C (0.3, $CHCl_3$). An X-ray structure of the rhodium complex is described in Ref. *8*.

(*R,R*)-4,5-Bis(*p*-tosyloxymethyl)-2,2-diphenyl-1,3-dioxolane (Compound 30). A mixture of 4.3 g (.01 mol) of (*R,R*)-1,2,3,4 butane tetrol-1,4-ditosylate and 2.4 g (.01 mol) of dichlorodiphenyl methane in 25 mL of *o*-dichlorobenzene was refluxed under nitrogen for 12 h with evolution of HCl. After evaporation of the solvent, the residue was crystalized from benzene/heptane to obtain 3.6 g (60.5% of product); mp 121°–122°C. $[\alpha]^{20}_{\text{D}}$ = −6.63 (1, C_6H_6).

(*R*,*R*)-4,5-Bis(diphenylphosphinomethyl)-2,2-diphenyl-1,3-dioxolane (Compound 23). A solution of 12.8 mmol of lithium diphenylphosphide in 65 mL of THF was prepared using 2-chloropropane to destroy the phenyllithium. This mixture was stirred for 16 h at 25°C with the above bis-tosylate. Evaporation of the solvent followed by hydrolysis and extraction gave a product which crystallized from ethanol; mp 135–137°C, $[\alpha]_D^{20} = -40.4$°C (1, C_6H_6).

Hydroformylations. The hydroformylation of vinyl acetate was run at 6 to 7 atm with a 44 : 56 CO/H_2 mixture at 80°–100°C using benzene as a solvent and an $8 \times 10^{-4}M$ concentration of rhodium and a 3.45*M* concentration of vinyl acetate. The ratio of ligand to metal was varied from 1 to 6 and the best results were obtained only at high ratios.

Isolated yields in the hydroformylation ran about 75–85%. Rotations were measured on distilled aldehyde and compared with a standard for pure α-acetoxypropanal: $[\alpha]_D^{20} = -34.9$°C (5.0, toluene).

This standard was obtained in two ways. An asymmetric hydroformylation was run and the aldehyde was oxidized to *S*-α-acetoxypropionic acid. A comparison of the rotation obtained with the literature (*19*) for pure *S*-acid $[\alpha]_D^{20} = -49.3$°C (7.2, $CHCl_3$) was used to arrive at the 34.9°C figure for pure *S*-aldehyde. This number was checked using an NMR shift reagent.

Results

The following asymmetric reactions were considered for this evaluation:

Z—AAC

$$\underset{C_6H_5}{\overset{H}{\diagdown}}C{=}C\underset{NHCOCH_3}{\overset{COOH}{\diagup}} \rightarrow C_6H_5CH_2{-}CH\underset{NHCOCH_3}{\overset{COOH}{\diagup}}$$

I

E & Z—BAC

$$\text{Z—}\ \underset{C_6H_5}{\overset{H}{\diagdown}}C{=}C\underset{NHCOC_6H_5}{\overset{COOH}{\diagup}} \qquad \text{E—}\ \underset{H}{\overset{C_6H_5}{\diagdown}}C{=}C\underset{NHCOC_6H_5}{\overset{COOH}{\diagup}} \rightarrow C_6H_5CH_2{-}CH\underset{NHCOC_6H_5}{\overset{COOH}{\diagup}}$$

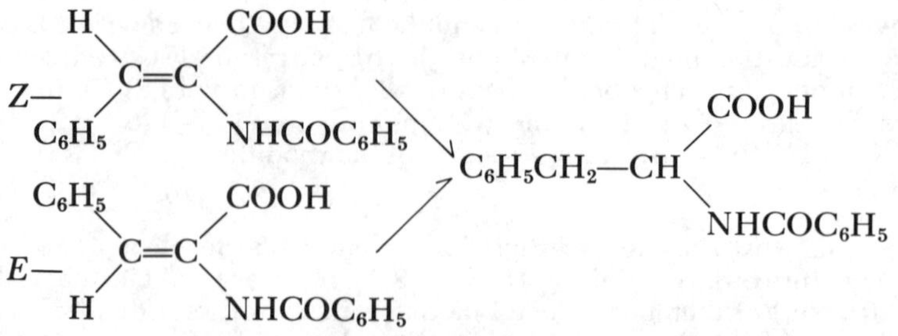

II

DHV

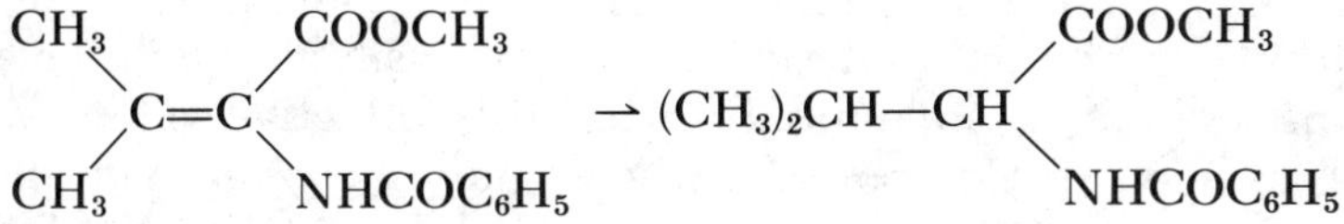

$$(CH_3)_2C{=}C(COOCH_3)(NHCOC_6H_5) \rightarrow (CH_3)_2CH{-}CH(COOCH_3)(NHCOC_6H_5)$$

III

α-Phenylacrylic Acid

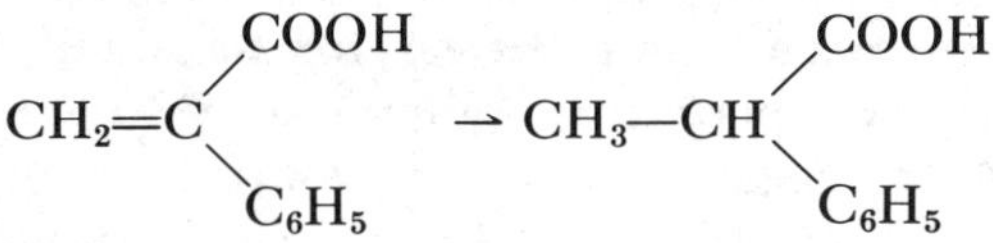

$$CH_2{=}C(COOH)(C_6H_5) \rightarrow CH_3{-}CH(COOH)(C_6H_5)$$

IV

Vinyl Acetate

$$CH_2{-}CHOCOCH_3 \xrightarrow[CO]{H_2} CH_3CH(OCOCH_3)(CHO)$$

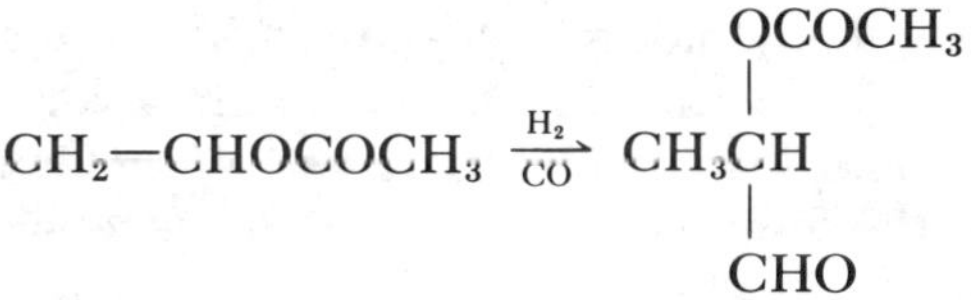

V

Of these reactions only **I** and **II-Z** and the corresponding ee's work with high efficiency. The best phosphine available to us for these conversions was (*R,R*)-1,2-ethanediylbis-[(2-methoxyphenyl)-phenylphosphine](DiPAMP) (3). Since a number of variants of this molecule were readily available from a common, resolved intermediate, it seemed worthwhile to make a few of these with the hope of achieving marginal gains as well as improving our understanding of this remarkable catalysis. Table I shows a number of bisphosphines that were prepared for this study.

Substituting aliphatic groups for the phenyl in DiPAMP, as in Compounds 2 and 3, gave a surprisingly inferior catalyst for reducing α-acetamidocinnamic acid (AAC) (*see* Reaction **I**). Substituting the phenyl with an electron-donating species as in Compound 4 or an electron-withdrawing group as in Compound 5 or 6 appeared to have no market effect. The 4-dimethylamino substituent in Compound 7 was designed to be an electron-donating species in base and an electron-withdrawing species in acid. It was surprising that this phosphine showed such a marked difference between the acid and base mode when Compound 4 or 5 (as did DiPAMP) showed only minor differences.

Substitution of the phenyl with a 2-naphthyl gave an identical catalyst (Compound 8). Replacement of the 2-methoxy with a 2-methylthio (Compound 9) destroyed the catalysis, although the corresponding sulfone (Compound 10) was fairly effective. The 2-fluorophenyl, as seen in Compound 11, gave only a moderately efficient catalyst. Since Variant 9 probably destroyed the catalysis by forming a tight complex with the metal, we expected that the free phenol, Compound 12 would behave in a similar manner. As it turned out, this compound was a fairly efficient, though sluggish, catalyst.

In our early work we found that varying the small, medium, and large groups on the phosphorus was unproductive (*4*). This led to our choice of the highly successful *o*-anisyl group. Perhaps other hetero atoms in the substituent might lead to catalysts with new and useful properties. We explored a few sulfone substituents and found that these were interesting but not superior to our present models. Table II shows that if we substitute the cyclohexyl group of CAMP ((*R*)-cyclohexyl-2-methoxyphenylmethylphosphine) (28) (*1*) with a methyl sulfonylmethyl, as in Compound 14, we can obtain a very respectable 92% ee on Z-AAC. The low results achieved with Compounds 15 and 17 show that the *o*-anisyl group is still necessary for good

Table I. Bisphosphines $(R_1)(R_2)P{-}CH_2CH_2P(R_3)(R_4)$ **(*R,R*)**

Compound No.	*R₁*	*R₂*	*R₃*	*R₄*	*Best ee %*[a]
1	C_6H_5	$2\text{-}CH_3OC_6H_4$	R_1	R_2 (DiPAMP)	96% *S*
2	C_2H_5	$2\text{-}CH_3OC_6H_4$	R_1	R_2	60 *S*
3	C_6H_{11}	$2\text{-}CH_3OC_6H_4$	R_1	R_2 (DiCAMP)	64 *S*
4	$4\text{-}CH_3OC_6H_4$	$2\text{-}CH_3OC_6H_4$	R_1	R_2	92 *S*
5	$3\text{-}ClC_6H_4$	$2\text{-}CH_3OC_6H_4$	R_1	R_2	96 *S*
6	$4\text{-}CH_3SO_2C_6H_4$	$2\text{-}CH_3OC_6H_4$	R_1	R_2	89 *S*
7	$4\text{-}(CH_3)_2NC_6H_4$	$2\text{-}CH_3OC_6H_4$	R_1	R_2	83 (acid) 15 (base)
8	2-naphthyl	$2\text{-}CH_3OC_6H_4$	R_1	R_2	95 *S*
9	$2\text{-}CH_3SC_6H_4$	C_6H_5	R_1	R_2	NR
10	$2\text{-}CH_3SO_2C_6H_4$	C_6H_5	R_1	R_2	73 *S*
11	$2\text{-}FC_6H_4$	C_6H_5	R_1	R_2	60 *S*
12	C_6H_5	$2\text{-}HOC_6H_4$	R_1	R_2	84 *S*[b]
13	C_6H_5	$2\text{-}CH_3COOC_6H_4$	R_1	R_2	63 *S*[b]

[a] The % ee was measured for the hydrogenation of AAC.
[b] The % ee was measured for the hydrogenation of BAC.

Table II. Sulfone Based Phosphines (R_1—P—R_2[a] with R_3 on P)

Compound No.	R_1	R_2	R_3	ee % for AAC: Base (Temp.)	ee % for AAC: Acid (50°)
14	2-$CH_3OC_6H_4$	CH_3	$CH_3SO_2CH_2$	92 (0°) 83 (50°)	79 *S* *S*
15	C_6H_5	CH_3	$CH_3SO_2CH_2$		19 *S*
16	2-$CH_3OC_6H_4$	CH_3	$(CH_3)_2NSO_2CH_2$	84 (50°)	78 *S*
17	C_6H_5	CH_3	$(CH_3)_2NSO_2CH_2$		26 *S*
18	2-$CH_3OC_6H_4$	CH_3	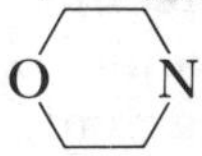 (morpholino)NSO_2CH_2	90 (0°)	79 *S*
19	*R,R*-1,2-Ethanediylbis[(2-methoxy-4-sodium sulfonylphenyl)phenylphosphine]				85 *S*
20	*R,R*-1,2-Ethanediylbis[2-methoxy-4-dimethyl-aminosulfonylphenyl)phenylphosphine]				79 *S*
21	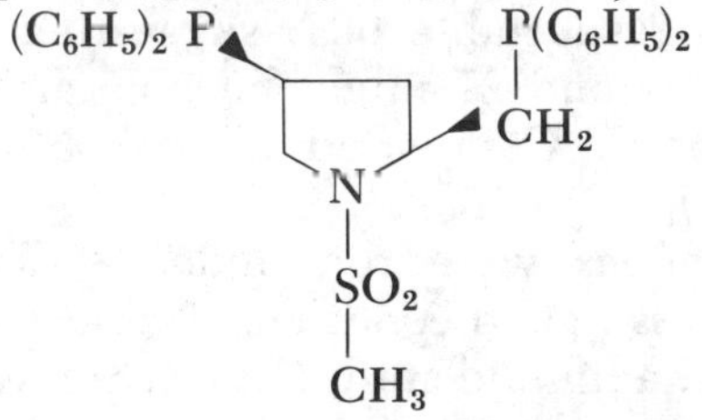 $(C_6H_5)_2P$–, $P(C_6H_5)_2$–CH_2– pyrrolidine, N–SO_2–CH_3 (Achiwa Sulfone)				91 *R* (IPA) 88 *R* (H_2O)

[a] Since displacement of $(R)_p$ menthyl ester and the Si_2Cl_6 reduction both occur with inversion, all products are presumably of the *R*-configuration.

catalysis. Other substituted sulfones such as Compounds 16 and 18 behave almost identically. One difference between these sulfones and our previous catalysts is their solubility in aqueous media. With a homogeneous catalyst it is only necessary for the substrate to be marginally soluble. In fact, it is very convenient to go from a slurry of reactant to a slurry of product. With sulfones unlike DiPAMP and DIOP one can use water as the slurrying means. We attempted to modify DiPAMP with a sulfonyl group as well as Achiwa's BPPM (*1, 5, 6*). Compounds 19, 20, and 21 were successful modifications, performing in water and giving only marginally poorer results than the parent compound.

Since DIOP (22) (*1, 7*) and its close derivatives are the most effective ligands for nonchelating olefins like α-phenylacrylic acid (*1, 7*) (Reaction **IV**) and for hydroformylations (Reaction **V**) (*1*), we decided to try a few variants of this versatile molecule as shown in Table III.

We were able to substitute the *gem*-methyls in the ketal ring with *gem*-phenyls. Compound 23 shows that this perturbation was quite harmful to both Reactions **I** and **V.** This result was unexpected since X-ray structure showed the *gem*-methyls to be well removed from the site of action (*8*) and one would expect *gem*-phenyls to be situated similarly.

Meta-substitution on the phenyl groups of DIOP gave marginal improvement (*1*), and the β-naphthyl moiety should simulate a meta substituent. Actually Compound 24 turns out to be better than DIOP since it gives a 83% yield under a condition where DIOP would give only a 76% yield (3 atm and 50°C). The α-naphthyl was, as might be expected for so hindered a molecule, quite inferior. The *m*-trifluoromethyl product (*see* Compound 26) gave significantly faster hydroformylations, but at best was only marginally better. Another DIOP derivative was prepared to check whether the rigidity of the acetonide ring was really necessary. The adjacent methyls in Compound 27 might be expected to influence a preferred conformation as in Bosnich's chiraphos (*1, 9*). Actually, under base conditions Compound 27 gave a respectable 66% suggesting that with sufficiently large groups Compound 27 might become an efficient ligand. With respect to Reaction **IV,** Compound 27 gave 55% ee as compared with 60% for DIOP (*7*).

Hydroformylations were run under a variety of temperature, pressure, and excess-ligand conditions. Table III shows only the best ee's obtained after a limited study. In no case were we able to exceed results published by others in other systems (*1*).

Table III. (*R,R*)-DIOP Derivatives $\left((R_1)_2C \begin{matrix} \diagup O{-}CH{-}CH_2P(R_2)_2 \\ | \\ \diagdown O{-}CH{-}CH_2P(R_2)_2 \end{matrix} \right)$

Compound No.	Structure *R*$_1$	Structure *R*$_2$	*Hydrogenation of AAC Best ee %*	*Hydroformylation*[a] *of Vinyl Acetate Best ee %*
22	CH_3	C_6H_5(DIOP)	83 *R*	40 S
23	C_6H_5	C_6H_5	38 *R*	29 S
24	CH_3	2-Naphthyl	83 *R*	39 S
25	CH_3	1-Naphthyl	NR	6 S
26	CH_3	3-$CF_3C_6H_4$	7*R*	42 S
27	$CH_3{-}CH{-}CH_2P(C_6H_5)_2$ / $CH_3{-}CH{-}CH_2P(C_6H_5)_2$ (S,S)[b]		49 S (acid) 66 S (base)	

[a] All hydroformylations were run with excess ligand (1–6 moles). Best ee's were obtained with 6 moles of ligand at the expense of slow reactions.

[b] Ref. *16*.

Table IV. Dehydrovaline (Reaction III)

Compound No.	*Ligand*	*ee %*
28	$(C_6H_{11})(2\text{-}CH_3OC_6H_4)(CH_3)P$(CAMP)	23
29	$(C_6H_{11})(\text{iso-PrOC}_6H_4)(CH_3)P$	44
30	$[(C_6H_{11})(2\text{-}CH_3OC_6H_4)PCH_2-]_2$(DiCAMP)	72

When one tries asymmetric hydrogenations on a hindered amino acid as in Reaction **III**, all of the well-known ligands (*1*) give very slow reactions and poor ee's. Hydrogenations generally don't finish DiPAMP. Table IV shows that using the isopropyl ether of CAMP (28) (*10*) gave much faster reactions and modest efficiency. The dimer of Compound 28, [DiCAMP(3)], a catalyst which was disappointingly poor for AAC (*see* Reaction I), gave 72% ee for Reaction III at acceptable rates under relatively forcing conditions (75°C and 4 atm partial pressure of hydrogen). This lead indicates that some further structural modification might solve this system if there was sufficient incentive.

Synthetic Methods

All of the bisphosphines in Table I were prepared by the reaction of the appropriate Grignard reagent on a resolved menthyl phosphinate ester as described in our previous work (*3*). R′ was either phenyl (*11*) or *o*-anisyl (*3*). R_1 corresponds to R_1 group in Table I.

$$R'\text{—}\overset{O}{\overset{\|}{P}}(CH_3)\text{—O Men} \xrightarrow{R_1MgX} R'\text{—}\overset{O}{\overset{\|}{P}}(CH_3)\text{—}R_1 \xrightarrow[2)\ CuCl_2]{1)\ LiN(iPr)_2} (R'\text{—}\overset{O}{\overset{\|}{P}}(R_1)\text{—}CH_2)_2$$

$$R'\text{—}\overset{O}{\overset{\|}{P}}(CH_3)\text{—O Men} \xrightarrow{(CH_3)_3O^+\ BF_4^-} R'\text{—}\overset{O}{\overset{\|}{P}}(CH_3)\text{—}OCH_3 \xrightarrow{R_1MgX} R'\text{—}\overset{O}{\overset{\|}{P}}(CH_3)\text{—}R_1$$

$$(R'\text{—}\overset{O}{\overset{\|}{P}}(R_1)\text{—}CH_2)_2 \xrightarrow{Si_2Cl_6} (R'\text{—}P(R_1)\text{—}CH_2\text{—})_2$$

When the Grignard reagent was too hindered for facile reaction, the menthyl ester was converted first to the methyl ester by the method outlined by DeBruin (*12*). Phosphines 9, 10, and 11 were prepared in this manner.

Coupling by preparing the anion with lithium diisopropylamide followed by $CuCl_2$ oxidation worked nicely in all cases (*1*).

Phosphine 10 was prepared by oxidizing Compound 9 as the phosphine oxide with $KMnO_4$ and then reducing the phosphine oxide with Si_2Cl_6. The sulfone was inert to this silane reagent.

Phosphine 12 was prepared by cleaving the methyl ether on DiPAMP with lithium diphenylphosphide in tetrahydrofuran (THF) at ambient temperatures.

Melting points and rotations of these phosphines and their corresponding oxides are reported in Table V. For catalytic work only a few milligrams of ligand were required for evaluation so that only NMR and MS analyses were used. Whenever possible the products were purified by crystallization to constant melting points. In the case of oils, there always will remain some doubt as to optical purity. The phosphines were air-sensitive and had to be crystallized under nitrogen. Actually, in much of our earlier work the phosphines were contaminated with 5–10% oxide which caused no change in ee's or rates as long as the weights were adjusted properly.

The sulfones were prepared by the following reactions:

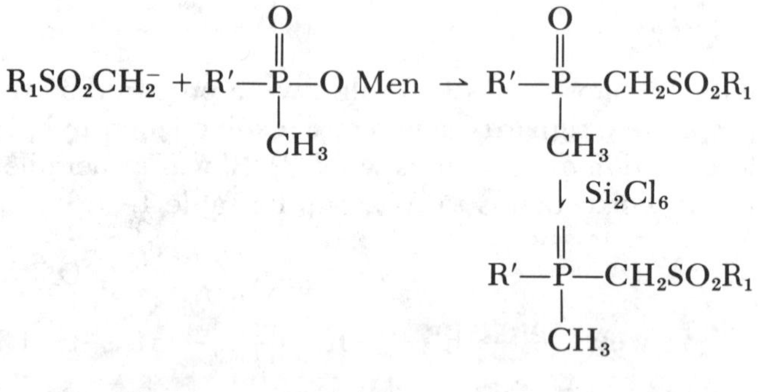

$$R_1SO_2CH_2^- + R'\!-\!\overset{\overset{\displaystyle O}{\|}}{\underset{\underset{\displaystyle CH_3}{|}}{P}}\!-\!O\ Men \rightarrow R'\!-\!\overset{\overset{\displaystyle O}{\|}}{\underset{\underset{\displaystyle CH_3}{|}}{P}}\!-\!CH_2SO_2R_1$$

$$\downarrow Si_2Cl_6$$

$$R'\!-\!\overset{\|}{\underset{\underset{\displaystyle CH_3}{|}}{P}}\!-\!CH_2SO_2R_1$$

R′ was either phenyl (*11*) or *o*-anisyl (*3*). R_1 was CH_3, $(CH_3)_2N$, or morpholino as in R_3 of Table II.

Compound 14 could be made this way or alternately from the anion of dimethylsulfoxide (*13*) followed by oxidation to the sulfone with $KMnO_4$. This compound, although nicely crystalline, was difficult to purify and we found it better to make the morpholino derivative, Compound 18, which handled much better.

Compound 19 was obtained by direct sulfonation of Compound 1 with concentrated H_2SO_4 at 25°C. The dimethylamino sulfone was prepared by chlorosulfonating and aminating DiPAMP as the bisoxide followed by silane reduction.

The sulfone derivative of Achiwa's BPPM (*5, 6*) was prepared from the free base and methane sulfonyl chloride.

Table V. Physical Data on Bisphosphine Oxides

Compound No.	(R,R)-Bisphosphine Oxide M.P. (°C)	(R,R)-Bisphosphine Oxide $[\alpha]_D^{20}$(1, MeOH)	(R,R)-Bisphosphine, m.p.(°C)
2	188–9	−76.1°	oil
3	222–4	−52.7°	84–6
4	oil	−30.4°	122–3
5	oil		89–91
6	oil		oil
7	222–5	+95°	175–80
8	217–220	−30.5°	133–4
9	179–80	−61.3°	127–9
10	noncrystalline		noncrystalline
11	193–4	−14.2°	128–30 (not pure)
12	327	+70.8° (CF_3COOH)	148–9 $[\alpha]_D^{20}$ = +56.7° (1, $CHCl_3$)
13			oil

The benzophenone ketal of DIOP (Compound 23) could not be made directly from the corresponding diol but was prepared as follows:

$$\begin{array}{l} HO—CH—CH_2OTs \\ \quad\;\; | \\ HO—CH—CH_2OTs \end{array} + (C_6H_5)_2CCl_2 \rightarrow \begin{array}{l} C_6H_5\;\;O—CH—CH_2OTs \\ \quad\;\;\; \times \quad\;\; | \\ C_6H_5\;\;O—CH—CH_2OTs \end{array}$$

30

The bis-tosylate (Compound 30) then was converted to Compound 23 in the usual manner by reacting with lithium diphenylphosphide (*7*).

The α- and β-naphthyl derivatives, Compounds 24 and 25, as well as Compound 26 were prepared from the bis-tosylate (*7*) and the corresponding lithium diarylphosphide.

The dimethyl derivative, Compound 27, was prepared from racemic 1,2 dimethylsuccinic acid by first resolving with quinine (*14, 15, 16*), preparing the dimethyl ester, and then proceeding exactly by the same sequence used for DIOP from dimethyl tartrate (*7*). Table VI gives melting points and rotations of the compounds in Table III.

Table VI. Physical Constants DIOP Derivatives

Compound No.	M.P. (°C)	$[\alpha]_D^{20}$	
23	135–137	−40.4°	(1, C_6H_6)(*R,R*)
24	110–125	13.9°	(2, C_6H_6)(*R,R*)
25	104–106	−.46°	(2, C_6H_6)(*R,R*)
26	oil	3.24°	(1, toluene)(*R,R*)
27	96–97	47°.	(1, $CHCl_3$)(*S,S*)

Conclusions

It is fairly well established that the success achieved with Reactions **I** and **II**-Z are facilitated by the ability of the substrate to chelate with the metal (*1, 3, 8*). The carbonyl of the amide and the olefin are situated perfectly for bond formation and the carboxyl or its equivalent forms a third point of tie-down. This three-point landing apparently forms a rigid structure that enables a rather simple ligand to achieve steric control. Fortunately, this special case is applicable to synthesizing α-amino acids where the importance of the compounds offsets the lack of generality. In Reactions **III**, **IV**, and **V** the tie-down is either highly hindered or nonexistent, and perhaps one can only achieve success in these more difficult systems with a ligand of much greater complexity.

In the hydroformylation reaction, **V**, the situation is even worse. Here there is no definite stereochemistry between the phosphine ligand and the metal. One of the reactants, carbon monoxide, competes so well with the phosphine for sites on the metal that it is difficult to insure that the chiral agent is present when the new asymmetric center is formed.

Literature Cited

1. Valentine, D.; Scott, J. W. *Synthesis* **1978**, *5*, 329.
2. Koenig, K. E.; Sabacky, M. J.; Bachman, G. L.; Christopfel, W. C.; Barnstorff, H. D.; Friedman, R. B.; Knowles, W. S.; Stults, B. R.; Vineyard, B. D.; Weinkauff, D. J. *Ann. N.Y. Acad. Sci.* **1980**, *333*, 16.
3. Vineyard, B. D.; Knowles, W. S.; Sabacky, M. J.; Bachman, G. L.; Weinkauff, D. J. *J. Am. Chem. Soc.* **1977**, *99*, 5946.
4. Knowles, W. S.; Sabacky, M. J.; Vineyard, B. D. *Ann. N.Y. Acad. Sci.* **1973**, *214*, 119.
5. Achiwa, K. *J. Am. Chem. Soc.* **1976**, *98*, 845.
6. Ojima, I.; Kogure, T.; Yoda, N. *J. Org. Chem.* **1980**, *45*, 4728.
7. Kagan, H. B.; Dang, T. P. *J. Am. Chem. Soc.* **1972**, *94*, 6429.
8. Knowles, W. S.; Vineyard, B. D.; Sabacky, M. J.; Stultz, B. R. "Fundamental Research in Homogeneous Catalysis", Plenum: New York, 1979; Vol. 3, p. 537.
9. Fryzuk, M. D.; Bosnich, B. *J. Am. Chem. Soc.* **1977**, *99*, 6262.
10. Knowles, W. S.; Sabacky, M. J.; Vineyard, B. D. In "Homogeneous Catalysis—II", *Adv. Chem. Ser.* **1974**, *132*, 274.
11. Korpiun, O.; Lewis, R. A.; Chickos, J.; Mislow, K. *J. Am. Chem. Soc.* **1968**, *90*, 4842.
12. DeBruin, K. E.; Perrin, D. E. *J. Org. Chem.* **1975**, *40*, 1523.
13. Corey, E. J.; Chaykovsky, M. *J. Am. Chem. Soc.* **1965**, *87*, 1345.
14. Korver, O.; Joberg, S. *Tetrahedron* **1975**, *31*, 2603.
15. Berner, E.; Leonardson, R. *Ann.* **1939**, *538*, 1.
16. Carnmalm, B. *Chem. Ind.* (*London*) **1956**, 1093.
17. Cramer, R. *J. Am. Chem. Soc.* **1964**, *86*, 217.
18. Schrock, R. R.; Osborn, J. A. *J. Am. Chem. Soc.* **1971**, *93*, 2397.
19. Cohen, S. G.; Crossley, J.; Khedouri, E.; Zand, R.; Klee, L. H. *J. Am. Chem. Soc.* **1963**, *85*, 1685.

RECEIVED July 10, 1980.

Asymmetric Catalytic Hydrogenation

B. BOSNICH and NICHOLAS K. ROBERTS

Lash Miller Chemical Laboratories, University of Toronto, Toronto, Ontario, Canada

Four cationic rhodium(I) catalysts incorporating structurally different chiral bidentate ditertiary phosphines are described. All of the catalysts hydrogenate (Z)-N-*acylaminoacrylic acids to give optically active* N-*acylamino acids. The optical yields generated by these four catalysts display a pattern that reflects the chiral and achiral conformations of the chelate rings and support the view that the major source of discriminatory interaction arises from the chiral array of phenyl groups bonded to the phosphorus atoms. Thus a prescription for phosphine design can be enunciated. The details of the mechanism of asymmetric hydrogenation are discussed.*

Wilkinson's (*1*) discovery that the soluble rhodium(I) phosphine complex, $[Rh(PPh_3)_3Cl]$, was capable of homogeneous catalytic hydrogenation of olefins immediately set off efforts at modifying the system for asymmetric synthesis. This was made possible by the parallel development of synthetic methods for obtaining chiral tertiary phosphines by Horner (*2*) and Mislow (*3, 4, 5*). Almost simultaneously, Knowles (*6*) and Horner (*7*) published their results on the reduction of atropic acid (*6*), itaconic acid (*6*), α-ethylstyrene (*7*) and α-methoxystyrene (*7*). Both used chiral methylphenyl-*n*-propylphosphine coordinated to rhodium(I) as the catalyst. The optical yields were modest, ranging from 3 to 15%.

Two major advances soon followed. The first was the discovery that rhodium(I) complexes incorporating bidentate chelating diphosphines also were capable of hydrogenating olefins. The other important development was the discovery that (Z)-*N*-acylaminoacrylic acids (*see* Figure 1), the precursors of amino acids, were hydrogenated at surprisingly high rates in view (*1*) of the fact that they were generally trisubstituted olefins. Moreover, these substrates seem to have characteristics that lead to high optical yields with a minimum of elabora-

0065–2393/82/0196–0337$05.00/0

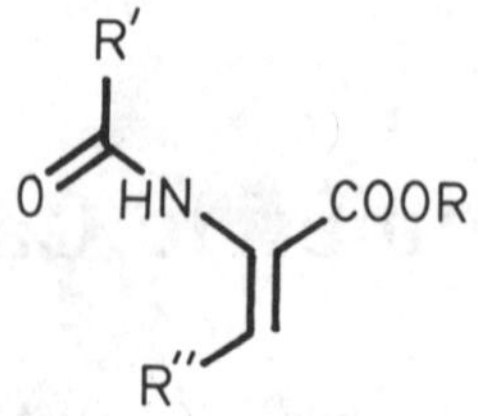

Figure 1. Structure of the amino acid substrates

tion of the phosphine. Thus Knowles (*8*) found that (Z)-*N*-acylaminoacrylic acids were reduced in optical yields up to 90% using the unidentate chiral phosphine, *o*-anisylcyclohexylmethylphosphine. This was surprising because unidentate chiral ligands are generally conformationally nonrigid and hence tend to give only modest discrimination. It is believed (*8*) that, in the case of the *o*-anisylcyclohexylmethylphosphine ligand, weak coordination of the oxygen atom to the rhodium(I) provides a restriction on the rotameric fluxionality of the ligand. Even so, many of the other unidentate phosphines, without the possibility of secondary coordination, gave respectable optical yields with the (Z)-*N*-acylaminoacrylic acids (*8*).

The next generation of catalysts investigated were those that incorporated chiral chelating diphosphines. Kagan (*9*) introduced *diop* and Knowles (*10*) extended his work to the bidentate ligand (*R*,*R*)-1,2-ethanediylbis-[(2-methoxyphenyl)phenylphosphine](1) (DiPAMP) (*see* Figure 2). Both of these ligands can generate cationic complexes during the hydrogenation cycle and have somewhat different mechanisms for reduction (*11*) compared with the original neutral species. Kagan's diop has chirality residing at the carbon atoms and has achiral phosphorus atoms whereas DiPAMP contains only chiral tertiary phosphine centers. Both, however, produce very efficient asymmetric hydrogenation catalysts for producing optically active amino acids. Following these studies numerous chiral chelating phosphines have been developed (*12–24*), all of which have been reasonably successful for asymmetric reduction of (Z)-*N*-acylaminoacrylic acids.

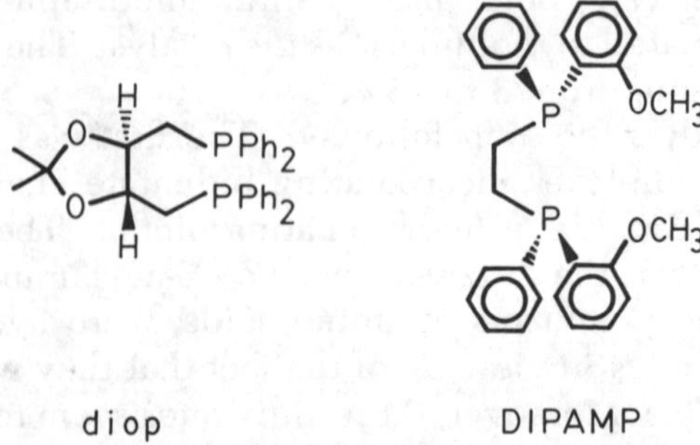

Figure 2. Structures of the diop and DiPAMP ligands

We believe that there is a rational explanation for the success of these chelating diphosphines. The purpose of this chapter is to adumbrate a general hypothesis for the chiral discrimination of these systems. We describe our work on a series of chelating diphosphines which we believe, as a whole, incorporate most of the structural features necessary to support our hypothesis.

Mechanism of Hydrogenation

The mechanism of hydrogenation of olefins by soluble cationic rhodium(I) complexes appears to involve: (a) olefin coordination; (b) hydrogen uptake; (c) addition of one hydrogen atom to the coordinated olefin to give a σ-alkyl-hydrido-rhodium intermediate; and (d) insertion of the remaining hydrido ligand into the σ-alkyl bond to produce the alkane (*11*). This sequence of events is shown in one of the columns of Figure 3; we shall deal with the additional details later in this chapter. Despite the presence of the σ-alkyl intermediate, there appears to be no scrambling of the β-hydrogen atoms of this intermediate as a result of β-elimination (*25, 26, 27, 28*). The hydrogen addition, at least for the substrates we deal with here, is completely stereospecific giving a cis-addition product, the hydrogen atoms being delivered to the olefinic face coordinated to the rhodium atom. Figure 4 shows the consequences of addition followed by β-elimination for Intermediate B using as an example the illustrated olefin substrate.

In order to achieve asymmetric hydrogenation, we need to incorporate a chiral phosphine that is capable of distinguishing between the enantiotopic faces of the prochiral olefin, so that one of these faces is coordinated preferentially to the rhodium atom. (It should be noted that the attachment of a prochiral olefin to a metal atom produces a chiral system irrespective of what else is attached to the metal.) Thus the catalyst solution will contain two diastereomeric species, each having the same phosphine chirality but with opposite chiralities of the metal olefin center. Such an equilibrium is shown as K_1 in Figure 3.

If K_2, k_1, k_{-1}, and k_2 had the same values as K'_2, k'_1, k'_{-1}, and k'_2, then the optical purity of the product, *RH, would be determined solely by the value of the diastereomeric equilibrium constant K_1. If, however, the primed and unprimed constants were different, the final optical yield could be determined by both thermodynamic and kinetic factors, and in one extreme could result in the observation that the preferred enantiomer of the product originated in the minor diastereomer. Clearly, kinetic factors can be important since the steric interactions of the initial two diastereomers are different and these could affect the rate constants of the reaction. Moreover, the σ-alkyl intermediate is chiral, as shown for one of the initial olefin diastereomers in Figure 4, and the rate of hydrogen addition and insertion

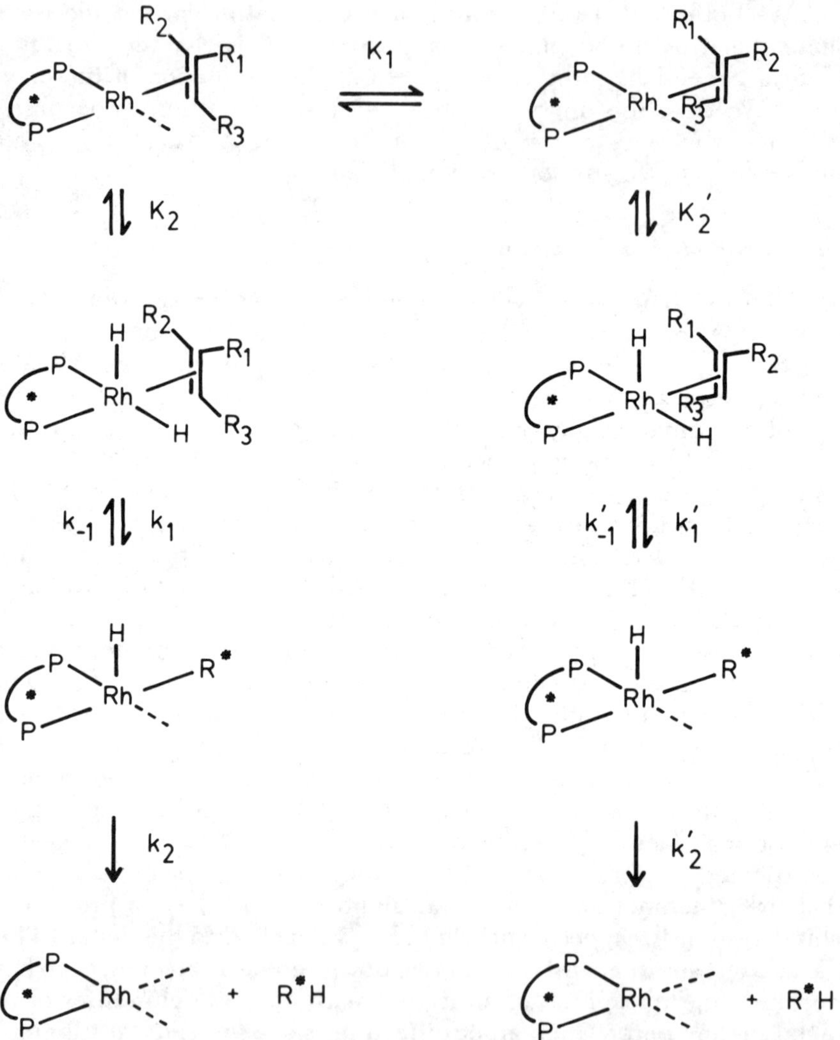

Figure 3. General mechanism for asymmetric catalytic hydrogenation using chelating phosphine rhodium complexes

could be affected by the different interactions of the chiral σ-alkyl intermediates with the chiral phosphine ligand. Thus, in another extreme, K_1 could represent unity and yet significant asymmetric reduction could be obtained because of the different diastereomeric interactions of the intermediates affecting the overall rates of reaction.

It is known, however, that K_1 for (Z)-*N*-acylaminoacrylic acids with a number of chiral chelating diphosphines or rhodium(I) corresponds closely to the values obtained for the optical yields (*29*). This in itself

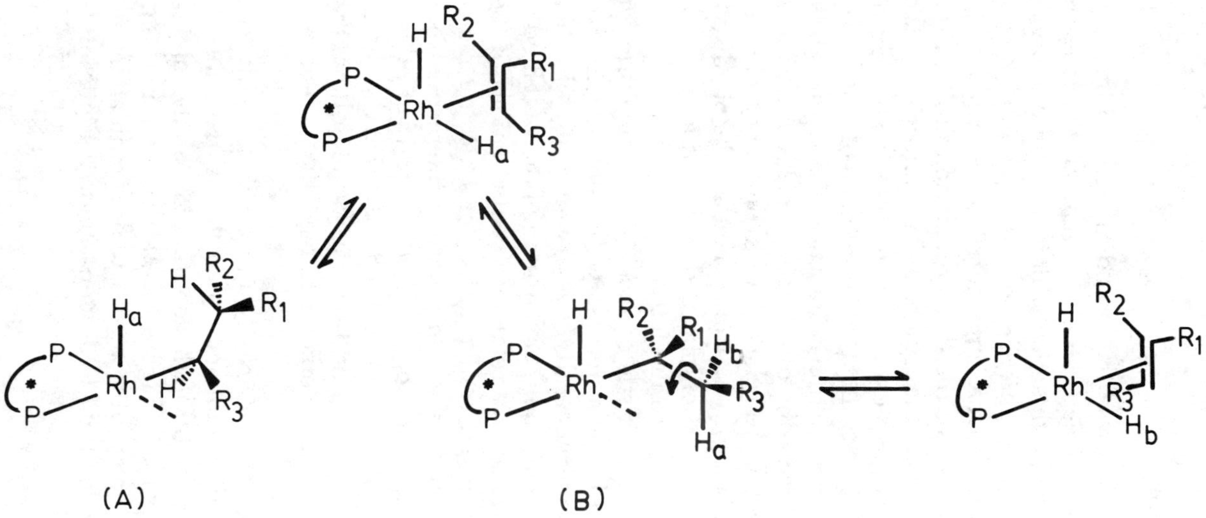

Figure 4. The possible consequences of β-elimination of various σ-alkyl-rhodium-hydrido intermediates

does not mean that the reductions proceed under thermodynamic control but that the chiral phosphines are capable of strong, asymmetric induction. (We shall discuss at the end of this chapter some of the surprising features of this induction.)

Asymmetric Induction

Internal diastereomers of the kind depicted in Figure 3 have two kinds of interaction between the chiral centers: the total interaction and that part of the total interaction which is discriminatory. The latter we call the diastereotopic interaction, and it is only this part of the interaction that causes either K_1 in Figure 3 to be displaced from unity or the activation energies of the diastereomeric transition states to be different. It follows that the object in asymmetric synthesis is to maximize the diastereotopic interaction. In the absence of a clear lock-and-key compatibility for the inducing molecule and the substrate, the stereochemical criteria for maximizing the diastereotopic interaction are not always obvious. For example, it is a commonly held view that an increase in the relative steric bulk of certain groups will increase discrimination. This may only increase the nondiscriminatory interaction unless such elaboration is directed at features of the substrate which control the discrimination.

However one characteristic of successful asymmetric synthesis is clear, that steric rigidity is a necessary (but not sufficient) condition. Labile, flexible systems generally give low optical yields because the net diastereotopic interaction is constituted of the weighted average of all of the conformational and rotameric configurations of the system. The result is that a given configuration (or inducing molecule and prochiral substrate) may induce one sense of interaction, another configuration may induce in the opposite sense, and yet others may not cause any discernible induction. The effect is that the discrimination tends to be washed-out in nonrigid systems (*30*). It follows that rigidity is only the first condition; the other is to have the correct stereochemical configuration.

For the case of the (Z)-*N*-acylaminoacrylic acids attached to rhodium(I) incorporating bidentate phosphines, the substrate is held in a fixed conformation through coordination of the olefin and the acyl-oxygen atom (*31*) (*see* Figure 5). As well as imparting rigidity, chelation increases the stability of substrate coordination and thereby contributes to the velocity of hydrogenation (*see* Figure 3).

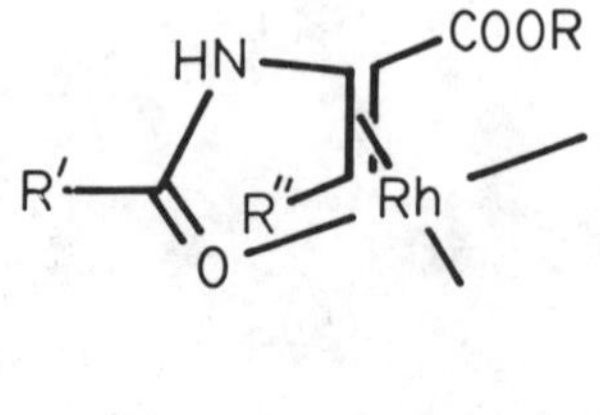

Figure 5. The mode of chelation of the amino acid substrates

Phosphine Design

Saturated bidentate ligands forming five-membered or higher-membered chelate rings adopt puckered conformations (*32, 33, 34*). As in saturated carbocyclic systems, the number of possible conformations increases with ring size although only a few are significantly populated. For the case of five-membered rings, two enantiomeric conformations exist. These rings rapidly interconvert from one enantiomeric conformation to the other (*see* Figure 6), provided no chiral centers exist on the ligand or about the metal center. In the process, the substituents on the donor atoms and on the carbon atoms of the ring interchange their quasiaxial and quasiequatorial dispositions (*see* Figure 6). If one or both of the carbon atoms is monosubstituted, and thereby chiral carbon centers are introduced, the conformational equilibrium depicted in Figure 6 will be displaced by the requirement (*32, 33, 34*) that the substituents be disposed equatorially. Thus the chiral carbon centers will induce the chelate ring to adopt one, preferred chiral conformation and this, in turn, will dispose the substituents on the donor atoms in a preferred chiral array. A prochiral olefin coordinated to a metal atom incorporating this chiral ring will be induced to adopt a preferred enantiomeric configuration by the chiral interactions generated mainly by the chiral array of substituents on the donor atoms.

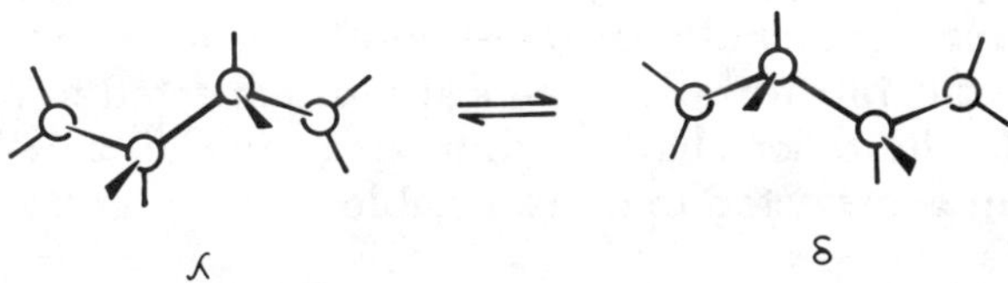

Figure 6. Conformational equilibria of five-membered chelate rings

Figure 7 shows the preferred conformations of *S,S*-chiraphos and *R*-prophos. According to our hypothesis, catalysts formed by these two ligands should give similar optical yields but of opposite chirality. This is because we assert that the chiral array of phenyl groups determines the optical induction and hence whether the ring conformations

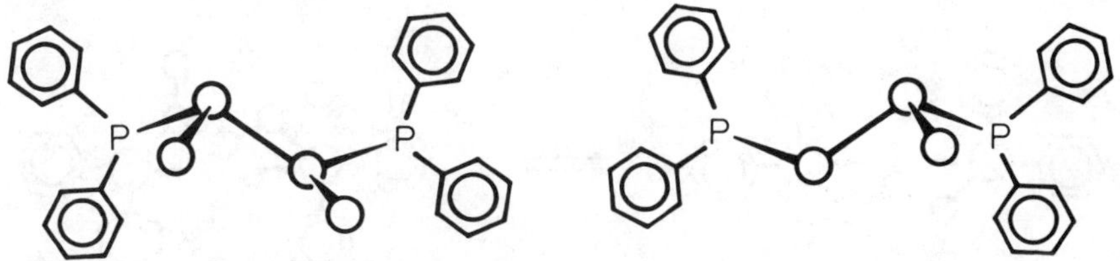

Figure 7. The preferred conformations of S,S-chiraphos and R-prophos chelate rings

are induced by one or two methyl groups is irrelevant. Moreover, the two phosphines stabilize enantiomeric ring conformations and hence should generate opposite optical inductions.

Six-membered chelate ring systems present a similar but somewhat more complicated situation. These rings potentially have many conformational arrangements. Of these, only two are important in the present context—the chiral skew and the achiral chair conformations—and of these two the latter is intrinsically more stable (*33, 34*). Substitution at the 2- and 4-positions of the ring can lead to stabilizing one or the other conformation.

Figure 8 shows some of the conformations of *S,S*-skewphos-(*S,S*-2,4-bis(diphenylphosphino)pentane). The chiral skew rings are shown with the methyl groups in both diaxial and diequatorial dispositions, whereas the chair conformation necessarily has one axial and one equatorial methyl group. Conformational studies (*35, 36*) on the corresponding diamine complexes suggest that the most stable conformation is the skew arrangement with two equatorial methyl groups. This preference, however, may not be overwhelming and stereochemical encumbrances on the metal atom may tilt the stereochemical balance to the chair conformation, even though this configuration possess a destabilizing axial methyl group.

Figure 9 shows some of the conformations of *S*-chairphos(*S*-1,3-bis(diphenylphosphino)butane). A chair and a skew conformation exist that have equatorially disposed methyl groups. The chair conformation that is intrinsically more stable is expected to be preferred (*33, 34, 37*). The other chair conformation with an axially disposed methyl group is expected to be less stable.

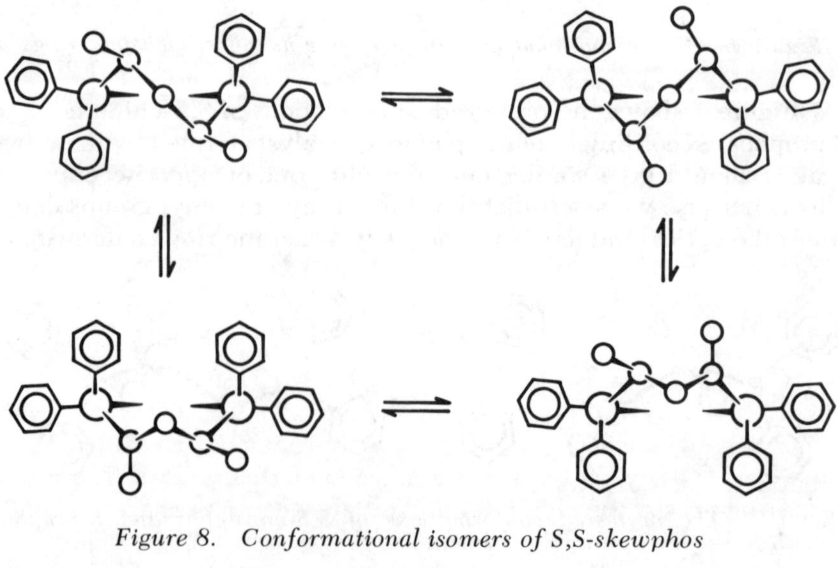

Figure 8. Conformational isomers of S,S-skewphos

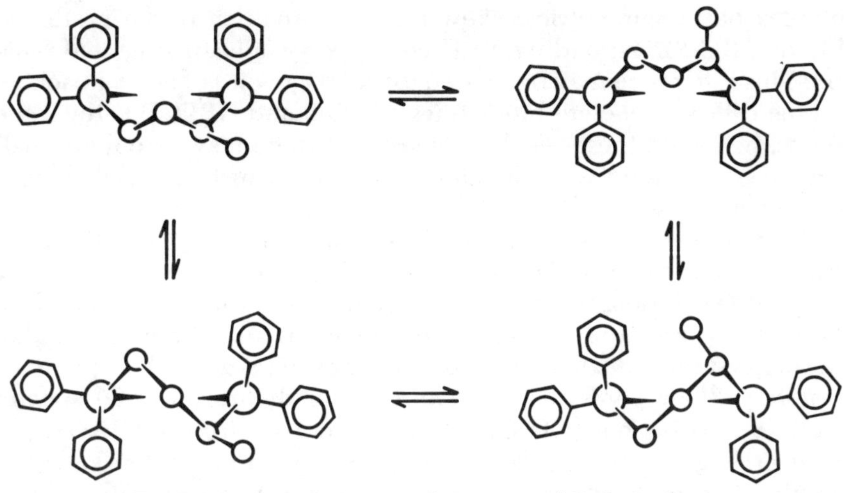

Figure 9. Conformational isomers of S-chairphos

Thus, if we assume that *S,S*-skewphos prefers the chiral skew conformation and that *S*-chairphos prefers the achiral chair conformation, we can predict the optical yields generated by catalyst formed by these two ligands. The skew conformation disposes the phenyl groups in a chiral array similar to that described for the five-membered ring systems, although there are differences in the orientations. The six-membered skew ring fixes the phenyl groups in more clearly defined axial and equatorial positions than for the five-membered rings, and the inclinations of the phenyl groups relative to the mean molecular plane of the complex are somewhat different in the two cases. Generally, however, we would expect that *S,S*-skewphos catalysts will give similar optical yields to those observed for chiraphos and prophos. The *S*-chairphos system in the chair conformation, on the other hand, is intrinsically achiral as far as the phenyl group orientations are concerned (*see* Figure 9). We thus expect a catalyst incorporating *S*-chairphos, although optically active, to give negligible optical yields.

These four ligands, therefore, provide an incisive test for our hypothesis concerning the origins of the optical induction.

Ligand Structures

Crystal structures of [Rh(*S,S*-chiraphos)COD]ClO_4 (*38*), [Rh(*R*-prophos)NBD]ClO_4 (*39*), [Rh(*S,S*-skewphos)NBD]ClO_4 (*39*), and [Rh(*S,S*-skewphos)COD]ClO_4 (*39*), (where COD is cycloocta-1,5-diene and NBD is norbornadiene) have been determined. The molecular structures of the chiraphos and prophos complexes are essentially those that are depicted in Figure 7 but, whereas the skewphos-COD complex has a symmetrical skew ring conformation in the predicted chirality, the corresponding NBD complex has a chair ring conformation. This difference may be related to the exigencies of crystal packing and the different chelate bite angles of COD and NBD. Whatever the origins, however, these results suggest that there may be only a small difference in energy for the skew and chair rings of *S,S*-skewphos complexes.

The phosphorus NMR of [Rh(*S,S*-skewphos)NBD]$^+$ in CD_2Cl_2 solution gives a two-line spectrum at 30°C and down to the freezing point of the solution, indicating that, on an NMR time scale, the phosphorus atoms are equivalent. Two explanations arre admissible: either the ring system is fluxional in this temperature range or the ring exists in a static skew conformation. The circular dichroism (CD) spectra of [Rh(*S,S*-skewphos)(CH_3OH_2)]$^+$ and [Rh(*S*-chairphos)($CH_3OH)_2$]$^+$, however, suggest that the preponderant conformation of the skewphos complex is skew (*40*). (A discussion of this will appear later in this chapter.)

A tentative inference may be drawn from these results; the skewphos complex is predominantly in the skew conformation in solution, but minor stereochemical encumbrances may tip the ring into the chair conformation.

Optical Yields

All hydrogenations were carried out under ambient conditions, and the general methods are described elsewhere (*25*). Chemical yields are quantitative in all cases. The results are listed in Tables I, II, III, and IV. The optical yields speak for themselves, but we emphasize a number of features. The *S,S*-chiraphos and *R*-prophos catalysts give similar optical yields, but they deliver the opposite chirality. The *S,S*-skewphos catalyst gives similar optical yields to those for the five-membered ring analogs, but the *S*-chairphos catalyst gives very low optical yields that vary in enantiomer with changes in solvent and substrate. All of these observations were predicted except for the low optical yields obtained with the skewphos catalyst for the leucine substrates.

The fact that the *S,S*-skewphos and *S*-chairphos catalysts give similar optical yields for the leucine substrates might suggest that the chelate ring is chair in both cases during the hydrogenation of the substrate. This does not appear to be the case at least for the initial substrate-binding step. The CD spectra of [Rh(*S,S*-skewphos)-($CH_3OH)_2$]$^+$ and [Rh(*S*-chairphos)($CH_3OH)_2$]$^+$ are similar in form but, as expected (*34, 35, 37*), the former spectrum is about five times larger than the latter. An identical relationship holds for the CD spectra of

Table I. (*S*,*S*)-Chiraphos

Amino Acid	*Substrate*	*Optical Yield (%)* THF	*Optical Yield (%)* EtOH
Alanine	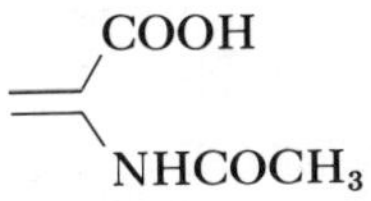 COOH, $NHCOCH_3$	88	92
Phenylalanine	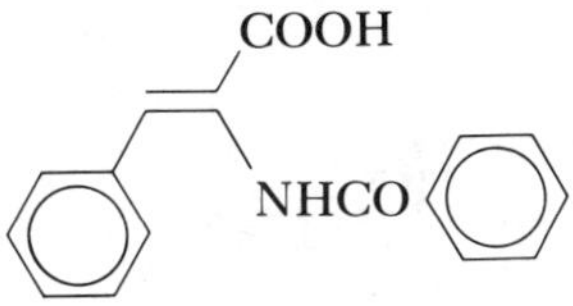 COOH, NHCO	99	95
	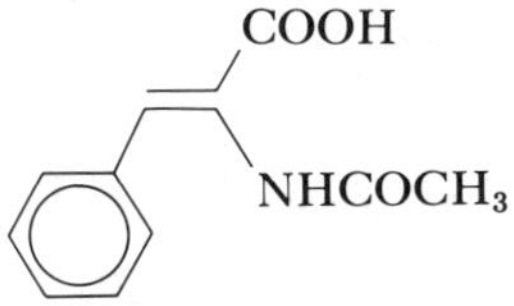 COOH, $NHCOCH_3$	74	89
	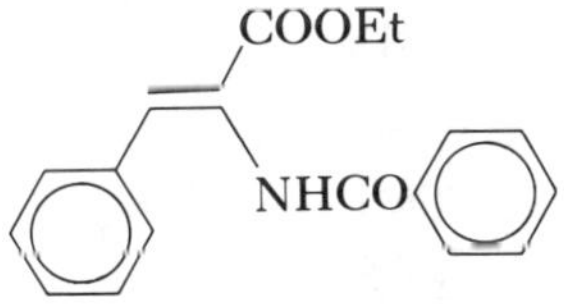 COOEt, NHCO	83	—
Leucine	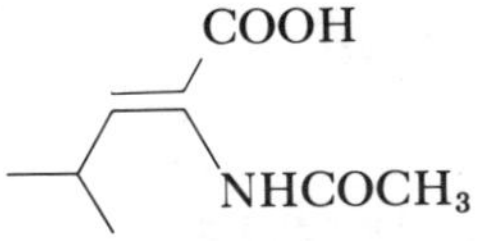 COOH, $NHCOCH_3$	100	93
	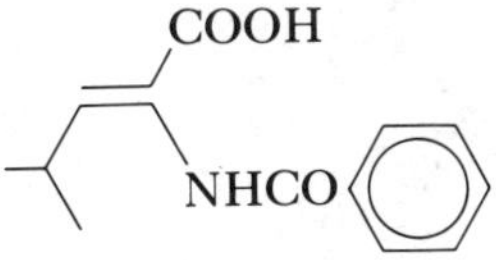 COOH, NHCO	87	72
Tyrosine	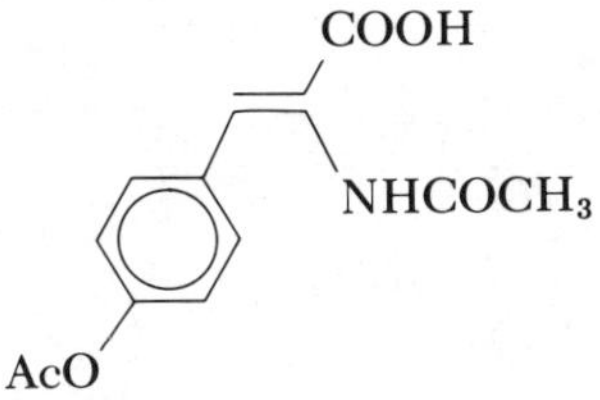 COOH, $NHCOCH_3$, AcO	74	88
DOPA	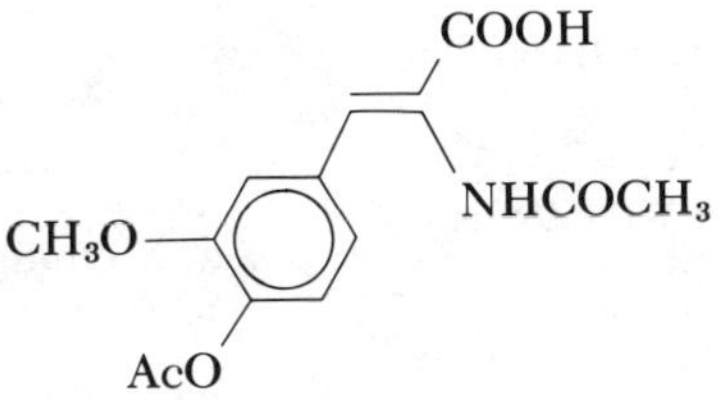 COOH, $NHCOCH_3$, CH_3O, AcO	80	83
		(all *R* configurations)	

Table II. (*R*)-Prophos

Amino Acid	*Substrate*	*Optical Yield (%)*	
		THF	*EtOH*
Alanine	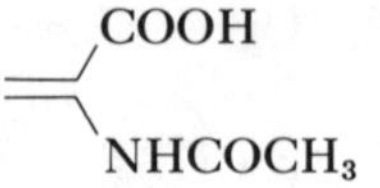 COOH, $NHCOCH_3$	87	90
Phenylalanine	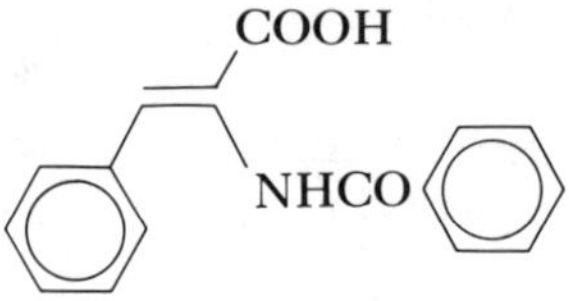 COOH, NHCO	93	91
	COOH, $NHCOCH_3$	91	90
	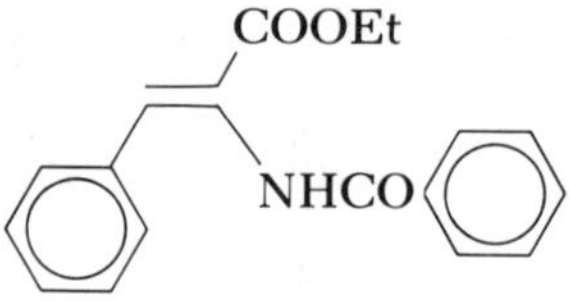 COOEt, NHCO	88	87
Leucine	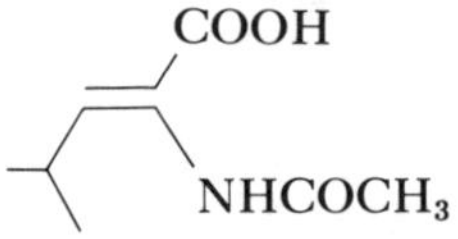 COOH, $NHCOCH_3$	87	87
	COOH, NHCO	—	—
Tyrosine	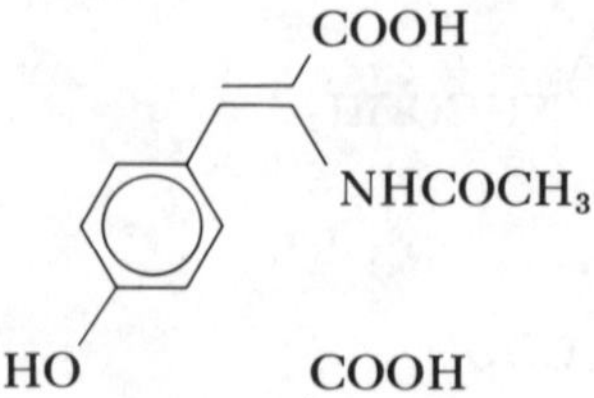 COOH, $NHCOCH_3$, HO	92	89
DOPA	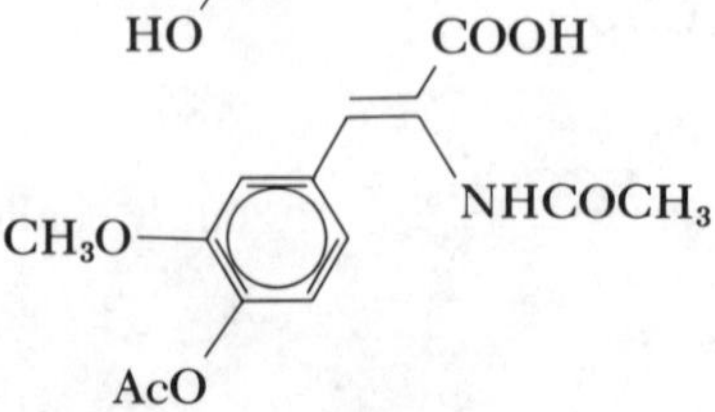 COOH, $NHCOCH_3$, CH_3O, AcO	89	87

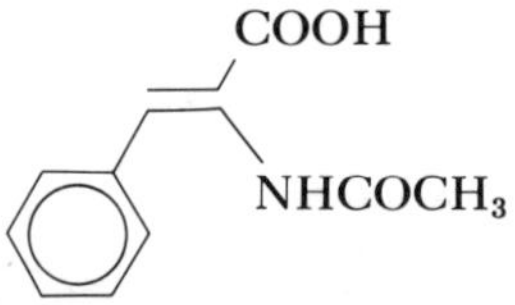

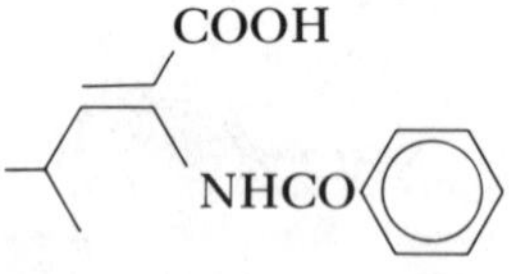

(all *S* configurations)

Table III. (*S*,*S*)-Skewphos

Amino Acid	*Substrate*	*Optical Yield (%)* THF	EtOH (95%)	MeOH
Alanine	$CH_2=C(COOH)NHCOCH_3$	95 (*R*)	98 (*R*)	96 (*R*)
Phenylalanine	$C_6H_5CH=C(COOH)NHCOC_6H_5$	90 (*R*)	80 (*R*)	
	$C_6H_5CH=C(COOH)NHCOCH_3$	93 (*R*)	92 (*R*)	
	$C_6H_5CH=C(COOEt)NHCOC_6H_5$	76 (*R*)	60 (*R*)	
Leucine	$(CH_3)_2CHCH=C(COOH)NHCOCH_3$	17 (*R*)	23 (*R*)	
	$(CH_3)_2CHCH=C(COOH)NHCOC_6H_5$	8 (*S*)	14 (*S*)	23 (*S*)
Tyrosine	p-$HOC_6H_4CH=C(COOH)NHCOCH_3$	90 (*R*)	84 (*R*)	
DOPA	3-CH_3O-4-AcO-$C_6H_3CH=C(COOH)NHCOCH_3$	93 (*R*)	92 (*R*)	

Table IV. (*S*)-Chiraphos

Amino Acid	*Substrate*	*Optical Yield (%)* THF	EtOH (95%)	MeOH
Alanine	COOH, $NHCOCH_3$	9 (*R*)	5 (*R*)	4 (*R*)
Phenylalanine	COOH, NHCO(Ph), Ph	12 (*R*)	10 (*S*)	
	COOH, $NHCOCH_3$, Ph	1 (*S*)	20 (*S*)	
	COOEt, NHCO(Ph), Ph	4 (*R*)	10 (*S*)	
Leucine	COOH, $NHCOCH_3$	5 (*S*)	3 (*S*)	
	COOH, NHCO(Ph)	11 (*S*)	16 (*S*)	24 (*S*)
Tyrosine	COOH, AcO	5 (*R*)	12 (*S*)	
DOPA	COOH, $NHCOCH_3$, CH_3O, AcO	1 (*R*)	9 (*S*)	

the corresponding complexes formed with the leucine substrates (*40*) and suggests that no major conformational change has occurred in replacing the methanol ligands for the bidentate leucine substrate. This, however, does not exclude the possibility that the dihydrido intermediates have chair rings. (*See* discussion on this later in this chapter.)

Discussion

Two recent studies (*41, 42*) have defined the mechanism of asymmetric catalytic hydrogenation with considerable clarity. Figure 10 shows an outline of the mechanism. The diastereomer equilibrium K_d gives a major and minor isomer depending on which (prochiral) olefin face is coordinated to the rhodium atom. In the case of chiraphos, for example, K_d is large enough that the minor isomer cannot be detected by NMR (*43*). What is remarkable, however, is that it is the minor isomer and not the major one that controls the asymmetric synthesis (*42, 43*). It now seems certain that the difference in velocity of hydrogenation of the two diastereomers is such that nearly all of the product is derived from the minor isomer. The difference in velocity is of the order of 10^3 for chiraphos (*43*) and the evidence suggests that other phosphine ligands behave in the same way. The origins of this velocity difference ultimately rest on the stereochemistry of the intermediates of reaction. It cannot be otherwise because asymmetric synthesis is essentially a stereochemical phenomenon that controls kinetics.

Low-temperature hydrogenation (*41, 42*) reveals that the rate-determining step is k'_3 below $-40°C$. Above $-40°C$, k'_1 is believed to be the rate-determining step. The σ-alkyl intermediate, as depicted in Figure 10, has been detected and its structure has been deduced by NMR. What seems to be established is that the acyl-oxygen atom and the α-carbon atom are bound to the rhodium atom; the participation of the carboxyl-oxygen atom in coordination is in dispute (*41, 42, 43*), although this may depend on whether an acid (*42*) or an ester (*41*) is involved. The dihydrido intermediate has not been detected, but clearly the origins of the asymmetric synthesis are described in an explanation for the different rates of dihydrido formation (k_1 vs. k'_1), and possibly to a lesser extent by the probably different rates of formation of the σ-alkyl intermediates (k_2 and k'_2).

Considering the k_1 vs. k'_1 step first, the two square planar diastereomers in principle can generate a total of eight chiral cis-dihydrido-bis(bidentate) octahedral complexes, four for each square planar diastereomer (*see* Figure 10). None of these octahedral complexes will be of the same energy because of the differing interligand interactions (*32, 44*). This is a well-known phenomenon in inorganic chemistry (*44*) where chiral chelate ligands stabilize a preferred chiral-

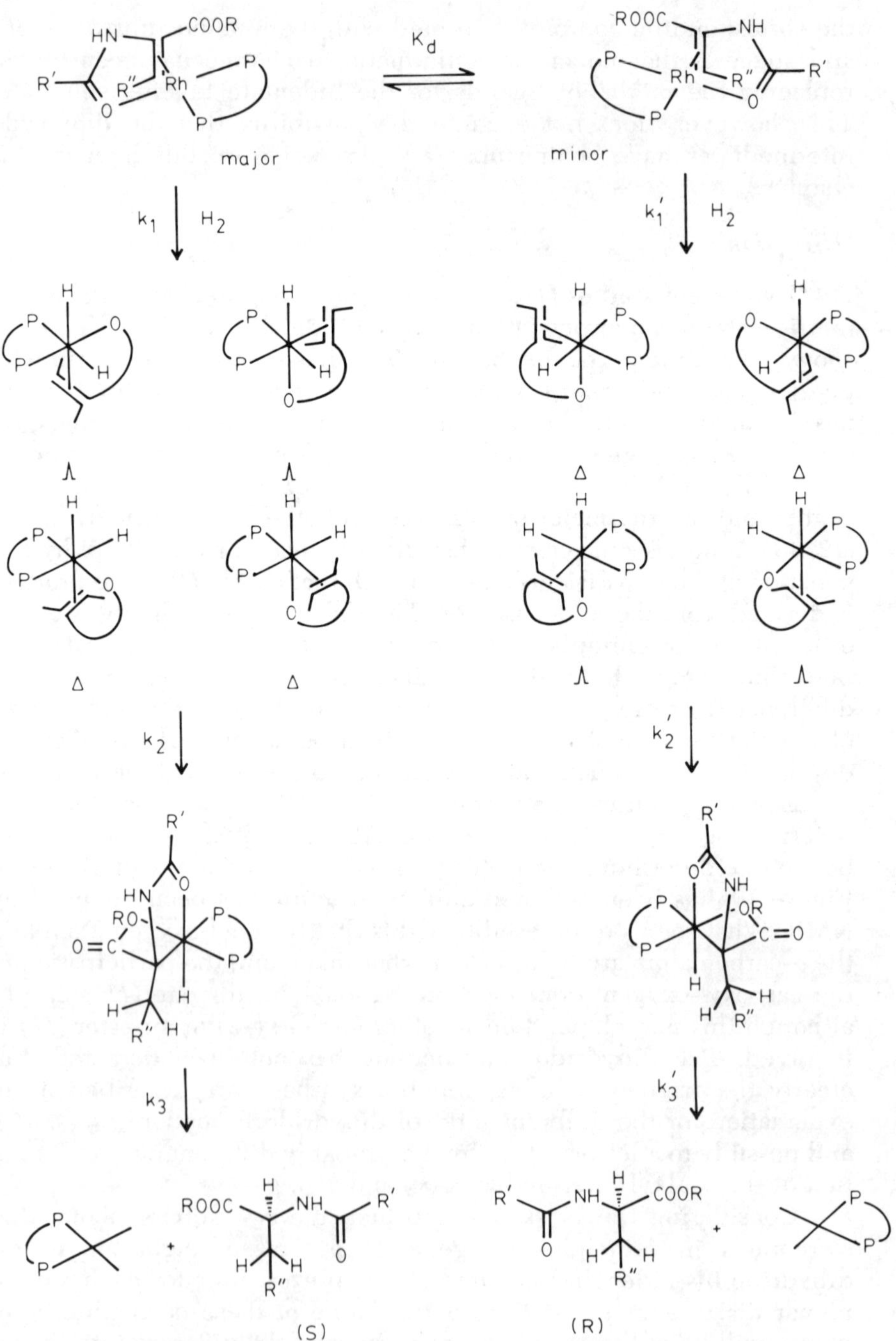

Figure 10. The mechanism of catalytic asymmetric hydrogenation of the amino acid substrates. This diagram depicts the structures of the intermediates. The absolute configurations refer to S,S-chiraphos.

ity about the metal (designated Δ or Λ). Thus the rate of asymmetric hydrogenation is governed by the most stable octahedral intermediate; the greater the stability of this species the greater the probability that hydride transfer will occur. Molecular models of these putative octahedral intermediates are not conclusive to us in deciding which configurational isomer should be the most stable, but what is clear is that the chiral array of phenyl groups is the major source of chiral discrimination. The bidentate olefinic substrate experiences the greatest amount of interaction from the phenyl groups. We may obtain a speculative explanation for the skewphos-leucine result if we assume that the phosphine ring is in a chair conformation in the dihydrido intermediate.

In addition to the steric differentiation of the constants k_1 and k'_1, the velocity of the hydride transfer steps (k_2 vs. k'_2) could also be different because the development of the diastereomeric transition states will require different energies. The contribution to the reaction velocity of this effect is difficult to assess.

If this is a general mechanism for asymmetric hydrogenation, then the question of designing effective ligands is addressed to the discrimination induced at the dihydrido intermediate. Reasonably reliable methods for assessing the stabilities have been documented (*32, 33, 34, 44*). The observation that rigid bidentate phosphines and rigid bidentate substrates have given the highest optical yields now is understood readily since it is the chiral chelating ligands, as opposed to chiral unidentate ligands, that impose the greater stereospecificity in octahedral complexes (*32, 33, 34, 44*).

Acknowledgments

This work was supported by the Natural Sciences and Engineering Research Council of Canada. Dr. Bosnich is pleased to acknowledge the Killam Foundation for awarding him a Killam Fellowship.

Literature Cited

1. Osborn, J. A.; Jardine, F. H.; Young, J. F.; Wilkinson, G. *J. Chem. Soc. A* 1966, 1711.
2. Horner, L.; Winkler, H.; Rapp, A.; Mentrup, A.; Hoffmann, H.; Beck, P. *Tetrahedron Lett.* 1961, 161.
3. Korpiun, O.; Lewis, R. A.; Chickos, J.; Mislow, K. *J. Am. Chem. Soc.* 1968, *90*, 4842.
4. Naumann, K.; Zon. G.; Mislow, K. *J. Am. Chem. Soc.* 1969, *91*, 7012.
5. Farnham, W. B.; Murray, R. K.; Mislow, K. *J. Am. Chem. Soc.* 1970, *92*, 5809.
6. Knowles, W. S.; Sabacky, M. J. *Chem. Commun.* 1968, 1445.
7. Horner, L.; Siegel, H.; Büthe, H. *Angew. Chem. Int. Edit. Engl.* 1968, *7*, 942.
8. Knowles, W. S.; Sabacky, M. J.; Vineyard, B. D. *Chem. Commun.* 1972, 10.
9. Dang, T. P.; Kagan, H. B. *J. Am. Chem. Soc.* 1972, *94*, 6429.

10. Knowles, W. S.; Vineyard, B. D.; Sabacky, M. J.; Bachman, G. L.; Weinkauff, D. J. *J. Am. Chem. Soc.* **1977**, *99*, 5946.
11. Halpern, J.; Riley, D. P.; Chan, A.C.S.; Pluth, J. P. *J. Am. Chem. Soc.* **1977**, *99*, 8055.
12. Cullen, W. R.; Yeh, E. S. *J. Organomet. Chem.* **1977**, *139*, C13.
13. Yamamoto, K.; Tomita, A.; Tsuji, J. *Chem. Lett.* **1978**, 3.
14. Achiwa, K. *J. Am. Chem. Soc.* **1976**, *98*, 8265.
15. Beck, W.; Menzel, H. *J. Organomet. Chem.* **1977**, *133*, 307.
16. Pracejus, G.; Pracejus, H. *Tetrahedron Lett.* **1977**, 3497.
17. Tanaka, M.; Ogata, I. *Chem. Commun.* **1975**, 735.
18. Hayashi, T.; Tanaka, M.; Ogata, I. *Tetrahedron Lett.* **1977**, 295.
19. Cullen, W. R.; Sugi, Y. *Tetrahedron Lett.* **1978**, 1635.
20. Tanao, K.; Yamamoto, H.; Matsumoto, H.; Miyake, N.; Hayashi, T.; Kumada, M. *Tetrahedron Lett.* **1977**, 1389.
21. Grubbs, R. H.; DeBries, R. A. *Tetrahedron Lett.* **1977**, 1879.
22. Hayashi, T.; Mise, T.; Mitachi, S.; Yamamoto, K.; Kumada, M. *Tetrahedron Lett.* **1976**, 1133.
23. Lauer, M.; Samuel, O.; Kagan, H. B. *J. Organomet. Chem.* **1979**, *177*, 309.
24. Brumer, H.; Pieronczyk, W. *Angew. Chem. Int. Ed. Engl.* **1979**, *18*, 620.
25. Fryzuk, M. D.; Bosnich, B. *J. Am. Chem. Soc.* **1977**, *99*, 6262.
26. Detellier, C.; Gelbard, G.; Kagan, H. B. *J. Am. Chem. Soc.* **1978**, *100*, 7556.
27. Koenig, K. E.; Knowles, W. S. *J. Am. Chem. Soc.* **1978**, *100*, 7561.
28. Fryzuk, M. D.; Bosnish, B. *J. Am. Chem. Soc.* **1979**, *101*, 3043.
29. Brown, M.; Chaloner, P. A. *J. Am. Chem. Soc.* **1978**, *100*, 4321.
30. Boucher, H. A.; Bosnich, B. *J. Am. Chem. Soc.* **1977**, *99*, 6253.
31. Chan, A. C. S.; Pluth, J. J.; Halpern, J. *Inorg. Chim. Acta* **1979**, *37*, L477.
32. Corey, E. J.; Bailar, J. C. *J. Am. Chem. Soc.* **1959**, *81*, 2620.
33. DeHayes, L. J.; Busch, D. H. *Inorg. Chem.* **1973**, *12*, 1505.
34. Nicketic, S. R.; Woldbye, F. *Acta Chem. Scand.* **1973**, *27*, 621.
35. Boucher, H.; Bosnich, B. *Inorg. Chem.* **1976**, *15*, 1471.
36. Kojima, M.; Fujita, M.; Fujita, J. *Bull. Chem. Soc. Jpn.* **1977**, *50*, 898.
37. Kojima, M.; Fujita, J. *Bull. Chem. Soc. Jpn.* **1977**, *50*, 3237.
38. Ball, R. G.; Payne, N. C. *Inorg. Chem.* **1977**, *16*, 1187.
39. Payne, N. C., personal communication.
40. Roberts, N. K.; Bosnich, B., unpublished results.
41. Chan, A. C. S.; Halpern, J. *J. Am. Chem. Soc.* **1980**, *102*, 838.
42. Brown, J. M.; Chaloner, P. A. *Chem. Commun.* **1980**, 344.
43. Halpern, J., personal communication.
44. Hawkins, C. J. "Absolute Configuration of Metal Complexes"; Cotton, F. A., Wilkinson, G., Eds.; Wiley–Interscience: New York, 1971; Chap. 3.
45. Sargeson, A. "Transition Metal Chemistry"; Carlin, R. L., Ed.; Marcel Dekker, Inc.: New York, 1966; Vol. 3, p. 303.

RECEIVED July 10, 1980.

Mechanistic Studies of Rhodium-Catalyzed Asymmetric Homogeneous Hydrogenation

JOHN M. BROWN, PENNY A. CHALONER, and DAVID PARKER
Dyson Perrins Laboratory, South Parks Road, Oxford OX1 3QY U.K.

The reaction pathway for rhodium-catalyzed asymmetric hydrogenation of enamides is described and intermediates are defined in solution by P-31, C-13, and H-1 NMR. The stereochemical relationship of bound enamide to rhodium alkyl and to the product of hydrogenation is demonstrated. Experiments involving the addition of HD to a variety of olefins in the presence of rhodium biphosphine catalysts suggest that a concerted addition of hydrogen to olefin and metal may occur in appropriate cases.

Since the discovery (*1*) of highly efficient catalysts for the asymmetric hydrogenation of dehydroamino acids, many groups have attempted to improve the range and selectivity of this reaction. Thus chiral biphosphines forming five- or seven-membered chelate rings have been widely used, with the site of asymmetry being phosphorus or more commonly one or more sites in the interphosphine chain. Hydrogenation normally is carried out close to atmospheric pressure in a polar solvent with the catalyst present as a cationic biphosphine rhodium complex most commonly derived from 1,5-cyclooctadiene or bicyclo[2,2,1]heptadiene.

Most published work in the last eight years has been concerned with synthesizing different types of chiral biphosphine and applying their complexes to reducing dehydroamino acids and related substrates. This has led inevitably to less emphasis on the detailed pathway of reaction, although a precise understanding of mechanism is vital to perfecting and extending asymmetric hydrogenation. In a general sense, asymmetric induction in organic reactions is understood

0065-2393/82/0196-0355$5.00/0

rather poorly (2) and relies heavily on empirical models of transition-state structure. It may be easier to understand asymmetric induction in organometallic catalysis because catalyzed reactions proceed through a number of stages linked by relatively low energy barriers and some of the intermediates are observable or even isolatable.

Three types of experiments have proved informative in the mechanistic study of asymmetric hydrogenation. These are, respectively, rate measurements and product analysis, X-ray crystallography and NMR-derived identification of stable and transient species involved in the catalytic cycle. The first two of these have been reviewed elsewhere (*2*, *3*, *4*); our own work has been concerned with NMR and has provided a surprising wealth of structural and mechanistic detail. A variety of chiral phosphine procatalysts have been used but the current discussion will be concerned largely with two. Thus (*R*,*R*)-1,2-ethanediylbis-[(2-methoxyphenyl)phenylphosphine] (1) (DiPAMP) (**1**) has been used by the Monsanto group for synthesizing L-Dopa (*4*) and forms a rigid five-ring rhodium chelate, whereas (*R*,*R*)-2,2-dimethyl-*trans*-4,5-(diphenylphosphinomethyl)dioxolane (DIOP) (**2**) forms a seven-ring chelate with a correspondingly greater degree of conformational flexibility.

The Catalytic Cycle: Stage One

The crystal structures of air-stable olefin Complexes 3 and **4** have been determined (3) and give an immediate insight into the factors that might control stereoselection. In both cases the *P*-phenyl rings are oriented in a specific conformation so that their nonbonded interactions with diolefin are different, pairs of aryl groups being related by a C_2 symmetry axis. Olefinic protons H_a and H_b are diastereotopic and will experience different ring-current contributions from *P*-aryl rings to their H-1 NMR chemical shifts. The expected separation is observed readily and is temperature dependent, as indicated in Figure 1. The proximity of olefinic protons to aryl C–H is demonstrated by a nuclear Overhauser experiment on Complex **5**. When the olefinic region is irradiated and the aromatic region is monitored, significant changes in intensity occur as indicated. Without a stereospecific labelling of bicyclo[2,2,1]heptadiene, these results cannot be interpreted in detail but they strongly suggest that intracomplex H–H repulsions help to determine the course of asymmetric catalysis.

When Complex **4** is dissolved in methanol the P-31 NMR spectrum of the resulting solution is a sharp doublet: $\delta = 17.1$ ppm and $J_{P-Rh} = 153$ Hz. Exposure of the solution to hydrogen leads to a quantitative change concomitant with hydrogenating the ligand. The new resonance at $\delta = 43.0$ ppm, and J_{P-Rh} 200 Hz is indicative of a solvent adduct 6 and no trace of the corresponding dihydride 7 is observed

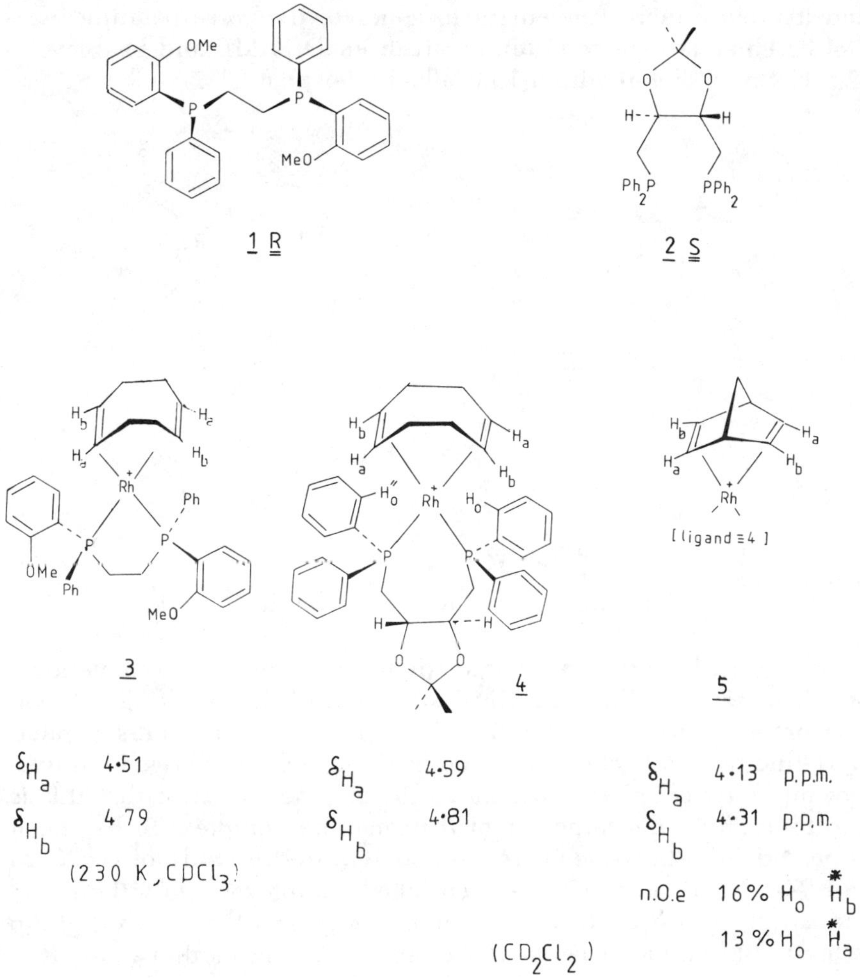

Figure 1. Proton NMR studies on olefin rhodium Complexes **1–5**. *The NOE effect is indicated in square brackets.*

here (5) or indeed with any other cis-chelating biphosphine. Dihydrides (8) are the normal stable state of the corresponding monophosphine complexes under 1 atm of hydrogen (6), although exceptions have been noted. The formation of Complex **6** rather than of Complex **7** is dictated by the destabilizing effect of mutually trans hydride and phosphine ligands; its lability is demonstrated by adding bicyclo[2,2,1]heptadiene when Complex 4 is reformed quantitatively. This immediately suggested that dehydroamino acid reduction could proceed by a similar ligand displacement pathway on Complex **6**.

Hydrogenation of Complex **3** under similar conditions is very slow and it proved more convenient to generate the corresponding bicyclo[2,2,1]heptadiene complex in situ from DiPAMP and bis-bicyclo-[2,2,1]heptadiene–rhodium[I]hexafluorophosphate.

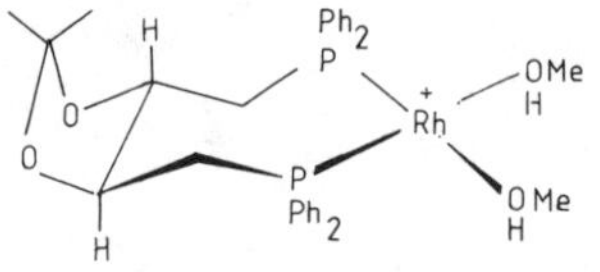

6

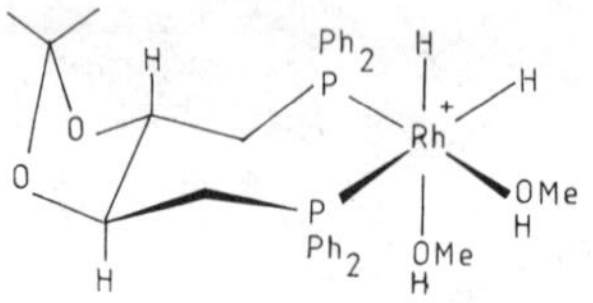

7

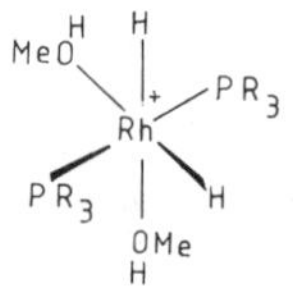

8

The Catalytic Cycle: Stage Two

Solutions of Complex **6** were stable for long periods of time in an argon atmosphere. Adding a solution of (Z)-α-benzamidocinnamic acid produced an immediate color change from yellow–brown to scarlet. The P-31 NMR spectrum of the resulting solution was broad at room temperature but gave a sharp, well-separated eight-line multiplet at 0°C, derived from a species with two inequivalent mutually coupled phosphorus nuclei. The evidence that this is due to a single diastereomer of the enamide complex (**9**) has been discussed in detail elsewhere (7) and required a study of the C-13 and P-31 NMR spectra of C-13-enriched complexes. Only the olefin and amide groups contribute to complexation and the carboxyl group is not metal bound here or in the corresponding methyl ester. It is interesting that *N*-vinylacetamide does not complex strongly other than at very low temperatures; additionally it is an inferior substrate, giving an optical yield of 24% when adding deuterium catalyzed by DIOP–rhodium complexes.

Since only one diastereomer of Complex **9** is observed, the NMR experiment does not yield any stereochemical information and it is presumptuous to assume that this is the species that participates in the catalytic cycle. Both diastereomers are observed in other chiral-chelating phosphine complexes and DiPAMP is particularly interesting. Here, under similar conditions, two diastereomers (Complexes **10a** and **10b**) are observed in a 10:1 ratio at room temperature in a spectrum showing no sign of exchange broadening, although Complex **10a** predominates very strongly at lower temperatures.

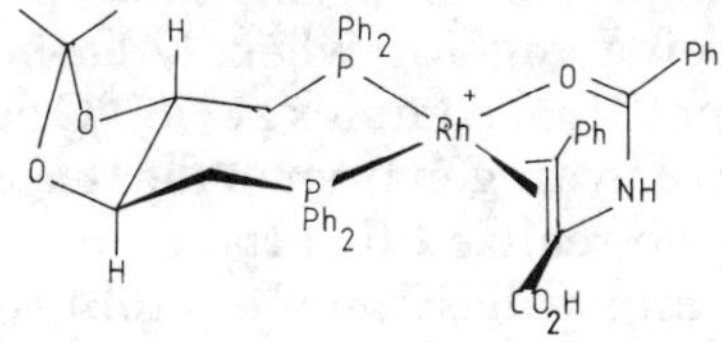

9

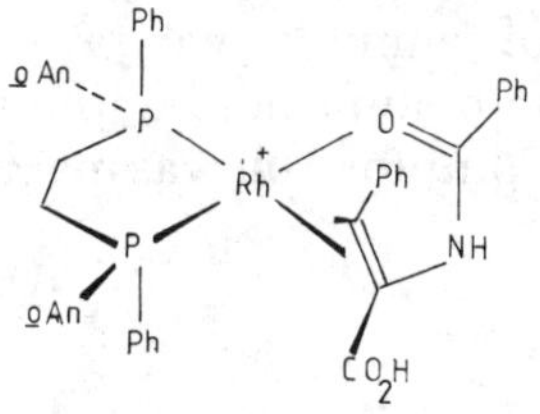

10a

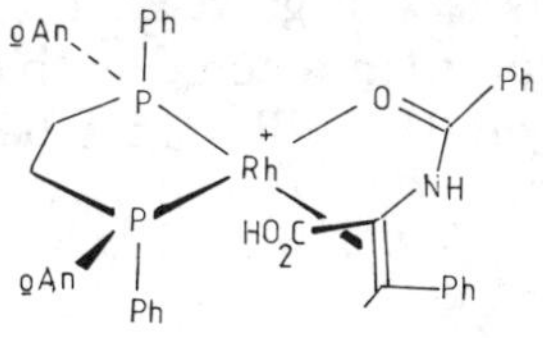

10b

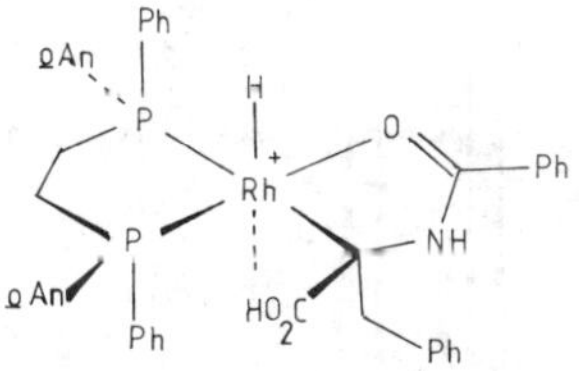

11

Both of these diastereomers are bound through olefin and amide groups on the basis of C-13 NMR studies (8). At low temperatures it is possible to prepare a nonequilibrated solution in which the ratio of **10a** to **10b** is about 2 : 1 because kinetic selectivity is considerably lower than thermodynamic selectivity. It may be that the first step in enamide binding is coordinating the amide group which is remote from the chiral center. In support of this we observe that exchange broadening is strongly dependent on the nature of the amide group and much less dependent on the acid or ester residue (9).

The Catalytic Cycle: Stage Three

Hydrogenating enamide (*see* Complex **10**) in methanol solution causes an extremely rapid reversion to the methanol complex with no observable intermediates. If the sample is cooled to −80°C before hydrogen is introduced and then agitated at −50°C for several minutes the P-31 NMR spectrum shows Complex **10a** and about 10% of a new species with similar chemical shifts but very different P–P and P–Rh coupling constants. The proportion of this new species was erratic in earlier experiments, being highest when (Z)-α-benzamidocinnamic acid was added to the methanol complex at −80°C and the sample then was sealed under 1 atm of hydrogen prior to the reaction at −50°C. With retrospective wisdom, we realized that these are precisely the conditions under which the minor diastereomer (**10b**) is formed at highest concentration and the absence of this species from samples equilibrated with hydrogen at low temperatures suggests that it is the precursor of the new complex. This supposition was confirmed by two experiments (*see* Figure 2). First, the methanol complex was prepared in situ and then reacted with (Z)-benzamidocinnamic acid with monitoring by P-31 NMR so that the proportion of Complex **10b** was maximized.

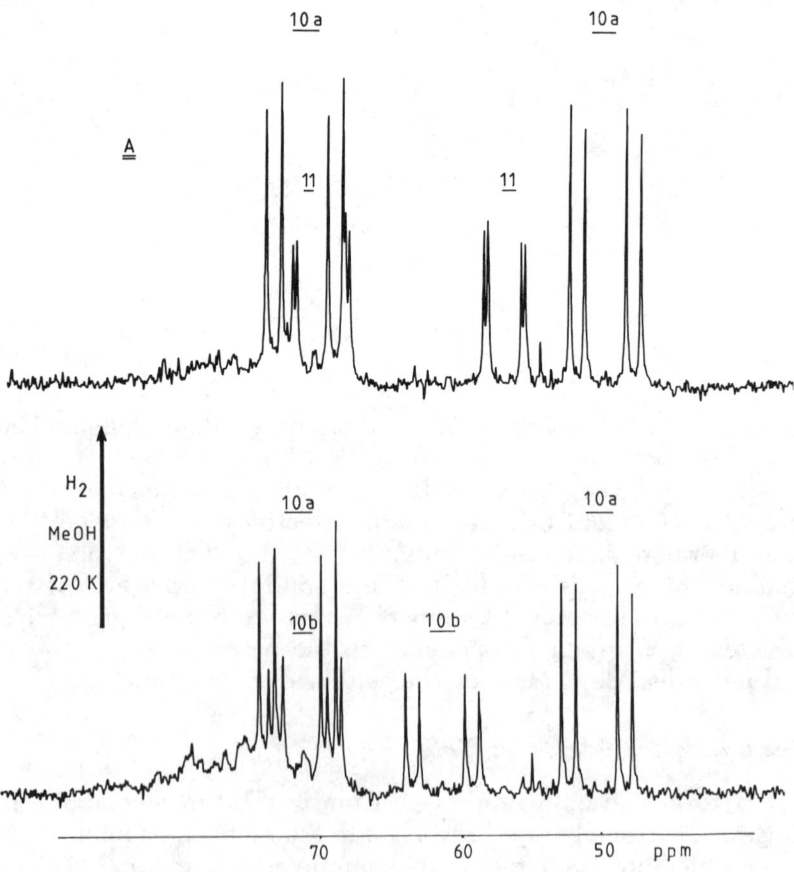

*Figure 2a. P-31 NMR spectra demonstrating formation of the alkyl (**11**) from the minor diastereomer (Complex **10b**) at 220 K: sample formed under kinetic control.*

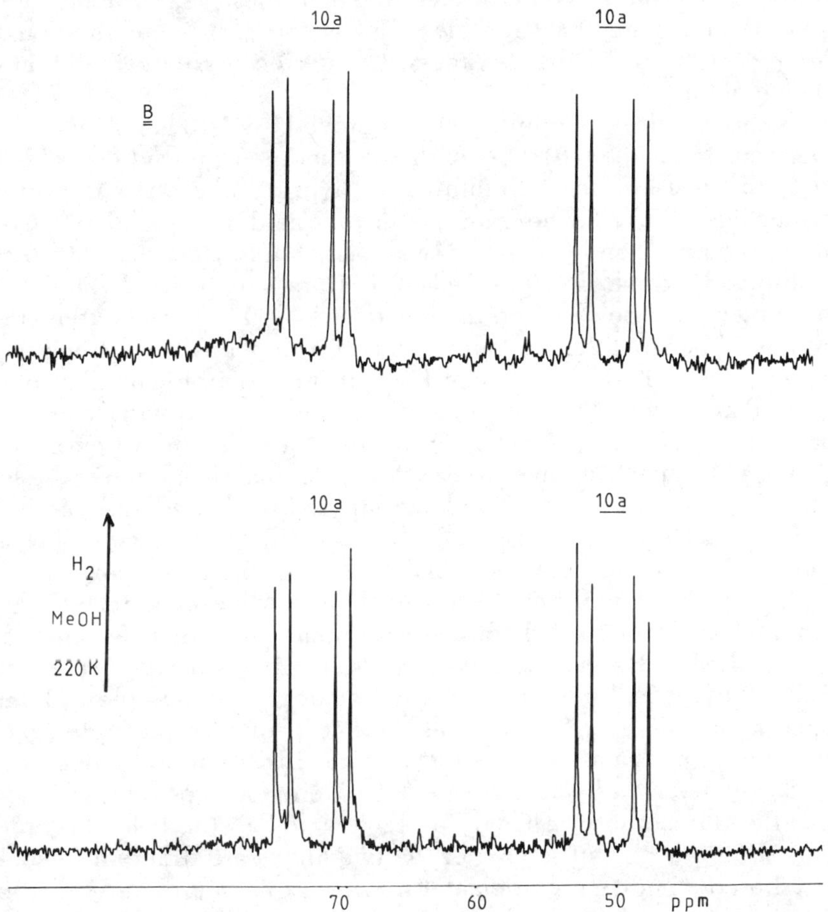

*Figure 2b. P-31 NMR spectra demonstrating formation of the alkyl (**11**) from the minor diastereomer (Complex **10b**) at 220 K: sample formed under thermodynamic control and predominantly Complex **10a**.*

The sample then was cooled to −80°C and sealed under hydrogen. It was warmed to −50°C in the probe, and the P-31 NMR spectrum was monitored at this temperature. The minor diastereomer (**10b**) disappeared concomitant with the formation of Complex **11**, while the concentration of Complex **10a** remained constant within experimental error. In a second experiment formation of the enamide complex was effected at −20°C, at which temperature the equilibration of Complexes **10a** and **10b** is rapid and cooling the sample to −50°C provides a spectrum from which the latter is effectively absent. When this solution is blanketed with hydrogen, little observable change occurs in the course of an hour at −50°C. Taken together, these experiments effectively demonstrate that Complex **11** has the same configuration as Complex **10b**, given that hydrogen transfer occurs with retention of configuration.

The structure of Complex **11** was proved by C-13 labelling experiments as described (*10*). One of these experiments introduced a technique that is useful for determining the hybridization state of a bound organic ligand. If (Z)-α-benzamidocinnamic acid is prepared with C-13 labels at both C1 and C3, then a two-bond coupling constant of 5.6 Hz is observed, which is attenuated to 5 Hz in Complex **10**. This C–C coupling cannot be detected in alkyl (Complex **11**) or in the reduction product *N*-benzoylphenylalanine, so that it depends very strongly on the C1–C2 bond order. It demonstrates that the double bond has been reduced already in Complex **11** and that this, coupled with the (eventual) observation of a metal hydride resonance at −19.3 ppm in the proton NMR spectrum and other NMR evidence (*10*), defines the hydridoalkyl structure. The amide group is bound and the carboxyl carbon experiences a considerable downfield shift in Complex **11** compared with Complex **10**. This may imply metal–carboxyl bonding but it alternatively could reflect the shielding experienced by a saturated acid with a β-rhodium substituent, for which few models are available. Halpern and Chan (*11*) have observed the 1,2-bis(diphenylphosphino)ethane (dppe) analogue of Complex **11** and suggest that the coordination site trans to hydride is occupied by a molecule of methanol, this being reinforced by the observation of its rapid displacement by acetonitrile at low temperatures. This contrasts with the unreactivity of Complex **11** towards acetonitrile at −50°C, suggesting that the structures of the two alkyls are different in some way. Based on current evidence it may be explained equally well by carboxyl binding, or coordination of one *o*-methoxy group trans to hydride in Complex **11**. Pertinent NMR information is summarized in Figure 3.

Since the formation of Complex **11** from **10b** is a second-order process and the formation of product from Complex **11** is a first-order process, their entropies of activation will be very different. A value of $\Delta S^{\ddagger} = -121\ \mathrm{J \cdot mol^{-1} deg^{-1}}$ has been reported for hydrogen addition to Vaska's compound, carbonylchloro bis-triphenylphosphineiridium(I) (*12*). As pointed out by Halpern (*11*), formation of the alkyl will be favored at low temperatures and it is observed to decay rapidly above −40°C. The observation and characterization of Complex **11** proved to be both fortunate and fortuitous, since we were unsuccessful in all of our attempts to form alkyls from chiraphos or DIOP, or from DiPAMP with itaconic acid derivatives.

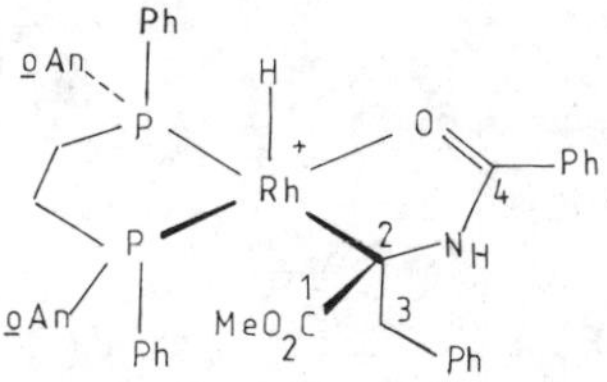

	$\delta\Delta_{ppm}$	J_{P_1C}, Hz	J_{RhC}, Hz
C1	12·1	3·5	2·5
C2	53·3	82	22·5
C3	91·3	3	
C4	9·2	8	

$J_{C_1C_3}$ <1 Hz

Figure 3. C-13 NMR data for the rhodium alkyl (11)

Probes for the Structure of the Transition State

The present understanding of the catalytic cycle in asymmetric hydrogenation may be illustrated in Figure 4. The reaction follows a pathway in which the original cationic rhodium diolefin complex first is hydrogenated to give a solvate, which then reacts rapidly with the dehydroamino acid substrate to give an enamide complex existing in two diastereomeric forms. These then traverse the rate-determining step by cis-specific addition of hydrogen, the difference in the free energy of the diastereomeric transition states being related to the optical yield (y) as $\Delta G\dagger = -RT \ln(100 + y/100 - y)$. This is because the equilibrium between the two enamide complexes which represent the resting state of the catalyst is fast relative to their hydrogenation. In the final post-rate-determining step of the reaction the rhodium alkyl decomposes rapidly by a cis-specific mechanism giving a product and the solvate complex.

At this point mechanistic studies have reached an impasse. All of the observable intermediates have been characterized in solution, and enamide complexes derived from diphos and chiraphos have been defined by X-ray structure analysis. Based on limited NMR and X-ray evidence it appears that the preferred configuration of an enamide complex has the olefin face bonded to rhodium that is opposite to the one to which hydrogen is transferred. There are now four crystal structures of chiral biphosphine rhodium diolefin complexes, and consideration of these leads to a prediction of the direction of hydrogenation. The crux of the argument is that nonbonded interactions between pairs of prochiral phenyl rings and the substrate determine the optical yield and that X-ray structures reveal a systematic relationship between *P*-phenyl orientation and product configuration.

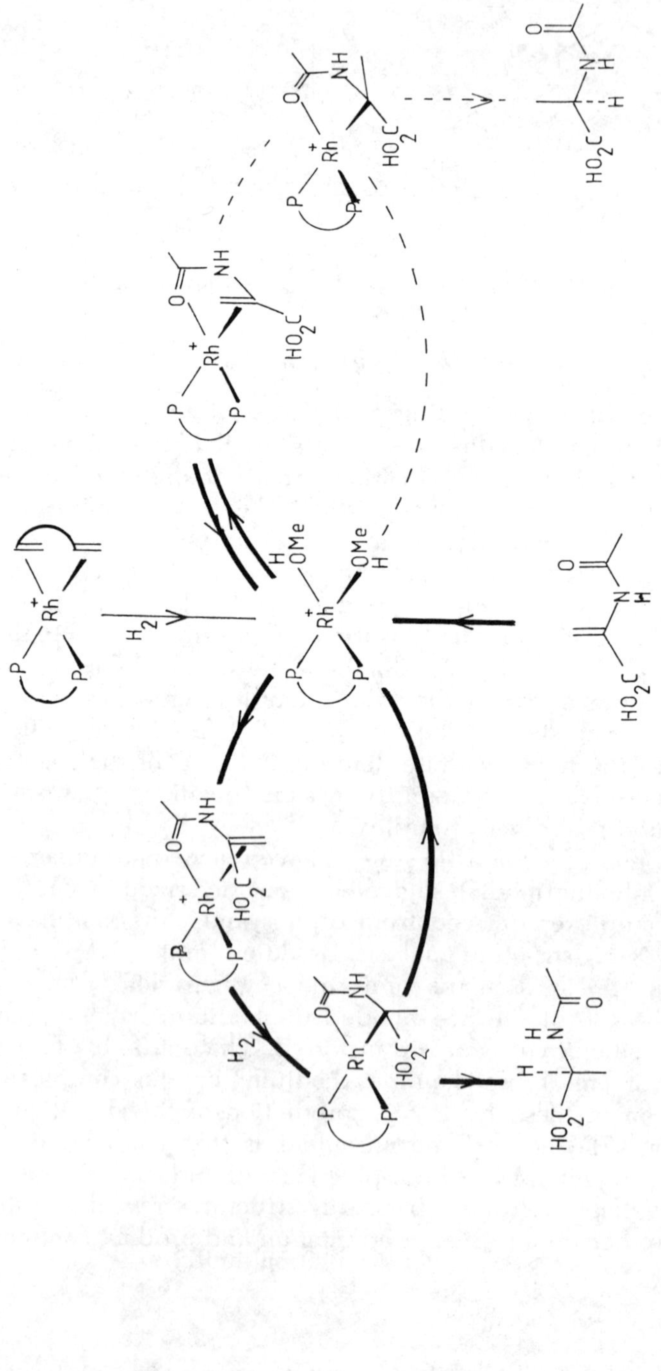

Figure 4a. Reaction cycle for asymmetric hydrogenation

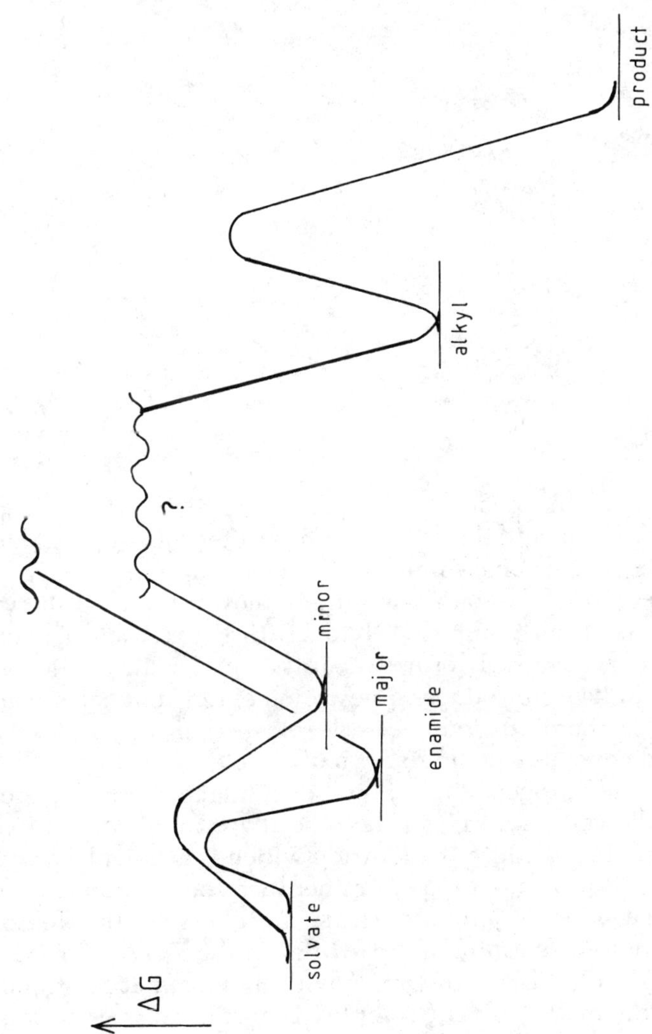

Figure 4b. Schematic energy barriers for the various steps

An alternative suggestion is based on the helicity of pseudoequatorial *P*-phenyl rings and explains the reversal of configuration between reactions catalyzed by (*R*,*R*)-*trans*-bis-1,2(*N*-diphenylphosphinamino)-cyclohexane complexes and (*R*,*R*)-*trans*-bis-1,2(*N*-methyl-*N*-diphenylphosphinamino)cyclohexane complexes (*13, 14, 15*).

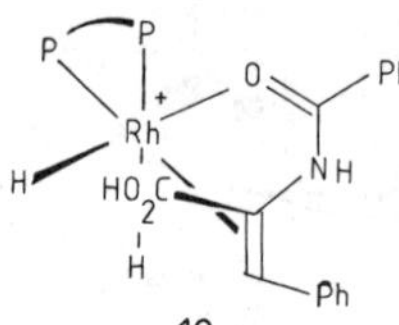

12

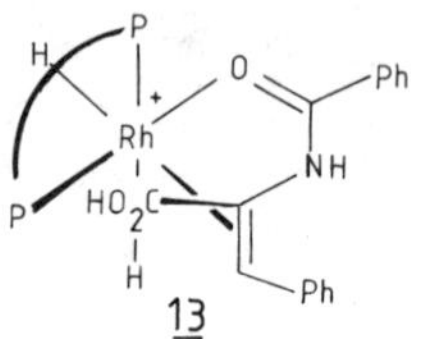

13

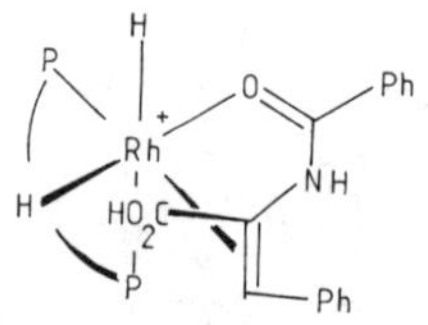

14

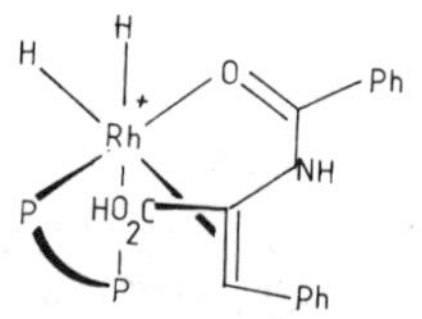

15

In general, the enamide complex is formed prior to the rate-determining stage, and the alkyl is formed after this step. All attempts to define the origins of asymmetric induction are pure conjecture unless the structure of the transition state is known. Figure 5 shows three possible routes from Complex **10b** to **11**. In Paths A and B, hydrogen addition occurs reversibly or irreversibly to give a dihydride intermediate. This in itself introduces a new element of stereochemical complexity, since there are four possible octahedral dihydrides derived from a single enamide complex, two of which (Complexes **12** and **13**) may transfer a hydrogen atom to the benzylidene site by a stereoelectronically allowed pathway; in the other two Complexes (**14** and **15**) this is impossible because the hydride which is syn-coplanar with the bound olefin is proximal to the β- rather than the α-carbon. Inspection of molecular models provides little insight into the reasons for stereoselectivity in forming a dihydride. In the DiPAMP case, minor enamide (Complex **10b**) must react with hydrogen at least 300 times faster than the major isomer (Complex **10a**) and yet there seems to be little steric barrier to the hydrogen approach to rhodium in either case. A discrimination of this order demands a rather tightly organized transition state, suggesting that the concerted pathway C (*see* Figure 5) is a serious possibility.

In the early papers (*16, 17*) on homogeneous hydrogenation by tris(triphenylphosphine)rhodium(I) chloride, Wilkinson and coworkers found the kinetic isotope effect for deuterium addition to be

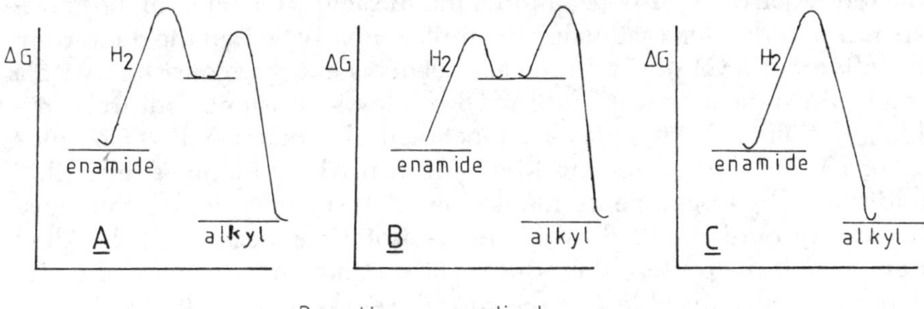

*Figure 5. Routes from Complex **10b** to Complex **11***

slightly less than unity in several cases (i.e., deuterium reacts faster than hydrogen) and ascribed this to a secondary isotope effect. It is now known (*18, 19*) that the addition of hydrogen is not rate limiting. Addition of hydrogen deuteride to olefins occurred without isotopic scrambling and gave an equimolar mixture of the two possible monodeuterated products (*16*). We expected that the concerted pathway C would give an unequal deuterium distribution when HD was added because its transition state has partial bonding of hydrogen to both rhodium and carbon. This means that Complex **16** will be preferred over **17** since the heavier rhodium atom will experience the smaller isotope effect on the basis of the reduced mass equation. For this reason we have carried out a series of reductions in methanol solution using HD (generated from $LiAlD_4$ and MeOH at −80°C) the results of which are recorded in Table I. The site of deuteration may be established very clearly from the proton and deuteron NMR spectrum and only two products are formed using Complex **4** as procatalyst in

Table I. Homogeneous Hydrogenation of Olefins with HD and Cationic Rhodium Complexes[a]

Phosphine	*Substrate*	*Selectivity[b]* (± 0.03)
Diphos	(Z)-α-acetamidocinnamic acid	1.37 : 1
Diphos	oct-1-ene	1.03 : 1
DiPAMP	(Z)-α-acetamidocinnamic acid	1.36 : 1
DIOP	(Z)-α-acetamidocinnamic acid	1.33 : 1
	1-octene	1.00 : 1
	styrene	1.01 : 1
	atropic acid; NEt_3	1 : 1.21
	acetamidoethylene	1.02 : 1

[a] Reaction of olefins with HD (600 mm, 20°C, MeOH; catalyst: substrate = 50 : 1). Deuterium distribution was estimated by proton and deuterium NMR.

[b] The isomer with deuterium at the more highly substituted site is listed first.

the reduction of (Z)-α-benzamidocinnamic acid. Here, and in the corresponding reduction by the dppe complex, nearly 60% of the deuterium is delivered to C1 of the amino acid. This is entirely consistent with a transition state such as Complex **18** but less consistent with irreversible addition of HD to give a rigid octahedral species such as Complex **12** or **13**. It does not rule out Mechanism B which involves reversible addition of hydrogen deuteride, followed by a step in which the transfer of protium is favored by a primary isotope effect so that the alkyl deuteride is the preferred product. The distinction between B and C is experimentally testable but the former seems less likely. Since it is impossible to observe a dihydride by NMR or other techniques under hydrogenation conditions, it is disfavored in the hypothetical equilibrium with its enamide precursor by a minimum factor of 10 : 1. This requires that reversal of the addition step has a rate constant at least ten times faster than that observed for a single cycle of catalytic hydrogenation and therefore in excess of 2 s^{-1}. This is much faster than is observed in other reversible hydrogenations (*20*).

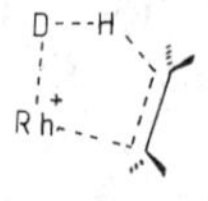

16

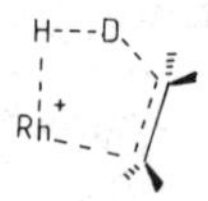

17

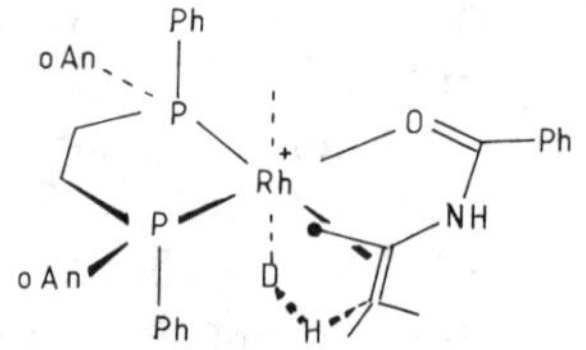

18 •= CO_2H

Thus concerted addition of hydrogen to the enamide complex must be considered as a possible mechanism. It has the manifold attraction of offering a simple explanation of the origin of asymmetric induction. The product-related enamide configuration is Complex **19**, this being normally present at a lower concentration than its isomer **(20)**. The orientation of aryl groups is shown for (*R*,*R*)-DIOP. The approach of hydrogen so that it bonds synchronously to the benzylidene carbon and to rhodium occurs with minimal steric hindrance in Complex **19**; in Complex **20**, however, a serious nonbonded interaction with the *o*-hydrogen of the axial phenyl ring occurs as hydrogen approaches. The synchronous addition may be accompanied by the occupation of the trans-coordination site by solvent or the carboxylate group. The same argument may be used to explain the course of asymmetric catalysis by DiPAMP complexes on the basis of available X-ray structural evidence, and since it has been claimed that all chiral-chelating diphosphines operate similarly, it affords a general explanation. Further experiments are in progress to test this possibility, which is admittedly speculative at present.

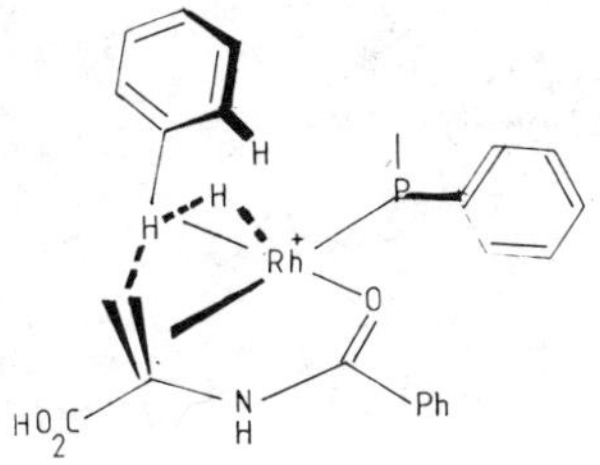

19 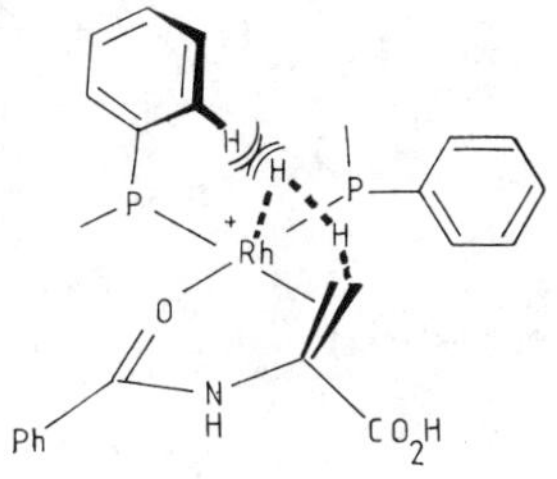 20

Since the submission of this article, further work by Halpern and co-workers (*21*) has led to the description of a chiraphos rhodium enamide crystal structure.

Literature Cited

1. Kagan, H. B.; Dang, T. P. *J. Am. Chem. Soc.* **1972,** *94,* 6429.
2. Valentine, D.; Scott, J. W. *Synthesis* **1978,** 329.
3. Knowles, W. S.; Vineyard, B. D.; Sabacky, M. J.; Stults, B. R. "Fundamental Research in Homogeneous Catalysis"; Tsutsui, M., Ed.; Plenum: New York; 1979, Vol. 3, p. 537.
4. Halpern, J. In "Inorganic Compounds with Unusual Properties—II," *Adv. Chem. Ser.* **1979,** *173,* 16.
5. Brown, J. M.; Chaloner, P. A. *J. Chem. Soc. Chem. Commun.* **1978,** 321.
6. Brown, J. M.; Chaloner, P. A.; Nicholson, P. N. *J. Chem. Soc. Chem. Commun.* **1978,** 646.
7. Brown, J. M.; Chaloner, P. A. *J. Chem. Soc. Chem. Commun.* **1979,** 637.
8. Brown, J. M.; Chaloner, P. A. *J. Am. Chem. Soc.* **1980,** *102,* 3040.
9. Brown, J. M.; Chaloner, P. A.; Glaser, R.; Geresh, S. *Tetrahedron* **1980,** *36,* 815.
10. Brown, J. M.; Chaloner, P. A. *J. Chem. Soc. Chem. Commun.* **1980,** 436.
11. Chan, A.S.C.; Halpern, J. *J. Am. Chem. Soc.* **1980,** *102,* 838.
12. Vaska, L. *Acc. Chem. Res.* **1968,** *1,* 335.
13. Fiorini, M.; Marcati, F.; Giongo, G. M. *J. Mol. Catal.* **1978,** *4,* 125.
14. Fiorini, M.; Giongo, G. M. *J. Mol. Catal.* **1979,** *5,* 103.
15. Onuma, K. I.; Ito, T.; Nakamura, A. *Tetrahedron Lett.* **1979,** 3163.
16. Jardine, F. H.; Osborn, J. A.; Wilkinson, G. *J. Chem. Soc.* (A) **1967,** 1574.
17. Osborn, J. A.; Jardine, F. H.; Young, J. F.; Wilkinson, G. *J. Chem. Soc.* **1966,** 1711.
18. Rousseau, C.; Evard, M.; Petit, F. *J. Mol. Catal.* **1977/1978,** *3,* 309.
19. de Croon, M. H. J.; van Nisselrooij, P. F. M. T.; Kuipers, H. J. A. M.; Coenen, J. W. E. *J. Mol. Catal.* **1978,** *4,* 325.
20. Tolman, C. R.; Meakin, P. Z.; Lindner, D. L.; Jesson, J. P. *J. Am. Chem. Soc.* **1974,** *96,* 2762.
21. Chan, A. S. C.; Pluth, J. J.; Halpern, J. *J. Am. Chem. Soc.* **1980,** *102,* 5952.

RECEIVED July 28, 1980.

Advances in Asymmetric Hydrocarbonylation

GIAMBATTISTA CONSIGLIO and PIERO PINO

Swiss Federal Institute of Technology, Department of Industrial and Engineering Chemistry, Universitaetstrasse 6, 8092 Zurich, Switzerland

Optically active oxygenated products have been obtained with up to 69% optical yield by reacting olefins with carbon monoxide and hydrogen or alcohols. Synthesizing optically active esters, including α-amino acid esters, from olefinic substrates, carbon monoxide, and alcohols is discussed in detail. Some regularities in the results obtained in enantiomer- and enantioface-discriminating reactions are emphasized. A simple model is discussed which provides for a correct qualitative prediction of prevailing antipode and regioisomer in the reaction products in more than 85% of the cases investigated.

The first experiments in asymmetric hydrocarbonylation were carried out in the early seventies when styrene and other substrates containing aromatic groups were hydroformylated with chiral catalytic systems formed by $[Rh(CO)_2Cl]_2$ and (*R*)-methylphenylpropylphosphine (*1*) or by $Co_2(CO)_8$ and [(*S*)-*N*-1-phenylethyl]salicylaldimine (2) (*see* Scheme I, *X* = H).

Subsequently, aliphatic 1-substituted, 1,1-, and 1,2-disubstituted ethylenes were investigated (3) using $RhH(CO)(PPh_3)_3$ in the presence of (-)DIOP (diisooctylphthalate) (*4*) as the chiral ligand, and asymmetric induction was shown to occur during or before the formation of a rhodium–alkyl intermediate (R–[Rh]). This formation appears to be practically irreversible under the reaction conditions used (5). As another consequence of this important feature, regioselection in hydroformylation is likely to occur during or before the formation of the rhodium–alkyl intermediates.

The regularities observed in the hydroformylation of different types of olefins have been rationalized using a simple model (5) based

0065–2393/82/0196–0371$5.00/0

Scheme I.

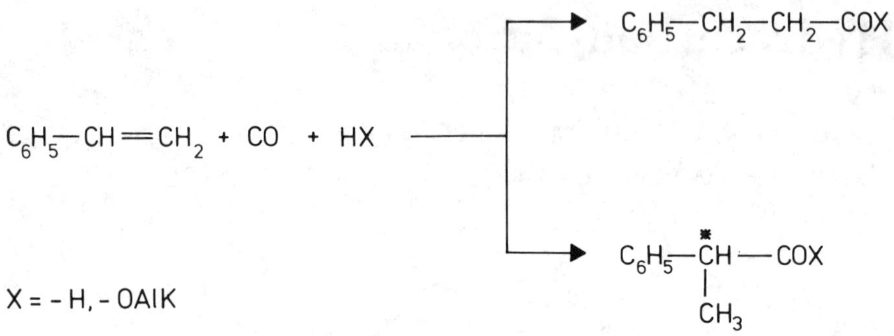

$$C_6H_5-CH=CH_2 + CO + HX \longrightarrow C_6H_5-CH_2-CH_2-COX$$

$$\longrightarrow C_6H_5-\overset{*}{C}H(CH_3)-COX$$

X = - H, - OAlK

on the presence of chiral metal atoms in the catalytic complex and on steric interactions between substrate and ligands bound to the metal atom.

The above results were reviewed in 1974 (*5*). Since then the main advances in the field have been the achievement of asymmetric hydrocarbalkoxylation (*see* Scheme I, *X* = -OR) using palladium catalysts in the presence of (-)DIOP (*6*), the use of other diphosphines as asymmetric ligands in hydroformylation by rhodium (*7*), and the achievement of the platinum-catalyzed asymmetric hydroformylation (*8, 9*). Further work in the field of asymmetric hydroformylation with rhodium catalysts has been directed mainly towards improving optical yields using different asymmetric ligands (*10*), while only very few efforts were devoted to asymmetric hydroformylation catalyzed by cobalt or other metals (*11, 12*) and it will be discussed in a modified form in this chapter.

Surprisingly, in view of the generally small enantiomeric excess obtained, the very simple model proposed (*5*) to explain regioselectiv-

Scheme II.

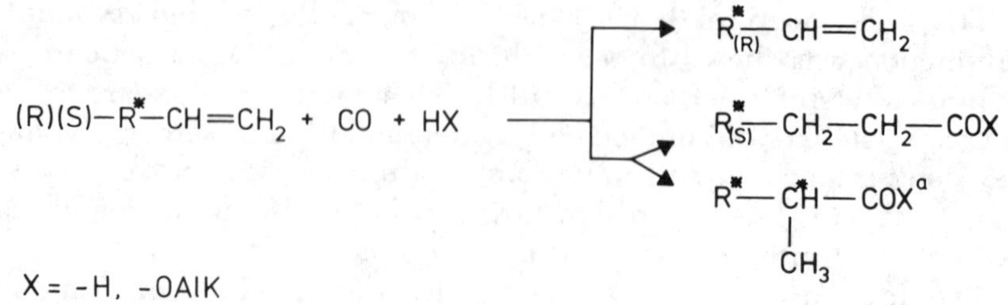

$$(R)(S)-R^*-CH=CH_2 + CO + HX \longrightarrow R^*_{(R)}-CH=CH_2$$

$$\longrightarrow R^*_{(S)}-CH_2-CH_2-COX$$

$$\longrightarrow R^*-\overset{*}{C}H(CH_3)-COX^{a}$$

X = -H, -OAlK

[a] MINOR ISOMER (UP TO 7%) IN THE CASES INVESTIGATED. THE TYPE OF CHIRALITY OF THE ASYMMETRIC CARBON ATOMS WAS NOT DETERMINED.

ity and asymmetric induction gives qualitatively correct predictions for isomeric and stereoisomeric composition of the reaction products in about 85% of the experiments. This model also makes it possible to logically relate the results of enantiomer-discriminating hydrocarbonylation reactions (see Scheme II) with those of enantioface-discriminating hydrocarbonylation.

In the present review, the results obtained in hydrocarbalkoxylation will be considered. The advances in asymmetric hydroformylation are discussed in another review (*13*).

Asymmetric Hydrocarbalkoxylation of Olefins

The hydrocarbalkoxylation of olefins with palladium catalysts has been known for a long time and has been reviewed adequately (*14*, *15*). Adding H– and –COOR to the olefin occurs with a cis-stereochemistry (*16*); the most likely mechanism proposed up to now is reported in Scheme III (*17*). A complete discussion of the mechanism will be published elsewhere.

Scheme III.

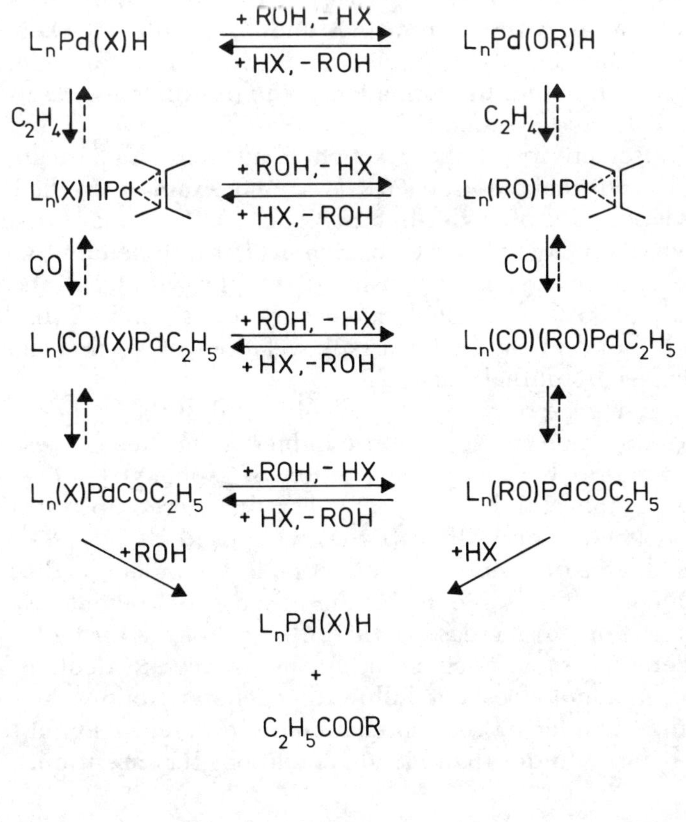

$$L_nPd(X)H \underset{+HX,\,-ROH}{\overset{+ROH,\,-HX}{\rightleftharpoons}} L_nPd(OR)H$$

$$\downarrow C_2H_4 \qquad\qquad \downarrow C_2H_4$$

$$L_n(X)HPd\langle \| \underset{+HX,\,-ROH}{\overset{+ROH,\,-HX}{\rightleftharpoons}} L_n(RO)HPd\langle \|$$

$$\downarrow CO \qquad\qquad \downarrow CO$$

$$L_n(CO)(X)PdC_2H_5 \underset{+HX,\,-ROH}{\overset{+ROH,\,-HX}{\rightleftharpoons}} L_n(CO)(RO)PdC_2H_5$$

$$\downarrow \qquad\qquad \downarrow$$

$$L_n(X)PdCOC_2H_5 \underset{+HX,\,-ROH}{\overset{+ROH,\,-HX}{\rightleftharpoons}} L_n(RO)PdCOC_2H_5$$

$$\searrow +ROH \qquad\qquad +HX \swarrow$$

$$L_nPd(X)H + C_2H_5COOR$$

Scheme IV.

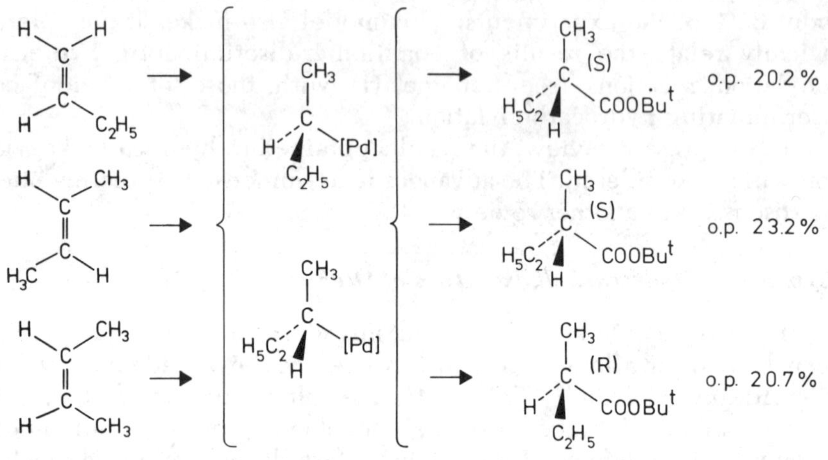

o.p. 20.2%

o.p. 23.2%

o.p. 20.7%

Similarly to the hydroformylation, under certain reaction conditions the formation of the intermediate palladium–alkyl complex can be practically irreversible as shown by the different prevailing chirality of the 2-methylbutanoic acid ester obtained from 1-butene and (Z)-2-butene, as well as from (*E*)- and (Z)-2-butene. Therefore, regioselection and enantioface selection must occur, as in hydroformylation, during or before the formation of the postulated palladium–alkyl intermediate (*see* Scheme IV).

Regioselectivity of the reaction is affected dramatically by the type of ligands that are used (*20, 21, 22*). For example, in the hydrocarbalkoxylation of 2-phenyl-1-butene using (-)DIOP, more than 96% of carbonylation takes place at Position 1 (*19*) independently from the alcohol used; on the contrary, using [(*S*)-2-phenylbutyl]diphenylphosphine, about 95% of carbonylation occurs at Position 2. A similar effect has been found in the hydrocarbalkoxylations of *N*-vinylsuccinimide (*22*) and *N*-vinylphthalimide (*23*).

The influence of the type of alcohol and of the carbon monoxide partial pressure on the asymmetric induction has been investigated in the asymmetric hydrocarbalkoxylation of 2-phenyl-1-propene using (-)DIOP as the chiral ligand (*18*). For the same substrate different chiral phosphines and diphosphines were used also (*17, 24*).

The degree of asymmetric induction is dependent on the structure of the alcohol (*see* Table I). In the absence of solvent, asymmetric induction seems to increase as the bulkiness of the alcohol increases; in benzene the same appears to be true for acyclic alcohols. In both cases *sec*-butanol does not follow the general trend. This is rather surprising considering that no enantiomer differentiation of this alcohol takes place under the reaction conditions that are used.

Table I. Influence of the Structure of the Alcohol on the Enantiomeric Excess in the Asymmetric Hydrocarbalkoxylation of 2-Phenyl-1-propene[a] in the Presence of (−)DIOP

Alcohol	*(S)-3-Phenylbutanoate Optical Purity, %* A[b]	B[c]
CH_3OH	19.3	3.0
C_2H_5OH	37.0	10.3
iso-C_3H_7OH	35.4	14.2
sec-C_4H_9OH	40.6	8.2
t-C_4H_9OH	38.9	19.6
c-$C_6H_{11}OH$	29.0	n.d.[d]
$C_6H_5CH_2OH$	30.6	n.d.[d]

[a] Reaction conditions: 100°C; 400 atm CO (initial pressure at 100°C); $PdCl_2$/(−)DIOP = 1:2.
[b] A: Benzene as the solvent (40 mL); olefin, 0.1 mol; alcohol/olefin = 1.5.
[c] B: Without solvent; alcohol, 50 mL; olefin, 0.1 mol.
[d] n.d. = not determined.

When $(PPh_3)_3PdCl_2$ is used as the catalyst precursor and optically active alcohols are used as the hydrogen donors, a small asymmetric induction occurs (*see* Table II) which could be due to a solvent effect (25). However, the large influence of the alcohol structure on regioselectivity suggests that the alcohol residue is present in (at least one of) the catalytic complexes.

Table II. Influence of the Structure of the Alcohol in the Hydrocarbalkoxylation of 2-Phenyl-1-propene in the Presence of $(PPh_3)_2PdCl_2$[a]

Alcohol	*Alcohol to Olefin Molar Ratio*	*3-Phenylbutanoate* %[b]	*Optical Purity, %*	*Absolute Configuration*
Ethanol	1	7	—	—
(−)-2-Butanol[c]	1	21	0.2	(*R*)
t-Butyl alcohol	1	68	—	—
(−)-2-Methylbutanol[d]	1	12	0.1	(*S*)
(−)-Menthol[e]	1	3	1.4	(*S*)
(−)-Menthol[e]	4	n.d.[f]	3.4	(*S*)

[a] Reaction conditions: 100°C; 400 atm CO.
[b] With respect to the sum of both esters.
[c] Optical purity 84%.
[d] Optical purity 98%.
[e] Under 260 atm CO.
[f] n.d. = not determined.

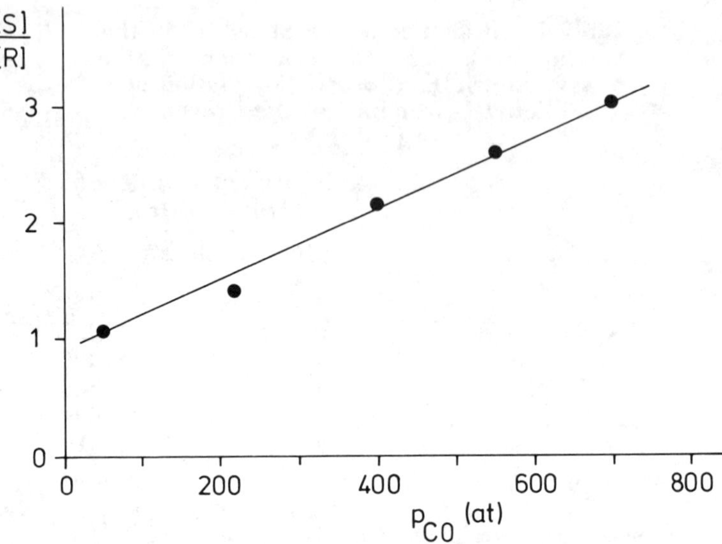

Figure 1. Influence of carbon monoxide pressure on the ratio between the enantiomers of t-*butyl-3-phenylbutanoate arising from asymmetric hydrocarbalkoxylation of 2-phenyl-1-propene by (−)DIOP/$PdCl_2$ in a benzene solution at 100°C*

The effect of the carbon monoxide pressure on the asymmetric induction in 2-phenyl-1-propene hydrocarbalkoxylation is opposite of what is expected (*see* Figure 1). In fact, the optical purity of the reaction product is increased as the carbon monoxide pressure is increased (*18*). This effect agrees with the fact that with the same substrate asymmetric induction is increased by lowering the (-)DIOP to the $PdCl_2$ molar ratio at a constant carbon monoxide pressure (*18*). In fact, in experiments in which mixtures of $(PPh_3)_2PdCl_2$ and [(-)DIOP]$PdCl_2$ were used to maintain the ratio P/Pd = 2 (*26*), the maximum optical yield has been obtained for a ratio(-)DIOP/Pd = 0.5. If a competition for coordination to palladium between carbon monoxide and phosphine exists (*see* Scheme V) as in the case of cobalt (*27*), the presence in the catalytic complex of only one phosphorus atom bound to pal-

Scheme V.

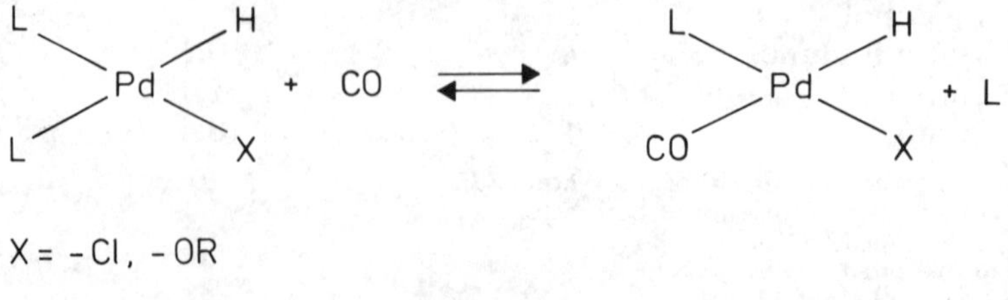

$$L_2Pd(H)(X) + CO \rightleftharpoons L(CO)Pd(H)(X) + L$$

X = -Cl, -OR

ladium as expected at high carbon monoxide pressure, appears to be more effective in inducing asymmetric synthesis than the presence of two phosphorus atoms in the cis position.

These data refer to the hydrocarbalkoxylation of 2-phenyl-1-propene and cannot be generalized. In fact, no effect of this type has been observed in the hydrocarbalkoxylation of *N*-vinylsuccinimide (*22*).

Some data are also available on the effect of the structure of the phosphine or diphosphine on the optical yield in the asymmetric hydrocarbalkoxylation of 2-phenyl-1-propene (*24*) (*see* Table III). The best optical yield was obtained using 2,2-dimethyl-4,5-bis(dibenzophosphol-5-ylmethyl)-1,3-dioxolane as the chiral ligand. Much lower optical yields were obtained using aminophosphines or monophosphines.

Different types of substrates have been hydrocarbalkoxylated using (-)DIOP as the chiral ligand (*see* Table IV). In addition to the examples of enantioface-discriminating reactions, two examples of enantiomer-discriminating reactions are reported also. The possibility of synthesizing optically active amino acids starting with *N*-vinylsuccinimide (*22*) or with other enamides or enimides (*23*) is remarkable.

From a stereochemical point of view, the results obtained in the hydrocarbalkoxylation of 2-phenyl-1-butene using [(*S*)-2-phenylbutyl]diphenylphosphine as the chiral ligand are particularly interesting. In this case two chiral products are obtained; the prevailing antipodes arise from the same enantioface but have different optical purities (*see* Scheme VI), showing that, at least formally, regioselection is different on the two prochiral faces of the olefin (*28*).

Scheme VI.

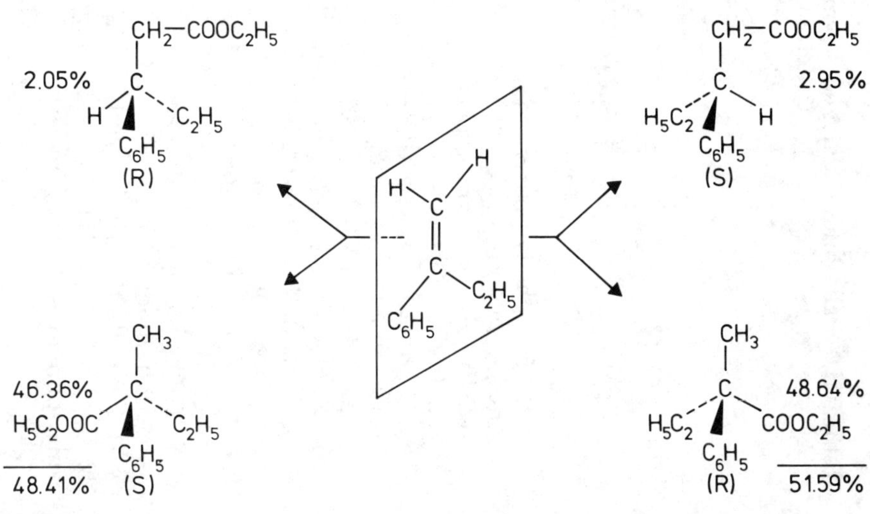

2.05% (R): $CH_2-COOC_2H_5$, C, H, C_2H_5, C_6H_5

2.95% (S): $CH_2-COOC_2H_5$, C, H_5C_2, H, C_6H_5

46.36% (S): CH_3, C, H_5C_2OOC, C_2H_5, C_6H_5

48.64% (R): CH_3, C, H_5C_2, $COOC_2H_5$, C_6H_5

48.41%   51.59%

Olefin: H, H, C=C, C_6H_5, C_2H_5

Table III. Palladium-Catalyzed Asymmetric Hydrocarbalkoxylation of 2-Phenyl-1-propene in the Presence of Different Phosphine Ligands (in Benzene at 100°C)

Chiral Ligand	*Hydrogen Donor*	P_{CO} *(atm)*	*P/Pd*	*Ester of the 3-Phenylbutanoic Acid*			*Ref.*
				%	*Absolute Configuration and Optical Purity*		
(*S*,*S*)1,2-Bis(diphenylphosphinomethyl)cyclohexane	iso-PrOH	230	2	92	(*R*)	22.0	*24*
(*S*,*S*)1,2-Bis(5*H*-dibenzophosphol-5-ylmethyl)cyclohexane	iso-PrOH	230	2	97	(*S*)	7.3	*24*
(*R*,*R*)2,2-Dimethyl-4,5-bis(diphenylphosphinomethyl)-1,3-dioxolane	*t*-BuOH	400	0.8	~ 100	(*S*)	58.6	*18*
(*R*,*R*)2,2-Dimethyl-4,5-bis(5*H*-dibenzophosphol-5-ylmethyl)-1,3-dioxolane	*t*-BuOH	240	0.8	not given	(*S*)	69.0	*24*
(*R*,*R*)1,2-Bis(diphenylphosphinomethyl)cyclobutane	iso-PrOH	230	2	99	(*S*)	9.3	*24*
(*R*,*R*)1,2-Bis(5*H*-dibenzophosphol-5-ylmethyl)cyclobutane	iso-PrOH	230	2	97	(*S*)	40.0	*24*
N[(*S*)-1-Phenylethyl]-*N*-[diphenylphosphino]methylamine	EtOH	400	2	17	(*R*)	2.2	*12*
[(*S*)-2-Phenylbuthyl]diphenylphosphine	EtOH	400	2	10	(*S*)	7.7	*17*
1,2:3,4-Di-*O*-isopropylidene-6-diphenylphosphino-D-galactopyranose	EtOH	400	2	44	(*S*)	7.0	*17*

Table IV. Palladium-Catalyzed Asymmetric Hydrocarbalkoxylation of Some Olefinic Substrates at 100°C in the Presence of (−)DIOP

Substrate	*ROH* R =	P_{CO} *(atm)*	Chiral Ester: *Name*	Chiral Ester: %	Chiral Ester: *Optical Purity (%) and Chirality*	
1-Butene	$t\text{-}C_4H_9$	700	2-methylbutanoate	22	20.2	(*S*)
3-Methyl-1-butene	CH_3	440–374	2,3-dimethylbutanoate	10	10.3	(*S*)
2,3-Dimethyl-1-butene	CH_3	405–326	2,3,3-trimethylbutanoate	5	2.0	(*S*)
Styrene	$t\text{-}C_4H_9$	700	2-phenylpropanoate	44	10.0	(*S*)
2-Methyl-1-butene	$t\text{-}C_4H_9$	390–350	3-methylpentanoate	~ 100	4.3	(*R*)
2,3-Dimethyl-1-butene	$t\text{-}C_4H_9$	385–355	3,4-dimethylpentanoate	~ 100	4.6	(*S*)
2,3,3-Trimethyl-1-butene	$t\text{-}C_4H_9$	396–365	3,4,4-trimethylpentanoate	~ 100	19.5	(*S*)
2-Phenyl-1-propene	$t\text{-}C_4H_9$	700	3-phenylbutanoate	~ 100	50.4	(*S*)
2-Phenyl-1-butene	$t\text{-}C_4H_9$	396–368	3-phenylpentanoate	~ 100	44.7	(*S*)
(*Z*)-2-Butene	$t\text{-}C_4H_9$	700	2-methylbutanoate	96	20.7	(*R*)
(*E*)-2-Butene	$t\text{-}C_4H_9$	700	2-methylbutanoate	98	23.2	(*S*)
(*E*)-1-Phenyl-1-propene	C_2H_5	380	2-phenylbutanoate	45[a]	10.0	(*S*)
Bicyclo[2.2.1]2-heptene	CH_3	380–370	bicyclo[2.2.1]heptane-1-carboxylate	~ 97 (exo)	4.5	(1*R*,2*R*,4*S*)
N-Vinylsuccinimide	C_2H_5	400	2-succinoimidylpropanoate	9	17.1	(*S*)
Methylmethacrylate[b]	CH_3	380–370	methylsuccinate	~ 100	49.0	(*S*)
3-Methyl-1-pentene	CH_3	495–450	4-methylhexanoate	93	1.2	(*S*)[c]
3,4,4-Trimethyl-1-pentene	CH_3	380–355	4,5,5-trimethylhexanoate	96	0.3	(*R*)[d]

[a] 50% 4-Phenylbutanoate is formed.
[b] 120°C.
[c] 50% conversion.
[d] 45% conversion.

A Model to Interpret the Regularities Observed in Asymmetric Hydrocarbonylation

Experimental Facts Leading to the Conception of the Model. Considering isomeric and stereoisomeric composition of the products obtained in the asymmetric hydrocarbalkoxylation investigated using the palladium–(−)DIOP catalytic system, some interesting regularities are apparent. This is also true of the rhodium–(-)DIOP-catalyzed hydroformylation using the same substrates (*13*).

This fact is surprising in view of the low enantiomeric excess experimentally found indicating a low difference in the free energies of the diastereomeric-activated complexes controlling asymmetric induction. However, from enantiomeric excess only a minimum value for that difference can be evaluated and, therefore, the actual value of the difference in the above free energies might be much larger.

In any case, while few results involving enantiomeric excesses smaller than 40% cannot be taken as a sound basis for speculating about the factors influencing asymmetric induction, the regularities that have become apparent in the numerous results concerning more than 90 independent experiments of hydrocarbonylation (including both hydroformylation and hydrocarbalkoxylation) in which either the substrate or the catalytic system is different, in our opinion make worthwhile an attempt to correlate the above results with a simple stereochemical model.

Hydrocarbonylation reactions are multistep catalytic processes (*14*). For this reason our first problem was to determine in which step asymmetric induction takes place, since models could be very different depending on the step in which asymmetric induction is determined (e.g. in the metal alkyl complex formation or in the final reductive elimination step).

We first established that hydrocarbonylation reactions occur with cis-stereochemistry (*29*, *16*) and that asymmetric induction occurs before or during the formation of the metal alkyl intermediate (*5*, *6*). This means that is either during the π-olefin complex formation between catalyst and substrate or during the insertion of the π-complexed olefin into the M–H bond. Therefore, the model should focus on the interactions between the substrate double bond and the catalytically active metal atom of the catalyst.

Since asymmetric induction decreases rapidly when the distance between the inducing asymmetric center and the new asymmetric center to be formed in the product increases (*25*), we assumed that the metal atom approached by the substrate (that is the atom of the catalyst directly interacting with the unsaturated atom of the substrate that becomes chiral during the reaction) is chiral (*5*). Complexes containing chiral metal atoms are known in the literature (*31*); furthermore, the presence of metal atoms having different chiralities in catalytic complexes containing different metal atoms offers a very simple explanation for the fact that, when the same chiral ligand((-)DIOP) is used, the same regularities, but opposite prevailing chiralities, are found mostly in the series of experiments with the same substrates in the palladium-catalyzed hydrocarbalkoxylation and in the rhodium-catalyzed hydroformylation (*see* Table V).

A Simple Model for the Transition State Leading to the Intermediate Metal–Alkyl Complex. The enantiomeric excess in the products reflects the relative free energies of the two diastereomeric-activated complexes controlling asymmetric induction and leading to either antipode. In the first attempts to settle a stereochemical model, the assumption was made that activation energies for the transformation of the diastereomeric π-olefin complexes in the metal–alkyl complexes leading to either antipode were very similar. In this case differences in free energies of formation of π-complexes would reflect, at least qualitatively, energy differences between transition states controlling asymmetric induction (*5*). However no suitable way has been found up to now to confirm experimentally this assumption. Therefore we prefer to schematize directly in this chapter the transition states controlling asymmetric induction, leaving unanswered the problem concerning the possible relationships between the relative free energy of the transition states leading to the metal–alkyl complexes and that of the corresponding π-complexes.

A very simple way to schematize these transition states is to represent in a plane the steric situation faced by the double bond when it approaches the M–H bond (*see* Figures 2a and b) and to superimpose the olefinic substrate (e.g., (Z)-2-butene) to this representation with the double bond parallel to the M–H bond (Figures 2a′ and b′).

If the transition state controlling asymmetric induction is the one leading from the π-olefin complex to the metal alkyl complex, the π-olefin complex can be conceived reasonably as a trigonal bipyramid in which the metal, situated in the center of the pyramid, is coordinated to a hydrogen atom, the olefinic double bond, and three other ligands, two of which must differ in size (e.g., a PPh_2R and a CO group for a monometallic complex). If the transition state controlling asymmetric induction is the one leading to the π-olefin complex, it is conceivable that there is a similar structure for the catalyst, where the metal still is bound to five ligands, one of which (e.g., a solvent molecule) dissociates immediately before the approach of the substrate. Unsaturated chiral metal complexes maintaining their chirality already have been postulated in the literature (*31*).

In Figures 2a and b L and S are two substituents having different sizes (L > S); in 2b Z is a substituent that is larger than H. The space available for accommodating the substituent(s) of the olefinic double

Table V. Influence of the Structure of the Substrate on the Prevailing Chirality and on the Maximum Optical Yield Obtained in Asymmetric Hydrocarbonylation with Rhodium or Palladium Catalysts in the Presence of the Same Asymmetric Ligand [(−)DIOP]

	Hydroformylation $RhH(CO)(PPh_3)_3$		*Hydrocarboxylation* $PdCl_2$	
Substrate	*Prevailing Chirality*	*Optical Yield, %*	*Prevailing Chirality*	*Optical Yield, %*
1-Butene	*R*	18.8	*S*	20
1-Pentene	*R*	19.4		
1-Octene	*R*	16.5		
3-Methyl-1-butene	*R*	15.2	*S*	10.3
3,3-Dimethyl-1-butene			*S*	2
Styrene	*R*	25.2	*S*	10
Allylbenzene	*R*	15.5		
(*Z*)-2-Butene	*S*	27.0	*R*	20.7
(*E*)-2-Butene	*S*	3.2	*S*	23.2
Norbornene	(1*S*,2*S*,4*R*)	5.3	(1*R*,2*R*,4*S*)	4.5
2-Methyl-1-butene	*R*	1.0	*R*	4.3
2,3-Dimethyl-1-butene	*S*	3.5	*S*	4.6
2,3,3-Trimethyl-1-butene			*S*	19.5
α-Methylstyrene	*R*	1.6	*S*	59
α-Ethylstyrene	*R*	1.8	*S*	49

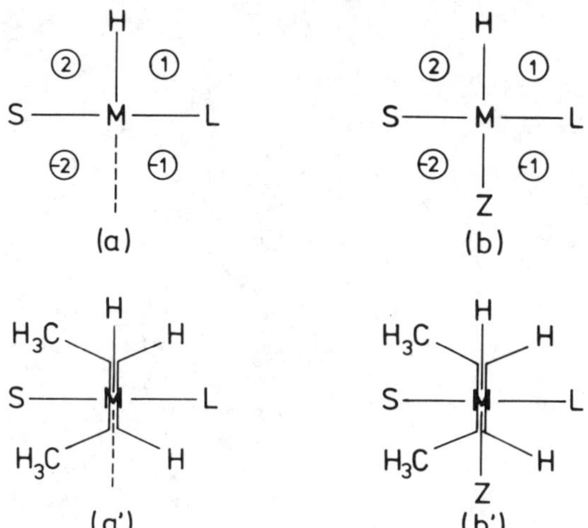

Figure 2. Models for the transition states controlling asymmetric induction in the hydrocarbonylation of olefins

bond of the substrate (parallel to the M–H bond) in each of the four quadrants—Q_1, Q_2, Q_{-1}, and Q_{-2}—(each one of which is defined by the axis connecting the metal atom with the double bond of the approaching substrate and by the projections of the bonds between two of the substituents and the central metal atom in a plane containing the M–H bond and perpendicular to the above axis) is different, depending on the size of the substituents, which in a rough approximation are considered as spheres, the centers of which are situated on the metal-to-ligand bond. It is expected that the space available in a quadrant for Figure 2b decreases in the following order: $Q_2 > Q_1 > Q_{-2} > Q_{-1}$. If the substrate is a monosubstituted ethylene, the substituent could be situated in either of the four quadrants, thus giving rise to four possible low-energy transition states.

If steric interactions are the main factors in determining the different free energies for the four possible transition states, the substituent would occupy preferentially that quadrant in which more space is available. If the quadrant preferred by the substituent is known, i.e., if the relative positions of L, S, Z, and H are known, we can predict the enantiomer that is formed prevailingly. The same model also allows us to predict which of the two unsaturated carbon atoms of the substrate will be bound to the metal and carbonylated eventually. For instance, in the case of a monosubstituted olefin (*see* Scheme VII), Isomer b will prevail over the sum of the antipodes of the other isomer (a + c) and Antipode a will prevail over Antipode c.

Scheme VII.

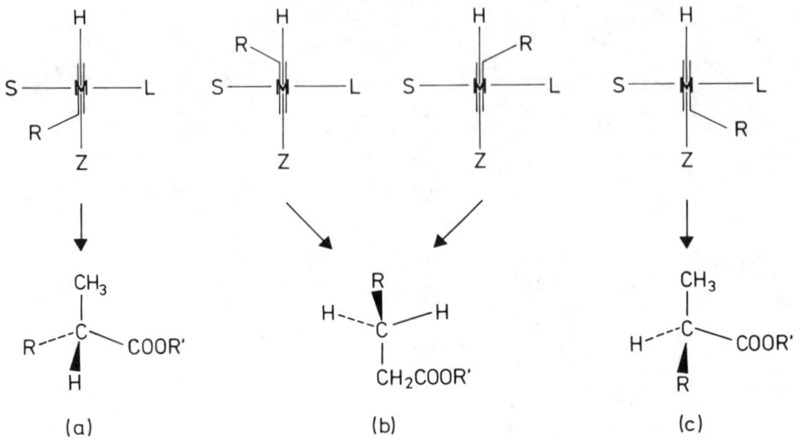

H H H H
R R
S—M—L S—M—L S—M—L S—M—L
R R
Z Z Z Z

CH_3 R CH_3
H---C—H
R---C—COOR' H---C—COOR'
H CH_2COOR' R

(a) (b) (c)

In order to decide if model 2a or 2b must be chosen and to establish the relative positions of L, S, and H in the model, the results of the hydrocarbalkoxylation of a substrate having C_s symmetry (e.g., 1-butene) have to be considered. The prevalence of the linear isomer over the branched one tells us that in Q_{-1} and Q_{-2} there is less space available than in Q_1 and Q_2; therefore Z must be present and Figure 2b is preferred to Figure 2a. Prevalence of the (*S*)-enantiomer in the branched isomer tells us that more space is available in Q_{-2} than in Q_{-1}; therefore the large ligand L must be settled in Figure 2b on the right side to the metal. The resulting situation is represented in Scheme VII.

Just as for the asymmetric hydroformylation experiments (*13*), the agreement between predicted and experimental results is very good (*see* Table VI). The only partial deviation is observed in the case of norbornene where the less hindered diastereoface is the only one reacted, but the prevailing enantiomer obtained is opposite to the one that is predicted. Another case in which prediction is uncertain is in the case of (*E*)-2-butene. In fact, the two methyl groups can occupy in either Q_1 and Q_{-2} or Q_2 and Q_{-1}. According to the predictions, space availability in Q_{-2} is larger than in Q_{-1}, and it is larger in Q_2 than in Q_{-1}. Furthermore, the results obtained with 1,1-disubstituted ethylenes show that the sum of the space available in Q_1 and Q_2 is much greater than the sum of the space available in Q_{-1} and Q_{-2}. The result with (*E*)-2-butene shows that the sum of the space available in Q_2 and Q_{-1} is greater than the sum of the space available in Q_1 and Q_{-2} contrary to what is observed in the hydroformylation by the rhodium–(-)DIOP catalytic system (*13*). Since we know that more space is available in Q_{-2} than in Q_{-1}, the difference in space availability in Q_2 with respect to Q_1 appears to be larger than that between Q_{-2} and Q_{-1}.

Table VI. Comparison Between Experiments and Forecast for the Prevailingly Carbonylated Unsaturated Carbon Atom and for Absolute Configuration in Asymmetric Hydrocarbalkoxylation by $PdCl_2/(-)$DIOP on the Basis of the Results of 1-Butene Hydrocarbalkoxylation

Substrate	*Prevailingly Formylated Unsaturated Carbon Atom*		*Prevailing Antipode in the Chiral Isomer*	
	Found	*Predicted*	*Found*	*Predicted*
1-Butene	1	—	(*S*)	—
3-Methyl-1-butene	1	1	(*S*)	(*S*)
3,3-Dimethyl-1-butene	1	1	(*S*)	(*S*)
Styrene	1	1	(*S*)	(*S*)
N-Vinylsuccinimide	1	1	(*S*)	(*S*)
2-Methyl-1-butene	1	1	(*R*)	(*R*)
2,3-Dimethyl-1-butene	1	1	(*S*)	(*S*)
2,3,3-Trimethyl-1-butene	1	1	(*S*)	(*S*)
2-Phenyl-1-butene	1	1	(*S*)	(*S*)
(*Z*)-2-Butene	2		(*R*)	(*R*)
(*E*)-2-Butene	2		(*S*)	
(*E*)-1-Phenyl-1-propene	[a]	2[a]	(*S*)	(*S*)[a]
Bicyclo-[2.2.1]-2-heptene	2		(1*R*,2*R*,4*S*)	(1*S*,2*S*,4*R*)
Methyl methacrylate	1	1	(*S*)	(*S*)

[a] 50% 4-Phenylbutanoate is formed.

Finally, the last uncertain case concerns the prevailing isomer arising from hydrocarbalkoxylation of (*E*)-1-phenyl-1-propene. In this case practically none of the predicted prevailing isomer 2-methyl-3-phenylpropanoate is formed. However, the prevailing ester (55%) is not the other expected isomer (2-benzylpropanoate) but rather the 4-phenylbutanoate which arises from isomerization of either the substrate ((*E*)-1-phenyl-1-propene to 3-phenyl-1-propene) or the catalyst–substrate complex.

The same transition-state model can be used to correlate the results for the enantioface-discriminating hydrocarbalkoxylation discussed above with the few results obtained in the enantiomer-discriminating hydrocarbalkoxylation, as will be discussed elsewhere (*32*).

Taking into account the experiments carried out with the same substrate but with different alcohols using the palladium–(-)DIOP catalytic system, the prediction of prevailing antipodes is not correct in one out of 21 experiments and the prediction of the prevailing isomer is not correct in two out of 24 experiments (*17*).

Unfortunately in hydrocarbonylation in the presence of chiral ligands other than DIOP, only 2-phenyl-1-propene was used as the substrate and therefore the model cannot be applied because the relative positions of the substituents L, S, H, and Z are unknown. However, the prevailing chirality and isomer observed in hydrocarbalkoxylation of 2-phenyl-1-propene with five other ligands, having structures similar to DIOP (*24*), indicate (*see* Table III) that in four cases out of five, the relative position of the substituents in the transition state should be the same as in the model of the transition state containing DIOP.

The situation is different with monophosphines as far as the prevailing isomer is concerned. In fact it appears (*see* Table III) that the branched isomer prevails over the linear one (*17*), contrary to the predictions of the model formulated for the palladium–(-)DIOP catalytic system. The available experimental data are not sufficient to allow us to formulate a model for the transition state in the case of palladium monophosphine–catalytic systems. To attempt a preliminary explanation of the isomeric composition found, we suggest a transition state having a geometry approximating a square pyramid in which the olefinic bond of the substrate interacts with three and not with four substituents (*see* Figures 2a and 2a′).

Final Remarks

Asymmetric hydrocarbonylation is a promising method for synthesizing optically active oxygenated compounds from prochiral olefins. Despite the reaction conditions, which include high carbon monoxide pressure and temperatures up to 120°C, the highest optical yields obtained up to now reach 47% (*33*) in the hydroformylation (1-butene with the Pt/DIOP system at 60°C) and 69% in the hydrocarbalkoxylation (2-phenyl-1-propene with a $PdCl_2$/DBP–DIOP catalytic system and *t*-butyl alcohol as the hydrogen donor). These results are far from the best optical yield achieved in asymmetric hydrogenation. However, asymmetric induction in hydrocarbonylation (as in the examples we have discussed) is based substantially on repulsive steric interactions, and improvements can be expected when polar groups in a suitable position, capable of attractive interactions with the metal atom of the catalyst, are present in the substrates.

The fact that a model for the transition state controlling asymmetric induction based on steric interactions allows us to correctly predict the type of prevailing regio- and stereoisomer for about 85% of the asymmetric hydrocarbonylation experiments (including hydroformylation and hydrocarbalkoxylation) is an indication that asymmetric induction in these catalytic reactions is based mainly on steric interactions. The data obtained so far do not allow us to establish whether the more stable or the less stable π-olefin complex intermediate is the one that reacts preferentially. However, the regularities that we observed indicate that the kinetic features are the same, at least in most of the experiments.

Researchers have assumed that the different prevailing chirality of the products arising from the same substrate in the hydrocarbalkoxylation with the palladium–(-)DIOP catalytic system and in hydroformylation with the rhodium–(-)DIOP catalytic system is connected with the different absolute configuration of the metal atom in the two different catalytic complexes. The different chiralities of the metal atom, where the asymmetric ligand is the same, should be connected with different nonchiral ligands in the metal complexes. In fact, the position of the diastereomeric equilibrium (*30, 31*) involving inversion of configuration at the metal atom (and hence the prevalence of metal atoms having the same chirality) generally depends on the nature of the ligands.

Acknowledgment

We thank the Schweizerischer Nationalfonds zur Förderung der wissenschaftlichen Forschung for financial support.

Literature Cited

1. Himmele, W.; Siegel, H.; Aquila, W.; Mueller, F. J. *Ger. Offen.* 1973, *2*, 132–414; *Chem. Abstr.* 1973, *78*, 97328.
2. Botteghi, C.; Consiglio, G.; Pino, P. *Chimia* 1972, *26*, 141.
3. Consiglio, G.; Botteghi, C.; Salomon, C.; Pino, P. *Angew. Chem.* 1973, *85*, 663.

4. Kagan, H. B.; Dang, T. P. *J. Amer. Chem. Soc.* **1972,** *94,* 6429.
5. Pino, P.; Consiglio, G.; Botteghi, C.; Salomon, C. In "Homogeneous Catalysis—II" *Adv. Chem. Ser.* **1974,** *132,* 295.
6. Botteghi, C.; Consiglio, G.; Pino, P. *Chimia* **1973,** *27,* 477.
7. Tanaka, M.; Ideka, Y.; Ogata, I. *Chem. Lett.* **1975,** 1115.
8. Hsu, C. Y., Ph.D. Dissertation, Univ. of Cincinnati, 1974; *Chem. Abstr.* **1975,** *82,* 154899.
9. Consiglio, G.; Pino, P. *Helv. Chim. Acta* **1976,** *59,* 642.
10. Hayashi, T.; Tanaka, M.; Ikeda, Y.; Ogata, I. *Bull. Chem. Soc. Jpn.* **1979,** *52,* 2605.
11. Piancenti, F.; Menchi, G.; Frediani, P.; Matteoli, U.; Botteghi, C. *Chim. Ind. (Milan)* **1978,** *60,* 808.
12. Tinker, H. B.; Solodar, A. J., *Can. Pat.* **1973,** 1,027,141; *Chem. Abstr.* **1978,** *89,* 42440.
13. Pino, P.; Consiglio, G., unpublished data.
14. Pino, P.; Piacenti, F.; Bianchi, M. In "Organic Syntheses via Metal Carbonyls"; Wender, I.: Pino, P., Eds.; John Wiley & Sons: New York, 1977.
15. Henry, P. M. "Palladium Catalyzed Oxidation of Hydrocarbons"; Reidel: Dordrecht, Holland, 1980; p. 193.
16. Consiglio, G.; Pino, P. *Gazz. Chim. Ital.* **1975,** *105,* 1133.
17. Wenzinger, F., Ph.D. Dissertation, Eidgenoessische Technische Hochschule, Zurich, 1978.
18. Consiglio, G.; Pino, P. *Chimia* **1976,** *30,* 193.
19. Consiglio, G. *Helv. Chim. Acta* **1975,** *59,* 124.
20. Consiglio, G.; Marchetti, M. *Chimia* **1976,** *30,* 26.
21. Sugi, Y.; Bando, K.; Shin, S. *Chem. Ind. (London)* **1975,** 397.
22. Consiglio, G.; Haelg, P.; Pino, P. *Rend. Accad. Naz. Lincei,* in press.
23. Becker, Y.; Eisenstadt, A.; Stille, J. K. *J. Org. Chem.* **1980,** *45,* 2145.
24. Hayashi, T.; Tanaka, M.; Ogata, I. *Tetrahedron Lett.* **1978,** 3925.
25. Morrison, J. D.; Mosher, H. S. "Asymmetric Organic Reactions"; Prentice-Hall: Englewood Cliffs, NJ, 1971.
26. Consiglio, G. *J. Organometal. Chem.* **1977,** *132,* C26.
27. Bianchi, M.; Benedetti, E.; Piacenti, F. *Chim. Ind. (Milan)* **1969,** *51,* 613.
28. Pino, P.; Stefani, A.; Consiglio, G.; In "Catalysis in Chemistry and Biochemistry, Theory and Experiment"; Pullman, B., Ed.; Reidel: Dordrecht, Holland, 1979.
29. Stefani, A.; Consiglio, G.; Botteghi, C.; Pino, P.; *J. Amer. Chem. Soc.* **1977,** *99,* 1058.
30. Pino, P.; Consiglio, G.; In "Fundamental Research in Homogeneous Catalysis"; Tsutsui, M., Ed.; Plenum: New York and London, 1978; p. 519.
31. Brunner, H. *Adv. Organometal. Chem.* **1980,** *18,* 151.
32. Pino, P.; Consiglio, G., unpublished data.
33. Haelg, P.; Consiglio, G.; Pino, P., unpublished data.

RECEIVED July 10, 1980.

24

P-31 NMR Comparisons of the Crystalline and Solution States of Rhodium(I) Diphosphine Catalysts

GARY E. MACIEL, DANIEL J. O'DONNELL, and RANDY GREAVES
Department of Chemistry, Colorado State University, Fort Collins, CO 80523

Six different rhodium(I) diphosphine catalysts and the corresponding ligands were investigated via P-31 NMR spectroscopy. For the solid samples cross polarization/magic-angle spinning (CP/MAS) was used. A comparison of the solid-state spectra with solution NMR spectra showed a general pattern of similarity for most of the compounds, but a different degree of phosphorus atom nonequivalence between the two states for some of the compounds. P-31 NMR data were obtained also on a polymer-supported diphosphine ligand and its corresponding rhodium(I) catalyst. Prospects and limitations for using NMR as a structural tool for comparing the solid and liquid states are discussed.

Several correlations have appeared in the recent literature (*1, 2, 3*) concerning the use of X-ray-determined crystal structures of rhodium(I) diphosphine catalysts to aid in predicting the chirality induced in the hydrogenation products obtained by using these homogeneous catalysts. Such solid-vs.-liquid-structure correlations always present an inherent uncertainty concerning the validity of the extrapolation of structural data obtained by solid-state methods to the solution state. Since most of the reliable, detailed structural data available to chemists have been obtained by diffraction experiments on solids and since most chemical processes are carried out in solution, this kind of uncertainty is not uncommon. NMR, with the advent of the cross polarization and magic-angle spinning (CP/MAS) methods for solids (*4, 5*), presents a clear opportunity to examine directly the relationship between the structures of the solid and liquid.

0065–2393/82/0196–0389$5.00/0

Until rather recently, high-resolution NMR techniques were limited to the liquid state, in which rapid, random molecular tumbling motions eliminate the splittings and line broadening effects of magnetic dipole–dipole interactions and the line-broadening effects of chemical shift anisotropy, and give rise to efficient spin–lattice relaxation mechanisms that make possible frequent repetitions of pulse NMR experiments. However with solid samples these motions are absent, and the application of standard liquid-state (pulse Fourier transform) approaches is largely futile. Recently the use of high-power H-1 decoupling (for minimizing line broadening due to dipolar interactions with protons) with cross polarization (to eliminate the bottleneck of long spin–lattice relaxation times) has been combined with MAS (which eliminates the line broadening due to chemical shift anisotropy) to provide an NMR technique for spin-one-half nuclei in solids that approaches the liquid-state techniques in resolution characteristics (*4, 5*). It is this combination that makes the detailed structural comparison of solid and liquid states possible.

Presented herein are the results of a P-31 NMR study of the solid states and liquid solution states of the following rhodium catalysts and their constituent diphosphine ligands: $(Rh(COD)diphos)^+$ $C10_4^-$, **I**; $(Rh(COD)DiPAMP)^+$ $C10_4^-$, **II**; $(Rh(COD)BPPM)^+$ $C10_4^-$, **III**; $(Rh(NBD)chiraphos)^+$ $C10_4^-$ and $(Rh(COD)chiraphos)^+$ $C10_4^-$, **IVa** and **IVb**; and $(Rh(NBD)prophos)^+$ $C10_4^-$, **V**. (COD = cyclooctadiene; NBD = norbornadiene; diphos = 1,2-ethanediylbis(diphenylphosphine); DiPAMP = (*R*,*R*)-1,2-ethanediylbis(*o*-methoxyphenylphenylphosphine); BPPM = (2*S*,4*S*)-*N*-*t*-butoxycarbonyl-4-diphenylphosphine-2-diphenylphosphinomethylpyrrolidine; chiraphos = (*S*,*S*)-2,3-butanediylbis(diphenylphosphine); and prophos = (*R*)-1,2-bis(diphenylphosphino)propane. The spectra show some distinct differences between the phosphine environments of the solid and solution states, implying that significant structural differences exist which may be important when considering the prediction of solution properties from solid-state structures. However, many uncertainties remain regarding the origins of nonequivalences observed in the solid-state NMR spectra. In any case, this chapter presents the first extensive collection of solid-state P-31 NMR data on phosphine ligands and/or their metal complexes.

Experimental

Preparation of the Ligands. *S*,*S*-chiraphos and *R*-prophos were prepared via methods given in the literature (*6, 7*). Diphos was obtained from Strem Chemicals. BPPM was generously supplied by K. Achiwa and *R*,*R*-DiPAMP also was generously supplied by W. Knowles.

Preparation of the Complexes. $[Rh(COD)Cl]_2$ and $[Rh(NBD)Cl]_2$ were prepared by Chatt and Venanzi's method (*8*) for $[Rh(COD)Cl]_2$. These were used to prepare Rh(COD)acac and Rh(NBD)acac by Cramer's method (9) for $Rh(C_2H_4)_2$acac. The complexes $[Rh(COD)diphos]^+$ ClO_4^-, **I**; $[Rh(COD)$-$DiPAMP]^+$ ClO_4^-, **II**; $[Rh(NBD)chiraphos]^+$ ClO_4^-, **IVa**; $[Rh(NBD)prophos]^+$ ClO_4^-, **V**; and $[Rh(COD)chiraphos]^+$ ClO_4^- were prepared via Schrock and Osborn's method (*10*) from the corresponding Rh(diene)acac precursor. The complex $[Rh(COD)BPPM]^+$ ClO_4^-, **III**, was prepared from $[Rh(COD)Cl]_2$ via the method for $[Rh(COD)P(C_6H_5)_2CH_{32}{}^*]^+$ ClO_4^- by Schrock and Osborn (*10*).

NMR Spectra. All solution P-31 spectra were obtained on a Nicolet NT-150 spectrometer operating at 60.7455 MHz. All solution samples were prepared in a K.S.E. glove box under a nitrogen gas atmosphere. Solution spectra of the ligands were obtained in $DCCl_3$. The solution spectra of the prophos and diphos complexes also were obtained in $DCCl_3$. Spectra of the complexes of chiraphos, BPPM, and *R,R*-DiPAMP were taken in acetone-d_6. The solid-sample spectra were obtained on a JEOL FX-60Q spectrometer modified to produce CP/MAS spectra at a frequency of 24.22 MHz (P-31) and on a home-built multi-nuclear CP/MAS spectrometer based on a Varian HR-60 magnet. Samples were packed in 10 or 13-mm (diameter) bullet rotors and were spun at the magic angle at speeds of about 2500 Hz.

Results and Discussion

The P-31 NMR spectra of solid and solution samples of the diphosphine ligands and complexes studied in this investigation are shown in Figures 1–11. The corresponding data are summarized in Tables I and II. The ligands show a chemical shift range in either the solid or solution states of about 23 ppm, roughly corresponding to the internal shift between the two chemically nonequivalent phosphorus atoms of prophos. The diphosphine complexes show a chemical shift range of 47.5 ppm in the liquid state and about 51 ppm in the solid state.

Inspection of the data in Tables I and II show for both ligands and complexes some apparent shifts between the solid and liquid states that appear to be large enough to reflect significant structural differ-

Table I. P-31 NMR Data of the Diphosphine Ligands

Ligand	*Solution*[a] δ_p*(ppm)*[b]	*Solid* δ_p*(ppm)*
Diphos	−11.9	−12.5
DiPAMP	−21.0	−25.6
BPPM	−7.3, −21.9	−14.8, −21.7
Chiraphos	−8.7	−11.9, −14.0
Prophos	2.3, −20.0 (J_{P-P} = 20.4)	−4.5, −27.3

[a] The solvent was $DCCl_3$.
[b] Chemical shifts given with respect to 85% H_3PO_4.

Table II. P-31 NMR Data of Rhodium(I) Diphosphine Catalysts

Catalyst	Solution[a]		Solid	
	δ_P(ppm)[b]	J_{Rh-P}(Hz)	δ_P(ppm)	J_{Rh-P}(Hz)
$[Rh(COD)diphos]^+ClO_4^-$ **I**	58.3	132.1	56.8	126
$[Rh(COD)DiPAMP]^+ClO_4^-$ **II**	52.3	150.9	51.6, 45.8	138, 145
$[Rh(COD)BPPM]^+ClO_4^-$ **III**	43.6[c]	146.4 (J_{P-P} = 36)	38.9	[d]
	43.4[c]	144.8 (J_{P-P} = 37)		
	14.7	139.5	12.7	
$[Rh(NBD)chiraphos]^+ClO_4^-$ **IVa**	59.2	154.0	62.2, 61.4	~ 107, ~ 133
$[Rh(COD)chiraphos]^+ClO_4^-$ **IVb**	56.6	146.2	57.6, 54.6	~ 117, ~ 125
$[Rh(NBD)prophos]^+ClO_4^-$ **V**	62.2	156 (J_{P-P} = 33)	64.0	162
	43.6	155 (J_{P-P} = 34)	43.4	124
$[Rh(NBD)diop]^+ClO_4^-$ **VI**	13.6	154	14.2	[d]

[a] The solvent was either $DCCl_3$ (*see* Complexes I and V) or acetone-d_6 (*see* Complexes II, III, and IV).
[b] Chemical shifts given with respect to 85% H_3PO_4.
[c] Assumed to represent two conformational forms of the $-CH_2-P(C_6H_5)_2$ phosphorus.
[d] Not observable.

ences. However, one should keep in mind that all of the chemical shifts were determined by substitution, which does not take differences in bulk susceptibilities into account. Such effects can be expected to amount to up to a few parts per million, as one is not concerned only with differences in compound identities, but also with different states of matter and major differences in density. For this reason, the solid-state chemical shifts should be considered to be uncertain by up to about ±3 ppm. Of special interest in this regard are the internal chemical shifts obtained for the nonequivalent phosphorus atoms within a given compound, i.e., BPPM and prophos and their complexes. Such internal shifts do not bear uncertainties associated with bulk susceptibility differences. The prophos ligand shows an internal shift difference of 22.3 ppm in solution and 22.8 ppm in the solid state; for the prophos complex (Structure V) the internal shift is 18.6 ppm in the solution sample and 20.6 ppm in the solid state. These differences are sufficiently small to indicate no major structural differences between the solid and solution states of each of these two compounds. For the BPPM ligand, the internal chemical shift difference is 14.6 ppm in the solution state and 6.9 ppm in the solid; the corresponding values for the BPPM complex are 28.8 ppm for the solution state and 26.2 ppm for the solid state. The fact that the internal shift difference for the BPPM ligand is nearly three times as large as for the corresponding rhodium(I) complex may be due to the constraining influence of rhodium–phosphorus bonding on the geometry of the ligand. In the absence of this constraining effect, the geometry of the ligand is more likely to respond to crystal packing influences, yielding a somewhat different structure (e.g., conformation) in the solid state than in solution.

In general the solid-state spectra in Figures 1–11 have considerably broader lines, and in some cases a related decrease in fine structure. These differences in linewidths are typical of the difference in resolution limits that one finds between the spectra taken on solids by CP/MAS and on solutions. In a few cases one actually sees additional fine structure (or apparently higher multiplicities) in the solid-state spectra in comparison with the corresponding compounds in solution. Such situations clearly correspond to cases in which there is greater structural diversity in the solid state than in solution, and they clearly show the great structural sensitivity of CP/MAS P-31 NMR of solids.

One can imagine several possible sources of the higher level of structural diversity observed for some of the solids in comparison with the liquids. All of these possibilities involve structural differences that are locked in by the rigidity of the solid state, but they are lost in the motional averaging made possible by the structural flexibilities of molecules in the liquid state. These structural differences can be dependent on chemically very consequential factors such as conformation. They also might be due to such considerations as crystallographic nonequivalences or even polymorphism, which are not related so directly to local structure or the molecular correlations between molecular structure and reactivity which are so interesting in chemistry. One does not expect nonlocal chemical nonequivalences to give rise to chemical shift differences of very large magnitude, but long-range effects of a few parts per million in P-31 NMR spectra cannot be ruled out. Furthermore, in order to rule out or evaluate phosphorus nonequivalences of nonlocal crystallographic origin and to unequivocally identify solid-state NMR data with solid-state structures based on X-ray diffraction, one must have the detailed diffraction study of the same sample on which the NMR experiments are carried out or at least on a crystallographically equivalent sample. That is, the small crystal used for the X-ray diffraction study must be representative of the large sample used in the P-31 CP/MAS NMR experiments. While these requirements have not been met in the present study, we have made every reasonable effort to use samples prepared in the same ways as those used by others in the relevant X-ray studies (exceptions are noted in the text). In this regard it may be pertinent to note that isolated structural disorder that might render a crystal useless for X-ray analysis would not contribute in a significant way to the bulk NMR properties of the sample and probably would not give an observable effect in the NMR spectra. Indeed it is quite likely that many of the crystals used in this study were "flawed" in the sense that they would be useless for X-ray diffraction analysis.

One interesting way to view the data given in Tables I and II is to consider the complexation shift for each ligand, i.e., the P-31 chemical shift of the ligand complex minus the P-31 shift of the ligand. These values are summarized in Table III for the solution- and solid-state data. One sees that for most of the complexes these shift differences

Table III. Complexation Shifts[a]

Ligand (Complex)	*Solution (ppm)*	*Solid State (ppm)*
Diphos (**I**)	70.2	69
DiPAMP (**II**)	73.3	74
BPPM (**III**)	50.8[b], 36.6	54, 35
Chiraphos (**IVa** and **IVb**)	67.9	75
	65.3	68
Prophos (**V**)	59.9, 63.6	68, 71

[a] P-31 chemical shift of complex minus P-31 chemical shift of ligand.
[b] Obtained by considering the average peak position of the multiplets at 43.4 and 43.6 ppm for the complex.

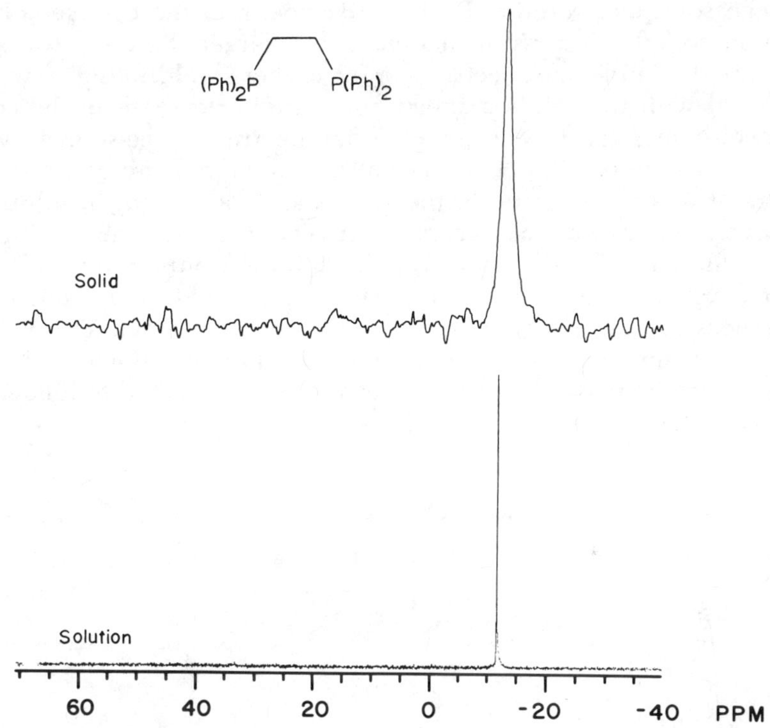

*Figure 1. Solid-state (*top*) and solution (*bottom*) P-31 NMR spectra of diphos. The solution spectrum was obtained in $DCCl_3$. The scale is in parts per million from external 85% H_3PO_4.*

correspond rather well between the solution- and solid-state data. The largest deviations (Complexes **IVa** and **V**) are 7 or 8 ppm, nearly within the experimental uncertainties of subtraction of two solid-state chemical shift values. From this point of view, it would appear that in general there are no dramatic structural differences between the solid and solution states of these rhodium(I) diphosphine complexes. Nevertheless, there are some indications described below, where each system is discussed briefly, in which significant solid-vs.-solution structural differences may exist.

Diphos

The solid-state and liquid-solution-state P-31 NMR spectra of the diphos ligand are presented in Figure 1. A small difference in shielding was observed between the solution spectrum of the diphos ligand (−11.9 ppm relative to external H_3PO_4) and the corresponding solid-state spectrum (−12.5 ppm).

The solid and solution P-31 NMR spectra of the corresponding rhodium complex are given in Figure 2. A larger shielding was observed in the solid-state spectrum for the higher-shielding signal when compared with the solution spectrum. A slight decrease in the coupling constant, J_{Rh-P}, was observed in the spectrum of the solid as well (132 Hz in solution, 126 Hz in the solid), which may indicate a small constraint on the molecule in the solid state not present in solution. However, the overall comparison between the solid and solution spectra for diphos and the derived rhodium(I) complex revealed no major differences between the two states that could be attributed to differences in the configuration about the rhodium atom. This is not very surprising since diphos is a symmetrical achiral ligand with no sterically positioned substituents that might be expected to influence drastically the structure in the solid state.

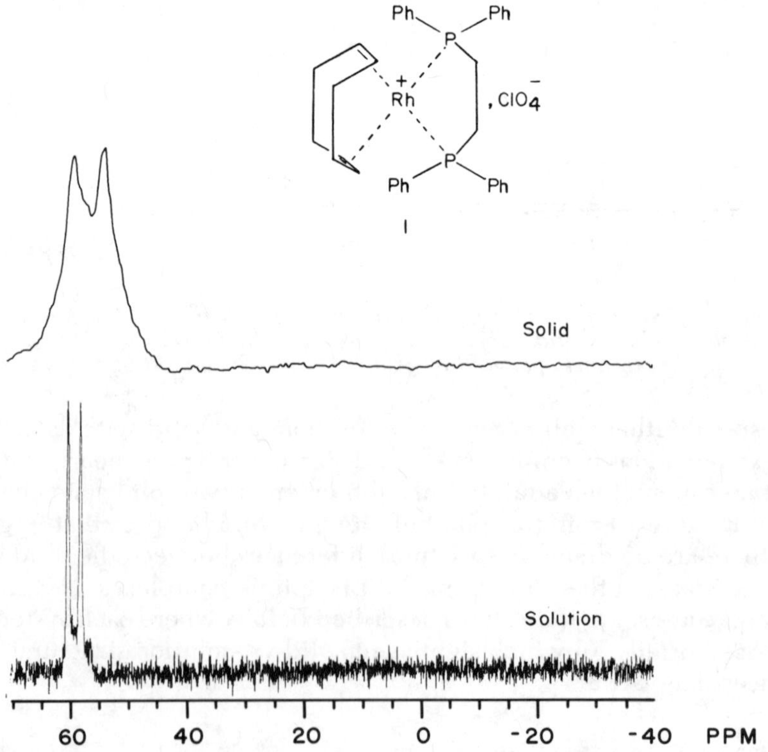

*Figure 2. Solid-state (*top*) and solution (*bottom*) P-31 NMR spectra of diphos complex **I**. The solution spectrum was obtained in $DCCl_3$. The scale is in parts per million from external 85% H_3PO_4.*

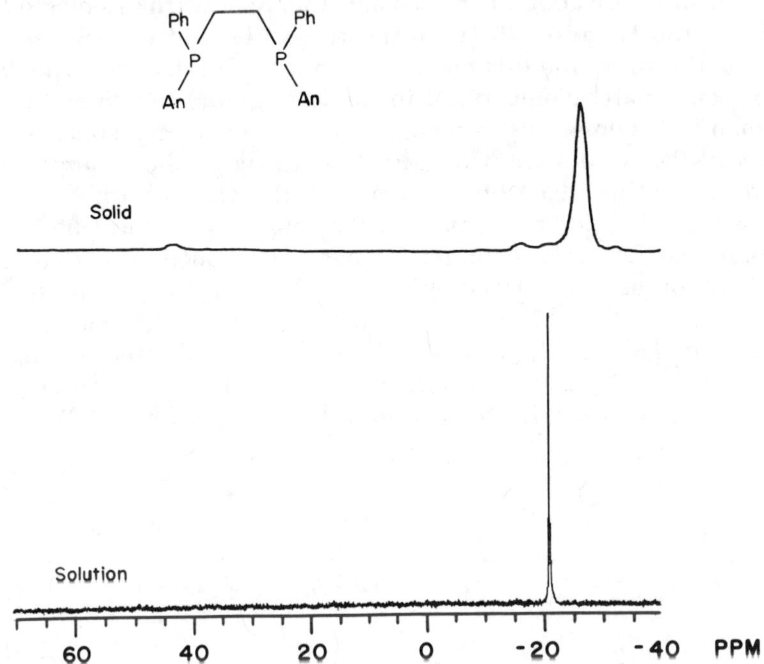

*Figure 3. Solid-state (*top*) and solution (*bottom*) P-31 NMR spectra of DiPAMP. The solution spectrum was obtained in $DCCl_3$. The scale is in parts per million from external 85% H_3PO_4.*

DiPAMP

The solid and solution P-31 NMR spectra of the DiPAMP ligand are presented in Figure 3. A shift of −4.5 ppm (increased shielding) was observed for the singlet observed in the solid (−25.6 ppm) compared with the solution spectrum (−21.0 ppm). Overall, the spectra resemble the corresponding spectra for diphos, and hence the same conclusions could be drawn—i.e., there are no major structural (including conformational) differences between the solid and solution states of the ligand. It should be noted that the DiPAMP ligand is chiral at each phosphorus atom, as opposed to the achirality of the diphos ligand.

The solid and solution P-31 NMR spectra for the corresponding rhodium–DiPAMP complex are given in Figure 4. The solution spectrum displayed the expected doublet centered at 52.3 ppm, with a J_{Rh-P} of 150.9 Hz. The solid, however, showed an apparent triplet centered at 48.8 ppm. It is proposed that this group of signals is in fact a set of two overlapping doublets, corresponding to nonequivalent phosphorus atoms, each of them with Rh-103–P-31 spin–spin coupling, in the solid state. According to this interpretation the two contributing chemical shifts are about 51.6 ppm and 45.8 ppm, with associated J_{Rh-P} values of 136 Hz and 145 Hz, respectively. This would suggest that very definite structural influences exist in the solid structure which are not present or are averaged out in solution, giving rise to constraints that manifest themselves as nonequivalences of the phosphorus resonances in the solid state. Comparison with the diphos complex (*see* Figure 2) implies that this constraint is associated with the orthomethoxy group and/or the presence of chirality at the phosphorus atoms in the DiPAMP complex. It has been concluded from X-ray studies (*3*) of the DiPAMP complex (with BF_4^- anion) that the specific arrangement of the chiral centers, and more exactly the phenyl substituents on the phosphorus atoms, impose chirality in the hydrogenation substrate. However, no significant structural difference between the phosphorus atoms is obvious from the X-ray results (*1, 3*).

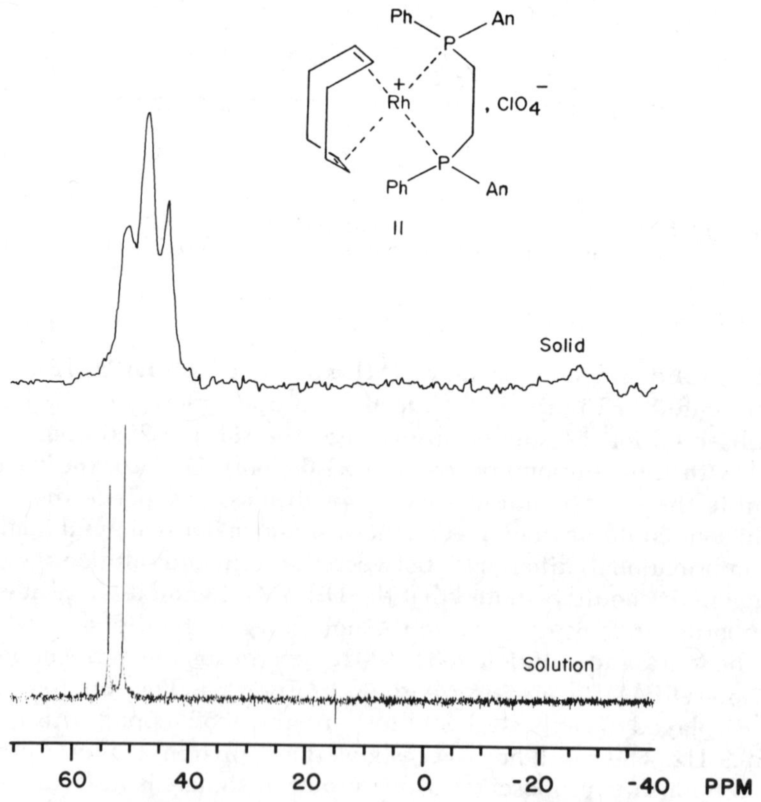

*Figure 4. Solid-state (*top*) and solution (*bottom*) P-31 NMR spectra of the DiPAMP complex **II**. The solution spectrum was obtained in acetone-d₆. The scale is in parts per million from external 85% H_3PO_4.*

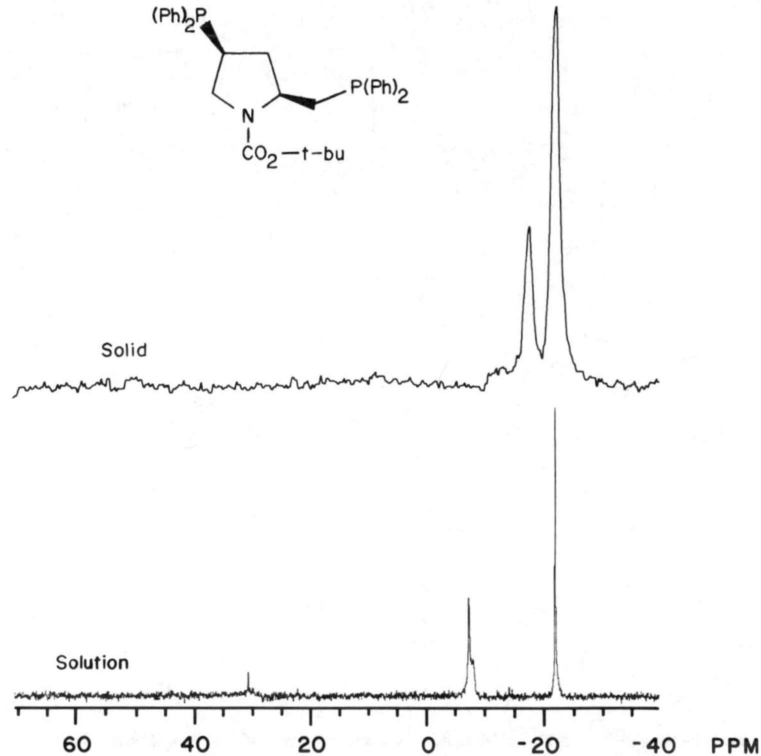

*Figure 5. Solid-state (*top*) and solution (*bottom*) P-31 NMR spectra of BPPM. The solution spectrum was obtained in $DCCl_3$. The scale is in parts per million from external 85% H_3PO_4.*

BPPM

The solid and solution P-31 NMR spectra of the BPPM ligand are given in Figure 5. The two signals in the solid-state spectrum of the ligand occur at −14.8 and −21.7 ppm, compared with −7.3 and −21.9 ppm for the solution spectrum. Thus, the lines in the solid-state spectrum were more than 7 ppm closer to each other in chemical shift than the signals in the solution spectrum, apparently because the shielding of the less shielded phosphorus nucleus is about 7 ppm higher in the solid than in the solution. Nevertheless, the two spectra clearly resembled one another qualitatively.

The solid and solution P-31 NMR spectra of the BPPM–rhodium complex are shown in Figure 6. The solution spectrum was obtained in acetone-d_6 as the solvent, and displays an unusual coupling pattern when compared with the previously published spectrum obtained in methanol-d_4 (*11, 12*). The spectrum shown in Figure 6 displays four sets of phosphorus signals: two sets of four signals are centered at 43.5

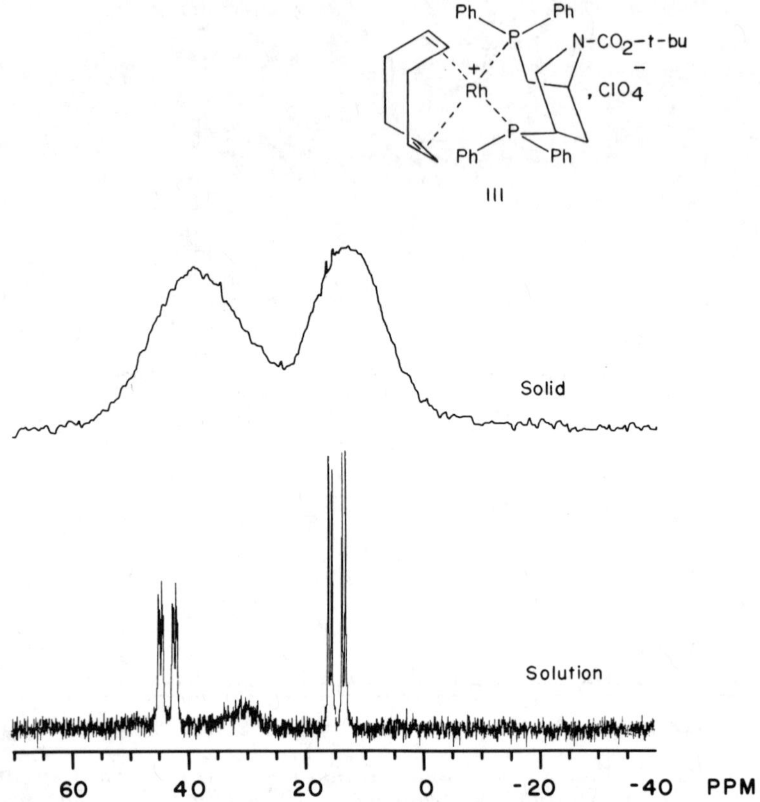

*Figure 6. Solid-state (*top*) and solution (*bottom*) P-31 NMR spectra of the BPPM complex **III**. The solution spectrum was obtained in acetone-*d_6*. The scale is in parts per million from external 85%* H_3PO_4.

ppm, arising from the diphenylphosphinomethylphosphorus (*see* Complex **III**) in two different structural conformers in solution, and two sets of doublets are centered at 14.7 ppm and assigned to the diphenylphosphinophosphorus (*see* Complex **III**). The spectra previously published by Achiwa (*11*) and Ojima (*12*) showed two sets of doublets (resulting from P-31–P-31 and Rh-103–P-31 splittings) centered at ca. 42 ppm (diphenylphosphinomethylphosphorus) and two sets of four signals (resulting from two phosphorus environments, each with P-31–P-31 and Rh-103–P-31 splittings) centered at ca. 15 ppm (diphenylphosphinophosphorus). The interchange of splitting patterns between the two phosphorus signals on changing the solvent must result from a change in the nature of solvent interaction with the complex. This alters the relative influence on the conformational populations of the local structures about the two phosphorus atoms in the ligand.

The solid-sample P-31 NMR spectrum of the complex (*see* Figure 6) displayed two broad signals that correspond to the two different phosphorus types in BPPM. The extreme broadness observed in this spectrum can result from both the presence of broadening due to the quadrupolar N-14 nucleus and the underlying multiple signal patterns evident in the solution spectra.

Chiraphos

The solid and solution P-31 NMR spectra of the chiraphos ligand are displayed in Figure 7. The solid-state spectrum of the ligand is notable since two signals, one at −11.4 ppm and one at −14.0 ppm, were observed, whereas the solution spectrum revealed only one signal at −8.7 ppm. This difference could result from crystal-packing forces in the solid state which could dictate a conformational arrange-

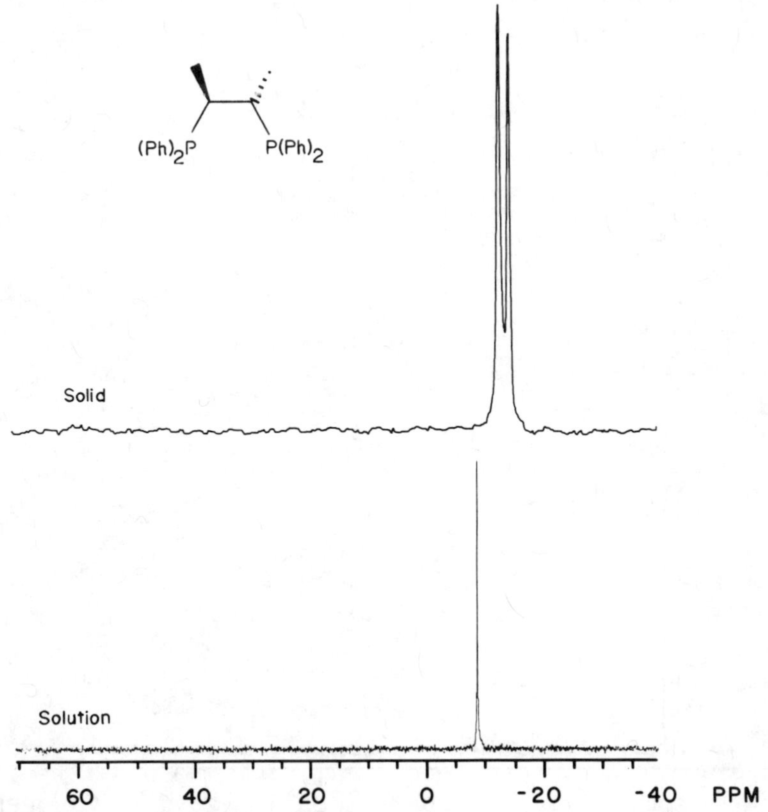

*Figure 7. Solid-state (*top*) and solution (*bottom*) P-31 NMR spectra of chiraphos. Solution spectrum was obtained in* $DCCl_3$*. The scale is in parts per million from external 85%* H_3PO_4.

ment that makes the phosphorus atoms nonequivalent. The absence of these forces and the presence of rotational motions in the liquid lead to only one phosphorus environment in that state. This feature may be associated with the chiral carbons of the butane backbone of the molecule, since this was the only instance in which nonequivalence occurred in the solid without a corresponding nonequivalence in the solution spectrum for any of the ligands.

The solid and solution P-31 NMR spectra of the chiraphos–rhodium complex are given in Figure 8. The solution spectrum showed the expected doublet centered at 59.2 ppm, arising from equivalent phosphorus atoms coupled to the rhodium atom (J_{Rh-P} = 154 Hz). The solid-state spectrum appears to display four overlapping signals at 64.6, 64.1, 60.0, and 58.6 ppm. It is suggested here that these signals arise from two nonequivalent phosphorus atoms with chemical shifts of 62.2 and 61.4 ppm, each coupled to the rhodium atom, with approximate coupling constants of about 107 Hz

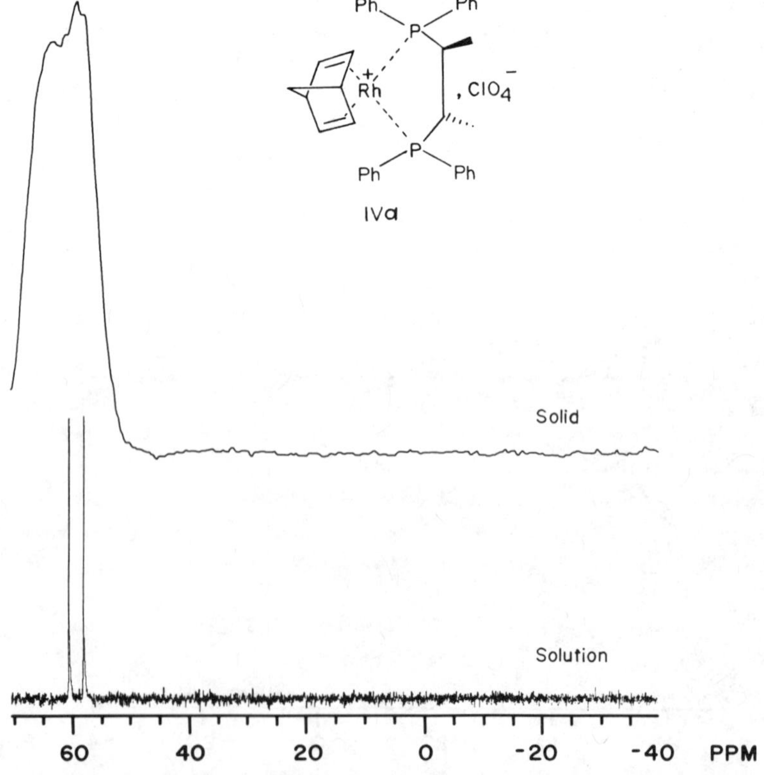

*Figure 8. Solid-state (*top*) and solution (*bottom*) P-31 NMR spectra of the chiraphos complex* **IVa***. The solution spectrum was obtained in acetone-*d$_6$*. The scale is in parts per million from external 85%* H_3PO_4.

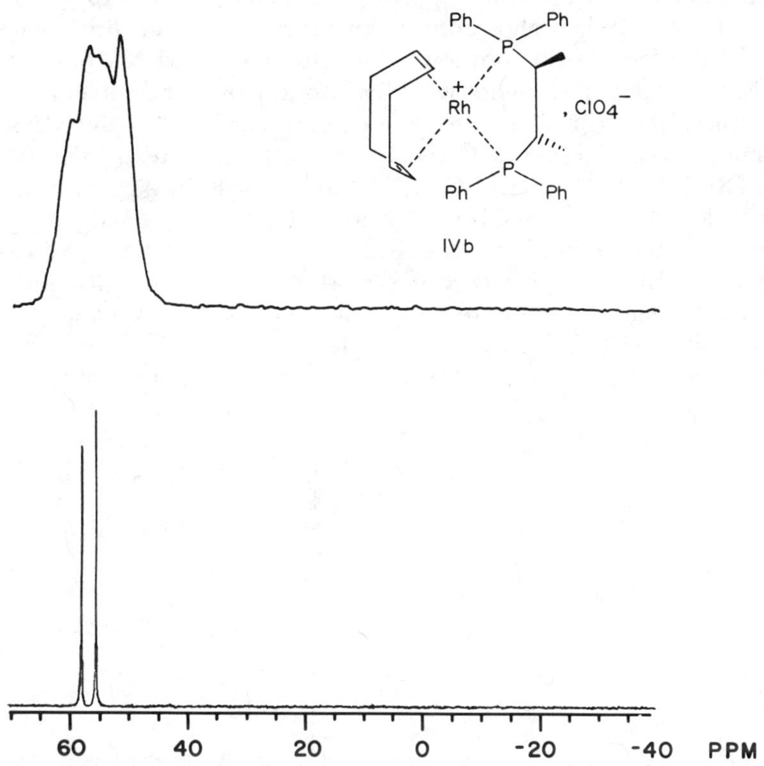

Figure 9. Solid-state (top) *and solution* (bottom) *P-31 NMR spectra of* [*Rh(COD)chiraphos*] $^{+}ClO_4^{-}$, *Complex* ***IVb***. *The scale is in parts per million from external 85% H_3PO_4.*

and 133 Hz. The breadth of the lines prevented an accurate determination of the coupling constants. However, the presence of nonequivalency between the phosphorus atoms in the solid state seems likely, indicating the occurrence of another solid-vs.-solution structural difference. It would appear that, whatever forces cause the nonequivalence in the solid state, they must be somewhat weaker in the case of chiraphos when compared with DiPAMP, for which a larger chemical shift difference was observed (*see* earlier discussion). The center of chirality of the chiraphos ligand is further away from the rhodium atom than is the center of chirality of the DiPAMP ligand, and this increase in distance may be acting to lessen the total effect seen for chiraphos.

In order to test the effect of the diene on the nature of the catalysts in the solid state, spectra of the solid and solution state of [Rh(COD)chiraphos]$^{+}$ $C10_4^{-}$ also were obtained and are presented in Figure 9. The solution spectrum again displayed the expected doublet, centered at 56.6 ppm, (J_{Rh-P} = 146.2). The differences between the chemical shifts and coupling constants in the solution spectra of the COD and NBD chiraphos complexes must arise from differences induced by the diene in solution. The solid-state P-31 NMR spectrum (*see* Figure 9) again showed two overlapping doublets, one centered at 57.6 ppm (J_{Rh-P} = 117 Hz) and one centered at 54.6 (J_{Rh-P} = 125 Hz). Comparing these values with those obtained from the spectrum of the solid [Rh(NDB)chiraphos]$^+$ $C10_4^-$ catalyst (*see* Figure 8), one can see very little effect in the solid spectra which could be assigned to an influence of the diene. Since the X-ray structure of this complex (2) has shown that the phosphorus atoms in this complex are situated very nearly identically, the differences between the solid-state and solution spectra of the COD–chiraphos complex must arise from very subtle constraints in the crystal, possibly arising from the nature of the chiral ligand.

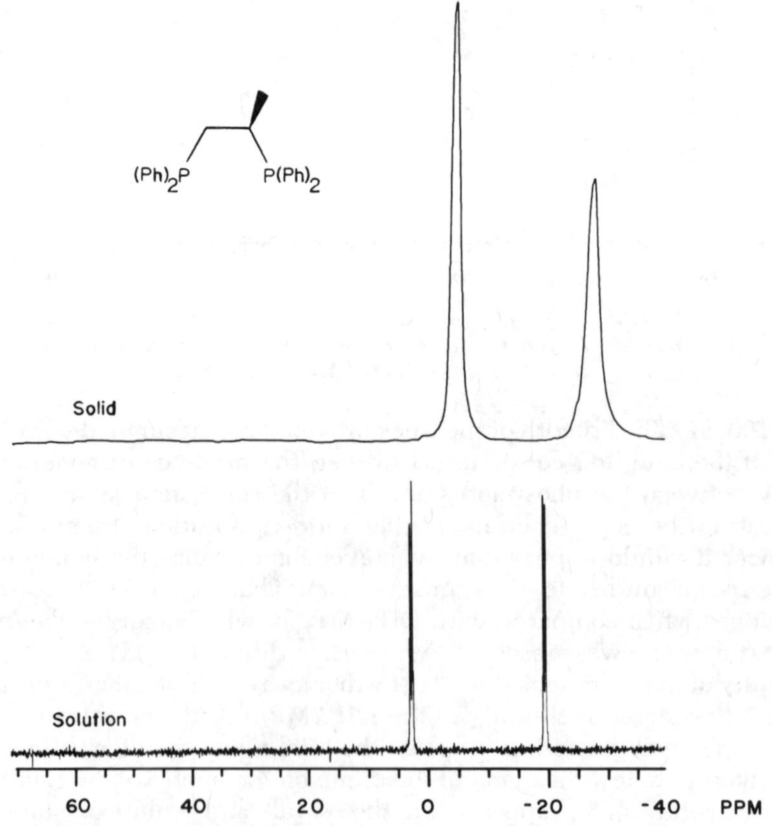

*Figure 10. Solid-state (*top*) and solution (*bottom*) P-31 NMR spectra of prophos. The solution spectrum was obtained in $DCCl_3$. The scale is in parts per million from external 85% H_3PO_4.*

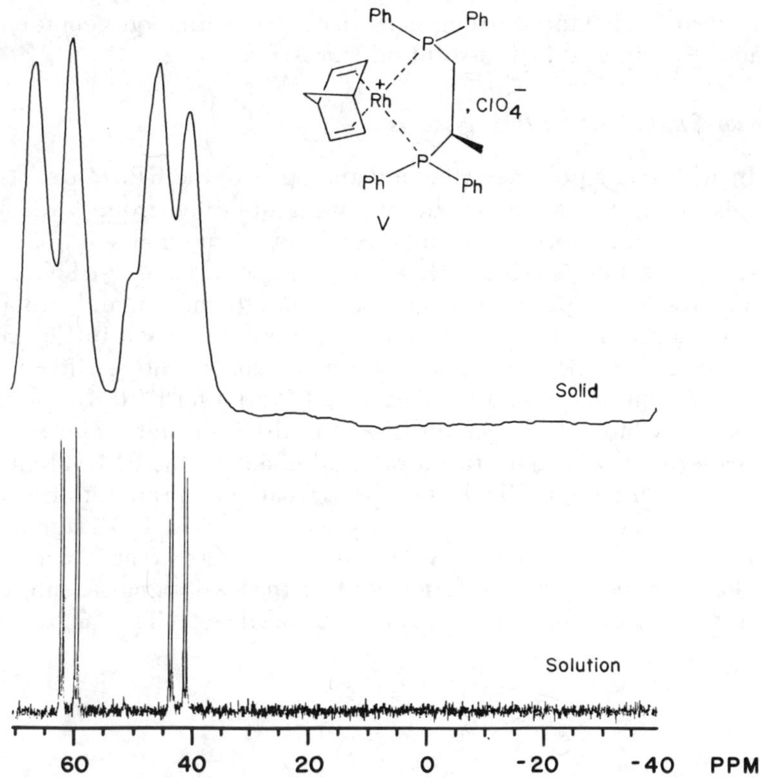

*Figure 11. Solid-state (*top*) and solution (*bottom*) P-31 NMR spectra of the prophos complex V. The solution spectrum was obtained in $DCCl_3$. The scale is in parts per million from external 85% H_3PO_4.*

Prophos

The solution and solid P-31 NMR spectra of the prophos ligand are presented in Figure 10. As seen from this figure, the expected nonequivalence of the phosphorus atoms was observed in both spectra. The signals in the solid-state spectrum showed increased shielding relative to the solution spectrum. The solution spectrum also revealed a P–P coupling (J_{P-P} = 20.5 Hz).

The solid and solution spectra of the prophos–rhodium complex are shown in Figure 11. In the solution spectrum, both Rh–P and P–P couplings are observed (J_{Rh-P} = 155.0 and 156.0 Hz; J_{P-P} = 33.4 Hz). The solid-state spectrum, with larger linewidths, showed only the Rh–P coupling (J_{Rh-P} = 162 and 124 Hz). A slight impurity was also evident from the NMR spectrum of the solid, but this should not interfere with the overall interpretation. The differences in the coupling constants observed between the solid- and liquid-state spectra may be the result of substantial differences in the coordination geometry between the solid and liquid-solution states.

Polymer-Supported BPPM Systems

In addition, a polymer-attached analogue of the BPPM ligand and a corresponding rhodium(I) catalyst were obtained from J. K. Stille. Because of the nature of the polymer, X-ray or normal-solution NMR analysis were not possible. However, the solid-state P-31 CP/MAS NMR analysis was performed on a polymer-attached ligand and a corresponding rhodium(I) complex. The spectra are shown in Figure 12. The polymer-attached ligand showed two signals centered at −16 and 25 ppm, as compared with −7.3 and −21.9 ppm for BPPM in solution and to −14.8 and −21.7 ppm for BPPM in the solid state. This strongly implies some substantial, structural differences for the BPPM ligand in these three situations. The rhodium(I) catalyst prepared from the polymer-attached ligand displayed only one broad P-31 signal centered at 25 ppm, as compared with multiplets centered at 43.5 and 14.7 ppm for Complex **III** in solution and to the two signals centered at 38.9 and 12.7 ppm for Complex **III** in the solid state. The fact that only

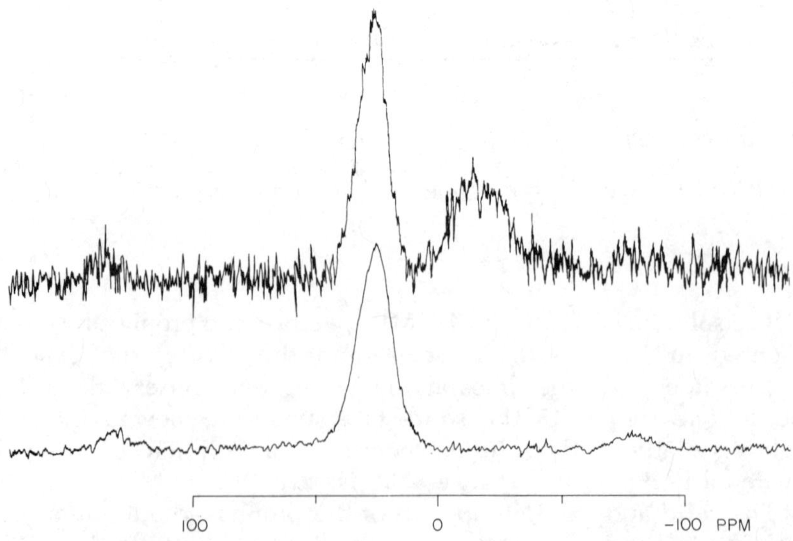

*Figure 12. Solid-state P-31 NMR spectra of a polymer-bound BPPM ligand (*top*) and the corresponding polymer-bound rhodium(I)COD catalyst (*bottom*). The scale is in parts per million from external 85% H_3PO_4.*

one P-31 band was observed for the polymer-bound catalysts implies that there are major structural differences between that catalyst and the simple catalyst (Complex **III**). More work clearly is needed with this and other polymer-attached species to determine the significance of these observations.

Conclusions

The results given here demonstrate the great structural sensitivity of solid-state CP/MAS P-31 NMR. A few of the results indicate that significant differences may exist between the solid and liquid-solution states of some of the compounds studied; but the nature and chemical significance of these differences remain in some respects in doubt because of possible influences of crystallographic factors that are as yet uncharacterized. Empirically these differences seem to be associated, at least in part, with the chirality of the diphosphine ligands. Until such differences are resolved, they must cast some doubt on the validity of extrapolating solid-state structures obtained by X-ray data to the solution state (*1, 2, 3*). Nevertheless, we do not wish to question the substance of correlations drawn from X-ray studies on the rhodium(I) diphosphine catalysts (*1, 3*). Such correlations have yielded in almost all of the cases the correct prediction of the chirality induced by the specific hydrogenation catalyst. However, the present study pleads for caution in the prevalent tendency in many chemical fields to assume that solid-state structures can be used to interpret stereochemistry in solution. More work is needed in the areas of X-ray-structure determination and solid-sample NMR spectroscopy. In any case, there is a general similarity between the P-31 NMR solid-state and the solution spectra of most of the compounds of this study.

It should be mentioned that other nuclei (C-13 and Rh-109) are available for further study of the systems discussed in this chapter. Preliminary results in our laboratory indicate that the solid-sample C-13 NMR spectra of these catalysts may be too complex to be routinely useful for these systems, but they may be useful in certain cases. Rh-103 experiments are planned.

CP/MAS NMR, with C-13, P-31, and a variety of spin-one-half nuclides can yield high-resolution NMR data that provide badly needed bridges between the liquid and solid states. These bridges can be used to estimate the extent to which diffraction-determined structural results on crystalline solids can be extrapolated into the liquid state. A substantial effort in this direction is underway in our laboratory. The potential for CP/MAS NMR in characterizing polymer-supported catalysts is also evident.

Acknowledgments

The authors wish to acknowledge partial support of this work by the National Science Foundation under Grant No. CHE74–23980 and the use of the Colorado State University Regional NMR Center, which was funded by the National Science Foundation Grant No. CHE78–18581. They are also grateful to J. K. Stille for providing the polymer-bound species, to K. Achiwa and W. S. Knowles for providing ligand samples, and to V. J. Bartuska for invaluable instrumental assistance. We also acknowledge helpful discussions with B. R. Stultz and N. Payne concerning the X-ray-determined structures of some of the catalysts.

Literature Cited

1. Vineyard, B. D.; Knowles, W. S.; Sabacky, M. J.; Bachman, G. L.; Weinkauff, D. J. *J. Am. Chem. Soc.* **1977,** *99,* 5946.
2. Ball, R. G.; Payne, N. C. *Inorg. Chem.* **1977,** *16,* 1187 and private communication from N. C. Payne.
3. Knowles, W. S.; Vineyard, B. D.; Sabacky, M. J.; Stultz, B. R. In "Fundamental Research in Homogeneous Catalysis"; Minoru; Tsutsui Eds.; Plenum: New York, 1978; Vol. 3, p. 537, and private communication from B. R. Stultz.
4. Schaefer, J.; Stejskal, E. O. "Topics in Carbon-13 NMR Spectroscopy"; Levy, G. C., Ed.; Wiley-Interscience: New York, 1979; Vol. 3.
5. Miknis, F. P.; Bartuska, V. J.; Maciel, G. E. *Amer. Lab.* **1979,** *19,* 11.
6. Fryzuk, M. D.; Bosnich, B. *J. Am. Chem. Soc.* **1977,** *99,* 6262.
7. Fryzuk, M. D.; Bosnich, B. *J. Am. Chem. Soc.* **1978,** *100,* 5491.
8. Chatt, J.; Venanzi, L. M. *J. Chem. Soc. A* **1957,** 4735.
9. Cramer, R. *J. Am. Chem. Soc.* **1964,** *86,* 217.
10. Schrock, R. R.; Osborn, J. A. *J. Am. Chem. Soc.* **1971,** *93,* 2397.
11. Achiwa, K.; Ohga, Y.; Iitaka, Y. *Tetrahedron Lett.* **1978,** *47,* 4683.
12. Ojima, I.; Kogure, T. *Chem. Lett. Tokyo* **1978,** 1145.

RECEIVED July 10, 1980.

APPENDIX

Papers Presented at Poster Session

1. **Steric and Electronic Effects in Cobaloxime Complexes of Phosphines and Phosphites**
   Joseph A. Kargol, Roger W. Crecely, and John L. Bumeister, Department of Chemistry, University of Delaware, Newark, DE 19711 and Paul J. Toscano and Luigi G. Marzilli, Department of Chemistry, The Johns Hopkins University, Baltimore, MD 21218

2. **Cone Angles and MINDO/3 Calculations of Phosphines**
   John A. Mosbo, Janice T. DeSanto, Paul L. Bock, and Bruce N. Storhoff, Department of Chemistry, Ball State University, Muncie, IN 47306

3. **A Novel Metal–Template Synthesis of Triphosphine Macrocycles**
   Arlan D. Norman and Bruce N. Diel, Department of Chemistry, University of Colorado, Boulder, CO 80309

4. **First Synthesis and X-Ray Structural Determination of the Macrocyclic Schiff Base Phosphineamine Complex, Ni([16]dieneN_2P_2)$(PF_6)_2 \cdot 0.5\ H_2O$**
   L. G. Scanlon, Y.-Y. Tsao, K. Toman, and S. C. Cummings, Departments of Chemistry and Geology, Wright State University, Dayton, OH 45435 and D. W. Meek, Department of Chemistry, The Ohio State University, Columbus, OH 43210

5. **New Ligand Systems for the Linkage of Dissimilar Transition Metals**
   N. E. Schore, L. Benner, and S. Sundar, Department of Chemistry, University of California at Davis, Davis, CA 95616

6. **Synthesis and Structural Properties of Phosphenium Ion Complexes**
   R. T. Paine, C. F. Campana, L. D. Hutchins, and R. W. Light, Department of Chemistry, University of New Mexico, Albuquerque, NM 87131

7. **Cyclopentadienylbis(ligand)nickel(I): Synthesis, Characterization, and Reactivity**
   E. Kent Barefield, David A. Krost, Rodney Trytko, Scott Edwards, and Shaun O'Rear, School of Chemistry, Georgia Institute of Technology, Atlanta, GA 30332 and Alex N. Williamson, Department of Chemistry, North Carolina State A&T University, Greensboro, NC 27411

8. **Controlling the Number of Metal Sites to Which a Poly(tertiary phosphine) Coordinates in Tungsten Carbonyls**
   R. L. Keiter, J. W. Brodack, and R. D. Borger, Department of Chemistry, Eastern Illinois University, Charleston, IL 61920 and L. W. Cary, Life Science Division, Stanford Research Institute, Menlo Park, CA 94025

9. **Kinetics and Mechanisms of Ligand Replacement Reactions in Trigonal Bipyramidal Nickel(II) Complexes of a Tetradentate Ligand Containing an Apical Phosphorus Donor Atom**
   Eugene Grimley and Nancy MacNeill, Department of Chemistry, Mississippi State University, Mississippi State, MS 39762

10. **Ruthenium Complexes with Polydentate Phosphines**
Bernardo Fontal, Chemistry Department, Universidad de Los Andes, Merida, Venezuela and Henry Taube, Chemistry Department, Stanford University, Stanford, CA 94305

11. **Application of High-Pressure IR Spectroscopy to the Study of the Interaction of Small Gaseous Molecules with V(II), V(III), and Ti(III) Phosphine Complexes**
V. Hall and C. David Schmulbach, Department of Chemistry and Biochemistry, Southern Illinois University, Carbondale, IL 62901

12. **Stereochemistry and Mechanism of Carbonyl Insertion on Octahedral Low-Spin Iron(II)**
C. R. Jablonski, Department of Chemistry, Memorial University of Newfoundland, St. John's, Newfoundland, Canada AIB 3X7

13. **A Copper(I)–Triphenylphosphine System to Reduce Dinitrogen**
B. D. James, R. Payne, and B. E. Green, LaTrobe University, Bundoora, Victoria, Australia 3083

14. **Enantioselective Carbon–Nitrogen Bond Formation**
G. E. Bossard and T. A. George, Department of Chemistry, University of Nebraska at Lincoln, Lincoln, NE 68588

15. **Studies in Asymmetric Synthesis, Preparation, Reactivity, and Solid-State Structures of Cationic Rhodium Complexes Containing a Chiral Aminophosphine or Aminoarsine Ligand**
Douglas W. Stephan, Jo-Ann E. Forrest, and Nicholas C. Payne, Department of Chemistry, University of Western Ontario, London, Ontario, Canada N6A 5B7

16. **Synthesis of a New Chiral Phosphine and Its Application to Asymmetric Catalysis**
Dennis P. Riley and Robert Shumate, The Procter & Gamble Company, Miami Valley Laboratories, P.O. Box 39175, Cincinnati, OH 45239

17. **Phosphine Nitrosyl Complexes of Rhenium(I)**
K. R. Grundy, K. Robertson, and E. Woods, Department of Chemistry, Dalhousie University, Halifax, Nova Scotia, Canada B3H 4J3

18. **Chemical and Catalytic Studies on Binuclear Cationic Complexes of Rhodium(I)**
Joel T. Mague and Steven H. DeVries, Department of Chemistry, Tulane University, New Orleans, LA 70118

19. **Complexes of Rhodium(I) with Carbon Monoxide, Bis(diphenylphosphino)-methane, and Various Anionic Ligands**
Alan R. Sanger, Alberta Research Council, 11315–87 Avenue, Edmonton, Alberta, Canada T6G 2C2

20. **Reactions of 2,6-Bis(di-*t*-butylphosphinomethyl)phenylrhodium(I)**
William C. Kaska and Craig Jensen, Department of Chemistry, University of California, Santa Barbara, CA 93106

21. **An Extremely Active and Selective Homogeneous Hydrogenation Catalyst, Acetyldicarbonyl-Bis(trimethylphosphite) cobalt(I), $CH_3C(O)Co(CO)_2(P(OCH_3)_3)_2$**
Thomas S. Janik, Michael F. Pysczcek, and Jim D. Atwood, Department of Chemistry, SUNY at Buffalo, NY 14214

22. PD(PPH_3)$_4$-Catalyzed Alkylation of Enol Phosphates with Trialkylaluminum Compounds
Kazuhiko Takai, Koichiro Oshima, and Hitosi Nozaki, Department of Industrial Chemistry, Kyoto University, Japan

23. Pentacoordinate Platinum(II) Phosphole Complexes
John H. Nelson and John Jeffrey MacDougall, University of Nevada, Reno, NV 89557 and Francois Mathey, The Institute National de Recherche Chimique Appliqueé, 91710 Vert-le-Petit, France

INDEX

A

A-frames, molecular 246*f*
A-frame species 247
Absorbance data for the reaction of *trans*-$[Rh(PR_3)_2Cl(CO)]$ with methyl iodide 288*t*
Acetyl chloride, reaction of $RhCl(PPh_3)_3$ ....196 sch
Acetylides
  μ_2-η^2, reactivity 184–188
  μ_3-η^2, reactivity 184–188
  μ_3-η^2 and nucleophilic attack ... 188
Acid yields
  from cyclic and acyclic alcohols 298*t*
  using $Rh_6(CO)_{16}$ with changing pressure of CO and O_2 ... 298*t*
  using $Rh_6(CO)_{16}$ vs. the pressure of CO/total pressure . 299*t*
α-(Acylamido)cinnamic acid derivatives, optical yields for the asymmetric hydrogenation .. 322*t*
Acyl chlorides, oxidative addition to $IrCl(PMePh_2)_3$ 198
Acyl chlorides, oxidative addition to $MClL_3$ 197–201
(Z)-*N*-Acylaminoacrylic acids. .. 338
Alcohol(s)
  cyclic and acylic, acid yields .. 298*t*
  oxidation 297
  structure, dependence of asymmetric induction 375*t*
  structure, influence on hydrocarbalkoxylation of 2-phenyl-1-propene 375*t*
Aldehydes
  and catalyst selectivity, steric effects 70*t*
  decarbonylation, diphosphine complexes as effective catalysts 72
  product yield and conversion .. 72*t*
Alkoxy groups, terminal, synthesis of polyphosphines 318–319
Alkyl migration in cationic rhodium(III) and iridium-(III) complexes 199
Alkylcobaloximes 87
Alkyliridium(III) complexes, secondary, containing tertiary phosphines, isomerization ..201–205
Alkyliridium(III) complexes, secondary, isomerization rates .. 204*t*
σ-Alkyl-rhodium-hydrido intermediates, η-elimination 341*f*
Allylic alcohols 224–225
π-Allylic palladium halide complexes 215
Aluminum alkyls, P-31{H-1} NMR data in solutions 261*t*
Amino acid substrate, catalytic symmetric hydrogenation 352*f*
Amino acid substrates, chelation . 342*f*
Arrhenius plot for PPh_3–Rh complex dissociation 54*f*
Arsinobenzaldehydes 304
Asymmetric
  homogeneous catalysts 325–336
  hydrocarbalkoxylation of olefins with palladium catalysts 373–379
  hydrocarbalkoxylation by $PdCl_2/(-)DIOP$ 385*t*
  hydrocarbonylation, advances 371–387
  hydrocarbonylation, model ..380–386
  hydrogenation
    catalytic 337–353, 340*f*
    energy barriers for steps 365*f*
    homogeneous, rhodium-catalyzed 355–369
    reaction cycle 364*f*
  induction 342
    dependence on alcohol structure 375*t*

B

Benzaldehyde
  decarbonylation 78
    catalytic 68*t*
    using cationic diphosphine complexes of rhodium(I) 67
Benzaldehydes, *para*-substituted, electronic effects 71*t*
Benzene production
  vs. benzaldehyde concentration at 158° for $Rh(dppp)_2$ ... 75*f*
  from neat benzaldehyde at 135°C as a function of $Rh(dppp)_2$ concentration . 74*f*
  rate, at 158°C 76*f*
Benzoyl chloride 79
Binding energies, ESCA, for platinum–tin complexes 38*t*
Binding energy of platinum, effect of $SnCl_3$ 37
Binuclear
  complexes, desirable functional groups 231–232
  organoplatinum complexes as catalytic intermediates ..231–240
  reductive elimination 236

Bis(diphenylphosphino)-
methane 232, 243
as bridging ligand 243–255
planar metal centers bridged
by, idealized structural
types 244f
Bisdiphosphine catalysts 70
Bisphosphine(s) 330t
oxides, physical data 335t
rhodium complex hydroformulation catalyst systems, P-31 NMR studies of equilibria and ligand exchange 43–64
synthesis 333–335
Bis-1,2-diphenylphosphinoethane 57
Bis-1,3-diphenylphosphinopropane 57
Bond length structure, influence
of *X* variation 90–92
Bond lengths in $LCo(DH)_2Cl$
and $LCo(DH)_2CH_3$ 91f
Bonding, metal–phosphorus and
P-31 NMR 1–22
Bonding modes in bridging
situations 232t
Bonding modes in terminal
situations 232t
1-Butene hydroformylation
studies 47

C

C-13 NMR (*see* NMR, C-13)
C-13 shift in $(CH_3O)_3PCo(DH)_2X$ vs. the P-31 shift of $(CH_3O)_3$-$PCo(DH)_2X(CDCl_3)$ 89f
Carbon dioxide complexes,
formation 156–157
Carbon dioxide, reduction 142
Carbonyl clusters
with μ_3-η^2 acetylides 171
binuclear 174–175
open triangular 180
phosphido-bridged iron
group 163–192
rhodium, complexes 291–300
structural parameters 173f
tetranuclear 179
Carbonyl derivatives of binuclear
organoplatinum complexes . 234–235
Carbonyl phosphines 304
Carbonyls, binuclear iron, P-31
chemical shift–bond angle
correlations 183–189
Catalysis, effect of ligand excess .. 47t
Catalysts, asymmetric homogeneous 325–336
Catalyst selectivity and aldehydes,
steric effects 70t
Catalytic
activity, effect of added
reagents 73t
cycle of olefins 356–363
hydrogenation, asymmetric .. 337–353
Catalytic (*continued*)
amino acid substrates 352f
intermediates, binuclear organoplatinum complexes 231–240
Chelate complexes
large-ring vs. open-chain 104
trimethylene bisphosphine ... 60f, 61t
derived from tris(triphenylphosphine)rhodium(I) carbonyl hydride and bisphosphines 57–63
Chelate ring(s)
conformational equilibria of
five-membered 343f
(*S,S*)-chiraphos 343f
large
coupling constants 108
stabilization 104–107
treatment with base 107
nine-membered 308
(*R*)-prophos 343f
stability 104–114
systems, six-membered 344
Chelates in tridentate en=P_2
complexes 308
Chelating agents, iminoarsine,
metal complexes 303–311
Chelating agents, iminophosphine,
metal complexes 303–311
Chelating phosphine rhodium
complexes assymetric catalytic hydrogenation using ... 340f
Chelation of amino acid
substrates 342f
Chirality, influence in asymmetric
hydrocarbonylation 382t
Chlorocobaloximes containing
phosphorus ligands, ORTEP
drawings 97f
Cobaloxime, $LCo(DH)_2X$ 86f
Cobalooximes containing phosphorus-donor ligands 85–98
Cobalt–phosphorus bond lengths
in $R_3PCo(DH)_2X$ compounds 91t
Cobalt–*X* bond lengths in the
$LCo(DH)_2X$ series 98t
Cobalt(I) compounds 86
Cobalt(II) compounds 86
Cobalt(III) compounds 87
Cobalt(III) reaction rates, influence of *X* variation 87–88
Cobalt(IV) compounds 87
Complexation shifts of ligands ... 394t
Coupling constant(s) 1
$[Ag(PBu_3)_n]^+(BF_4)^-$
against 1/n 5f
with bond length, correlation .. 6–10
bonds to the ligand
cis-$[PtCl_2L(PEt_3)]$ 18f
trans-$[PtCl_2L(PCy_3)]$ 18f
$[W(CO)_5L]$ 19f
bonds to PPh_3 in platinum(II)
complexes 18f

Coupling constant(s) (*continued*)
cis and *trans*
$[Pt(SCF_3)X(PEt_3)_2]$ 12*f*
cis-$[PtCl(CX_n)(PPh_3)_2]$ ...13*f*, 14*f*
$[PtCl(SnX_3)(PPh_3)_2]$ 15*f*
trans-$[PtCl(SnX_3)(PPh_3)_2]$. 14*f*
relationship to varying ligand *X* 11–16
dependence on *cis* ligands 10–16
interpretations 2
large chelate ring 108
NMR 31–36
one-bond, for platinum–tin complexes 35
PPh_3 *trans* to tin 16*f*
trans-$[PtCl_2(NH_2R)L]$ against that for *trans*-$[PtCl_2(PBu_3)L]$ 3*f*
phosphonato ligand in *trans*-$[PtX\{P(O)(OPh)_2\}(PBu_3)_2]$ against that for *trans*-$Pt(CH_3)X(PEt_3)_2$ .. 2*f*
phosphine ligands 17
phosphite ligands 17
protonated phosphorus donors for platinum(II) complexes 20*f*
Pt–P bond lengths in platinum(II) and (IV) 7*t*
Coupling, two-bond 35–36
Crystalline structure of $Os_3(CO)_9(COOEt)(OH)(Ph_2PC{\equiv}CiPr)$ 181*f*
Cyclohexanol, oxidation to adipic acid, solvent effects 297*t*
Cyclohexanol, reaction pathway in autoxidation 296
Cyclohexanone, oxidation 294–297
to adipic acid, solvent effects .. 294*t*
to adipic acid, temperature and oxygen pressure effects ... 295*t*
$Rh_6(CO)_{16}$-catalyzed, pressure effects 299–300
Cyclohexanone, reaction pathway in autoxidation 296
Cyclohexyl ring, dehydrogenation 151–152
Cyclometallation reactions 107–111
stereopopulation control 104–114

D

Decarbonylation
aldehyde, diphosphine complexes as effective catalysts 72
benzaldehyde 78
catalytic 68*t*
using cationic diphosphine complexes of rhodium(I) 67
$RhCl(PPh_3)_3$ 71
reaction(s)
homogeneous catalytic, effect of chelating diphosphine ligands 65–82
Decarbonylation (*continued*)
reactions (*continued*)
mechanism using Rh(bis-1,3-diphenylphosphinopropane)$_2^+$ 73
mechanism using $RhCl(PPh_3)_3$ 66–67
synthetic usefulness 69–73
Dehydrogenation of a cyclohexyl ring 151–152
Dehydrovaline 333*t*
Dialkylamino groups, terminal synthesis of polyphosphines containing 318–319
Diiminodiphosphines 306–308
Diiminodiarsines, chiral 308–310
Dinuclear metal complexes, novel reactions 243–255
Diphos–rhodium complex 396*f*
Diphos, solid-state P-31 NMR spectra and solution .395*f*, 395–397
Diphosphine
catalysts, rhodium(I), P-31 NMR comparisons 398–408
complexes as effective catalysts for aldehyde decarbonylation 72
ligands, chelating, effect on homogeneous 65–82
ligands, P-31 NMR data 391*t*
Diplatinum(I) complexes 232–234
Dirhodium tetrahydridotetrakis-(triisopropylphosphine)
angles 128*t*
bond distances 128*t*
crystal data and intensity collection 120*t*
fragmentation 125
hydrido hydrogen atom locations 129–131
molecule 127
parameters derived for the rigid group atoms 124*t*
positional and thermal parameters for the nongroup atoms 122–123
preparation 118
reactions 121
solution and refinement 119
spectra 121
structure 126–129, 131
synthesis 121
as tetrahydride 126
unit-cell packing 127*f*, 129

E

Electron-donating ligands 141
β-Elimination of σ-alkyl-rhodium-hydrido intermediates 341*f*
Enamide complex 366
addition of hydrogen 368

Energy barriers for the steps of asymmetric hydrogenation .. 365*f*
ESCA binding energies for platinum–tin complexes 38*t*

F

Fermi contact expression 32
Fermi contact term 2
  evaluation 3
Fragmentation of $Ru_3(CO)_{11}(Ph_2PC{\equiv}CR)$ .167–168

H

H-1 NMR (*see* NMR, H-1)
Halogens, addition to $Pd_2(dpm)_3$. 252
1-Hexene, vinylic substitution productions 220*t*
Homogeneous catalysts, asymmetric 325–336
Homogeneous hydrogenation, asymmetric, rhodium-catalyzed 355–369
Homogeneous hydrogenation of olefins with HD and cationic rhodium complexes 367*t*
Hydration of unsaturated compounds 138
Hydride derivatives of binuclear organoplatinum complexes .235–238
Hydride fluxionality processes 110
β-Hydride elimination 214
trans-Hydridoalkyls or hydridoaryls of platinum(II) 207
Hydrido-bridged rhodium dimer
  angles 128*t*
  bond distances 128*t*
  crystal data and intensity collection 120*t*
  fragmentation 125
  hydrido hydrogen atom locations 129–131
  molecule 127
  derived parameters for the rigid group atoms 124*t*
  positional and thermal parameters for the nongroup atoms 122–123
  preparation 118
  reactions 121
  solution and refinement 119
  spectra 121
  structure 126–129, 131
  synthesis 121
  as tetrahydride 126
  unit-cell packing 127*f*, 129
Hydrido complexes of platinum(II) 205–210
Hydrido–hydroxo complexes of platinum(II) 206
Hydroxo complexes, inert 208
Hydroxo complexes of platinum(II) 205–210
Hydrocarbaloxylation
  asymmetric, of olefins 373–379
    palladium-catalyzed 379*t*
  assymetric, by $PdCl_2/(-)$DIOP 385*t*
  effect of carbon monoxide pressure 376
  palladium-catalyzed asymmetric, of 2-phenyl-1-propene . 378*t*
  2-phenyl-1-propene, influence of alcohol structure 375*t*
  substrate having C_s symmetry . 384
  (*E*)-1-phenyl-1-propene 386
Hydrocarbons, H–D exchange reaction 138
Hydrocarbonylation, asymmetric
  advances 371–387
  influence of chirality 382*t*
  influence of optical 382*t*
  model 380–386
Hydrocarbonylation of olefins, transition states controlling asymmetric induction 383*f*
Hydroformylation 332
  olefin, energetics of catalytic intermediates 56*f*
  studies, 1-butene 47
Hydrogen
  addition, to enamide complex ... 368
  consumption vs. time for the hydrogenation of 1-octene at constant pressure 266*f*
  evolution from water by rhodium complex 140–142
  –D exchange reaction of hydrocarbons 138
  molecular exchange reaction with D_2O 139
  peroxide, decomposition 297–299
    catalyzed 298*t*
  pressure, effect on hydrogenation of olefins 266*t*
Hydrogenation
  asymmetric
    α-(acylamino)cinnamic acid derivatives, optical yields 322*t*
    catalytic 337–353, 340*f*
      amino acid substrates 352*f*
    energy barriers for steps 365*f*
    homogeneous, rhodium-catalyzed 355–369
    reaction cycle 364*f*
  catalysis of olefins by rhoduim hydride complex 257–271
  catalyst species, PPh_3–Rh complex 44
  homogeneous, of methyl linoleate ...280*t*, 282*t*, 284*t*, 285*t*

Hydrogenation (*continued*)
homogeneous, of olefins with HD and cationic rhodium complexes 367*t*
low-temperature 351
1-octene 267*f*
olefin 279–282
effect of hydrogen pressure .. 266*t*
mechanism 339–342
rhodium catalyzed, probes for structure of transition state 363–369

I

Imines, chiral 305
Iminoarsine chelating agents, metal complexes 303–311
Iminophosphine chelating agents, metal complexes 303–311
Induction, asymmetric 342
Iron carbonyls, binuclear, P-31 chemical shift-bond angle correlations 183–189
Iron group, phosphido-bridged, carbonyl clusters 163–192
IR data for platinum–tin complexes 39*t*
Iridium
complexes 148–151
tertiary phosphine 195–212
tricyclohexylphosphine ...145–161
–diphosphine chemistry 111
hydride, stereochemistry 112
trichloride 111
Iridium(III) complexes, cationic, alkyl migration 199
Isomerization
rates
$IrClI\{CH(CH(CH_3)_2\}-(CO)L_2$ 203*t*
$IrClI\{CH(CH_3)_2\}(CO)-(PMe_2Ph)_2$ 202*t*
various secondary alkyl-iridium(III) complexes . 204*t*
secondary alkyliridium(III) complexes containing tertiary phosphines 201–205
tetrahydrofuran 202

K

Ketones, cyclic acid yields 296–297
Ketone oxidation 294

L

Ligand(s)
bis(dimethylphosphino)methane 233
bridging, palladium complexes containing 243–255
complexation shifts 394*t*

Ligand(s) (*continued*)
complexes containing $R_2(O)P^-$ .96–97
cis, dependence of coupling constants 10–16
diphosphine, P-31 NMR data .. 391*t*
effect of excess on catalysis 47*t*
en= P_2 306
angular parameters for 309*f*
exchange
rate variation for *X* ligands in $(CH_3O)_3PCo(DH_2X)$ vs. Taft's polar substituent constant 88*f*
reaction, log K_{obs} vs. electron-donor ability 93*f*
reaction, log K_{obs} vs. Tolman cone angle 93*f*
phosphine–rhodium complex catalyzed hydroformylation of olefins 45*f*
PPh_3 compounds with chelating bisphosphines . 58*f*
PPh_3–Rh complex with trimethylene bisphosphine at different temperatures 62*f*
PPh_3–Rh complex with ethylene bisphospine ... 59*f*
influence on NMR spectra 95–96
L-donor 96
structures 345–346
tertiary phosphine 274
variation on reaction rates and equilibria, influence .. 92
variation, influence on structure 96

M

Melting points of *cis*-$[Pt\{P(n-C_mH_{2m+1})_3)\}_2Cl_2]$ complexes 277*f*
Metal
complex(es)
dinuclear, novel reactions .243–255
iminoarsine chelating agents 303–311
iminophosphine chelating agents 303–311
–alkyl, transition states leading to intermediate 381
oxidation state and phosphorus ligand constant 4–6
–phosphorus bonding, applications of P-31 NMR 1–22
O-Metallation 102
Methoxo complexes of platinum(II) 205–210
Methyl
halides, addition to $Pd_2(dpm)_3$ 252–255
linoleate, homogeneous hydrogenation ..280*t*, 282*t*, 284*t*, 285*t*
migration 201

Methylplatinum complexes of binuclear organoplatinum complexes 238–240
Michael addition 314
Monoiminophosphines 305

N

NMR
C-13, data for rhodium alkyl .. 363*f*
C-13, spectrum of RhCl(ttp) and $AlEt_2Cl$ 265
coupling constants 31–36
H-1
data for platinum–tin complexes 34*t*
high-field, for $H_2IrCl(CO)$-$(PCy_3)_2$ 151*f*
olefin–rhodium complexes ... 357*f*
parameters for palladium compounds 253*t*
parameters of RhCl(ttp) and $AlEt_3$ or $AlEt_2Cl$ 263*t*
spectrum(a)
[Cu (en=As_2)ClO_4] 310*f*
[Cu(R-pn=As_2)ClO_4] ... 310*f*
$Pd_3(dpm)_2(\mu\text{-}CH_2)I_2$ 252*f*
RhCl(ttp) plus Et_3Al in toluene-d_8 .... 262*f*
H-1{P-31}, NMR spectroscopy . 112
P-31
comparisons of rhodium(I) diphosphine catalysts .389–408
coordination shifts of phosphines 96
data
diphosphine ligands 391*t*
platinum–tin complexes 33*t*
rhodium(I) diphosphine catalysts 392*t*
isomers of $Ru_2(CO)_5$-(C≡CiPr(PPh_2)-(Ph_2PC≡CiPr) 183*f*
parameters
phosphido-bridged iron-group carbonyl complexes 178*t*
for phosphinoacetylene iron-group carbonyl complexes 178*t*
trends and implications ... 180
spectrum(a)
chiraphos 401–405
demonstrating formation of rhodium alkyl .360*f*–361*f*
DiPAMP 397*f*
(Nme)(C_6H_5)PCH_2CH_2-$P(C_6H_5)_2$ 321
polymer-bound BPPM ligand 406*f*
prophos 404–406, 404*f*
proton-decoupled, of [$Pd_2(Ph_2PCH_2CH_2$-$PPhCH_2PPh_2)_2$][PF_6]$_2$ 249*f*

NMR (*continued*)
P-31 (*continued*)
spectrum(a) (*continued*)
[Rh(COD) chiraphos]$^+$-ClO_4^- 403*f*
rhodium–DiPAMP complex. 398*f*
solid-state, of diphos 395*f*, 395–397
solution, of diphos .395*f*, 395–397
study(ies)
crystalline and solution status of rhodium(I) diphosphine complexes 390
equilibria and ligand exchange in bisphosphine rhodium complex hydroformylation catalyst systems 43–64
equilibria and ligand exchange in PPh_3–Rh complex 43–64
ligand exchange in PPh_3–Rh complexes at various temperatures 49*f*, 49–53
metal–phosphorus bonding . 1–22
platinum–tin homogeneous hydrogenation catalysts 23–39
P-31{H-1} data for solutions containing aluminum alkyls 261*t*
P-31{H-1} spectrum(a)
as a function of temperature for [$Rh(dppb)_2$]BF_4 ... 81*f*
as a function of temperature for [$Rh(dppp)_2$]BF_4 ... 77*f*
trans [$PtCl(SnCl_3)PEt_3)_2$] and *trans*-[$Pt(SnCl_3)_2$-$PEt_3)_2$] mixture 29*f*
RhCl(ttp) and $AlEt_3$ 259–260
Pt-195, data for platinum–tin complexes 34*t*
Pt-195 studies on platinum–tin homogeneous hydrogenation catalysts 23–39
Sn-119
data for platinum–tin complexes 34*t*
spectrum of *trans*-[$Pt(SnCl_3)_2$-$PEt_3)_2$] 27*f*
spectrum of *trans*-[PtH-$(SnCl_3)PPh_3)_2$] 30*f*
studies on platinum–tin homogeneous hydrogenation catalysts 23–29
spectra, influence of *X* variation 89*f*
time scale 113
Neomenthyl groups, terminal, synthesis of polyphosphines containing 319–320

O

1-Octene, hydrogenation 267*f*
hydrogen consumption vs. time . 266*f*

Olefin(s)
catalytic cycle 356–363
concentration, effect on reaction rate 268t
five-carbon 227
four-carbon 226
functionalized
reactions 222–228
three-carbon, additions ...222–226
for vinylic substitution reaction 217t
with HD and cationic rhodium complexes, homogeneous hydrogenation 367t
hydrocarbakoxylation
asymmetric 373–379
palladium-catalyzed 379t
hydroformylation, energetics of catalytic intermediates .. 56f
hydrogenation 279–282
catalysis by rhodium hydride complex 257–271
effect of hydrogen pressure .. 266t
mechanism 339–342
rhodium complexes, proton NMR studies 357f
six-carbon 228
π-Olefin complex 381
Optical yield(s)
(S)-chiraphos, catalysts 346–351
influence in asymmetric hydrocarbonylation 382t
(S,S)-skewphos catalysts 346–351
Organic compounds, oxidation ..293–294
Organoplatinum complexes, binuclear
carbonyl derivatives 234–235
catalytic intermediates 231–240
hydride derivatives 235–238
methylplatinum complexes ..238–240
ORTEP drawing
$[Pd_2Cl_4\{tBu_2P(CH_2)_7$-$PtBu_2\}_2]$ 105f
$[Rh(dppb)_2]BF_4$ 82f
$[Rh_2(CO)_4(dppb)_3]^{2+}$ 80f
ORTEP II plot of $Ru_3(CO)_8$-$(C{\equiv}CtBu)(PPh_2)$ 174f
ORTEP II plot of $Ru_3(CO)_9$-$(C{\equiv}CiPr)(PPh_2)$ 172f
Oxidative addition
acyl chlorides to $IrCl(PMePh_2)_3$ 198
acyl chlorides to $MClL_3$ 197–201
reactions of $Pd_2(dmp)_3$, two-center, three-fragment ..250–252
trans-$[Rh(PR_3)_2Cl(CO)]$ ...283–289
water to low-valence transition metal compounds 136
Oxidation
cyclohexanol to adipic acid, solvent effects 297f
cyclohexanone 294–297
to adipic acid, solvent effects . 294t
Oxidation (*continued*)
cyclohexanone (*continued*)
to adipic acid, temperature and oxygen pressure effects 295t
$Rh_6(CO)_{16}$-catalyzed, pressure 299–300
ketone 294
organic compounds 293–294
PPh_3 292–293

P

P-31
chemical shift-bond angle correlations in binuclear iron carbonyls 183–189
NMR (*see* NMR, P-31)
shift of $(CH_3O)_3PCo(DH)_2X$ vs. that of $Bu_3PCo(DH)_2$-$X(CDCl_3)$ 90f
Palladium
catalysts, asymmetric hydrocarbalkoxylation of olefins with 373–379
-catalyzed asymmetric hydrocarbalkoxylation of 2-phenyl-1-propene 378t
-catalyzed asymmetric hydrocarbalkoxylation of olefins . 379t
complexes containing dpm as a bridging ligand 243–255
compounds, H-1 NMR parameters 253t
–palladium bond in Pd_2-$(dpm)_2X_2$ 244–250
–triarylphosphine complexes as catalysts for vinylic halide reactions 213–230
Phosphido
bridge, characteristics 164
bridge reactivity 188–189, 191
-bridged
complexes 165
iron group carbonyl clusters 163–192
iron group carbonyl complexes, P-31 NMR parameters 178t
groups bridging nonbonding metals 182
groups, synthesis of complexes containing 166–170
Phosphine
complexes, tertiary, of rhodium, iridium, and platinum ..195–212
design 343–345
ligands, tertiary 274
long-chain tertiary, transition-metal complexes 273–289
P-31 NMR coordination shifts .. 96
sulfone-based 331t
tertiary, isomerization of secondary alkyliridium(III) complexes containing ...201–205

Phosphinoacetylene iron group carbonyl complexes, P-31 NMR parameters 178t
Phosphinobenzaldehyde 304
Phosphorus
atoms, nonequivalence 405
-donor ligand, variation 92–96
donors, protonated 20f
–hydrogen bonds, polyphosphines containing, synthesis 315–318
–H coupling of methyl resonance in $[LCo(DH)_2CH_3OH]^+$ vs. electron-donor ability 95f
ligand constant and metal oxidatioin state 4–6
ligand, dependence 16–21
Platinum
binding energy 37
chemical shifts 37
–L bond lengths in platinum(II) complexes 20f
–P bond length plotted against coupling constants for platinum(II) complexes of trialkylphosphines 9f
–P bond lengths in platinum(II) and (IV), coupling constants and 7t
–P against coupling constants for *cis*-$PtCl_2L(PEt_3)$ complexes 11f
–tin complexes 23–39
ESCA binding energies 38t
IR data 39t
P-31 NMR data 33t
Pt-195, Sn-119, and H-I NMR data 34t
tertiary phosphine complexes 195–212
Platinum(II)
complexes
coupling constants of protonated phosphorus donors 20f
hydrido 205–210
hydrido–hydroxo 206
hydroxo 205–210
methoxo 205–210
mononuclear 205–210
Pt–L bond lengths 20f
of trialkylphosphines 8
trans-hydridoalkyls or hydridoaryls 207
and platinum(IV), coupling constants and Pt–P bond lengths 7t
preparation of *cis*-hydridomethyls 209
Polymer-bound rhodium(I) COD catalyst 406
Polymer-supported BPPM systems 406–407
Polyphosphines synthesis
containing phosphorus–hydrogen bonds 315–318
Polyphosphine synthesis (*continued*)
containing terminal dialkylamino or alkoxy groups .318–319
containing terminal neomenthyl groups 319–320
recent advances 313–323
Phosphos–rhodium complex 405
Pt-195 NMR (*see* NMR, Pt-195)
Pyridine, aqueous, at 20°C, pH values 136t

R

Reduction of carbon dioxide 142
Rhodium
alkyl, C-13 NMR data 363f
alkyl, P-31 NMR spectra demonstrating formation 360f–361f
carbonyl clusters, complexes .291–300
-catalyzed asymmetric homogeneous hydrogenation ..355–369
-catalyzed hydrogenation, probes for structure of transition state 363–369
complexes 155
chelating phosphine 340f
H_2 evolution from water by 140–142
concentration, effect on reaction rate 268t
concentration vs. rate of reaction 269f
dimer, hydrido-bridged
angles 128t
bond distances 128t
crystal data and intensity collection 120t
fragmentation 125
hydrido hydrogen atom locations 129–131
molecule 127
derived parameters for the rigid group atoms 124t
positional and thermal parameters for the nongroup atoms 122–123
preparation 118
reactions 121
solution and refinement 119
spectra 121
structure 126–129, 131
synthesis 121
as tetrahydride 126
unit-cell packing 127f, 129
–DiPAMP complex 397
P-31 NMR spectra 398f
–diphos complex 396f
hydride complex, hydrogenation catalysis of olefins by ...257–271
hydride complex, solutions ...264–265
tertiary phosphine complexes 195–212
tricyclohexylphosphine complexes 145–161

Rhodium(I)
complex catalysts 337–353
complexes of (Nmen)(C_6H_5)-$PCH_2CH_2P(C_6H_5)_2$ 321
diphosphine catalysts, P-31 NMR comparisons 389–408
diphosphine catalysts, P-31 NMR data 392*t*
diphosphine complexes, P-31 NMR study of crystalline and solution status 390
Rhodium(III) complexes, cationic, alkyl migration 199
Ruthenium complexes 146–148
tricyclohexylphosphine 145–161

S

Sn-119 NMR (*see* NMR, Sn-119)
$SnCl_3^-$ ligand, interaction with platinum 24
Spectroscopic data for *trans*-PtH(R)L_2 complexes 208
Stabilization of large chelate rings 104–107
Stereochemistry of [RhCl-$(COCH_3)(PMe_2Ph)_3]^+$ cation 200*f*
Stereopopulation control 103
on cyclometallation reactions 101–114
Steric rigidity 342
Substrates, hydrogenated 270–271
Synthesis of $Fe_2(CO)_6$-(C≡C*t*Bu)(PPh_2) 166–167

T

Tertiary phosphines, long-chain, transition-metal complexes .273–289
Tetrahydrofuran, isomerization ... 202
Thorpe-Ingold effect 102–104
Tolman cone angle 92
Transition
metal complexes of long-chain tertiary phosphines 273–289
metal compounds, low-valence, oxidative addition of water . 136
states controlling asymmetric induction in the hydrocarbonylation of olefin 383*f*
states leading to intermediate metal–alkyl complex 381
Tris(PPh_3)rhodium(I) carbonyl hydride and bisphosphines, chelate complexes derived from 57–63
Tris(PPh_3)rhodium(I) carbonyl hydride plus triphenylphosphine systems 48–57
Trialkylphosphine(s)
complexes, splitting of water molecule by 135–144
coupling constants for platinum(II) complexes 9*f*
platinum(II) complexes 8
Triarylphosphine–palladium, complexes as catalysts for vinylic halide reactions 213–230
Tricyclohexylphosphine complexes of ruthenium, rhodium, and iridium 145–161
Tricyclohexylphosphine complexes, spectroscopic data 147*t*
Trimethylene bisphosphine, chelate complexes 60*f*
parameters 61*t*
Triphenylphosphine
dissociation 48–54
–rhodium catalyzed 1-butene hydroformylation, Arrhenius plots 55*f*
–rhodium complex
dissociation, Arrhenius plot .. 54*f*
ligand exchange, with ethylene bisphosphine .. 59*f*
P-31 NMR studies of equilibria and ligand exchange 43–64
at various temperatures, P-31 NMR studies of ligand exchange 49*f*, 49–53
oxidation 292–293

U

Unsaturated compounds, hydration 138

V

Vinylic
halide reactions 214
olefin–amine reaction 216
palladium–triarylphosphine complexes as catalysts for 213–230
substitution productions from 1-hexene 220*t*
substitution reaction
chain extension and functionalization 216–221
functionaliezd olefins 217*t*
functionalized vinylic bromides 217*t*

W

Water
–adduct formation 137
H_2 evolution by rhodium complexes 140–142
molecule, splitting by trialkylphosphine complexes ...135–144
oxidative addition to low-valence transition metal compounds 136
reversible splitting 153–155

X

X-ray crystal structure of Fe_2-$\{(CO)_6[C(CN$*t*$Bu)CPh]\}$-(PPh_2) 189*f*
X-ray structure of $Ru_3(CO)_{11}$-$(Ph_2PC{\equiv}CPh)$ 168*f*